Springer Series
in Physical Environment

13

Otto Fränzle

Contaminants in Terrestrial Environments

With 75 Figures and 3 Fold-Outs

Springer-Verlag
Berlin Heidelberg New York
London Paris Tokyo
Hong Kong Barcelona
Budapest

Professor Dr. rer. nat. OTTO FRÄNZLE
Christian-Albrechts-Universität zu Kiel
Geographisches Institut
Projektzentrum Ökosystemforschung
Olshausenstraße 40
24118 Kiel
Germany

ISSN 0937-3047
ISBN-13:978-3-642-77368-6   e-ISBN-13:978-3-642-77366-2
DOI: 10.1007/978-3-642-77366-2

Typesetting: Macmillan India Ltd., Bangalore-25
32/3145/SPS-5 4 3 2 1 0 – Printed on acid-free paper

To my wife Ursula

Preface

Environments are highly complex systems, whose evolution is governed by complicated networks of feedback loops. In the course of history, man has exerted an ever-increasing influence on the various types of terrestrial and aquatic environments, and, as times change, environmental problems appear to change with them. This applies in particular to our eventful and troubled century. In the 1950s, lake eutrophication was among the key issues, and then the focus of biological concern shifted to the "silent spring", i.e. the poisoning of birds by DDT and mercury-dressed seeds. In the course of the last two decades, one environmental alarm bell has sounded after another: lake and soil acidification, the dioxin threat, heavy metals and a bewildering number of more or less persistent organics in soil, damage of crops and forests by pollutant gases, material damage by accelerated corrosion, seas over-enriched with nutrients, the ozone hole in the stratosphere, and climatic changes caused by man's inadvertent interference with the composition of the atmosphere.

Thus today's society is concerned with a multitude of events affecting the highly diversified ecosystems of our planet, ranging from natural processes such as weather and climate systems to man-made phenomena such as ecotoxicology, acid rain and the greenhouse effect. To better understand the impact of these events on ecosystems and to resolve the rapidly growing number of environmental problems, knowledge of the basic earth and life sciences and how they interact is essential, and considering the bewildering number of interrelationships from a unified system's point of view appears indicated.

The way followed in this book to unravel, to a certain extent, this complex network of relationships and feedback loops which have to be considered in such an analysis is model building. It has the advantage of admitting a variety of broadly different approaches, ranging from empirical models for practical purposes, to rather abstract ones aiming at qualitative general insights. At one end of this spectrum there is a detailed and pragmatic description of specific systems such as soil horizons or more precisely defined adsorbents in interaction with one or a few pure chemicals in aqueous solutions. At the opposite end of the spectrum are relatively general models which have to sacrifice numerical precision for the sake of general principles. Such conceptual models, formulated in this book in the form of fairly comprehensive graphs, need not correspond in detail to any single "real world" process, but aim to provide a framework for the discussion of broad classes of pnenomena or simply of contentious issues. Rationally handled, these different approaches mutually reinforce each other, thus providing reciprocally new and deeper insights.

Following this general strategy, the book is organized into three distinct parts. The first (Chaps. 1 and 2) is devoted to fundamentals of environmental chemistry and ecology, which must be understood before attempts can be undertaken to follow the way of environmental chemicals through the major compartments of terrestrial ecosystems. The second part (Chap. 3.1–5) deals with the complex atmospheric pathways of anthropogenic chemicals, starting with a description of the manifold sources of potential pollutants, then following their fate via transport and physical and chemical transformation processes to deposition onto natural or anthropogenic surfaces. The last part of the book (Chap. 4.1–4) is mainly devoted to the manifold interactions of environmental chemicals deposited onto or introduced by other mechanisms into the soil–vegetation complex of ecosystems. It includes chapters on pollutant impact on materials and a review of chemical fate modelling.

Unfortunately, the information explosion and the necessarily limited space have prevented an all-inclusive integration. I tender my apologies to all investigators whose important contributions have not been included. My only excuse is that I could devote only a limited amount of time to this activity, as I have been continuously involved with ongoing ecological research.

Parts of the tables and the diagrams for this book were prepared by Mrs. D. Busch while the typing was done by Mrs. M. Weller, assisted by Mrs. I. Müller, K. Otto, A. Peper and T. Schramm. Many thanks are due to them for both their skill and patience.

I am particularly indebted to my family for sympathetically suffering the frame of mind into which one sinks when writing a book which one hopes is worthwhile.

Kiel, August 1993   OTTO FRÄNZLE

Acknowledgements

The following individuals and organisations are thanked for permission to reproduce text illustrations and tables:

Figures 1.1, 1.2, 1.3 reproduced from *Umweltgutachten* 1978 by permission of Statistisches Bundesamt; Fig. 2.3 reproduced from *Umweltbundesamt-Berichte 10/78* (H. J. Hueck et al.) by permission of Erich Schmidt Verlag; Figs. 3.1, 3.2, 3.3, 4.1 and Tables 4.1, 4.2, 4.10 reproduced from *Environmental Systems* (I. D. White et al.) by permission of Chapman and Hall; Fig. 3.4 reproduced from *Umweltwissen* (H. Bossel) by permission of Gesamthochschule Kassel; Figs. 3.5, 3.23, 3.32 and Tables 1.1, 2.6, 2.9, 3.5, 3.16, 3.41 reproduced from *Jahresberichte* by permission of Umweltbundesamt Berlin; Fig. 3.6 and Table 3.9 reproduced from *Bochumer Geographische Arbeiten 36* (W. Kuttler) by permission of Ferdinand Schöningh GmbH Verlag; Fig. 3.7 reproduced from *Geowissenschaften in unserer Zeit 4/86* by permission of Bruno Rudolf; Figs. 3.8, 4.18, 4.19, 4.25 and Table 4.17 reproduced from *Scientific Basis for Soil Protection in the European Community* (H. Barth, P. L'Hermite eds.) by permission of Elsevier Applied Science Publishers Ltd., © 1987 Elsevier Science Publishers BV; Table 4.23 reproduced from *Aspects of Degradation and Stabilization of Polymers* (H. H. G. Jellinek) by permission of Elsevier Applied Science Publishers Ltd., © 1977 Elsevier Science Publishers; Fig. 3.9 reproduced from *Boundary Layer Climates* (T. R. Oke) by permission of Routledge Publisher Ltd.; Fig. 3.10.1 and Tables 3.4, 3.5, 3.13, 3.17. 3.18, 3.20, 3.21, 4.18 reproduced from *The Character and Origin of Smog Aerosols* (G. M. Hidy et al. eds.) by permission of John Wiley and Sons Inc., © 1980 John Wiley and Sons; Table 4.21 reproduced from *Sulfur in the Environment* (J. Nriagu) by permission of John Wiley and Sons Inc., © 1978 John Wiley and Sons; Fig. 3.10.2 and Table 3.1 reproduced from *Planets and Their Atmospheres. Origin and Evolution* (J. S. Lewis, R. G. Prinn) by permission of Academic Press Inc.; Table 3.3 reproduced from *Proceedings of the 2nd International Clean Air Congress* (G. M. Hidy, S. K. Friedlander) by permission of Academic Press Inc.; Figs. 3.11, 3.12, 3.15, 3.24, 3.25, 3.26, 3.31, 3.33, 3.35, 3.36, 3.37 and Tables 3.15, 3.28, 3.33, 3.34, 3.35, 3.36, 3.39, 3.40 reproduced from *Deposition of*

Atmospheric Pollutants (H. W. Georgii, J. Pankrath eds.) by permission of Kluwer Academic Publishers B.V.; Figs. 3.22.2, 3.27, 4.11 and Tables 3.14, 3.24, 3.25, 3.26, 3.27, 3.29, 3.30, 3.31, 3.32 reproduced from *Saure Niederschläge – Ursachen und Wirkungen –* (VDI-Berichte 500) by permission of VDI-Verlag GmbH; Fig. 3.28 reproduced from *Saure Niederschläge – eine Trendanalyse* (P. Winkler) by permission of Deutscher Wetterdienst; Figs. 4.2, 4.3, 4.8, 4.9 and Table 4.10 reproduced from *Plant–Water Relationships in Arid and Semi-Arid Conditions* (UNESCO) by permission of United Nations Educational Scientific and Cultural Organization, Tables 4.45, 4.46, 4.49 reproduced from *Soil Biology* (M. Alexander) by permission of UNESCO; Fig. 4.10 reproduced from *Ein Methodenvergleich zur Laborbestimmung des K_f-Wertes von Sanden* (A. Pekdeger, H. D. Schulz) by permission of Geologisch-Paläontologisches Institut der Universität Kiel; Fig. 4.13 reproduced from *Umweltbereich Luft* (Folienserie) by permission of Fonds der Chemischen Industrie; Fig. 4.20 and Tables 2.7, 4.47 reproduced from *Einfluß von Pflanzenschutzmitteln auf die Mikroflora von Böden* by permission of J. C. G. Ottow; Fig. 4.21 reproduced from *Structure–Activity Relationships in Toxicology and Ecotoxicology: An Assessment* by permission of ECETOC – European Chemical Industry Ecology & Toxicology Centre; Fig. 4.22 reproduced from *Verölung und Ölabbau im Lebensraum Meer* permission of Bruno P. Kremer; Fig. 4.23 and Tables 2.1, 2.4 and 2.5 reproduced from *Handbook of Environmental Data on Organic Chemicals* (Verschueren) by permission of van Nostrand Reinhold Company Inc.; Tables 1.2, 3.11 reproduced from *Produktionsmengen und Verwendung chemischer Stoffe* (G. Rippen) by permission of Battelle Europe; Table 2.11 reproduced from *Amer Soc for Testing and Material, Spec Tech Publ 657* (Cairns et al.) by permission of American Society for Testing and Material; Table 3.2 reproduced from *Umwelt 1/82, 2/82* (M. Häberle) by permission of VDI-Verlag; Tables 3.6, 3.7, 3.8 reproduced from *Räumliche Erfassung . . . Materialienband 1* by permission of NUKEM GmbH; Table 3.10 reproduced from *Berichte Dt. Wetterdienstes Bd. 4 Nr. 26* (R. Bögel) by permission of Deutscher Wetterdienst; Table 3.22 reproduced from *Environ Sci Tech 9* (W. L. Dilling et al.) by permission of American Chemical Society; Table 3.23 reproduced from *Environ Sci Tech 19* (K. R. Darnall et al.) by permission of American Chemical Society; Table 3.42 reproduced from *Taschenbuch für Umweltschutz Bd. 1* (W. Moll) by permission of Steinkopff; Table 4.3 reproduced from *Das Klima der bodennahen Luftschicht* (R. Geiger) by permission of Friedrich Vieweg & Sohn Verlagsges. mbH; Table 4.11 reproduced from *Monitoring of Air Pollutants by Plants* (H.-J. Jäger) by permission of Dr. W. Junk Publ.; Tables 4.24, 4.24 reproduced from *NATO CCMS Pilot Study on Air Pollution Control*

Strategies and Impact Modelling 139 by permission of NATO; Table 4.22 reproduced from *Forschungsbericht 85 106 08 010* (F. Jörg et al.) by permission of Fraunhofer Ges.; Tables 4.2, 4.26 reproduced from *Lehrbuch der Bodenkunde* (F. Scheffer, P. Schachtschabel) by permission of Ferdinand Enke Verlag; Tables 4.31, 4.32, 4.33 reproduced from *Mitt. Dt. Bodenkdl. Ges. 53* (H. P. Blume, G. Brümmer) by permission of Dt. Bodenkdl. Ges.; Tables 4.36, 4.37, 4.38, 4.39 reproduced from *Soil Science 134* (Boyd) by permission of The Williams & Wilkins Co.; Tables 4.40, 4.41 reproduced from *Environ Sci and Tech 15* (R. Schwarzenbach, J. Westall) by permission of American Chemical Society; Tables 4.6, 4.7 reproduced from *Z. f. Pflanzenernährung und Bodenkunde* (Bücking, Krebs) by permission of VCH; Tables 4.43, 4.44 reproduced from *Zeitschrift für Pflanzenernährung und Bodenkunde* (Wilke) by permission of VCH; Tables 4.50, 4.51, 4.52, 4.53, 4.54, 4.55, 4.56 reproduced from *DVWK – Merkblätter zur Wasserwirtschaft* by permission of Deutscher Verband für Wasserwirtschaft und Kulturbau e.V.

Contents

1 Introduction

1.1 Man and Environment

In the perspective of the last two million years of human evolution, culture and biology have been inextricably interwoven. Our early Palaeolithic or Acheulian ancestors, radiating from East Africa and inhabiting much of the three Old World continents by 500,000 B.P., already attempted to circumvent environmental limitations, via culture, in expressing dietary preferences different from the youngest australopithecines and in colonizing both drier and colder environments. Culture, instead of obviating the role of biological evolution, acted as a catalyst (Brain 1981).

Environmental changes, in particular the Mid-Pleistocene episodes of cold climate in higher latitudes and corresponding aridity in tropical regions, triggered cultural responses that favoured biological capacity for culture. The peculiar human tendency to defy environmental constraints was one of the essential prerequisites to the development of increasingly flexible subsistence settlement systems.

Specifically, by 35,000 B.P., man had acquired the linguistic and social skills essential to the increasingly rapid evolution of our cultural heritage. New continents such as the America and Australia were taken possession of; brilliant artistic expression merged from the darkness of prehistoric ages, manipulation of plants and animals led eventually to their domestication, and agriculture and pastoralism began to provide supplementary, then alternative, modes of subsistence. Intensified agriculture allowed the creation of permanently settled villages; and a further intensification, supplemented by artificial irrigation, began to support multitiered economies and urbanized centres. Finally, complex stratified societies allowed the development of industry and industrialization, and the present technical world civilization emerged.

In the course of his adaptive biological and, in particular, cultural radiation, man has exerted an ever-increasing influence on the ecosystems that ensured his livelihood. In Acheulian times, *Homo erectus* was dispersed over large geographical ranges. The repetitiveness of the archaeological residues suggests a strongly patterned if rudimentary human way of life (Butzer 1977). It is probable that the food preferences of these hunters and gatherers required larger territories, and thus comparatively few Acheulians could be supported in any one area; mobility and the periodicity of seasonal activities are likely to have

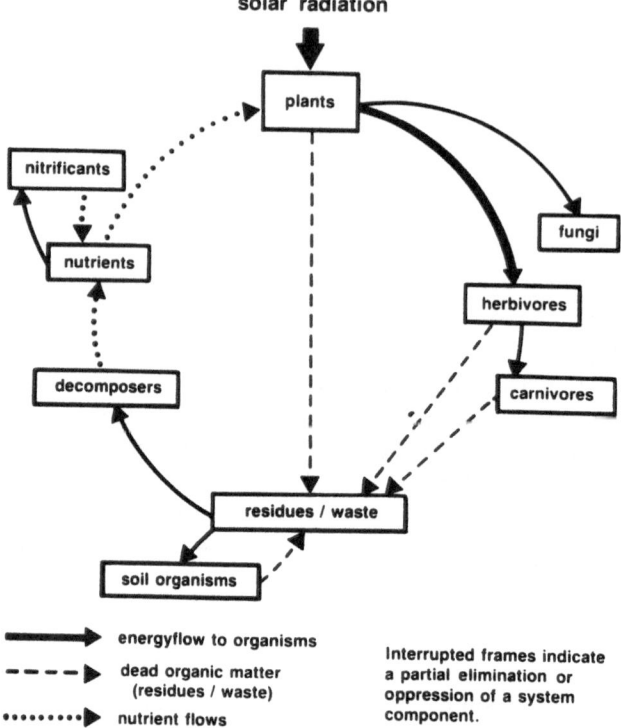

Fig. 1.1. Simplified model of a natural ecosystem. (After R.S.U. 1978)

been considerable (Freeman 1973; Leroi-Gourhan 1983). The surviving material culture in Africa, Europe and Asia exhibits little geographical variation of a functional rather than stylistic nature, and the sites indicate a preference for open, grassy environment, with large herds of gregarious herbivores. This means that human influence on ecosystems was by no means different from that of other hunting or gathering animals; a specific formal distinction of man from coeval herbivores and carnivores in an ecosystem model is consequently not indicated (Fig. 1.1).

Things changed drastically when agricultural societies arose, which was probably the case in much of southwest Asia and the Aegean sphere during the tenth millenium B.P. The new subsistence system, based on crop-planting, livestock-raising and essentially permanent settlements, villages first and towns only a few millenia later, had been successfully developed to the point that long-term population increases were possible. Significant ecosystem changes resulted as native faunas and floras began to be replaced by domesticated stock and cultigens. Not infrequently, the hydrological cycle was disturbed and soil erosion initiated or enhanced.(Seymour and Girardet 1985).

From the viewpoint of ecosystems, this means that the largely prevailing agrarian and woodland ecosystems were not only profoundly changed in species composition, as the two models of agrarian ecosystems from mediaeval and present times indicate (Figs. 1.2, 1.3), but that re-cycling processes constantly lost importance, while the intensity of chemical impacts grew at an ever-increasing rate.

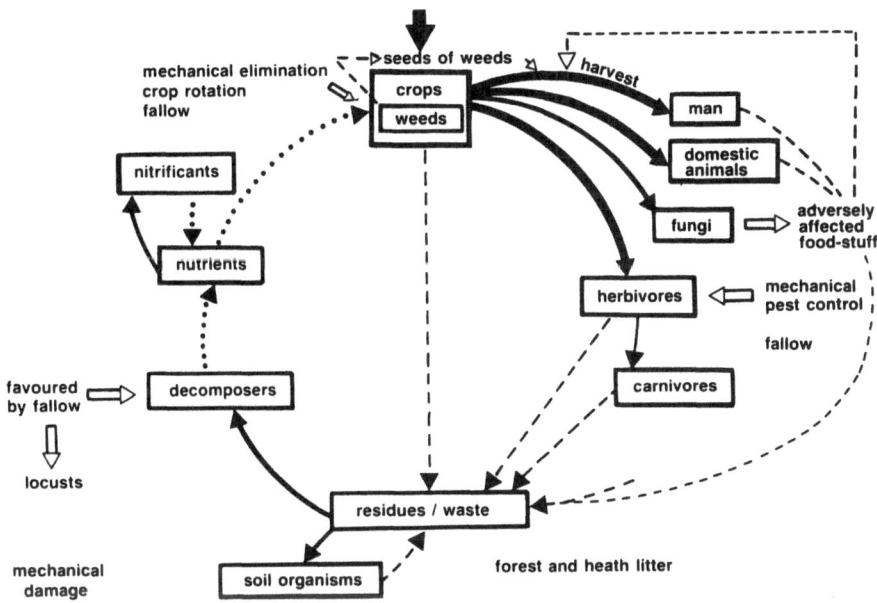

Fig. 1.2. Agrarian ecosystem in medieval Europe. (After R.S.U. 1978)

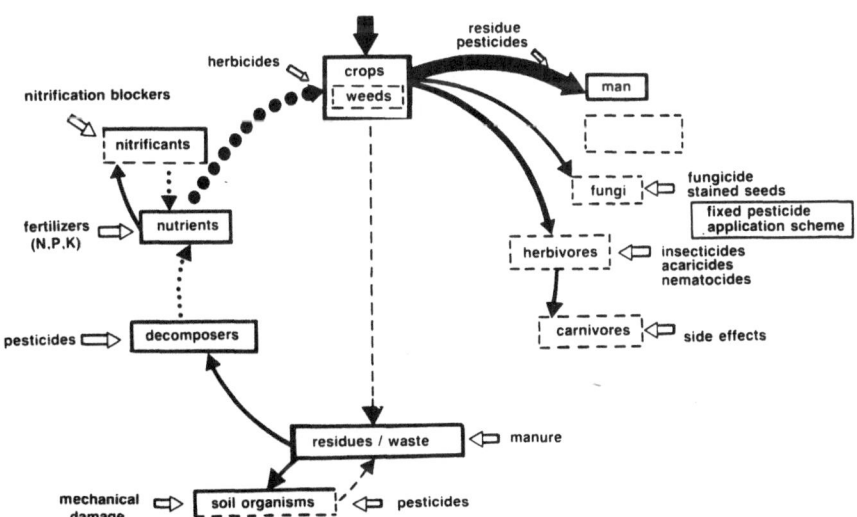

Fig. 1.3. Modern high-yield grain production system. (After R.S.U. 1978)

In addition to these structural changes of agrarian ecosystems, the fact that the formerly "isolating" woodlands or uncultivated areas around them were constantly reduced, merits particular attention. It means that the buffering capacity of these non-agrarian areas, which had played a very important role in the intricate ecosystem interplay of the past, no longer exists in the densely populated parts of the world. Consequently, both agrarian and non-agrarian types of ecosystems are now open to human impacts on the regional and continental levels, and present-day analytical knowledge provides ample evidence that even the most remote areas on earth bear more or less marked imprints of human activity. Among them the chemical imprints merit particular attention, because their importance has grown since the middle of the last century with an ever-increasing speed, and the development is not likely to change in principle, albeit sectoral differentiations are evident.

1.2 Technical Chemistry and Ecology

Since 1960, the chemical industry has grown at about twice the rate of the overall industry sector, and it plays a critical role in economic growth and industrial development throughout the world. In 1978, the chemical industry of OECD member countries manufactured products worth over 350 billion dollars, an estimated two-thirds of world production. It ensures the livelihood of many more people in those "down-stream" industries whose very existence depends on the products the highly diversified chemical industry manufactures.

Just because these products are now an essential part of both our economic and social life, governments and industry alike have become increasingly concerned about potential unintended consequences which the use of man-made chemicals could have on both human health and the environment. The numbers are striking: of some 4 million known chemical substances, some 100,000 are produced in commercial quantities, and it is estimated that as many as 1000 new substances reach the market every year.

A closer analysis of these figures in the light of production volume information as available from the US Environmental Protection Agency (EPA) Toxic Substances Control Act inventory and national and international production statistics (cf. Rippen et al. 1987a) shows that an estimated maximum of only some 20.000 substances out of the universe of commercial chemicals are of technical importance. A first comparison on the basis of production volume correlations indicates that the USA and the member states of the European Community produce one-third of the total world production of chemicals each, while the Federal Republic of Germany, in turn, accounts for about one-third of the EC production, and consequently somewhat less than 10% of the total world production. More revealing are sub-divisions of the high-volume materials in terms of product categories and production ranges. Table 1.1 indicates for the USA as a representative economic unit that a relatively small number of materials accounts for the bulk of the production volume.

Table 1.1. Volume distribution of commercial chemicals in USA. (After Blair 1981, on the basis of the entire EPA inventory)

Production range (lbs/yr)	Number of materials	%	Total production (million lbs/yr)	%	Cumulative production %
$> 10^{11}$	1	< 0.1	102,000	2.5	2.5
10^{10}–10^{11}	95	0.2	3,119,000	76.5	79.0
10^9–10^{10}	216	0.5	656,000	16.1	95.1
10^8–10^9	436	1.1	155,000	3.8	98.9
10^7–10^8	1065	2.7	33,800	0.8	99.7
10^6–10^7	1983	5.0	8140	0.2	99.9
10^5–10^6	3798	9.7	1720	0.04	99.98
10^4–10^5	4689	11.9	225	0.01	99.99
$< 10^4$	27,010	68.7	28	< 0.01	100.00

Dividing all those substances produced in quantities of 1 million pounds per year ($450 \, t \, a^{-1}$) or more into major categories such as organics, inorganics, polymers etc. shows that petroleum derivatives (gasoline, kerosene, distillation cuts etc.) represent 10% of the total number of entries in the inventory, but account for 55% of the total production. The inorganics represent both 12% of the materials and 12% of the production. Another 6.9% of the production is due to materials which are residues from the processing of ferrous metals. The saturated hydrocarbons are responsible for 6.7%. Well-defined organic substances, the materials which cause major concern in testing, are the most numerous, as they represent about 34% of the inventory sample, but they account for only 6% of the total production. Polymers and plastics represent 24% of the number of materials and 3% of the total production.

Table 1.2 provides an exemplary comparison of selected high-volume chemicals in the Comecon, Federal Republic of Germany (D), European Community (EC), USA and the world. The figures listed indicate that man, owing to the accelerated growth of the chemical industry during the past 40 years in all continents except Antarctica, produces and uses a great many synthetic compounds in quantities which appear, in localized exposure at least, no longer negligible in comparison to natural fluxes and concentrations. Thus the ecological role of man has changed profoundly in little more than one century of his long cultural history, since he is now capable, by introducing new substances or increasing the concentration of natural ones such as CO_2, CH_4 and CO, of changing structure and function of the terrestrial ecosystems to an extent unthought of in the past (cf. Table 1.3).

Because the products of the chemical industry are also a most essential part of both our economic and social life, governments and industry alike have become increasingly concerned about potential unintended consequences which the use of man-made chemicals could have on both human health and the environment. Therefore a number of key chemical-producing countries have

Table 1.2. Comparative volume distribution of technical chemicals (t/a). (After Rippen et al. 1987a)

Product		World	USA	D	EC	Com
H_2SO_4	(1985)		3.593×10^7	4.199×10^6		
HCl	(1985)		2.519×10^6	0.94×10^6		
Cl_2	(1985)		9.5×10^6	3.5×10^6		
HNO_3	(1985)		7×10^6	2.9×10^6		
NaOH	(1985)		9.9×10^6	3.7×10^6		
Na_2CO_3	(1985)		7.7×10^6	1.4×10^6		
Na_2SO_4	(1981)		1.1×10^6	9×10^3		
P_2O_5	(1985)	4.4×10^6	1.04×10^6	1.9×10^5	6×10^5	6×10^5
KOH	(1975)	5×10^5	1.3×10^5	1×10^5		
$CaO/Ca(OH)_2$	(1981)	1.17×10^8	1.7×10^7	0.5×10^6	1.4×10^7	3.5×10^7
CaC_2	(1976)	6×10^6	2.5×10^5	4.6×10^5		
Al	(1981)	1.6×10^7	4.7×10^6	7.3×10^5		2.4×10^6
Al_2O_3		3.2×10^7	6×10^6	1.4×10^6		5×10^6
$Na_2Cr_2O_4$	(1976)	3.8×10^5	1.3×10^5	6×10^4	1.3×10^5	1×10^5
Plastics			1.57×10^7	7.6×10^6		
Fertilizers		7.42×10^7	1.11×10^7	1.16×10^6		
Synthet, caoutchouc		9×10^6	1.8×10^6	0.45×10^6		
Synthet, fibres		1.25×10^7	2.8×10^6	0.8×10^6		
CH_3COOH		3×10^6	1.1×10^6	0.266×10^6		
CH_3CHO		2.4×10^6	0.45×10^6	0.36×10^6		
C_6H_6		1.44×10^7	3.77×10^6	0.89×10^6	4.4×10^6	
$C_2H_4Cl_2$		1.28×10^7	4.2×10^6	1.3×10^6	5.3×10^6	
C_2H_3Cl		7.7×10^6	2.1×10^6	1.1×10^6	3.5×10^6	
C_2H_4		2.5×10^7	1.3×10^7	3.6×10^6	8.7×10^6	
C_6H_5OH		3×10^6	1.3×10^6	0.27×10^6		
C_2HCl_3		0.6×10^6	0.14×10^6	0.38×10^5	0.24×10^6	

passed, or are enacting, general substance control legislation (cf. Johnson 1981; BUA 1989). Common to all these legal instruments is the preventive aspect, i.e. the notification of chemicals prior to marketing, which entails the presentation of data derived from laboratory investigations together with additional information permitting the evaluation of potential hazards. In addition to this prospective notification procedure, the German Chemicals Act of 1980 (Chemikaliengesetz – ChemG) stipulates that also existing chemicals must be reported to the competent federal authority, if they exhibit properties which indicate they may be hazardous, either alone or in combination with other chemicals [§4 (6) ChemG]. To this end, an Advisory Committee on Existing Chemicals of Environmental Relevance (BUA) was established by the Society of German Chemists (GDCh) in 1982. Since then it has been selecting and examining existing chemicals for environmental and health purposes on the basis of consecutive priority lists (BUA 1989).

Table 1.3. Reference chemicals in order of estimated global release in kt/yr (ECETOC 1988c)

1,2,3-Trichlorobenzene	0.09	1,2-Dichloropropane	15
1,3,5Trichlorobenzene	0.09	Diethlyphthalate	23.1
1,2,4,5-Tetrachlorobenzene	0.15	Pentachlorophenol	27
4-Chloroaniline	0.16	p-p'-DDT	28.5
2-Chloroaniline	0.17	Aniline	32.7
1,2-Dichloroethene	0.3	Dichloroisopropyl ether	45
1,1,2,2-Tetrachloroethane	0.45	Nitrobenzene	46.5
4-Nitroaniline	0.55	Malathion	47.4
Hexachlorobutadiene	0.6	o-Xylene	63
N,N'-Dimethylaniline	0.72	1,4-Dichlorobenzene	66
3,4-Dichloroaniline	1.2	Hexachlorobenzene	80
4-Methylphenol	1.2	Tetrachloromethane	84
Dimethoate	1.8	p-Xylene	100
1,1,2-Trichloroethane	2.4	Atrazine	111
1,3-Dichlorobenzene	3	Alachlor	123
Dieldrin	3	Phenol	180
Tri-N-Butylphosphate	3	Styrene	210
1,2,4-Trichlorobenzene	3.3	Di-N-Butylphthalate	230
m-Xylene	3.6	Trichlorofluoromethane	291
Endrin	6	Trichloroethene	326
Chlorobenzene	6.6	Di(2-Ethylhexyl)phthalate	402
Di-Methylphthalate	6.6	Tetrachloroethane	470
Aldrin	7.5	1,1,1-Trichloroethane	537
Heptachlor	8.1	1,2-Dichloroethane	547
1,2-Dibromoethane	9	Ethylbenzene	600
Parathion (methyl and ethyl)	9.6	Cyclohexanol	1200
Chlorpyrifos	10	Toluene	1800
1,1-Dichloroethane	12	Benzene	11000
γ-HCH (Lindane)	14.1		

A comprehensive assessment of the potential ecotoxicity of new and existing chemicals implies the determination of their persistence and distribution potential on the one hand, and the analysis of the respective modes of immission and the structure of the ecosystems exposed on the other. The way followed in this book to unravel, to a certain extent, the bewildering number of interrelationships which have to be considered in such an analysis is model building. It has the advantage of admitting a variety of broadly different approaches, ranging from empirical models for practical purposes, to rather abstract ones aiming at qualitative general insights.

At one end of this spectrum there is a detailed and pragmatic description of specific systems such as soil horizons or their defined adsorbents in interaction with one or a few pure chemicals in aqueous solutions. This "tactical" approach, as Holling (1966, 1968) has it, may be of considerable importance both for particular projects of resource management and as a method of grouping and codifying masses of experimental data. At the opposite end of the spectrum are relatively general "strategic" models, again in Holling's terminology, which have to sacrifice numerical precision for the sake of general principles. Such models, formulated in the present book in the form of fairly comprehensive graphs, need not correspond in detail to any single "real world" process, but aim to provide a conceptual framework for the discussion of broad classes of phenomena or simply of contentious issues. Rationally handled, tactical and strategic approaches mutually reinforce each other, thus providing reciprocally new and deeper insights.

In ecotoxicology, it may be surmised, tactical models of the systems analysis kind, applied to specific problems of hazard assessment or environmental management, are likely to remain more fruitful in the near future than general theory. However, in the long run, a more evolved science embedded in a comprehensive long-term ecosystem research (Ellenberg et al. 1978) will draw upon the wide spectrum of theoretical models, from the very general or abstract to the very particular or specific ones. They should reflect the distribution of potentially toxic substances in whole ecosystems or selected compartments, their short- and long-term effects on the structure and functioning of the systems affected, and accumulation and degradation processes via the food chains.

2 Basic Chemical and Ecological Principles

Ecotoxicology may be defined as "a science which is concerned with the toxic effects of chemicals and physical agents on living organisms, especially on populations and communities within defined ecosystems; it includes the transfer pathways of those agents and their interactions with the environment" (Butler 1978).

In formal analogy with the objectives of classical toxicology which has evolved as an aut- or demecologically oriented discipline, ecotoxicology aims at defining toxic effects in terms of reactions of ecosystems or representative compartments of such systems. In view of the complicated structure of ecosystems and their tremendous variability in time and space, however, particular problems arise. While many experimental approaches exist to predict hazards of drugs, food additives and contaminants for man which are based on extrapolation from animal or other data, the analogous ecotoxicological procedures are much more complicated and hitherto by far less satisfactory. This is, for instance, well exemplified by the valuable efforts of OECD working groups dealing with various aspects of the Chemicals Testing Programme of this organization.

The group on degradation and accumulation and the group on ecotoxicology have proposed test methods to identify a number of potential properties of chemicals which indicate a possible hazard for functionally important organisms. They are, in essence, mostly laboratory experiments, although it has been recognized that "where appreciable environmental concentrations of the chemicals are likely to be involved, and/or some indication of possible environmental hazards exists, it may be necessary to assess the effect in experimental systems more closely approaching something like natural conditions, especially with regard to interspecific relations and the functioning of multispecies systems".

The very crux of the ecotoxicological extrapolation problem is the difficulty that an eventual decision-making process will lead either to "false negatives or to false positives" (Koeman 1982). In this connection a false negative implies that a chemical looking acceptable on the basis of laboratory data may yet cause considerable damage under practical circumstances. If, for instance, the prediction of possible environmental exposure levels on the basis of standardized laboratory tests turns out to underestimate the actual levels occuring in "real world" environments, populations of certain species will be much more at risk than expected. Thus species with mono- or oligomaniacal feeding behaviours, i.e. food specialists, run a much higher risk of falling prey to bioaccumulation effects if specific chemicals are applied against their favourite or only prey animals than less specialized predators. Eider ducks, to give another example, appear to be much more in danger from certain organochlorine pesticides than most other bird species living in the same environment, because the females do not feed during the incubation period. The resulting partial atrophy of their tissues mobilizes lethal concentrations of the compounds (Koeman and van Genderen 1972).

In contrast, the false positive involves rejection of chemicals because laboratory tests indicate intolerable hazards for the environment, while in fact such undesirable effects are not likely to occur. Pesticides may become labelled as toxic to fish on the basis of standardized laboratory trials, although this does not necessarily imply that they will always cause mortality in fish when applied in, or in the immediate neighbourhood of, their natural aquatic habitat. In the framework of tsetse fly control in West Africa, it could be amply demonstrated that the potentially fish-toxic insecticides endosulfan and deltamethrin (Industrieverband Pflanzenschutz 1982) could actually be used without causing any noticeable effect on fish (Takken et al. 1978). The effective dose rate for killing tsetse flies was minimalized by a careful selection of the insecticide formulation on the one hand and application methods on the other.

These examples are to indicate that the hazards of pesticides usually depend largely on such factors as the amount applied, the formulation, ways and times of application, and intensity of use, which are, in their extremely variable specific combinations, virtually impossible to simulate. Attempts are necessary, therefore, to consider the ecological characteristics of target areas as well as possible non-target effects in connection with the use pattern envisaged in order to avoid unrealistic predictions. A compound which appears unacceptable in a certain application in one place need not necessarily be so also in another. Hence basic physical–chemical data like those described in the *Collection of Minimum Pre-Marketing Sets of Data* of the OECD Working Group on Exposure Analysis (Umweltbundesamt 1982) have to be matched to data on the relevant properties of typified environments into which the compounds are ultimately released. It is under this particular perspective that the essential structural and dynamic characteristics of both chemicals and ecosystems which are relevant to ecotoxicology are described in the following chapters.

2.1 Exposure and Effect Criteria of Chemicals

Methodologies for estimating fluxes, transformations and resultant environmental concentrations of chemicals depend on the knowledge of how rapidly they are discharged to a specific compartment in the environment and how rapidly they are transformed or removed by physical, chemical and biological processes. Apart from the specific structure of the receiving compartment, and in particular its physical and chemical characteristics (cf. Chaps. 4.1, 4.2, 4.3 and 4.4), these processes are largely controlled by the inherent molecular properties of the chemicals released. These may be appropriately classified as sets of equilibrium and kinetic constants which define the exposure potential, and a complementary set of data which characterizes the effect potential, i.e. chemical, biotic and, in particular, ecotoxicological effects of a substance.

In the following sections the essential exposure and effect criteria are described. Other chemical and physical properties like flash points, flammability limits and autoignition temperature are omitted because they are not of direct concern to the environmentalist. The interested reader may find a comprehensive review of these and other dangerous properties of chemicals in *Dangerous Properties of Industrial Materials* by Sax (1975). With regard to the following quantitative specifications, it must be noted, however, that chemicals are never 100% pure, and that the very nature and quantity of the impurities can have a significant impact on many environmental qualities. In particular, the following parameters are very sensitive to the presence of impurities: water solubility, odour characteristics and threshold values, biological oxygen demand and toxicity (Verschueren 1983).

2.1.1 Structure–Activity Relationships

Structure–activity relationships rationalize the relations between selected aspects of the environmental behaviour of chemicals, e.g. distribution in air, soil and water or biological activity, and their structure. Structure–activity relationships may be quantitative, providing a specific value for a rate constant (Chapman and Shorter 1972) or qualitative, providing only an upper or lower limiting value for it. Thus the quantitative estimate of the constant gives the investigator a basis for estimating the range of values of the rate process, whereas the qualitative estimate frequently is a limiting value that makes it possible either to eliminate the process from further investigation or define research needs more precisely.

Over the past two decades there has been a rapid increase in the development of structure–activity relationships whose analysis started with the classical contribution of Hammett and Pfluger (1933). One reason is that a large number of chemicals has come under scrutiny regarding their potential toxic effects on humans or environmental species, and in the absence of adequate experimental data techniques for predicting effects from sets of easily measurable or easily calculable properties would be an attractive alternative. Qualitative and quantitative structure–activity relationships (customarily abbreviated as QSARs or simply SAR) have been of interest, mainly in the USA (Arcos 1983; Moore 1984), for the preliminary assessment of substances notified to the regulatory authorities for the control of chemicals.

2.1.1.1 Chemical Structure Descriptors

Any structure–activity relationship is characterized by three components:

- chemical structure descriptors, i.e. data on certain properties of the molecule, or physical or chemical potentials of the substance
- the environmental (for instance, biological) activity
- the technique used to define the relationship.

Many parameters have been applied as descriptors to express indirectly some or other aspects of chemical structure in QSARs. The following compilation lists the main ones (cf. ECETOC 1986a):

Physico-chemical descriptors

1. *General*

Melting point	Activation energy
Boiling point	Heat of reaction
Vapour pressure	Reaction rate constant (k)
Dissociation constant (pK_a)	Reduction potential

2. *Hydrophobicity*

   Partition coefficient (P)
   R_M coefficient from reverse-phase chromatography
   Solubility in water (S)
   Parachor

3. *Electronic*

Hammett constant (σ)	Dielectric constant
Taft polar substituent constant (σ^*)	Dipole moments
	H-bonding (HB)

   Ionization potential

4. *Quantum–chemical* (including indices derived from molecular orbital calculations)

   Molecular orbital indices such as
   - atomic charge   – bond energy
   - bond indices   – resonance energy
   - electron-donating character (energy of the highest occupied MO)
   - electron-accepting character (energy of the lowest unoccupied MO)
   - electrostatic potential distribution
   - Dewar numbers

   Electron density
   π-Bond reactivity
   Electron polarizability

Steric

Molecular volume   Substructure shape
Molecular shape   Taft steric substituent constant (E_s)
Molecular surface area
Molecular refractivity (MR)   Verloop STERIMOL constants
                                               $(L; B_1-B_5)$

Structural

Atom and bond fragments   Number of rings (in polycyclic
                                               compounds)

Substructures
Substructure environment   Molecular connectivity
                                               (extent of branching)

Number of atoms in a given
substructural element

Except for elementary ones, a description of these descriptors is given in the
following Section 2.1.1.2, while environmentally relevant structure–activity relationships are described in greater detail in Sections 2.1.2 and 2.1.3.

2.1.1.2 Techniques for Defining Structure–Activity Relationships

A variety of techniques has been used to define structure–activity relationships
(cf. Golberg 1983). The simplest ones are two-dimensional graphical plots of
a single chemical descriptor as related to a specific activity, e.g. scatter diagrams
or the equations derived from the elementary data by means of curve-fitting
procedures. More elaborate are different kinds of regression analysis involving
linear or exponential, simple or multiple regressions (Lambert 1967; Hance
1969; Karickhoff et al. 1979; Moreale and van Bladel 1980; Fränzle 1982, 1986;
Mingelgrin and Gerstl 1983; di Toro 1985; Topp 1986; Fränzle et al. 1987a, b).
    While the QSARs served to describe the relative importance of sorption-
relevant soil constituents such as organic matter, clay minerals, hydrated weakly
crystalline iron, aluminium, manganese oxides in relation to environmental
chemicals in terms of chemical structure (cf. Chap. 4.4.1) the widely used Hansch
(Hansch and Fujita 1964) method relates biological effects to hydrophobic,
electronic and steric properties of a chemical.
    Hydrophobicity (or lipophilicity) is generally defined by octanol/water parti-
tion coefficients (Sect. 2.1.2.6), and its influence on biological activity seems to be
related to the transport of a chemical into and through lipid membranes to its
site of action. It is also a descriptor for hydrophobic interactions, i.e. interactions
where the association of nonpolar regions in molecules with each other is
stronger than with water.
    Fujita et al. (1964) showed that within various series of aromatics the
contribution of a particular substituent to the *n*-octanol/water partition co-

efficient (P_{ow}) was a characteristic of the substituent. Thus the hydrophobic parameter for a substituent X can be defined as:

$$\pi_X = (\log P_{ow})_{RX} - (\log P_{ow})_{RH} , \tag{2.1}$$

where RX = substituted aromatic and RH = unsubstituted parent aromatic molecule. The predictability of P_{ow} has been extended to aliphatic compounds, and other techniques have been developed for calculating P_{ow} from molecular fragment values (Rekker 1977; Hansch and Leo 1979).

π values proved to be relatively constant from one system to another as long as there are no steric or electronic interactions of the substituents. The following Table 2.1 shows that, with the exception of the CH_2 and NH_2 functions, π values for aliphatic positions are lower than for aromatic ones.

It should be noted, however, that the additive–constitutive character of log P is influenced by:

– steric effects which may induce shielding of an active function by inert groups,
– inductive effects of substituents on each other,
– intra- and intermolecular hydrogen bonding,
– branching,
– conformational effects, for instance "balling up" of an aliphatic chain.

The inherent difficulties to assess the steric, inductive and conformational effects

Table 2.1. Comparison of aromatic and aliphatic π values. (Verschueren 1983)

Function	Aromatic π $\log P_{C_6H_5} - \log P_{C_6H_6}$	Aliphatic π $\log P_{RX} - \log P_{RH}$
NH_2	− 1.23	− 1.19
I	1.12	1.00
$S-CH_3$	0.61	0.45
$COCH_3$	− 0.55	− 0.71
$CONH_2$	− 1.49	− 1.71
$COOCH_3$	− 0.01	− 0.27
Br	0.86	0.60
CN	− 0.57	− 0.84
F	0.14	− 0.17
Cl	0.71	0.39
COOH	− 0.28	− 0.67
OCH_3	− 0.02	− 0.47
OC_6H_5	2.08	1.61
$N(CH_3)_2$	0.18	− 0.30
OH	− 0.67	− 1.16
NO_2	− 0.28	− 0.85
CH_2	0.50	0.50

are such that calculated $\log P$ values of complex molecules not infrequently differ from the true (i.e. experimental) ones by one or two orders of magnitude (Verschueren 1983). Nevertheless the n-octanol/water partition coefficient has proved useful as a simple means to predict biological uptake, lipophilic storage and biomagnification factors (cf. Sect. 2.1.3.5), while extrapolations of the multifactorial soil adsorption processes are much more difficult (cf. Chap. 4.4.1.2).

Electronic properties play a part in the specific interactions between a chemical compound and a biological or other receptor, e.g. electrostatic interactions between centres of opposite charge. In analogy to the above additive π value, the effect of a substituent on the electronic properties of the molecule can be characterized by the Hammett and Taft constants. The Hammett constant,

$$\sigma_X = \log K_{RX} - \log K_{RH} \, ,$$

is derived from the dissociation constants (K) of the substituted and unsubstituted benzoic acid, RX and RH, respectively (ECETOC 1986a). If σ_X is positive, then the group X is more electron-attractive than is H, and vice versa. The Taft polar substituent constant σ^* is similarly derived from substituted acetic acids.

Steric properties influence the approach or binding of a molecule to a biological or other site. As a parameter of the steric effect of a substituent in a molecule, Taft (1956) proposed a constant E_s determined from the rate constant (K) of the hydrolysis of the ester $(XCOOC_2H_5)$ according to equation:

$$(E_s)_X = \log K_{RX} - \log K_{RH} \, , \tag{2.2}$$

where K_{RX} and K_{RH} denote, respectively, the rate constants of the hydrolysis of the substituted and unsubstituted esters RX and RH. Values of σ and E_s for many substituents have been tabulated by Hansch and Leo (1979).

The above physico-chemical properties may be combined into one equation to describe more comprehensively the biological activity of chemicals:

$$\log BA = a\pi^2 + b\pi + c\sigma + dE_s + e \, , \tag{2.3}$$

where BA is the biological activity of a member of a series of compounds RX, with varying substituents X, and π, σ, E_s have the above meaning while a–e are coefficients determined by regression analysis. The first term was included because of the possibility of a parabolic dependence of biological activity on hydrophobicity.

In practical application, the Hansch equation requires values of the biological activity and the chemical descriptors for each of an initial series of compounds. These values are put into the equation and a general equation fitting all compounds, i.e. the structure–activity relationship is determined by regression analysis. The goodness of fit (possibly after emission of "outliers") is expressed by statistical quality criteria such as the product–momentum correlation coefficient, the standard error, and the F-ratio. Values of the activity of structurally related compounds can then be predicted from their chemical descriptors.

Free–Wilson and related group-contribution techniques are based on the assumption that in a series of structurally related chemicals each particular substituent adds a constant contribution to the activity of the parent molecule (Free and Wilson 1964). Hence

$$f(BA) = S_A + S_B + \cdots + \mu , \tag{2.4}$$

where BA denotes the biological activity, and μ is the contribution of a hypothetical parent compound to the biological activity, while $S_A, S_B, \ldots$ are the individual contributions added by substituents A, B, etc.

Because each compound yields an equation of the above type the substituent contributions are found by solving a system of multiple simultaneous equations (Purcell et al. 1973). The goodness of fit can be determined by regression or related analyses.

Specific physico-chemical properties are not ascribed to the substituent group in the Free–Wilson approach; they are contained within the additive group contributions. Therefore information on the mechanism of biological action is not usually obtained.

Enslein and Craig (1978) suggested an extension of the above approach by dividing each molecule into sub-structures, where the contribution of the individual sub-structures is calculated by regression and discriminant analysis.

Pattern recognition methods are applied to seek qualitative correlations between a set of descriptors of atomic or a molecular structure and the presence or absence of specific biological (or other) activities in a series of structurally related compounds. Most variants of this technique which is wholly empirical and predictive (i.e. a priori without a biological basis) have the following common features (ECETOC 1986a):

- The training set of chemicals is normally large in comparison to that of other QSAR techniques.
- A training set of atomic or molecular descriptors expressing chemical structure is chosen
- Each chemical structure is represented by a point in n-dimensional space, where n is the number of chemical descriptors chosen.
- If the structure-descriptors have been chosen appropriately, chemicals possessing a specific biological or environmental activity will cluster in one part of the n-dimensional space, while those lacking this activity will cluster in a separate part or may not cluster at all.
- If there is no adequate separation of active and inactive compounds, different sets of structural descriptors are sequently tested until a valid structure–activity relationship is found (or the attempt abandoned).
- A valid structure–activity relationship can then be used to predict whether a structurally related compound is likely to possess the environmental activity under consideration by using the appropriate structure descriptors and determining whether the chemical clusters with the active or inactive groups.

A variety of statistical methods is available for analyzing the data sets in the framework of pattern recognition techniques. The most common ones are factor analysis, nearest-neighbour analysis and discriminant analysis which are described in detail in all major textbooks of statistics (see, for instancce, Lachenbruch 1975 or Trampisch 1986). Therefore, particular reference is made only to the integrated software program ADAPT (i.e. automated data analysis using pattern recognition techniques). It performs data storage and manipulation, automatic construction of the molecular representation, and the statistical analysis. The representation of the molecule may include structural fragments, steric and physico-chemical parameters, partition coefficients (see Sect. 2.1.2.6) and atomic volumes (Stuper and Jurs 1976).

Another type of pattern recognition procedure was suggested by Fränzle (1986) involving cluster techniques, biplot analysis (Gabriel 1971) for metrically indicated chemicals, and entropy analysis (Wishart 1984) in the case of chemicals characterized by scores. Metric descriptors such as density, vapour pressure, molar mass, melting and boiling points, water solubility or n-octanol/water partition coefficients, dissociation constant, volatility and viscosity define the environmental behaviour of n chemicals of a training set in the form of an $n \times m$ matrix, or a geometrically equivalent point set in an m-dimensional space, where m is the number of descriptors. This equivalence means that any matrix with $n \times m$ elements may be represented by one vector for a row and another vector for each column such that the elements of the matrix are the inner products of the vectors. When the matrix is of rank 2 or 3, or can be closely approximated by a matrix of such rank, the vectors may be plotted and the resulting matrix representation, i.e. the so-called biplot inspected visually, which is of considerable practical interest for the analysis of large matrices.

A biplot can be made unique by introducing a particular metric for either row or column comparisons. To approximate the original $n \times m$ matrix of rank r by a matrix of rank 2 or 3, the singular-value decomposition is used. This approximate biplot not only permits viewing the individual data and their differences, but further allows scanning the standardized differences between units and inspecting the variances, covariances and correlations of the variables. In the present context, the biplot is a most useful graphical aid in interpreting the multivariate matrix of chemicals as defined by the above characteristic physical and chemical properties, and to test the mathematical validity of subsequent numerical classification strategies. In the form of cluster algorithms which are eventually optimized by means of the iterative procedure RELOCATE (Wishart 1984), they define homogeneous classes of chemicals in the precise operational sense of the term.

If the primary characterization of the training set chemicals is limited to scores, biplot analysis is not possible but entropy analysis (Wishart 1984) is indicated. In either case, a graphic display of the classes defined should be made on the basis of their T-ratios, i.e. the standardized deviations of the defining variables (= physical and chemical properties) of each class from the global mean values.

In the final step of analysis the individual classes of chemicals thus defined are interpreted in terms of chemical structure.

2.1.2 Equilibrium Constants for Chemicals Distribution

2.1.2.1 Aggregate State and Related Properties

Hazardous substances occur in all three aggregate states; and consequently their impact on the soil–vegetation complex (IN in foldout models I and II) may be due to gravitational settling or dry and wet deposition (cf. Chap. 3.5). In addition, lateral transport by either Horton overland flow, saturation overland flow, throughflow or groundwater flow is a very common phenomenon (Chap. 4.1.2.2). In dependence on the interceptive qualities of the vegetation stand-affected *solid particles* reach the ground in highly variable quantities. Deposition on leaves may facilitate a return into the lower atmosphere, otherwise gravitational settling or washing-down processes may follow. In the latter case, the water solubility is of decisive importance for the ultimate distribution on the ground. Provided the water solubility or the amount of water present are low, the transport of the solid particles into the soil will be controlled by the pore space spectrum of the soil.

A network of macropores which are typically 0.2–10 µm in diameter, but may be as narrow as 10 nm in clay soils lies between individual particles or between microaggregates within the peds. Between major soil aggregates, macropores (i.e. > 10 µm in diameter) occur as well as larger voids and cracks. Sometimes, interconnecting tubular passages may also occur, whose significance is referred to in Chap. 4.1.2.5. Together, the kind, size and distribution of soil aggregates and soil voids and pores determine the fabric of the soil and consequently its penetrability for solid particles.

A predominantly vertical translocation by gravity and, mostly, by percolating water comes to an end where pore size is equal to or somewhat inferior to the particle diameter. This purely mechanical retention of particles constitutes the filtering capacity of a soil. As far as it depends on the amount and continuity of coarser macropores, it can be positively or negatively influenced by soil tillage or by management measures. In long-term zero-tillage experiments in loess soils, for instance, the development of a relatively dense network of macropores due to earthworms has been observed (Ehlers 1975), while in intensively used soils, the continuity of macropores has been drastically reduced in the deeper topsoil layers by an increase in mechanical load. The reduced oxygen supply also impairs the appropriate decomposition of plant residues, which causes a distinct increase in the organic C and N contents (Nieder and Richter 1986). This, in turn, may have consequences for the fate of chemicals introduced into such a deepened topsoil.

In comparison with solid particles, the fate of *liquids* is largely controlled by the parameters: relative density, kinematic viscosity, volatility and relative vapour density in the framework of transport process through the unsaturated zone, while relative density (i.e. density in comparison to water), solubility or miscibility in water are of decisive importance for transport in the saturated zone. These parameters are described in greater detail in the following sections.

Table 2.2. Density, kinematic viscosity and water solubility of selected hydrocarbons and mineral oil products. (After Schwille 1981)

Name, formula	Density at 20 or 15°C (*) ρ (g/ml)	Kinematic viscosity at 20°C ν_{cSt} (mm^2/s)	Solubility in water or saturation concentration at 10 or 20°C (*) (mg/kg)
Aliphatic chlorohydrocarbons:			
Trichloroethylene, C_2HCl_3	1.46	0.40	1070
Tetrachloroethylene, C_2Cl_4	1.62	0.54	160
1,1,1-Trichloroethane, $C_2H_3Cl_3$	1.32	0.65	1700
Methylenechloride, CH_2Cl_2	1.33	0.32	13200*
Chloroform, $CHCl_3$	1.49	0.38	8200*
Carbontetrachloride, CCl_4	1.59	0.61	785*
Water	1.00	1.01	
Mineral oil products:			
Medium distillates (light heating oil, diesel fuel)	0.082–0.86*	2–8	3–8
Petroleum, jet fuel	0.77–0.83*	2–4	
Gasoline (carburettor fuels)	0.72–0.78*	0.5–0.7	150–300
Crude oils (Africa, Near East, North Sea)	0.80–0.88*	3–35	10 to 25

Bertsch et al. (1979) studied seepage velocities of mineral oils in identical porous media at constant temperatures as a function of kinematic viscosity and found excellent correlations, which indicates the suitability of this parameter for liquid-specific comparisons. Table 2.2 gives some illustrative figures.

The fate of *gases*, finally, is largely influenced by their density in relation to that of the ambient air on the one hand, and their reactivity with regard to the constituents of natural or polluted air on the other (cf. Chap. 3.5).

2.1.2.2 Specific Gravity, Molar Mass and Structure

The specific gravity of a substance, in particular in comparison to that of water or air, is of special environmental importance. Hydrocarbons are usually lighter than water. Figure 2.1 illustrates the relationship of molar mass and specific gravity.

The curves show that the specific gravities of the normal alkyl chlorides are somewhat higher than those of the hydrocarbons. Both are lighter than water and the specific gravities decrease as the number of carbon atoms is increased.

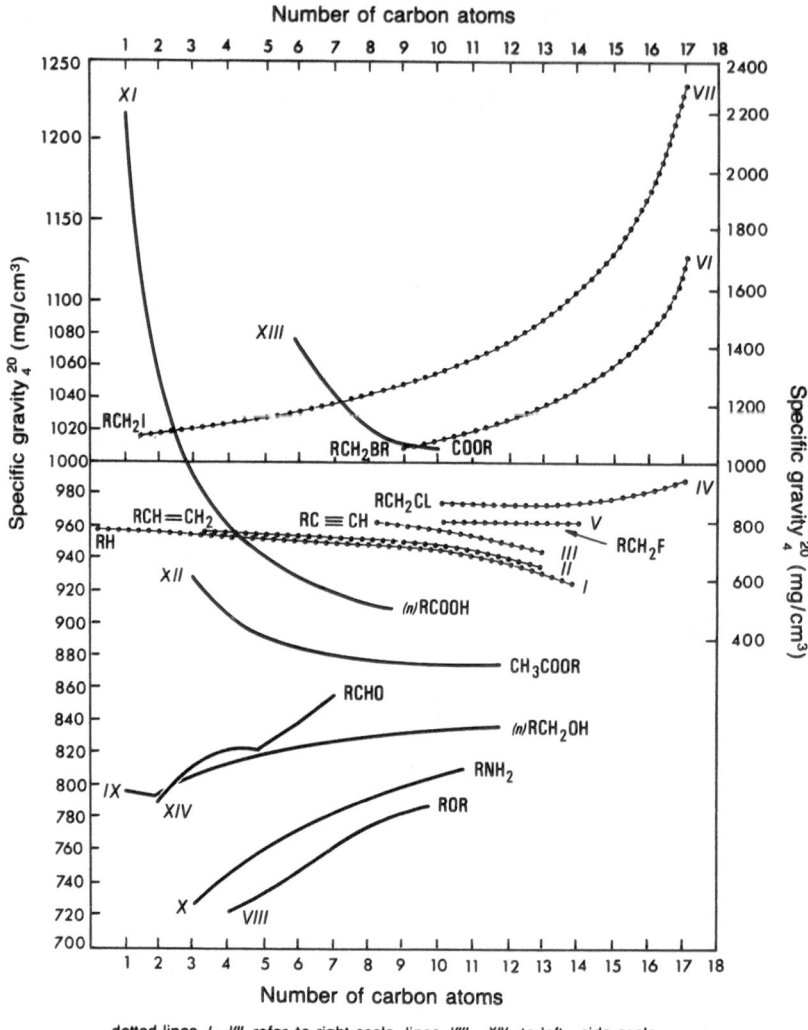

Fig. 2.1. Relationship of molar mass and specific gravity. (After Verschueren 1983)

The same applies to the primary alkyl bromides and iodides. The downward slope of Curves IV, VI and VII results from the fact that the halogen atoms constitute a smaller and smaller percentage of the molecule as the total molar mass is increased by increments of methylene radicals. Considering alkyl halides with the same carbon skeleton and of the same class specific gravity increases in the order

$$RH < RF < RCl < RBr < RI.$$

Similar relationships are found for secondary and tertiary halides and for aryl halides.

As Fig. 2.1 shows, the introduction of functional groups containing oxygen causes an increase in specific gravity. As a rule, compounds with two or more chlorine atoms or one Cl atom together with an O atom or an aryl group have a specific gravity greater than unity (Verschueren 1983). Curve VIII shows that ethers are the lightest of all organic oxygen compounds, while the aliphatic alcohols (Curve IX) are heavier than the ethers but lighter than water. The specific gravity of the alcohols exceeds unity if a chlorine atom, a second hydroxyl, or an aromatic nucleus is introduced.

The amines (Curve X) are less dense and less associated than the alcohols. Association causes the specific gravity of formic acid and acetic acid to be above that of water, while the higher fatty acids are lighter than water. In general, compounds with several functional groups, and in particular those promoting association, will have a specific gravity greater than unity.

2.1.2.3 Melting and Boiling Points

The melting (or freezing) point is the temperature at which a solid changes to a liquid (or vice versa). Normally, melting involves an increase in volume by about 10%; exceptions are water, bismuth, and gallium. According to the sign of this change in volume, melting temperature rises or falls by 0.01–0.1 K/bar, which means that, under environmental conditions, pressure has only little effect on melting point. Table 2.3 summarizes melting and boiling points of selected organic chemicals.

From a comparative analysis of these figures and related data collections, it ensues that the melting and boiling points of the members of homologous series increase with increasing molar mass. When comparing different halogens, however, it must be noted that the mass effect can be (partly) counterbalanced by polarization effects, e.g. in the case of CH_3F and CH_3Br.

A substitution of a hydrogen in an alkane by another atom or group results in an elevation of the boiling point. Thus alkyl halides, alcohols, aldehydes, ketones, acids boil at higher temperatures than the "elementary" hydrocarbons with the same carbon skeleton. The introduction of groups promoting association brings about a marked rise in boiling point, which is especially pronounced in the alcohols and acids, since hydrogen bonding can occur.

2.1.2.4 Vapour Pressure, Vapour Density and Volatility

The *vapour pressure* of a liquid or solid is the pressure exerted by the gas in equilibirum with the liquid or solid at a given temperature. The evaporative loss of a substance, volatilization, is a very important source of material for

Table 2.3. Melting and boiling points of selected hazardous organic chemicals. (BUA-Stoffberichte 1–25)

Compound	Molecular mass g/mol	Melting point °C	Boiling point °C (1013 hPa)
Methane, chloro-	50.49	− 97.7	− 24.2
Methane, dichloro-	84.93	− 94.9	39.6
Methane, trichloro-	119.38	− 63.2	61.3
Methane, bromo-	94.95	− 93.7	3.6
Ethanol, 2-chloro, phosphate (3:1)	285.49	− 62	—
2,2′-Dichloro-ethylether	143.02	− 50	178
NTA	191.10	241.5	—
Benzene	78.0	5.4	80.4
Benezene, 1,3-dichloro	147.01	− 24.8	173
Benzene, 1,2,4-trichloro	181.45	16.1	213.5
Benzene 1,3,5-trichloro	181.45	63.5	208.4
Benzene, 1-chloro-2-nitro	157.6	33	244
Benzene, 1-methoxy-2-nitro	153.14	9.4–10.45	276.8
Benzene, 1-methoxy-4-nitro	153.14	54.5	259
Benzene, 1,3,5-triethyl	120.19	− 44.7	164.7
Benzenamine, 2-nitro	138.13	72	284.1
Benzenamine, 4-nitro	138.13	149.5	331.7
Pentachlorophenol	266.4	189	309
Dinitrotoluenes:			
2,4 DNT	182.14	69.9	319.5
2,5 DNT	182.14	52.5	—
2,6 DNT	182.14	65.9	287.4
Anthracene	178.24	218	340
Dibenz (a, h) anthracene	278.36	264.5	524
Benzo(a)pyrene	252.3	180	496

airborne transport and subsequent transformation and deposition processes (cf. Chaps. 3.2, 3.3, 3.4, 3.5). Vapour pressure data permit, in combination with solubility data, calculations of rates of evaporation of dissolved compounds from water using Henry's law constant H_c which is normally defined as

$$H_c = PS^{-1},\qquad\qquad(2.5)$$

where P is the equilibrium vapour pressure of the pure chemical and S is saturation concentration or solubility at a fixed temperature. Mackay et al. (1985), who summarized current theory and methods for estimating H_c from structure, have shown that P may be expressed as a function of normal boiling point and melting point of a chemical (cf. Chap. 2.1.2.3). Chemicals with low vapour pressures, and high solubility or high adsorptivity onto solids are less likely to vaporize and become airborne than chemicals with high vapour pressures or with less tendency for solution in water or adsorption to soils or sediments. The latter substances are less likely to biodegrade or hydrolyze, but are prime candidates for photolysis and for involvement in adverse atmospheric processes such as smog formation etc. (cf. Chaps. 3.3 and 3.4).

Vapour density, as compared with the density of air (= 1), indicates whether a gas will be transported along the ground, possibly subjecting ecosystems to high exposure, or whether it will disperse rapidly. It is related to equilibrium vapour pressure through the fundamental equation of state for a gas:

$$PV = nRT .$$ (2.6)

If the mass of the substance and its molecular mass are substituted for the number of moles n, this yields (Verschueren 1983):

$$\text{Vd (Vapour density)} = \frac{PM}{RT} ,$$ (2.7)

where P = equilibrium vapour pressure (atm), R = 0.082 liter atmospheres/mol/K, M = molecular mass (g/mol), T = absolute temperature (K).

Volatilization of chemicals from water is an important transport process for a number of substances, which have low solubility and low polarity. In spite of very low vapour pressure, many compounds can volatilize rapidly owing to their very high activity coefficients in solution.

Following mathematical approaches by Liss and Slater (1974) and Mackay and Leinonen (1975) the rate constant for volatilization from water is given by the relation

$$k_{vw} = \frac{A}{V}\left(\frac{1}{K_L} + \frac{RT}{H_c K_G}\right)^{-1} ,$$ (2.8)

where A = evaporating surface (cm^2), V = liquid volume (cm^3), H_c = Henry's law constant (torr M^{-1}), K_L = liquid film mass transfer coefficient (cm h^{-1}), K_G = gas film mass transfer coefficient (cm h^{-1}), R = gas constant (torr K^{-1}M^{-1}), T = temperature (K).

Volatilization processes from soil surfaces are complicated by variable contribution from volatilization of the chemical from the water at the surface, evaporation of water itself and the related capillary rise (cf. Chap. 4.1.2.5) which transports more water and dissolved chemical to the surface (Spencer et al. 1973; Spencer and Farmer 1980). Initially, volatilization from the surface water as in Eq. (2.8) will be rate-controlling, but the model will fail as a concentration gradient is established through the soil column, and no simple laboratory experiment will reliably determine the process in a way that can be extrapolated to a complex field situation.

The *relative volatilization rate* is defined as the ratio of the rate of volatile removal by air contact to the rate water (or another liquid, e.g. ether) is evaporated under corresponding conditions. If the experimental value is greater than unity, stripping with air may be a feasible treatment operation for this wastewater.

On the basis of desorption experiments in a packed column, Thibodeaux (1974) developed a mathematical model which yields the following solution for fairly diluted pollutant concentration in water:

$$\log X_t = \log X_0 - (Ka^{-1} - 1)\log \frac{M_0}{M_t}, \tag{2.9}$$

where
X_t = concentration of pollutant (mg/l) in the desorption column at sample time t,
X_0 = initial concentration of pollutant (mg/l) in thr column at start of experiment,
M_t = quantity of water (g) in the desorption column at sample time t,
M_0 = quantity of water (g) in the desorption column at start of experiment,
K = specific desorption rate of the organic pollutant in the packed column,
a = specific desorption rate of water in the packed column.

The relative volatilization rate k/a can be derived from a log–log plot of X_t and M_0/M_t. The slope of a straight line through these raw data points yields $(K/a - 1)$.

Experimental K/a values for pure compounds at room temperature and initial concentrations of about 100–1000 mg/l are: methanol 7–13, n-butanol 10–18, acetaldehyde 171–207, acetone 36–100, benzene 107, acetic acid 1–5, propionic acid 0–1.5, phenol 0.5–1.9, sucrose 0.26 (Verschueren 1983). Substances whose relative volatilization rates are not significantly different from unity will not be removed appreciably by air stripping, and sucrose even concentrates in the solution upon stripping.

2.1.2.5 Water Solubility

Water solubility is an important parameter for the environmental assessment of all solid or liquid chemicals. When a substance (the solute) dissolves in a liquid (the solvent), the molecules of the solute must force their way between molecules of the solvent. This means that the two types of molecules have similar attractive forces, which leads to the rule that "like dissolves like", where "like" refers to similar characteristics of polarity.

In general, molecules with highly polar bonds will be relatively strongly attracted to one another; consequently, water and alcohol are potent solvents for ionic compounds such as salts, as well as polar non-ionic molecules such as sugar. On the other hand, non-polar molecules such as tetrachloromethane, benzene or gasoline are more weakly attracted to one another (Phadke et al. 1945). They mix easily and are good solvents for non-polar molecules, such as fats, greases, and paraffines, which are not soluble in water. Some solids are so firmly bonded that they are not significantly soluble in any common solvents except those with which they react chemically, whereby the nature of the molecule being dissolved is altered. Quartz is of this type: its solubility in pure water is of the order of 5 mg/l, but if the water contains a 0.01 molar concentration of ortho-diphenols quartz is transformed into soluble esters and the solubility is increased to 40 mg/l (Fränzle 1971b).

Influence of the Composition of Natural Waters

The composition of natural waters varies considerably in space and time (see, e.g. Garrels and Christ 1965; Hem 1970; Dutil and Durand 1974; Stumm and Morgan 1981; Mattheß 1990). Variables such as pH, water hardness, cations, anions, and organic substances such as gelatine, fulvic and humid acids, all affect the solubility of chemicals in water (Kruyt and Robinson 1927; Brintzinger and Beier 1933).

The solubility of lower n-alkanes in salt water is by about one order of magnitude higher than in fresh, distilled water, which is due to both physical and chemical factors. In contrast, the higher n-alkanes (i.e. C_{10} and higher) proved to be less soluble in seawater than in distilled water. In accordance with the McDevit–Long theory the magnitude of the salting out effect increases with increasing molar volume of the alkanes.

This means that freshwater saturated with respect to normal alkanes due to pollution will undergo sequential salting out when entering an estuary. The salted-out molecules might then either adsorb on suspended minerals or on particulate organic matter (Chandra et al. 1984) or rise to the surface as slicks. The same applies to all natural or pollutant organic molecules whose solubilities are decreased by the addition of electrolytes. Thus estuaries may limit the amount of dissolved organic carbon entering the ocean appreciably, while the amount of particulate organic carbon entering the marine environmental is increased (Gerlach 1976; Duinker and Hillebrand 1979; Duinker 1981; Verschueren 1983).

Structure–Solubility Relationships

Because of the marked polarity of the H_2O molecules, water is only a poor to very poor solvent for hydrocarbons. Since alkenic and alkinic linkages or benzenoid structures do not affect polarity considerably, unsaturated or aromatic hydrocarbons do not differ much from alkanes in their water solubility. Also the introduction of halogen atoms does not alter the polarity of the molecule appreciably, but it increases the molar mass, and for this reason solubility always falls off. Salts, as extremely polar compounds, exhibit a correspondingly high solubility in water, while other compounds such as alcohols, aldehydes, esters, ethers, acids, amines, amides, ketones and nitriles lie between salts and the more or less non-polar hydrocarbons.

Amines owe their comparatively high solubility, e.g. benzenamine, 4-nitro: 0.568 g/l (Collett and Johnston 1926), to their tendency to form hydrogen-bonded complexes with water molecules. Consequently, the solubility of amines diminishes as the basicity decreases. Furthermore, the fact that many tertiary amines are more soluble in cold than in hot water indicates that at higher temperatures the hydrate is unstable and the solubility measured is that of the free amine, while at lower temperatures the distinctly higher solubility of the hydrate itself is involved.

Monofunctional ethers, esters, ketones, alcohols, aldehydes, nitriles, amides, amines and acids display comparative behaviour with respect to water solubility (cf. Gross and Saylor 1931; Zoeteman et al. 1980; Kunte and Pfeiffer 1985; Eurocop-Cost 1984). In homologous series the hydrocarbon (non-polar) part of the molecule increases, whereas the polar function remains essentially unchanged. This involves a corresponding trend to a decrease in solubility in polar solvents such as water.

Table 2.4 illustrates the influence of functional groups on the aqueous solubility of representative benzene derivatives. An increase in molar mass is normally coupled with an increase in intermolecular forces in a solid. Therefore polymers generally exhibit low solubilities in water or ether, which is illustrated by the following examples. While formaldehyde is readily soluble in water, paraformaldehyde is insoluble. Glucose and many amino acids are soluble in water, but the polymers – starch, glycogen, cellulose, and proteins – are insoluble.

Table 2.4. Influence of functional groups on aqueous solubility of benzene derivatives. (Verschueren 1983)

	Functional group	S mg/l solubility mg/l (temp. °C)	$\log S$ mg/l	$\Delta \log S$ mg/l $\log S_{C_6H_5X}$ $- \log S_{C_6H_6}$
Aniline	$-NH_2$	34,000 (20)	4.53	1.28
Phenol	$-OH$	82,000 (15)	4.91	1.66
Benzaldehyde	$-COH$	3300	3.52	0.27
Benzoic acid	$-COOH$	2900	3.46	0.21
Nitrobenzene	$-NO_2$	1900	3.28	0.03
Benzene	—	1780	3.25	0.00
Fluorobenzene	$-F$	1540 (30)	3.19	-0.06
Thiophenol	$-SH$	470 (15)	2.67	-0.58
Toluene	$-CH_3$	515	2.71	-0.54
Chlorobenzene	$-Cl$	448 (30)	2.65	-0.60
Bromobenzene	$-Br$	446 (30)	2.65	-0.60
Iodobenzene	$-I$	340 (30)	2.53	-0.72
Diphenylether	O–⬡	21 (25)	1.32	-1.93
Diphenyl	–⬡	7.5 (25)	0.88	-2.37

From a correlation between the logarithm of hydrocarbon solubility in water and the molar volume of the hydrocarbons, the following conclusions can be drawn (Verschueren 1983):

- Branching increases solubility for alkanes, alkenes and alkynes, but not for cycloalkanes, cycloalkenes and aromatic hydrocarbons.
- For a given carbon number, ring formation increases water solubility.
- Double bond addition to a chain or ring structure increases solubility; the addition of further double bonds to a hydrocarbon of given carbon number further increases solubility proportionally.
- A triple bond in a chain molecule augments water solubility to a greater extent than two double bonds.

In the light of these generalized statements, it is worth noting, however, that the following effects cannot be assessed by this additive–constitutive character of solubility:

- steric effects which induce shielding of an active function,
- intra- and intermolecular hydrogen bonding,
- inductive effects of substituents on each other,
- conformational effects, for instance, "balling up" of an aliphatic chain.

Solubility of Mixtures

Mixtures of compounds behave differently from the single compounds when brought into contact with water, since each component will partition between the aqueous phase and the mixture. Compounds with a high solubility will tend to move into the aqueous phase while the less soluble components will have a greater tendency to remain in the mixture phase. Thus the fractional composition of the water-soluble fraction will differ from the original composition of the mixture, and the concentrations of the individual components of the water-soluble fraction are generally lower than their maximum solubilities (Lee et al. 1979).

2.1.2.6 n-Octanol/Water Partition Coefficient

The n-octanol/water partitiom coefficient (P_{ow}) is defined as the ratio of the equilibrium concentration C of a dissolved substance in the two-phase system n-octanol and water (two solvents which are largely immiscible):

$$P_{ow} = \frac{C_{octanol}}{C_{water}} .$$

Usually the decadic logarithm of P_{ow} is given, abbreviated $\log P_{ow}$.

It should be noted that P_{ow} is a dependent variable with regard to water solubility; therefore the simultaneous use of both in structure–activity relationships or multivariate classification procedures will give spurious correlations. In

fact, the experimental determination of both values of aqueous solubility and
n-octanol/water partition coefficient led to the following regression (ECETOC
1986a):

$$\log P_{ow} = 5.00 - 0.670 \log S \, , \tag{2.10.1}$$

where S = aqueous solubility in μmol/l.
    Expressing solubility in mg/l, Eq. (2.10.1) becomes:

$$\log P_{ow} = 4.5 - 0.75 \log S \, . \tag{2.10.2}$$

The regression covers many classes of both liquid and solid compounds with
highly different polarities from hydrocarbons and organic halides to aromatic
acids, pesticides and PCBs. The corresponding correlation coefficient is 0.97,
which normally allows an estimation within one order of magnitude of the
partition coefficient of a given chemical from its solubility in water. But the
scatter increases considerably for solubilities above 100 mg/l, as the example of
l-tyrosine may show, where the experimental $\log P_{ow}$ is -2.26 while calculations by means of Eq. (2.10.1) yield values of $+2.7$ or $+2.5$, respectively.
    Furthermore it appears questionable that the above equations would also
apply to chemicals with high dissociation constants such as salts and strong
acids or bases, since their activities in solutions cannot by approximated by their
concentrations. Basically, the same restriction applies to aliphatic acids or bases
whose partition coefficients can vary drastically with changes in pH.
    The most comprehensive compilation of data is due to Leo et al. (1971),
who frequently used relationships between values in different solvent systems for
calculation purposes which have the general form:

$$\log P_X = a \log P_o + b \, , \tag{2.11}$$

for instance,

$$\log P_{cyclohexanone} = 1.035 \log P_{octanol} + 0.896 \tag{2.12}$$

$$(n = 22; \quad r = 0.98; \quad s = 0.194)$$

$$\log P_{toluene} = 1.135 \log P_{octanol} - 1.777 \tag{2.13}$$

$$(n = 10; \quad r = 0.972; \quad s = 0.340) \, .$$

A comparative review of n-octanol/water partition coefficients shows that
$\log P_{ow}$ in a homologous series increases by approximately 0.5 per CH_2 group.
Furthermore, functional groups such as chlorine atoms, hydroxyl groups,
methyl groups, benzene etc. showed additive effects on $\log P_{ow}$. The analogous
additive effect of functional groups on $\log P_{ow}$ values of benzene derivatives is
summarized in Table 2.5.
    Values of complex molecules not infrequently differ from the true (i.e.
experimental) ones by one to two orders of magnitude. Nevertheless the
n-octanol/water partition coefficient has proved useful as a simple means to
predict biological uptake, liphophilic storage and biomagnification factors (cf.

Table 2.5. Influence of functional groups on n-octanol/water partition coefficient of benzene derivatives. (Verschueren 1983)

Product	Functional group	$\log P_{oct}$	$\Delta \log P_{oct}$ $\log P_{C_6H_5X} - \log P_{C_6H_6}$
Benzenesulfonic acid	$-SO_3H$	-2.25	-4.38
Benzenesulfonamide	$-SO_2NH$	0.31	-2.44
Aniline	$-NH_2$	0.90	-1.23
Phenol	$-OH$	1.46	-0.67
Benzaldehyde	$-COH$	1.48	-0.65
Benzonitrile	$-CN$	1.56	-0.57
Benzoic acid	$-COOH$	1.87	-0.28
Nitrobenzene	$-NO_2$	$1.85/1.88$	-0.28
Nitrosobenzene	$-NO$	2.00	-0.13
Benzene	—	2.13	—
Fluorobenzene	$-F$	2.27	$+0.14$
Thiophenol	$-SH$	2.52	$+0.39$
Toluene	$-CH_3$	2.80	$+0.67$
Chlorobenzene	$-Cl$	2.84	$+0.71$
Bromobenzene	$-Br$	2.99	$+0.86$
Iodobenzene	$-I$	3.25	$+1.12$
Diphenyl	$-C_6H_5$	3.6	$+1.47$
Diphenylether	$-O-C_6H_5$	4.21	$+2.08$

Sect. 2.1.3.5), while extrapolations of the multifactorial soil adsorption processes are much more difficult (cf. Sect. 4.4.1.2).

2.1.3 Kinetic Constants for Environmental Processes

Every chemical reaction takes place at a definite rate depending on the specific conditions under which it occurs. These boundary conditions are the concentrations or pressures of the reacting substances, temperature, presence or absence of a catalytic agent and radiation.

Some reactions are so rapid that they appear to be instantaneous, for instance, the neutralization of an acid by a base, while others are so extremely slow at ordinary temperatures that no change would be detected in many years, e.g. the combination of hydrogen and oxygen to form water in the absence of a catalyst. Between these extremes are many reactions which take place at measurable rates at "normal" temperatures under ordinary conditions. Chemical kinetics deals with the rates and mechanisms of these reactions whose understanding is rendered difficult by the fact that both time and intermediate products are involved as compared with the subject of chemical equilibrium, which is concerned only with the initial and final states of reactants.

Because of the complexity of these subjects (cf. elements PHOTO and CH in foldout model II), the following presentation is limited to the determination of the essential rate constants for environmental processes from structure alone, using structure-activity relationships. For a comprehensive treatment of chemical kinetics the interested reader is referred to *A Guidebook to Mechanism in Organic Chemistry* by P. Sykes (1976).

2.1.3.1 Hydrolysis

Hydrolysis of compounds usually results in the introduction of a hydroxyl function (–OH) into a chemical, and is most commonly associated with the loss of a leaving group (–X):

$$RX + H_2O \rightarrow ROH + HX \ . \tag{2.14}$$

In water the reaction is catalyzed mainly by H^+ and $(OH)^-$ ions, whereas in moist soil, loosely complexed metal ions such as copper or calcium may also constitute important catalysts for several types of chemical structures. In addition, also sorption (cf. Chap. 4.4.1) of the chemical may increase its reactivity toward hydronium or hydroxyl ions (Mill 1980; Wolfe 1980; Mattheß et al. 1985).

The general rate equation for hydrolysis in water is:

$$R_H = K_H[C] = K_A[H^+][C] + K_B[OH^-][C] + K'_N[H_2O][C] \ , \tag{2.15}$$

where K_H is the measured first-order rate constant at a given pH, $[C]$ is the concentration of chemical C, while the last term denotes the neutral reaction with water (K'_N being the second-order constant) which in water can be expressed as a pseudo-first-order constant K_N. The equation shows that the total rate of hydrolysis in water is pH-controlled unless K_A and $K_B = 0$.

Equation (2.15) can be modified with regard to the interaction of bound or free metal-ion catalysis in soil or sediments by introducing one or more terms for the form:

$$K_M K_A[M]_T \cdot (K_A + [H^+])^{-1} \ , \tag{2.16}$$

where K_M is the metal-ion catalysis constant, $[M]_T$ is the total metal ion concentration, and K_A is the equilibrium constant for dissociation of the hydrated ion complex.

Mabey and Mill (1978) reviewed kinetic data for hydrolysis of a variety of organics in aquatic systems in relation to the chemical characteristics of most freshwater systems. Quantitative structure–activity relationships, used as predictive test methods to develop the essential kinetic data, generally take the form of linear free-energy relationships such as the Hammett and Taft

$$\log K_X/K_0 = \rho\sigma \tag{2.17}$$

or Brönsted

$$\log K_X/K_0 = mpK_a + C , \tag{2.18}$$

where ρ and σ have their usual significance (cf. Sect. 2.1.1.2), K_0 and K_X are the rate constants for the unsubstituted and substituted structures, respectively, while the dissociation constant pK_a refers specifically to the acidity of the leaving group OR in the following reactions (Mill 1981):

$$HO^- + RO - Y = RO^- + HO - Y[Y = \quad C(O) \text{ or } \quad P(O)]$$

$$RO^- + H_3O^+ = ROH + H_2O . \tag{2.19}$$

A comparative evaluation of the validity of linear free-energy relationships for hydrolysis of several classes of compounds such as esters, phosphates, alkyl halides and carbonates indicates that for reactive chemicals estimates of $\log K_n$ are reliable for a specific structure with an error bound of 0.5 log units (Wolfe 1980).

Highly variable hydrolysis reactions have been observed on soils and sediments, which appears to be due to unusual pH relationships at the surface of soil constituents, by possible incursion of metal-ion catalysed processes and by general acid- and base-catalysed reactions possibly involving phenolic, amine or sulfide groups in soils (Delvigne 1965; Helling et al. 1974; Saltzman et al. 1974; Mill 1981; Mattheß et al. 1985). In some cases, e.g. in the presence of clay minerals, hydrolysis rates of organic chemicals were considerably accelerated compared to bulk solution, while adsorption to humic components not infrequently impedes hydrolysis. A detailed understanding of the mechanisms involved is limited, however, and structure–activity relationships appear to be available for only a few compounds (Mill 1980; Wolfe 1980).

2.1.3.2 Photolysis

Natural *direct photolysis* results from absorption of photons by the chemical in the solar spectrum above 290 nm. The specific absorption rate $(K_{a,\lambda})$ is a measure of the overlap between the solar spectrum and the electronic absorption spectrum of the compound, i.e. an inherent property of the photoreactive chemical (ECETOC 1983):

$$K_{a,\lambda} = 2.303 \, E_0(\lambda)\varepsilon_\lambda , \tag{2.20}$$

where $E_0(\lambda)$ = scalar irradiance and ε_λ = molar absorptivity at wavelength λ. Absorption of light energy in terms of photons results in the excitation of an electron from a lower to a higher orbital, which can have several possible transitions. Generally, an increase in the number of conjugated double bonds in the molecule will decrease the energy required for an electronic transition (Moore and Ramamoorthy 1984).

Since under most environmental conditions pollutant concentrations are low, photolysis can be described by a first-order expression, that is, the rate constant of direct photolysis k_d is directly proportional to pollutant concentration. From Eq. (2.20) then follows that

$$k_d = 2.303 \ \Phi_d \int E_0(\lambda) \ \varepsilon \ d\lambda \ , \tag{2.21}$$

where the integration is over the range of $E_0(\lambda)$ which is absorbed by the compound, and Φ_d denotes the quantum yield which may attain a maximum value of 1. It should be noted, however, that rapid direct photolysis can occur even with compounds that react with very low quantum efficiencies (ECETOC 1983).

Generalizations about the relation of chemical structure to ε_λ are found in Hendry and Mill (1980), but no simple structure–activity relationships are available to predict ε_λ or Φ (Mill 1981). ε_λ as an inherent property of a chemical can be readily computed from kinetic data obtained in the laboratory. According to Kasha–Vavilov's law, quantum yields in solution are generally wavelength-independent; however, the lower k_d is in distilled water, the greater is the probability that it may exhibit wavelength dependence or may be enhanced in natural waters.

Indirect photolysis may occur if a natural material can absorb solar photons and transfer part of the energy to a chemical. Thus compounds that do not absorb light directly can still undergo mediated photolysis (cf. Chap. 3.4.1.), and chemicals that photolyze directly may photolyze in natural waters containing organic substances in dissolved or particulate form much more rapidly.

Since in most cases the molecular structures of these natural (or anthropogenic) substances or photosensitizers have not been identified, it is difficult to define the action spectra for photosensitized reactions. Another complication results from the fact that a variety of photochemical processes may be involved in indirect photolysis. In spite of these difficulties, general rate expressions for sensitized photolysis can be defined.

The rate of light absorption is

$$I^s_{a,\lambda} = 2.3 \ \varepsilon_{\lambda,s} \ E_0(\lambda)[S] \ , \tag{2.22}$$

where $\varepsilon_{\lambda,s}$ is the molar absorptivity of the sensitizer and $[S]$ is its concentration (ECETOC 1983). It ensues from Eq. (2.22) that the rate constants for indirect photolysis should increase with increasing $[S]$ if $E_0(\lambda)$ is held constant. When the concentration of the pollutant P is very dilute, the quantum yield is directly proportional to it. Denoting C_λ the proportionality constant the total rate expression becomes

$$(Rate)_\lambda = 2.3 \ E_0(\lambda) \ \varepsilon_{\lambda,s} \ C_\lambda[S][P] \ . \tag{2.23}$$

Provided the sensitizer concentration is constant, as is usually the case in nature, the rate expression is of first-order form.

In most natural waters light attenuation effects must be taken into account. To this end, the scalar irradiance $E_0(\lambda, Z)$ at depth Z may be approximately equated to $E_0(\lambda, 0) \exp(- K, Z)$, where $E_0(\lambda, 0)$ denotes scalar irradiance just below the water surface and K_λ the attenuation coefficient of the water at λ. The average irradiance in a well-mixed water body $E_0^{av}(\lambda, Z)$ is

$$E_0^{av}(\lambda, Z) = E_0(\lambda, 0)\frac{F_\lambda}{K_\lambda Z},$$ (2.24)

where F_λ denotes the fraction of light attenuated at depth Z which equals
$1\text{-exp}(-K_\lambda Z)$. This yields the average rate constant for sensitizied photolysis

$$K_{s,\lambda} = E_0^{av}(\lambda, Z)X_{s,\lambda},$$ (2.25)

where $X_{s,\lambda} = 2.3\varepsilon_{\lambda,s}C_\lambda[S]$.

Various methods have been developed to compute $E_0(\lambda)$ for sunlight as
a function of season, latitude and depth in a water body. In summary it may be
said, however, that the possibilities of predicting the rates of photolysis of many
chemicals in air or water are severely limited by the lack of structure–activity
relationships for ε_λ, and in particular by the lack of detailed understanding of
the indirect processes. Even worse is the situation with regard to photolytic
reactions on soils where the processes remain largely unexplored.

2.1.3.3 Oxidation and Reduction

Direct and indirect phototransformation as described in the preceding section
are possible removal pathways only for those chemicals which, either by their
very nature or due to fixation onto light-absorbing carrier substances, can
absorb in the region of solar radiation. The main transformations, however,
leading to the removal of chemicals from atmosphere and water involve reactions with photochemically generated *oxidants* such as the hydroxyl radical
(OH·), ozone (O_3), hydroperoxyl radical ($HO_2^·$), organic peroxyl radicals
($RO_2^·$), singlet oxygen ($O_2 \, \Delta_g$) and nitrate radical ($NO_3^·$).

Based on the experimental assessment of indirect phototransformation processes in air and water (Mill et al. 1981; ECETOC 1983), structure–activity
relationships for oxidation by each of these species are available. The rate
constant for the reaction of a substance with hydroxyl radicals, which are
generally of prime importance (cf. Chap. 3.4.1), is computed as an additive
property. Assuming that the rate constant of the total molecule with OH· is the
sum of the rate constants for the reactivities of each portion of the molecule:

$$k_{OH} = \sum_{i=1}^{n} n_i \cdot \alpha_{H_i} \cdot \beta_{H_i} \cdot k_{H_i} \qquad \text{(I)}$$

$$+ \sum_{j=1}^{m} n_j \alpha_{E_j} \cdot k_{E_j} \qquad \text{(II)}$$

$$+ \sum_{l=1}^{o} n_l \cdot \alpha_{A_l} \cdot k_{A_l}. \qquad \text{(III)}$$ (2.26)

The first term (I) describes the abstraction of hydrogen atoms, k_{H_i} being the
reactivity of the ith H atom which depends on the degree of substitution on the

adjacent atom. α_{H_i} and β_{H_i} account for the effects of substitution in the α and β positions; n_i is the number of equivalent hydrogen atoms. The second term (II) corresponds to addition to a double bond, k_{E_j} being the reactivity of the jth carbon atom double bond, while α_{E_j} is taken as unity except when a hydrogen atom is attached to the double bond. n_j accounts for the repetition of each unique double bond. The third term (III) then corresponds to addition to aromatic rings, k_{A_l} being the reactivity of the lth aromatic group while α_{A_l} accounts for the effects of halogen atoms on the ring. The various rate constants k_{H_i}, k_{E_j}, k_{A_l} and the corrective factors have been determined in preliminary experiments on typical compounds such as alkanes, alkenes, benzene etc.

For substituted alkanes the above approach has been extended by Heicklen (1981) in order to include the temperature dependence of k_{OH}:

$$k(T) = (8kT/\mu)^{0.5} \, \sigma_R^2 \cdot \sum_{i=1}^{n} \gamma_i \exp \{ - a(D_i - D_0) \, T/RT \}, \tag{2.27}$$

where

$k(T)$	= reaction rate coefficient at absolute temperature T,
k	= Boltzmann's constant,
μ	= reduced mass of reacting species,
σ_R.	= intrinsic reaction radius for an OH· radical reaction with a C–H bond of bond length 1.5×10^{-8} cm,
γ_i	= number of equivalent C–H bonds of each type in the molecule,
D_i	= bond dissociation enthalpy for each type of C–H bond,
$D_0 (T)$	= a temperature-dependent energy parameter,
R	= ideal gas constant.

If D_i is taken as the C–H bond dissociation enthalpy at 298 K, then α becomes 0.323 and D_0 is derived from the empirical formula

$$D_0^{-1} = 1.062 \times 10^{-2} + 3.52 \times 10^{-6} T, \tag{2.28}$$

where D is in kcal/mol and T in K, which is valid for T between 200 and 400 K.

An analogous approach has been developed for the estimate of k_{0_3} as the sum of rate constants for the reactivities of the various reactive groups such as aliphatic or aromatic C=C, etc.

Table 2.6 summarizes the reactivity of different classes of organics toward each of the above oxidants. It is important to realize that for many chemicals only qualitative or semi-quantitative structure–activity relationships are needed in order to establish the relative importance of these processes compared with others.

The *reduction* of organic compounds requires an environment of low redox potential, i.e. eutrophic or deeper levels of aqueous systems (cf. Golterman 1975) and depths of soil where oxygen is limited (see Lindsay 1979; Schachtschabel et al. 1989). Various mechanisms for reducing organics may be distinguished, such as hydrogenation, halogenation, coupled oxidation-reduction etc., many of which require catalysts. Rates of reductive alteration of organics, like those of

Table 2.6. Rates of oxidation in air and water at 298 K. (After Mill 1981)

Class or structure	Half-life
Air: HO˙ Radical	
n-, iso-, Cycloalkanes	1–9 days
Alkenes	0.05–1 day
Halomethanes	0.2–100 years
Alcohols, ethers	1–3 days
Ketones	0.2–6 days
Aromatics	0.1–3 days
Water: $RO_2^˙$ Radical	
Alkanes, alkenes	220–2000 years
Alkyl derivatives	220–2000 years
Phenols, arylamines	1 day
Hydroperoxides	150 min
Polyaromatics	10 days
Water: $O_2 \Delta_g$	
Aliphatic compounds	100 years
Substituted or cyclic alkenes	8–40 days
Alkyl sulfides	1 day
Dienes	19 h
Eneamines	15 min
Furans	1 h

many other chemical reactions, are temperature and pH-sensitive (Fränzle et al. 1987).

As in the case of oxidative reactions, also geostatistically oriented in situ studies of straight chemical reduction are urgently needed. Since the most common mechanism for the reduction of organic molecules in the environment is via microbial attack, separating biologically induced from straight chemical reduction in situ is difficult. From the viewpoint of prospective hazard assessment, however, it is more important to identify those compounds which are potentially susceptible to reduction and the transformation products which might be deposited in the environment as a result.

2.1.3.4 Biotransformation

Biotransformation is more accurate and general a term than biodegradation, because many chemicals are transformed to products of comparable molecular complexity instead of being degraded. Chemically speaking, biotransformations in water and soil systems include the above hydrolysis, oxidation cleavage, and reduction, even in aerobic systems (Mill 1981). Kinetics of biotransformation are normally based on the Monod model (cf. Faust and Hunter 1978) which couples the rate of loss of the chemical to growth of the organisms involved. In the natural environment, however, growth is usually controlled by nutrients present in comparatively constant amounts, and the kinetic expression for biotransformation processes simplifies to a second- or first-order reaction:

$$d[C]/dt = k_{bt}[C][B] = k'_{bt}[C] ,$$ (2.29)

where $d[C]/dt$ = rate, $[B]$ = number of organisms per unit volume and k_{bt} = second-order rate constant for biotransformation or k'_{bt} = pseudo-first-order rate constant. The half-life of the chemical under transformation ($t_{1/2}$ at a given cell concentration) can then be calculated as:

$$t_{1/2} = \frac{\ln 2}{k_{bt}[B]} .$$ (2.30)

In deriving this equation, it is assumed that the microbial community in water, sludge or soil has already been acclimated to the chemical and there is no lag time involved in the production of the necessary level of biotransforming organisms or appropriate mutants or enzymes.

A comparison of the transformation rates of organic compounds in sterilized soil or water samples with comparable samples which have not received a sterilization treatment provides ample evidence for the amount of metabolism or microbial destruction of many organic molecules (Huber 1982; Wildförster 1985). In many instances, however, no microorganism can be isolated which is able to grow on the organic compound and use it as a source of energy, carbon or some other nutrient (Alexander 1967). Yet microorganisms can be found which transform the molecule, although not being able to use it as a nutrient source, which is a reflection of cometabolism (cf. Chap. 4.4.2).

The likeliest explanation of cometabolic processes is that the parent molecule is transformed by enzymes of low substrate specificity to yield a product which is not a substrate for other enzymes of the microorganism, and consequently accumulates. Hence products of cometabolism are structurally fairly similar to the original compounds, and therefore they may be as toxic, too. Molecules with chlorine, nitro and some other substituents are quite frequently subject to cometabolism.

Owing to the remarkable versatility of microorganisms in their catabolic activity, a great number of substances are subject to microbial mineralization or cometabolism in water and soil. If, however, microorganisms have no enzymes active in a needed catabolic sequence or if the substrate is protected from microbial attack because of special properties of the environment, the compound tends to persist. In some instances, persistence may be attributed to the absence of the requisite enzymes or to the lack of permeability of microorganisms to the substrate (Alexander 1980). Other mechanisms of recalcitrance may be inherent in the molecule itself. In other cases, biotransformable compounds are degraded quickly in one environment but are partially or totally refractory in another (cf. Chap. 4.4.2.2).

It ensues from the complicated interrelationships that reliable conclusions about biotransformation are not generally possible on the basis of structure alone. Among the more environmental variables affecting both the rate and the extent of aerobic and anaerobic transformation processes are: temperature, pH, salinity, dissolved oxygen, concentration of test substance, concentration of viable microorganisms, microbial species, quantity and quality of nutrients (other than test substances), trace metals and vitamins, and time. With these reservations, Table 2.7 summarizes exemplary results of experimental estimates of biotransformation. As a result of the comparative consideration of the below

Table 2.7. Persistence of various groups of pesticides. (After Ottow 1985)

Persistence	Chemical group	Use pattern
1–15 years	a) Chlorinated hydrocarbons, polychlorinated benzenes and phenols (e.g. hexachlorobenzene, pentachlorophenol, DDT)	I
	b) Other xenobiotics with several nonphysiological C–Cl groups (including metabolites)	
2–12 months	a) Phenol thiourea (e.g. diuron, linuron)	H
	b) Triazines (e.g. atrazines, simazine)	H
	c) Benzoic acids (e.g. dicamba)	H
	d) Bipyridyles, nitriles (e.g. paraquat, picloram)	H
1–6 months	a) Phenoxyacetic acids (e.g. 2,4-D, 2,4,5-T)	H
	b) Nitriles, polychlorinated and nitrified phenyls (e.g. metabolites like chloronitrophenols and chloroanilines)	H
	c) Acylanilides	H
	d) Dinitroanilines (e.g. nitralin, trifluralin, treflan)	H
2–12 weeks	a) Carbamates (e.g. carboryl, carbofuran, mesurol, propham)	H
	b) Organic esters (e.g. diazinon, malathion, mevinphos parathion)	I
	c) Thiocarbamates and dithiocarbamates (e.g. EPTC, diallate, thiram, ferbam, maneb)	I, F
	d) Benzimidazoles (e.g. benomyl, carbendazime)	F
	e) Phthalimide derivatives (e.g. captan, merpam, folpet)	F
	f) Dinitrophenols (e.g. dinosebacetate, DNOC) and aliphatic acids (e.g. dalapon, trichloroacetic acid)	H

H = herbicide, I = insecticide, F = fungicide.

set of chemical and biotic transformation processes, some conclusions are indicated:

1. Aromatic rings with chlorine or nitro substitution can be degraded in soils only by O_2 and oxygenases. Consequently, compounds of this type and their aromatic metabolites are relatively persistent under anaerobic conditions of water saturation. Benzene and the related phenoxyacetic acid compounds are also relatively resistant to nitrate which normally acts as a hydrogen acceptor (Fournier 1980).

2. On the other hand, relatively persistent chlorinated hydrocarbons (DDT, lindane, dieldrin, heptachlor) and chlorinated phenyl compounds (γ-HCB, pentachlorophenol, pentachloronitrobenzene) normally undergo a remarkably accelerated de(hydro)chlorination under anaerobic conditions with a biochemically reduced redox potential of -200 to -400 mV such as prevails part-time in flooded paddy soils (Ottow 1985). Consequently, a regular change of aerobic and anaerobic conditions proves particularly favourable to the mineralization of chlorinated aromatics and their metabolites.

2.1.3.5 Bioconcentration, Bioaccumulation and Ecological Magnification

Bioconcentration is the concentration of a chemical in an organism due to direct uptake from the environment, e.g. ambient air or water, neglecting contaminated food. *Bioaccumulation* includes the latter pathway of concentration, while *biomagnification* considers only concentration processes via food uptake. *Ecological magnification* defines the increase in concentration of a substance in an ecosystem or a foodweb when passing from a lower trophic level to a higher one (Korte 1987).

The bioconcentration or bioaccumulation process is the result of both kinetic (diffusional transport and biodegradation) and equilibrium (partitioning) processes. It thus involves the following set of fundamental events (Verschueren 1983):

- partitioning of the xenobiotic between the environment and some surface of the organism,
- diffusional transport of the molecules across cell membranes,
- transport processes mediated by body fluids, such as exchange between blood vessels and serum lipoproteins,
- concentration of the xenobiotic in various tissues depending upon its affinity for specific biomolecules, such as nerve lipids,
- metabolism, co-metabolism or biodegradation of the xenobiotic.

Bioconcentration kinetics of an aquatic organism can be defined as:

$$\frac{dc_A}{dt} = k_1 \cdot c_W - k_2 \cdot c_A \tag{2.31}$$

$$c_{At} = \frac{k_1}{k_2} \cdot c_W (1 - e^{-k_2 t}) \tag{2.32}$$

$$c_{As} = \frac{k_1}{k_2} \cdot c_W = BCF \cdot c_W, \tag{2.33}$$

where

c_A = concentration of chemical in an organism (ng g^{-1}),
t = time in days (d),
k_1 = rate constant for uptake (d^{-1}),
c_W = concentration of chemical in water (ng g^{-1}),
k_2 = rate constant of excretion (d^{-1}),
c_{At} = concentration of a chemical in an organism at time t (ng g^{-1}),
c_{As} = concentration of chemical in an organism under equilibrium conditions (ng g^{-1}),
BCF = bioconcentration factor.

The relative importance of the above processes is likely to differ considerably as a result of both the physico-chemical properties of the xenobiotic and the morphological and physiological structure of the target organisms (cf. Chap. 2.2.3). Therefore it is an essential ecotoxicological requirement to complement the determination of bioconcentration factors (BCFs) by an assessment of biomagnification values. Under "real-world" conditions, they would provide deeper insight into the complex bioaccumulation processes of xenobiotics as a result of transport and partitioning into the higher levels of the food web, which may ultimately result in toxic concentrations (cf. Metcalf et al. 1971; Cole et al. 1976; Ramade 1979; Gruttke et al. 1987).

Metcalf and Lu (1973) and Metcalf et al. (1975) have correlated the bioconcentration values for pentachlorobiphenyl, tetrachlorobiphenyl, trichlorobiphenyl, DDE, chlorobenzene, benzoic acid, anisole, nitrobenzene, and aniline from the fish of model ecosystems with both water solubility and with the octanol/water partition coefficient. In these cases the correlation between physical properties and biomagnification proved excellent, and the corresponding regressions were

$$\log BM = 4.48 - 0.47S \tag{2.34}$$

$$\log BM = 0.75 + 1.16 \log P_{ow}, \tag{2.35}$$

where BM = biomagnification and S = aqueous solubility (ppb). Könemann (1981a, b) and Könemann and Musch (1981) reported the occurrence of a nonlinear relation between $\log P_{ow}$ and $\log BCF$ in the fat of guppies (*Poecilia reticulata*) exposed to chlorinated benzenes. Bioaccumulation increased with $\log P_{ow}$ until reaching a maximum at $\log P_{ow} = 6.5$. The correlations between biomagnification and water solubility or the related octanol/water partition coefficient are only valid for compounds which are not liable to significant biodegradation. Sixty-two chemicals, too diverse to group into classes and including many pesticides, were studied by Veith et al. (1979). The authors determined BCFs of 30 of the 62 chemicals by standard method on the fathead minnow (*Pimephales promelas*), and therefore the values should be internally consistent. Of the remaining 32 chemicals, BCFs were taken from various papers relating to different fish species and exposure times. The comparability of this data set with that of the authors is therefore in some doubt, and only limited evidence could be provided that BCFs varied between species by only a factor of 3 at most. The results were summarized in the form of a linear regression ($n = 59$, $r = 0.947$)

$$\log BCF = 0.85 \log P_{ow} - 0.70 , \tag{2.36}$$

which, according to the authors, is a satisfactory structure–activity relationship yielding estimates of BCF to within an order of magnitude for chemicals of $\log P_{ow}$ ranging from 1 to nearly 7.

It is worth mentioning that most of the correlation equations between octanol/water partition coefficient or water solubility and biomagnification have been established on biorefractive compounds or on homologous series of compounds with comparable biodegradability characteristics (Frische et al. 1979; Verschueren 1983; ECETOC 1986b). Therefore the equations are not universally applicable as can be derived, for instance, from the relationship between water solubility of DDT, DDE, DDD and the DDT analogues and bioaccumulation in fish of terrestrial–aquatic model ecosystems. From these and similar findings, it further ensues that correlations between physico-chemical properties of compounds and biomagnification will become less meaningful in the future unless degradation velocity is taken into appropriate account. This applies in particular to pesticides, which are now engineered to be less persistent.

2.2 Stability of Ecosystems

In ecology, the discussion of stability has traditionally tended to equate stability with systems behaviour. Questions of persistence and the probability of extinction are of particular importance in this connection, and hence these measures are also frequently defined as aspects of stability. It should be kept in mind, however, that stability is the ability of a system to return to an equilibrium state after a temporary disturbance, while persistence is a measure of resilience in the sense defined by Holling (1976), i.e. the capacity to absorb changes of state and driving variables and parameters. The more rapidly a system returns to equilibrium, and with the least fluctuation, the more stable it is, which means, thermodynamically speaking, that stability is coupled with a (relative) minimum of entropy production (Prigogine 1976).

2.2.1 Different Notions of Stability

The above, and most common, meaning of stability, i.e. *stability in the vicinity of an equilibrium point* in a deterministic system is not only the most tractable mathematically, but also relates to more general stochastic situations, or to large amplitude disturbances (Lewontin 1969; May 1974). For population models in such deterministic environments, community equilibria are defined by situations where all net growth rates of species' populations are zero. The graphical visualization of the corresponding solutions of the equations of population dynamics for an m-species community may be represented on some m-dimensional surface, where each point marks a set of populations. The equilibria are in principle those points where the surface is flat; equating its configuration with that of a landscape, this means in customary geographical notation on hilltops and in valley bottoms. The hilltop equilibria, however, are obviously labile, i.e unable to survive the smallest displacement; only the valley bottoms are stable configurations.

Looking beyond the realm of a linearized stability analysis limited to the immediate neighbourhood of equilibrium points, for which straightforward mathematical tools exist, the situation becomes more complicated. The appropriate representation of the *global stability* of a system implies recourse to non-linear equations of population biology. In geographical terms, such a global analysis aims at comprehending the stability of the entire landscape, and not only just that of the immediate vicinity of equilibrium points. This involves an appropriate recognition of the fact that real environments are uncertain, stochastic, i.e. the corresponding environmental parameters in the model equations exhibit random fluctuations.

This leads to a yet more general meaning of stability which May (1974) termed *structural stability*. This refers to the qualitative effects upon solutions of the model equations which are due to gradual variations in the model parameters. A system is considered as structurally stable if these solutions change in a continuous manner. Conversely, a system is structurally unstable if gradual changes in the system parameters, such as alterations in site factors of a community, produce qualitatively discontinuous effects (cf. van der Maarel 1976; Rohde 1980; Fränzle 1981; Gigon 1983).

For a more thorough review of the meanings that may be given to stability, see Lewontin (1969), Holling (1973, 1976), May (1974) and Gigon (1983). Comparative accounts of mathematical models relating to stability of populations are given by Thom (1969), Maynard Smith (1971), May (1974), Freedman (1980) and Jørgensen (1988), for instance.

For practical purposes, reference is made to Stöcker's stability index (1974) which was conceived for the differential assessment of herbicide effects. Let $\Delta u = |u_0 - u_j|$ be the difference between the original state u_0 of a plant community and its present one u_j (i.e. after application of a herbicide), then the index of stability S_j is defined as

$$S_j = 1 - \frac{|u_0 - u_j|}{u_0} \quad \text{for } u_0 \geq u_j \qquad (2.37)$$

or

$$S_j = 1 - \frac{|u_0 - u_j|}{u_j} \quad \text{for } u_0 \leq u_j . \qquad (2.38)$$

This index is a dimensionless figure, $S_j = 0$ indicating a complete change of the system with respect to the quality measured, for instance biomass production,

Table 2.8. Average biomass production $(g/0.25\,m^2)$, u_0 = control, u_j = effect of $0.45\,g$ DCP-Na/m². (After Stöcker 1974)

Species	u_0	u_j	Δu	S_j
Gramineae	21.41	6.62	14.79	0.31
Cirsium acaulon	8.23	5.95	2.28	0.72
Ononis spinosa	4.27	9.60	5.33	0.44
Euphorbia cyparissias	1.33	0.72	0.61	0.54
Viola hirta	1.25	0.68	0.57	0.54
Teucrium chamaedrys	3.52	1.76	1.76	0.50
Fragaria viridis	1.34	0.77	0.57	0.58
Sanguisorba minor	1.21	1.73	0.52	0.70
Centaurea jacea	2.67	2.33	0.34	0.87
Salvia pratensis	0.97	2.76	1.79	0.35
Primula veris	0.23	0.62	0.39	0.37
Total	46.43	33.54	28.95	5.93

$S_j = 1$ conversely meaning stability in the sense of invariability of the structural characteristics considered.

Frequently, the arithmetic mean of the elementary stability indices s_j is more informative:

$$S_j = \frac{1}{m} \sum_{j=1}^{m} s_j , \tag{2.39}$$

as Table 2.8 shows. Application of index 2.37 yields the global S_j figure 0.72, while 2.39 yields a figure of 0.54, which is indicative of the selective response of the individual species to the herbicide.

In principle, comparable measures of stability may be derived from the comparison of diversity indices describing communities before and after a temporary disturbance. In practice, diversities, like the frequencies obtained from traditional quadrat sampling, have an essential dependence on sample size. It has proved useful, therefore, to define a diversity number to be the reciprocal of a mean proportional abundance (Rényi 1961; Hill 1973). Different means – harmonic, geometric and arithmetic – correspond to different measures of diversity. Entropies which are logarithms of diversity, are equivalent, but sometimes less easy to visualize although not less suitable for general use (Fränzle and Bobrowski 1983).

2.2.2 Stability and Diversity of Ecological Communities

Diversity is of considerable theoretical interest because it can be related to stability, evolutionary time, maturity, spatial heterogeneity, productivity, and predation pressure. For instance, MacArthur (1955) has argued cogently that stability is proportional to the number of links between species in a trophic web, while Müller (1977a) and Ellenberg (1979) emphasized that less diversified communities might exhibit a higher stability against external disturbances than others richer in species. Conversely, May's (1974) comparative mathematical analysis of large randomly connected cybernetic networks serving as ecosystem models showed that "too rich a web connectance or too large an average interaction strength leads to instability. The larger the number of species, the more pronounced the effect". Since ecosystems are likely to have evolved to a very small subset of all possible sets of open systems, however, MacArthur's and analogous conclusions might still apply in the real world.

2.2.2.1 Diversity Analyses

In fact, McNaughton's field test (1978) of May's inference, carried out in 17 grassland stands in Tanzania's Serengeti National Park showed "that both average interaction strength and connectance declined as species richness of the grassland increased. The correlation was somewhat stronger for interaction strength than for connectance". Hence it may be surmised, assuming the grassland communities analyzed are representative biocenoses (Fränzle 1984a), that species-poor stands are likely to be characterized by strong interactions among these species, while species interacting with many others do so only relatively weakly (Margalef 1968). Diffuse competition obviously increases with the number of species (cf. MacArthur 1972). In addition, McNaughton's findings suggest that communities have a block-like organization consisting of species interacting among themselves but only little with species in other blocks.

It ensues from a part of these analyses and studies on species diversity of tropical rain forest ecosystems in particular (Fränzle 1976a, 1977a), that a community is the more likely to have both low fluctuations in composition and low resilience, the more homogeneous its environment in space and time is. This positive relationship between diversity and stability is appropriately illustrated by the Hylaea associations of South America and Malesia (cf. van Steenis 1948/49; Aubréville 1961; Prance 1978; Stein 1978). Here, the entropy of the impoverished mature soils of the ferralsol, acrisol and podzol groups is high in terms of low content and equal distribution of nutrients (cf. van Wambeke 1978); consequently, the net entropy rates of the vegetation cover must be proportionally low in order to maintain stability of the total system in the most efficient way. In other words: species diversity of phytocenoses and nutrient content of the sites are negatively correlated provided energy flux rates are high enough to ensure stability by means of efficient entropy flux processes.

The realm of validity of this statement may be defined more precisely by means of comparative diversity analyses of ectropical plant communities (Fränzle 1979). The composition of 132 associations from all parts of the continental USA except Alaska evaluated in relation to climate and soils shows

that the amount of water available for evapotranspiration is of decisive importance for species diversity. This is further corroborated by the highly significant
correlation between diversity and distribution of rainfall in time. Conversely, the
analysis of 194 plant communities in northern Germany led to the conclusion
that, most frequently, high diversity coupled with medium to low proportional
abundances occurs on sites with high nutrient supply and medium to low soil
moisture, whereas the combination of both high diversity and high abundance
with high nutrient and medium to low moisture status is comparatively rare.
These inverse diversity-nutrient relationships in tropical and ecotropical environments on the one hand, and the fact that soil moisture controls diversity in
the USA but not in northern Germany on the other, lead to the final question of
whether a unifying interpretation of these apparently incoherent results is
possible. In this connection it should be recalled that an increase of net primary
production or biomass with growing soil moisture and elevated CO_2 level of the
microclimate does not necessarily imply a corresponding increase in structural
diversity (cf. mono- or oligospecific associations). Hence the following section
does not refer to well-known physiological plant–water relationships but stresses some more general biophysical aspects which seem to have played a lesser
role in relevant discussions hitherto.

2.2.2.2 Ecological Communities as Dissipative Structures

Biophysical aspects of plant and animal life are basically related to specifications
of the second law of thermodynamics as applied to open systems (Prigogine
1967). Because these systems are in exchange of energy and matter with the
outside world, their entropy production, dS, comprises the terms d_iS and d_eS,
where d_iS denotes the entropy production within the system while d_eS is a flux
term describing entropy "export" into the environment:

$$dS = d_iS + d_eS . \tag{2.40}$$

Only $d_iS > 0$, but d_eS can also be negative. Identifying entropy with disorder, it
ensues from Eq. (2.40) that an isolated system can only evolve towards greater
disorder. For an open system, however, the "competition" between d_eS and d_iS
permits the system, subject to certain boundary conditions, to adopt new states
or structures. These are stationary if

$$dS = 0 \tag{2.41a}$$

or

$$d_eS = -d_iS < 0 , \tag{2.41b}$$

respectively.

d_iS can be expressed in terms of thermodynamic "forces", X_i, and rates of
irreversible phenomena, J_i (Prigogine 1967). X_i may be gradients of temperature
or concentration; the corresponding "rates" are then heat flux and chemical

reaction rate. Hence

$$\frac{d_i S}{dt} = \sum_{i=1}^{n} J_i X_i .$$ (2.42)

Around equilibrium there is a linear relationship between fluxes and forces

$$J_i = \sum_{j=1}^{m} L_{ij} X_j ,$$ (2.43)

where L_i are specific coefficients, e.g. coefficients of thermal conductivity or diffusion. Provided the reservoirs of energy and matter in the environment of the open system are sufficiently large to remain essentially unchanged, the system can tend to a non-equilibrium stationary state far beyond the domain of linear thermodynamics. This state may be associated with dissipative structures (Glansdorff and Prigogine 1971), i.e. structures resulting from a dissipation of energy rather than from conservative molecular forces.

Considering phytocenoses from the point of view of stationary dissipative structures, the relationship of $d_i S$ and $d_e S$ as expressed in Eq. (2.41) and the specific boundary conditions controlling entropy production and flux rates appear to be particularly important. It is a consequence of Eq. (2.41b) that, thermodynamically speaking, stability or the capacity to maintain a non-equilibrium steady state is coupled with a (relative) minimum of total entropy production dS. Clearly this can be accomplished by either minimizing $d_i S$ or maximizing $d_e S$, or a combination of both strategies.

Concentration processes involved in the normal metabolic activities of living systems play an important role in this connection, as can be seen from the following equation

$$\Delta G^0 = RT \ln \frac{C_2}{C_1} ,$$ (2.44)

where ΔG^0 = difference in standard free energy, $R = 8.31 \, J \, mol^{-1} K^{-1}$, T = temperature in Kelvin and C_2, C_1 = higher or lower thermodynamic concentration, respectively. Changes in concentration are a physical prerequisite for the production of a great many compounds, and an absolutely cogent one if substances are produced whose free energy is higher than that of the corresponding "raw materials".

In Amazonia, for example, forest stands developed specific adaptations facilitating these concentration processes. A dense root mat is formed over the soil in intimate contact with litter, and root tips growing upwards into the fallen litter. Mycorrhizal associations seem to be active in the transfer of some nutrients directly from litter to roots. The dense fabric of predominantly fine roots also plays an important part in exchange and adsorption of nutrients from throughfall water. In addition, the structure of the foliage favours the use of nutrients by a long active life, the retransport of certain nutrients before leaf shedding and a high polyphenol content and coriaceous nature, which reduces herbivory. Leaves shed are decomposed slowly, thus reducing leaching losses. Another factor of high importance is the multilayered structure of the forest and the activities of epiphytes and microorganisms on much of the exposed surfaces, which acts as an efficient filtering system scavenging nutrients from rainwater. Finally biological nitrogen fixation in the root–humus–soil interface seems to play an important role in the nutrient budget (Herrera et al. 1981; Fittkau 1983; Klinge 1983).

Taking into account that even in the simplest cells the normal metabolic pathways imply several thousand complex chemical reactions which must be coordinated by means of an extremely sophisticated functional network means that biological order in both functional and spatio-temporal respect constitutes a further and most powerful negentropic factor. It characterizes every living system from the sub-microscopic level up to gigantic rainforest biomes like the Amazonian Hylaea.

The effectiveness of these negentropic processes is further enhanced by most efficient entropy fluxes related to the transpiration and nocturnal respiration of plants. The molal entropy of H_2O increases from $63\ \mathrm{J\,mol^{-1}\,K^{-1}}$ (liquid) to $189\ \mathrm{J\,mol^{-1}\,K^{-1}}$ (gas) in the course of evaporation, and CO_2 has a molal entropy of no less than $214\ \mathrm{J\,mol^{-1}\,K^{-1}}$. Consequently, also the reverse process, the photosynthetic fixation of CO_2, is of comparable importance for the negentropy balance.

In the light of these mechanisms the results of the above comparative diversity analyses may be given a unified interpretation:

1. Species diversity is not a monotonous function of the nutrient status of soil. If soils form on basic or intermediate parent materials pedogenic nutrient supply and species diversity may both increase during periods of 10^3–10^4 (10^5) years, provided the water and energy factors are not limiting. In Fig. 2.2, the period of ascending evolution is termed phase I. Shorter fluctuations of variable but generally decreasing amplitude are likely to be superimposed on this long-term trend.

In phase II, which can last for several hundred thousand years, soils degrade in regard to nutrient supply but species diversity keeps increasing for negen-

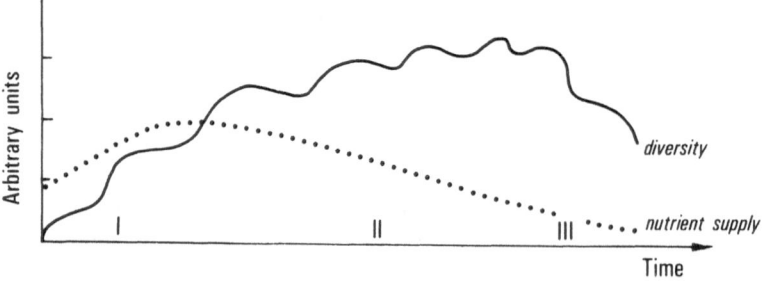

Fig. 2.2. Model of pedogenic nutrient supply and species diversity in the course of ecosystem evolution

tropic reasons (cf. the tropical rain forest stands in comparison to the ectropical associations described here). Clearly such an evolution towards and including the following phase requires a sufficient degree of climatic and geomorphic stability (as has been the case in central Amazonia and considerable parts of Malesia) or else the continuity of pedogenesis will be interrupted by truncation processes that rejuvenate the soil (as happened in the major part of the Congo Basin). Phase III marks the eventual and comparatively acute decrease in diversity of phytocenoses once the nutrient status of soil has fallen below a critical level.

2. Spatial heterogeneity as a characteristic feature of the sites during phase I results in variability in numbers of species and plant individuals. With this variability the associations can simultaneously retain genetic and behavioural types that can maintain their existence in low populations together with others which can capitalize on opportunities for dramatic increase.

The more homogeneous the environment in space and time (i.e. during phase II, and probably also phase III), the more likely is the system to have low fluctuations, i.e. high stability. Tropical rain forests represent climatically buffered and (largely) self-contained systems with relatively low (natural) variability on highly impoverished soils of the ferralsol, acrisol and podzol classes. Temporary external disturbances are consequently likely to affect a proportionally higher number of species than in phase-I populations, where the interspecific differences in ecological potency are usually distinctly higher. Hence resilience of mature biocenoses is low during phase II. On the contrary, biocenoses may be highly resilient in phase III of evolution, although unstable.

The highly reduced resilience of phase II stands is particularly well illustrated by the influence of clearing activities on tropical forests. The soil, with its low retention capacity, is irrevocably impoverished after a brief period of fertilization due to the sudden release of nutrients in the biomass. This brief period of fertility is used by shifting cultivators to produce crops; but even this system, developed by small native populations "in close contact with the environment" (Herrera et al. 1981), can induce severe losses if forced to fit a market-oriented economy or to feed larger populations (cf. Nye and Greenland 1960; Fränzle 1976b; Weischet 1977; Alvim 1978; Sioli 1980b; Klinge et al. 1981).

Even more illuminating in the framework of ecotoxicological considerations is the example of Vietnam's inland forests, which still bear the scars of a 10-year herbicide spraying programme by the US Air Force during the Vietnam war. According to official US figures, some 10.3% of these forests, in addition 36.1% of mangrove forests, 3% of cultivated land and 5% of other land was sprayed. Some 90,000 tons of herbicides and anti-plant chemicals were used between 1962 and 1971 (Hay 1983). Most of the spraying was with Agent Orange (i.e. a 1:1 mixture of 2,4-D and 2,4,5-trichlorophenoxy-ethane acid), Agent White (2,4-D and picloram) and Agent Blue (cacodylic acid).

Twenty one years after the spraying ended in 1971, there has been little regrowth in areas sprayed three or four times. The fertility of the soils is rapidly further reduced if forests are replaced with grassland or bamboo, as has happened in much of the sprayed region. In particular, there has been a loss of minerals and nitrogen and in some cases a fall in soil pH, which all will inhibit recolonization.

Defoliation has clearly affected the faunal assemblages, not only in the areas totally defoliated and subsequently converted to grassland, but also in those regions less frequently sprayed where the plants in the upper reaches of the forest were damaged. In both areas, the number of animals and birds has dropped dramatically, and animals are also at considerable risk in some of the forest areas which have been isolated from the main body by herbicide application. Furthermore there is concern that Agent Orange contained a high concentration of the highly toxic contaminant 2,3,7,8-tetrachlorodibenzo-p-dioxin, which is both teratogenic and carcinogenic in animals, and highly toxic to humans, too.

2.3 Biota Resilience Versus Chemical Persistence

2.3.1 Chemical Persistence

Chemical persistence can be defined as "the property of a compound to retain its salient physical, chemical and functional characteristics in the environment through which it is transported and/or distributed for a finite period following its release" (Stern 1980). The determination of this complex property plays an important role in the overall evaluation of a substance's environmental hazard. It should include concentration as a function of time and location as well as all major compounds resulting from the more important transformation processes. (In agreement with Mill 1981, transformation is the term preferred to describe any process producing a change in molecular structure. Terms such as hydrolysis, oxidation, photolysis etc. then refer to specific transformation processes and are used as appropriate.)

A kinetic model for the sum of fate processes or reactions, respectively, which define persistence assumes that the net rate of loss of a chemical (R_T) is equal to the sum of all loss processes and that each one can be appropriately described by a first order-process (Mill 1981).

$$R_T = \frac{d[C]}{dt} = \sum k_L'[C][E] \tag{2.45}$$

$$k_L'[E] = k_L$$

$$R_T = \sum k_L[C]$$

$$t_{1/2} = \frac{\ln 2}{\sum k_{L,}}, \tag{2.46}$$

where k'_L and k_L are second- and first-order rate constants, [E] denoting the environmental parameter in compatible units of concentration, and $t_{1/2}$ the half-life of the compound.

Since many chemicals enter the various environmental compartments air, water, soil and biota continuously, persistence is then better defined as the steady-state concentration C_{SS}, rather than in terms of half-life; it results from equal rates of input (R_I) and loss (R_L).

$$\frac{d[C]}{dt} = R_I - R_L = 0$$

$$R_I = \sum k_L[C]$$

$$[C]_{SS} = \frac{R_I}{\sum k_L}. \qquad (2.47)$$

The consideration of persistence in this comprehensive sense involves an evaluation of the major processes in each environmental compartment; this is done in

Table 2.9. Environmental processes influencing chemical persistence

Atmosphere	Convective and advective transport
	Sorption
	Nucleation, coagulation
	Rainout, washout, dry deposition
	Hydrolysis
	Photolysis
	Oxidation
Soil/sediment	Leaching, connection dispersion, diffusion
	Sorption
	Volatilization
	Hydrolysis
	Oxidation, reduction
	Photolysis
	Bio-uptake
	Biotransformation
	Erosion
Water	Transport by currents, turbulence
	Diffusion
	Sorption
	Volatilization
	Hydrolysis
	Photolysis
	Oxidation, reduction
	Bio-uptake
	Biotransformation

Chaps. 3 and 4. In the framework of this introductory review it may consequently suffice to list the essential processes (Table 2.9) and to comment on inherent problems in a general way.

As may be inferred from the list of influential processes in Table 2.9, an ecotoxicological measure of persistence in the comprehensive sense of the introductory definition is still far beyond present possibilities. Consequently, a number of methods have been tentatively selected to determine persistence in the narrower sense. Yet the results obtained pose problems which should be kept in mind; some of these can be summarized as follows:

– Transformation products may pose more serious ecological problems than the parent compounds from which they were derived.
– Where multiple metabolic pathways exist (cf. Chap. 4.4.2), it is difficult to ascertain whether their effects are simply additive, super-additive or mutually exclusive in the real world. Associated with this is the usual problem of extrapolating results of isolated laboratory tests to in situ events.
– In addition, there is always the difficulty of designing in an experiment the pertinent environmental conditions found in nature.

Furthermore, there are issues that have to be addressed for each mechanism of transformation. In biodegradation, for instance, there have been many arguments concerning:

– the use of mixed or pure cultures;
– the use of adapted or unadapted inocula;
– the selection of chemical concentrations;
– the influence of cometabolites on an organism's response to chemicals;
– the effects of synergism, symbiosis, antagonism, commensalism etc.;
– the distinction of biotic from abiotic transformation processes if natural materials are utilized as inocula.

2.3.2 Resilience of Natural Biota

In the light of such isolated, monospecific or media- and process-oriented measures of persistence, it is difficult to establish generalized relationships with ecosystem or biota resilience. Yet some broader issues of the above analysis of existing concepts can be deduced in combination with empirically meaningful definitions of resilience.

While theoretically measures of this quality of biocenoses can be derived from the behaviour of model systems in phase space, it is probably more difficult to find practically useful measures. In theory, two components are important (Holling 1976): "one that concerns the cyclic behaviour and its frequency and amplitude, and one that concerns the configuration of forces caused by the positive and negative feedback relations". A full development of this concept, which is beyond the scope of this introductory chapter, would show that for any fixed set of parameters, equilibrium populations tend to be recovered only from perturbations into a limited domain of population space, and that these equilibria change with the environmental parameters (Jørgensen 1988).

This approach requires an immense amount of knowledge of any particular system, taking into account the manifold mechanisms of stress avoidance, stress evasion and stress tolerance, which each species – more precisely, each ecotype – has developed (cf. Precht 1967; Levitt 1980). These adaptations to site qualities may be either stable, having arisen by evolution over a large number of generations, or unstable, i.e. depending on the developmental stage of the organism and the environmental factors to which it has been exposed. Also the unstable adaptation has arisen by evolution, but the hereditary potential is wide enough to permit larger changes during the growth and development of the organism.

Considering the complexity of multi-species systems resulting from the countless adaptations to site structures which may also vary tremendously in time and space, it is obvious that gaps in our knowledge to date usually prohibit the practical application of the above comprehensive approach. However, first feeble steps in this direction are represented, for example, by the work of Hassel and May (1974) on insect host-parasite systems, and by Lloyd's cicada-fungus model (Lloyd and May 1974).

It merits, therefore, all the more attention from the practical point of view that Hughes and Gilbert (1968) suggested a promising approach to measuring resilience by means of deriving probabilities of species or ecotype extinction. On a primarily purely descriptive basis they showed in a stochastic model that the distribution of surviving population sizes can be appropriately approximated by a negative binomial. This permits estimating the very small probability of zero, i.e. of extinction, from the observed mean and variance. The parameter values of the negative binomial, or any other distribution function, covariate with the configuration of the phase space such that they can be viewed as relative measures of resilience. It is important to develop this technique with a number of theoretical models in order to determine the appropriate distributions and their behaviour (cf. Bobrowski 1982) adequately incorporating the evolutionary aspects outlined in Section 2.2. It should then be feasible to sample populations in defined areas, apply the appropriate distribution, and use the parameter values as measures of resilience.

With a view to the frequently limited ecological meaning of traditional measures of chemical persistence, it appears useful to formulate the interesting relationships between biota resilience and persistence in the form of a correlation matrix:

$$[M(R_i, P_j)]_{1 \leqslant i, \, j \leqslant n} , \tag{2.48}$$

where R_i denote resilience measures of biocenoses and P_j measures of chemical persistence. Both the R_i and P_j are index numbers resulting from multidimensional combinations of the parameters of the binomial distribution or the different persistence measures of each chemical considered, respectively, by means of numerical classification procedures. Among these, an amalgam of biplot analysis (Gabriel 1971), hierarchical clustering algorithms, and optimized iterative procedures proved most useful (Fränzle et al. 1980).

The individual R_i, P_j-combinations, represented by the $\times$-symbols of the correlation matrix, denote specific resilience and persistence-oriented inter-relationships.

	R_1	R_2	R_3	. . . . .	R_n
P_1	$\times$		$\times$		$\times$
P_2	$\times$	$\times$			$\times$
P_3		$\times$			
.					
.					
.					
.					
.					
P_n	$\times$	$\times$	$\times$		$\times$

Biplot analysis not only permits viewing of the individual data of such a matrix and their differences, but further allows scanning of the standardized differences between units and inspecting the variances, covariances and correlations of the variables involved.

2.4 Model Ecosystems and Related Definitions of Environmental Toxicity of Chemicals

Hitherto the predictive assessment of dose–effect relationships between chemicals and biotic elements in the environment has usually started with laboratory tests. Their results need extrapolation to the complex "real-world" situation, which, as a rule, proves to be rather difficult, since the experiments are run with defined amounts of chemicals in contact with a single species or a restricted group of organisms. In contrast, the integral approach to assessing the effects of potential pollutants in the natural environment has to deal with the manifold interactions and complicating factors or boundary conditions which are described in the following chapters, and which are carefully avoided or controlled in the laboratory situation.

The most prominent of these complicating factors in the natural environment are (Hueck et al. 1978):

– Uneven distribution of the pollutant in an ecosystem compartment due to influence of gradients or currents, dilution, volatilization, sorption processes etc. Bioaccumulation in successive elements of the food web may cause considerable variations in the individual load of different species.

- Differences in specific sensitivity may lead to shifts in the abundance and diversity proportions of the biocenoses affected. Interspecific relationships, synergistic relations and food web effects can be analyzed only in multispecies systems.
- In a natural ecosystem the effect of a potential pollutant on the biocenosis will be greatly influenced by: weather, diurnal and seasonal variations, and simultaneously occurring variable loads such as thermal and chemical loads along with intermittent discharges.

To a certain extent, the influence of these complications can be accounted for by simulation techniques in the laboratory, but, necessarily, practical limitations exist. On the other hand, it is impossible to rely on retrospective field experiments only since the doses are hardly known exactly, intentional pollution for experimental purposes is frequently inadmissible and biological observations are time-consuming and expensive. As a consequence, it appears useful to carry out comparative tests with selected model chemicals under both laboratory and natural conditions in order to determine the relative representativity of the test procedure. This validation of the tests should also pay appropriate attention to the important question in how far the test system or the experimental field plot are representative from the regional point of view (Fränzle 1984a).

In Table 2.10 a compilation of experimental models is given in a decreasing order of resemblance to the "real-world" situation. The gradual build-up and increasing structural complexity of such systems is schematically represented in Fig. 2.3. Comparing this comprehensive, normative list of ecotoxicological models with the (still) predominant present-day practice, there is a fairly marked tendency to favour the bioindicators approach and aquarium tests or technically comparable procedures (cf. Müller 1977b; Bick and Neumann 1982; Steubing and Jäger 1982). This has the appreciable advantage of relative simplicity and assures a fairly high amount of reproducibility, but also poses manifold problems as to rationally and reliably extrapolating to the "real-world" situation.

Table 2.10. Survey of ecotoxicological models

---

1. Pollutants already present
   - Observation of individaul bioindicators
   - Gradient tests
   - Analysis of calamities
   - Survey of total stock (for instance abundance spectra, diversities, biomass production)

2. Pollutants intentionally applied
   - Aquarium tests with artificial communities
   - Compartment tests with separate trophic levels
   - Basin tests with natural communities

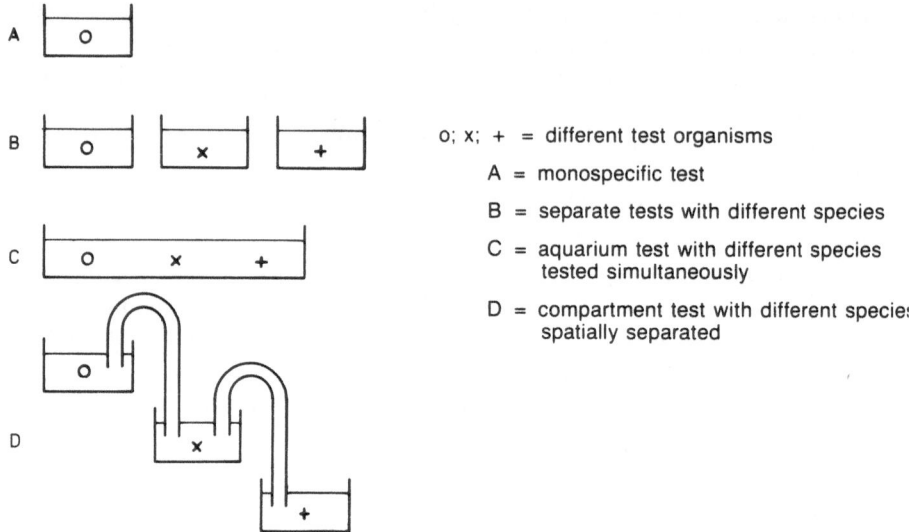

Fig. 2.3. Increasing structural complexity of ecotoxicological test systems. (After Hueck et al. 1968)

2.4.1 Ecotoxicological Tests

Toxicology studies designed to protect human health or the environment normally begin with the determination of direct toxicity of a compound due to ingestion, inhalation or adsorption through the skin. This level of direct toxicity is mainly measured by assessing an LD_{50} or LC_{50} (lethal dose or concentration 50), i.e. that single dose of a compound apt to kill half the number of organisms of a test group within 24 h. Clearly this figure provides useful information in cases of accidental exposure to high dose levels or for defining appropriate levels in additional toxicity experiments. From the ecotoxicological point of view, however, it must be recalled that the knowledge of such a dose level in one species can hardly give any information about eventual effects to be expected from long-term exposure to low concentrations of a substance to man and the environment.

Consequently, the more accurate assessment of so-called no-toxic-effect levels involves longer dosing periods; and the combination of a dose-range-finding study, coupled with a sub-acute toxicity test, can provide reliable insight of predictive value into the possible dangers of a compound. For the sake of greater safety it may be adequate to base the experimental determination of no-toxic-effect levels on a deliberately isolating consideration of critical organs or organ systems, respectively, of an animal species most representative for the situation. In addition to defining organs or organ systems most at risk. toxicological studies also have to make due allowance for determining the "high-risk groups" of the monospecific community tested, i.e. the group of animals within a species which proves most susceptible to the chemical. Such particularly sensitive groups can be: the young (males or females), the adult (males or females), the pregnant females, the unborn or the aged.

To this end, fertility studies can be conceived which investigate the effects of a chemical compound on the generative behaviour of both young males and females. It may suffice to indicate the relatively recent recognition that the entire biotransformation and pharmacocinetics can be changed during pregnancy, which also means that detoxification mechanisms may be altered considerably. This can affect not only the pregnant female, but also the unborn, although in the majority of cases the foetus seems to be more at risk than the female herself. Therefore a relatively simple teratological study with animals can give most useful insights into risks to be expected for the unborn. As a result of experiments based on the most critical organs or organ systems and properly defined high-risk groups, it may then be comparatively easy to measure the distance between the first toxic effects observed (in mg/kg body weight) and the maximum dose to which the human may be exposed, if the compound is used or released as foreseen.

Table 2.11 summarizes in a way the presentation of test procedures, although limiting the scope of a joint expert evaluation to aquatic toxicity tests. The relative importance of each of the six criteria (such as ecological significance, scientific and legal defensibility etc.) is rated numerically with 1 being the criterion of lowest significance and 6 being the criterion of highest significance. Then each type of test is rated (0–10 in increasing value) under each evaluation criterion. The numerical rating in the table is a weighted mean representing the product of the mean rating for each type of test and the rank of the criterion. Present relative utility is determined by summing the weighted means for each type of test and expressing these summations as a percent of the highest value attributed to acute lethality tests. Since a particular test can have a low or high rating under a single, several, or all criteria, these summations may not necessarily be indicative of overall value and their utility. For instance, residue accumulation tests have a relatively high overall rating because of the high ratings given to the ecological importance and legal defensibility, which are both considered the most important of all the criteria. Similarly, field studies have a fairly high overall rating due to the ecological significance of such studies.

2.4.2 Comparative Appraisal and Research Needs

A comparative analysis of ecotoxicological laboratory tests shows that the basic problem of relating exposure to effects data is the definition of reasonable safety margins between the "no observed effect levels", derived from single species tests, and environmental concentration. This is due to the elementary fact that neither tests nor models can accurately represent the complexity of a natural

Table 2.11. Consensus evaluation of the present relative utility of simple and complex toxicity tests used in assessing the risk associated with the occurrence of chemicals in aquatic environments. (Cairns et al. 1978)

Test	Evaluation criteria						Present relative utility
	Ecological significance	Scientific and legal defensibility	Availability of routine methods	Predictive utility	General applicability	Simplicity and cost	
Simple systems							
Acute lethality	33	36	35	22	24	26	100
Embryo/larval	33	30	23	20	25	14	82
Reproduction	36	30	21	27	24	8	82
Residue accumulation	27	32	22	24	17	14	77
Algal assay	30	23	26	18	14	20	74
Organoleptic	14	25	25	27	13	15	67
Structure activity	18	21	23	17	16	22	66
Behavioural	21	13	10	9	16	9	44
Histological	10	10	20	10	13	10	41
Physiological and biochemical	12	10	13	8	19	8	40
In vitro	4	5	9	4	11	11	25
Complex systems							
Field	34	24	18	21	21	3	69
Diversity	26	19	25	15	20	10	65
Benthic	21	18	12	18	17	12	56
Microcosm	19	10	15	14	17	9	48

atmosphere–plant–soil ecosystem, in particular not the immense temporal and spatial variability of weather conditions and climate, and in natural plant and animal populations. Consequently, the results of tests and models should not be assigned the same amount of precision normally associated with laboratory experiments and in the habitual terms of physical and chemical usage (cf. also in addition to the examples given in Table 2.11 the particularly useful discussions by Draggan and Giddings 1978; Cairns 1979; Klöpffer and Rippen 1983).

An incontestable advantage of laboratory tests is that they permit standardization, with the reservations made, for instance, in regard of acute toxicity tests for fish. They can be easily replicated in different laboratories with a set of chemicals for comparative purposes, and are usually less expensive than more sophisticated test procedures. In order to provide a relative optimum of information, laboratory tests should cover the whole range of parameters controlling the fate and behaviour of a compound in a natural ecosystem in addition to the relevant physico-chemical and biological properties such as vapour pressure, solubility, acute toxicity etc. Furthermore, an ecotoxicological test profile should quantify the range of significant disappearance factors.

Such a comparative approach implies the inclusion of environmental chemicals already better studied and appropriately documented on behalf of their past release, usage and presence in ecosystems. The results thus obtained allow a certain prediction on the basis of reference chemicals, although the underlying test procedure merely simulates some features of a natural ecosystem. An improvement of the methodology, therefore, must make better allowance for the vital role of the plant root system and the vast range of soil organisms in determining the fate of soil chemical residues. More reliable, yet still simple future laboratory tests should at least characterize overall field soil biomass and its activity, e.g. on the basis of $^{14}CO_2/CO_2$ respirometry, ATP levels etc. (Kokke and Winteringham 1980). In order to further improve the reliability of test data, particular attention has to be paid to regionally representative sampling, which is indispensable for subsequent areal extrapolations or interpretations, respectively, of the test results (Fränzle 1984a; Wildförster 1985). Tests of this kind could obviate the danger of a false negative decision in the above sense of the term when an apparently biodegradable compound proves unexpectedly persistent in nature because of its appearance in an otherwise untested soil of different composition.

Another largely neglected provision of laboratory tests is that for the capacity of bacteria, fungi, weeds, insects and even rodents to evolve, in the comparatively short time of a few generations, populations resistant to an otherwise toxic chemical.

Having established priorities by means of laboratory tests, the need for more sophisticated approaches on the microcosm level can be considered. These can indicate the specific distribution potentials for intercompartmental transfer processes, but their decisive disadvantage is due to the fact that they can never represent more complex natural biocenoses, genetic pools and climate fluctuations etc. (see Winteringham 1977, 1981; Frissel 1981).

Also the whole idea of mathematical modelling involves a sufficiently precise identification of only those pathways that account for the larger part of the phenomena observed and not of every possible pathway (cf. Fränzle 1984b; Chap. 4.2). Thus the main value of a mathematical model lies in the very exercise of stating how much is really known about individual pathways naturally operating in a number of combinations, and in the possibility of identifying the very type of research which is most urgently needed when significant contributions to the knowledge of the overall system behaviour are desired.

At present, our knowledge of the structure and functions of entire communities and ecosystems, whether stressed or not, is still at a fairly low level (cf. Ellenberg et al. 1978; Fränzle 1990). Therefore, the input of scientifically based information about environmental effects of chemicals on the definition of "safe" levels of substances is very small, and an arbitrary application factor must be introduced to estimate some recommended "safe" level from laboratory test results. Hence the reliable determination of such factors through sufficiently comprehensive ecotoxicological studies must be given high priority in future research.

The development of time-dependent models of ecosystems will entail, as a first step, definitions of ecosystems response variables ranging from simple measures of the population size for a particular species, to measures of both population size and mix of a number of species (Jørgensen 1988). In addition, major efforts should be devoted to studying the interactive effects of combined chemicals and the variable modifying influences of otherwise non-toxic environmental conditions. Statistically speaking, this implies in terms of dose–effect relationships, the definition of "equivalent dose" rules in the framework of intra-species comparisons. Since this extrapolation rule between species will not be known exactly, however, some measure of the error in estimating the toxicologically relevant species' properties must be incorporated into the final extrapolation. "Therefore, the estimation of risk to any population in the ecosystem will have some measure of uncertainty which is a combination of sampling errors in the data, choice of a particular mathematical model among the many possible models when extrapolating outside the observable range, and the uncertainties inherent in extrapolating between different . . . species. In general, the weakness of current techniques used to measure the relations of doses and responses is that they are aimed at single species under controlled conditions. We must begin to consider an ecosystem as a single entity, albeit a complex one, and formulate new methods of estimating dose response of the system as a whole" (Brown 1978).

3 Fluxes of Anthropogenic Chemicals in the Atmosphere

According to the Technische Anleitung zur Reinhaltung der Luft 1974 (= Technical Instruction Air of the Federal Republic of Germany 1974, Amendment 1985) pollutants are defined as "alterations of the natural composition of air, especially by smoke, soot, dust, gases, aerosols, steam or odorous substances". Other definitions stress the effects upon potential receptors, including modifications due to both natural and anthropogenic sources, for instance: "Air pollutants are substances which, when present in the atmosphere under certain conditions, may become injurious to human, animal, plant or microbial life, or to property, or which may interfere with the use and enjoyment of life or property". In the present context emphasis will be more on the physical and chemical processes and boundary conditions leading to a more or less polluted atmosphere at a location than on the effects evoked after exposure, and attention will be mainly paid to pollutants generated by human activities rather than by natural events.

The amount and type of pollution at any given site is determined by the nature of relevant emissions, the state of the atmosphere, and the physical and chemical configuration of the area affected. Clearly the rate of emission and the properties of the pollutants have a central bearing, but also other characteristics of the source are important such as shape of the emission area, duration and mode of releases and the effective height at which the injection of the pollutants occurs. After release, the disposition of pollutants is controlled by wind speed and temperature stratification of the atmosphere. While the former determines both the distance of downwind transport and the pollutant dilution, and establishes the intensity of mechanical turbulence as related to surface roughness, the latter defines the atmospheric stability which, in turn, regulates the intensity of thermal turbulence and the depth of the surface mixed layer. In the atmosphere pollutants may undergo manifold transformations which are related to air temperature, intensity of solar radiation, amount and distribution of water in form of vapour or droplets, and the presence or absence of other substances. Furthermore the eventual removal of pollutants by precipitation-related processes, by gravitational deposition, impaction or adsorption, is closely related to the state of the atmosphere.

The foldout model I *Fluxes of environmental chemicals in the atmosphere*, shows the interesting relationships in the form of a comprehensive relational graph, and this chapter is organized in basically the same sequence. First the input of pollutants to the boundary layer is described, then the essential atmospheric mechanisms controlling the dispersal, transformation and eventual removal of pollutants from the air are dealt with. In this connection a comparative evaluation of transport models is aimed at with an emphasis on application problems.

3.1 Natural and Anthropogenic Sources of Potential Pollutants

Air is composed of nitrogen, oxygen, argon, carbon dioxide and traces of such gases as krypton, helium, methane and hydrogen. In addition, air can hold water in vapour and particulate form, depending upon temperature. Conventionally, the presence of other materials, such as dusts, other gases, fumes, and vapours, is regarded as pollution. Since air is subject to pollution from both man-made and natural sources, it is practically impossible to find unpolluted outdoor air even in the most remote areas.

In order to provide for an introductory overview, Table 3.1 summarizes the abundances of a number of gaseous constituents of the atmosphere, with mixing ratios ranging down to 6×10^{-20} for radon. (It should be kept in mind, however, that mixing ratios below 1×10^{-9} are mostly extrapolations on the basis of measurements which lack spatial representativity.) The known particulate constituents, both solid and liquid, have mass fractions ranging from 10^{-3} for droplets in raining cloud to about 10^{-6} for large hydrated ions. These components are involved in manifold cyclic processes which, in addition to the atmosphere, may also involve the hydrosphere, biosphere, lithosphere, as is indicated in each foldout model.

Both for notional and technical reasons it appears useful to distinguish pollution as a practically universal phenomenon from the more regionalized or localized contamination. When a pollutant occurs in the atmosphere in sufficient concentrations and long enough to become a hazard to health it is called a contaminant. For instance, when the carbon dioxide level in the air reaches 500 parts per million (ppm) the gas is considered a pollutant; when a level of 5000 rpm is reached, however, the normal respiratory functions of man are affected and the compound must consequently be considered a contaminant.

There may be hundreds of polluting substances in the air, and they usually come from natural and anthropogenic sources in varying proportions. As is indicated in Table 3.1, the compounds found in the atmosphere generally contain sulphur, nitrogen, oxygen, the halogens (bromine, chlorine, fluorine and iodine), and an extremely complex number of organic materials, such as hydrocarbons. Table 3.2 gives an introductory listing of the most important types of air pollutants and their principal sources, which forms the basis for the following description of the nature of emissions.

Table 3.1. Composition of tropospheric air (from Lewis and Prinn 1984)

Gas	Volume mixing ratio
Nitrogen (N_2)	7.81×10^{-1} (dry air)
Oxygen (O_2)	2.09×10^{-1} (dry air)
Argon (^{40}Ar)	9.34×10^{-3} (dry air)
Water vapour (H_2O)	$\leq 4 \times 10^{-2}$
Carbon dioxide (CO_2)	$2\text{--}4 \times 10^{-4}$
Neon (Ne)	1.82×10^{-5}
Helium (^4He)	5.24×10^{-6}
Methane (CH_4)	$1\text{--}2 \times 10^{-6}$
Krypton (Kr)	1.14×10^{-6}
Hydrogen (H_2)	$4\text{--}10 \times 10^{-7}$
Nitrous oxide (N_2O)	3.0×10^{-7}
Carbon monoxide (CO)	$0.1\text{--}2 \times 10^{-7}$
Xenon (Xe)	8.7×10^{-8}
Ozone (O_3)	$\leq 5 \times 10^{-8}$
Ammonia (NH_3)	$\leq 2 \times 10^{-8}$
Sulfur dioxide (SO_2)	$\leq 2 \times 10^{-8}$
Hydrogen sulphide (H_2S)	$0.2\text{--}2 \times 10^{-8}$
Formaldehyde (CH_2O)	$\leq 1 \times 10^{-8}$
Nitrogen dioxide (NO_2)	$\leq 3 \times 10^{-9}$
Nitric oxide (NO)	$\leq 3 \times 10^{-9}$
Hydrochloric acid (HCl)	$\leq 1.5 \times 10^{-9}$
Nitric acid (HNO_3)	$\leq 1 \times 10^{-9}$
Hydrogen peroxide (HOOH)	$\sim 1 \times 10^{-9}$
Methyl chloride (CH_3Cl)	$\sim 6 \times 10^{-10}$
Carbonyl sulphide (COS)	$\sim 5 \times 10^{-10}$
Freon-12 (CF_2Cl_2)	2.8×10^{-10}
Sulfuric acid (H_2SO_4)	$\sim 1 \times 10^{-10}$
Freon-11 ($CFCl_3$)	1.7×10^{-10}
Carbon tetrachloride (CCl_4)	1.2×10^{-10}
Methyl chloroform (CH_3CCl_3)	1.2×10^{-10}
Freon-12 ($CHCl_2F$)	$\sim 1.4 \times 10^{-11}$
Methyl iodide (CH_3I)	$\sim 1 \times 10^{-11}$
Chloroform ($CHCl_3$)	$\sim 9 \times 10^{-12}$
Methyl bromide (CH_3Br)	$\sim 5 \times 10^{-12}$
Radon	$\sim 6 \times 10^{-20}$

3.1.1 Natural Atmospheric Cycles

The adequate assessment of the relative importance of anthropogenic inputs involves a comparative consideration of the fundamentals of the carbon, oxygen, nitrogen, phosphorus and sulphur cycles in nature. Together with

Table 3.2. Types, sources and quantities of atmospheric pollutants. (After Rasmussen and Went 1965; Robbins et al. 1968; Kellog et al. 1972; Rasmussen 1972; Robinson et al. 1973; Oke 1978; Woodwell 1978; Bolin et al. 1979; Häberle 1982)

Type	Natural source	10^6 t/a	Anthropogenic source	10^6 t/a
Particulates	Sea spray	1100	Combustion	110
	Wind action	365	Agriculture	1
	Forest fires, volcanoes, meteors	150		
Sulphur compounds				
H_2S, mainly	Bacteria	268	Burning fossil fuels	150
SO_4^{2-}, mainly	Sea spray	130		
H_2S, SO_2, SO_4^{2-}	Volcanoes	2		
Carbon monoxide	Oxidation of methane	1500–5000	Combustion engines, burning fossil fuels	450–460
	Decomposition of chlorophyll	100–500		
	Forest fires, oxidation of hydrocarbons	120		
Carbon dioxide	Plants	600,000	Burning fossil fuels and wood	22,000
	Animals			
	Forest fires			
	Volcanoes			
Nitrogen compounds	Bacteria		Burning coal	27
NH_3		1160		
NO_x		502		
N_2O		145		
Hydrocarbons			Burning coal	3
CH_4	Bacteria	1600	Petrol processing and consumption	48.5
Terpene, ethylene etc.	Plants	1000	Solvent consumption	10
			Burning rubbish	25
			Other uses	2

hydrogen, calcium, potassium, silicon and magnesium these elements account for practically 100% of the atoms of the biosphere (cf. Bolin and Cook 1983). The dominant process transferring hydrogen, carbon and oxygen, which structurally account for 99.6% of the organisms already, from the environment to living systems, is the photoreduction of carbon dioxide to carbohydrate with the liberation of molecular oxygen from water. From autotrophic organisms, carbon, hydrogen and oxygen are passed to heterotrophs along the food web. From living plants and animals CO_2 is released to the atmosphere during respiration, and O_2 during photosynthesis. Some bacteria which use nitrates or sulphates as an oxygen substitute in respiration reduce them to N_2 or H_2S respectively, both of which are released to the atmosphere. Other organic and inorganic compounds are excreted and decomposed together with dead organic matter. These processes of decay and decomposition are oxidation reactions proceeding with a net loss of free energy and ultimately releasing the constituent elements of the organic molecules back into the environment as simple inorganic compounds (cf. Likens 1981; Woodwell 1984).

The processes of carbon and oxygen transfer between the living systems of the biosphere and their immediate environment as coupled to the transpiratory transport of water and energy form the ecosphere loop of circulation depicted in Fig. 3.1 (for further details see foldout model I).

In the above combination, the carbon and oxygen cycles have been of predominant importance in the evolution of both eukaryotic life and the contemporary atmosphere (cf. for instance Schidlowski 1971; Lewis and Prinn 1984). In the present atmosphere the CO_2 content is largely controlled by evaporation from and dissolution into the oceans, where the observed CO_2 amounts are about 50–60 times those of the atmosphere. Owing to control by temperature and acidity of the seawater on the one hand, and the biological production rate on the other, the CO_2 pressure varies seasonally with an average atmospheric residence time of about 5 years (Junge 1963; Bolin and Cook 1983). Through the biosphere CO_2 is cycled at a distinctly slower rate, i.e. once in about 20–30 years.

In addition to CO_2, also minor amounts of CH_4, CO and CH_2O play a role in the atmospheric carbon cycle (see Table 3.2). Methane as a result of anaerobic decay of organic matter and the product of archaebacteria (Woese et al. 1978) such as *Methanobacterium*, *Methanococcus*, *Methanogenium*, *Methano-microbium* and *Methanosarcina* is oxidized to formaldehyde and then to carbon monoxide (Ekhalt 1974). The resultant life-time of CH_4 is about 2 years (Levy 1971) while the CO produced is further oxidized to CO_2 with a time constant of only a few months.

Like that of carbon, also the nitrogen cycle plays a significant role in the biosphere; its principle elements are illustrated in Fig. 3.2.

The fastest cycles in which atmospheric nitrogen compounds participate involve a variety of nitrifying and denitrifying bacteria (see, e.g. Mishustin and Shilnikova 1969; Alexander 1971; Burns and Hardy 1975; Duxbury et al. 1982). Under usually anaerobic conditions in poorly aerated soils or in seawater,

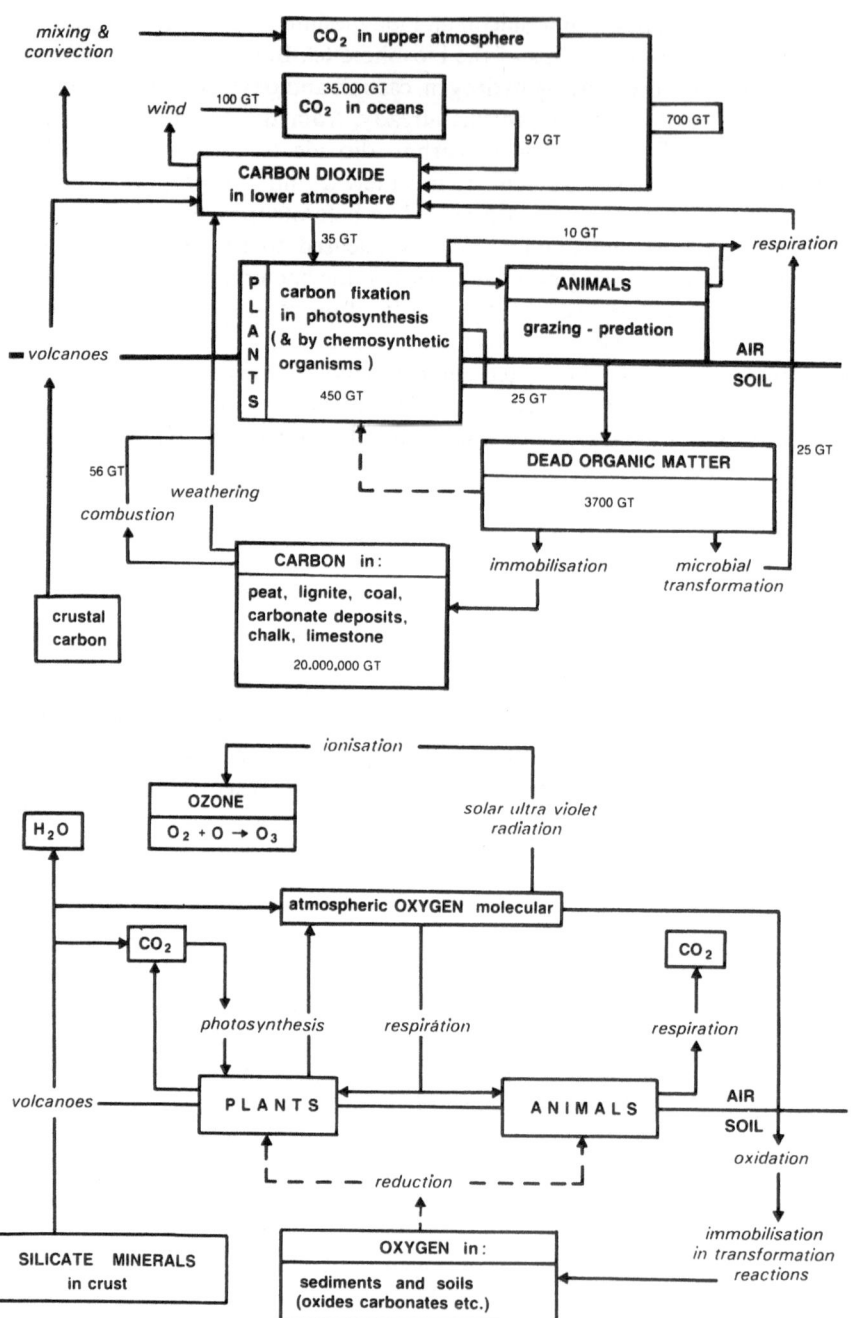

Fig. 3.1. The carbon and oxygen cycles. (After White et al. 1984)

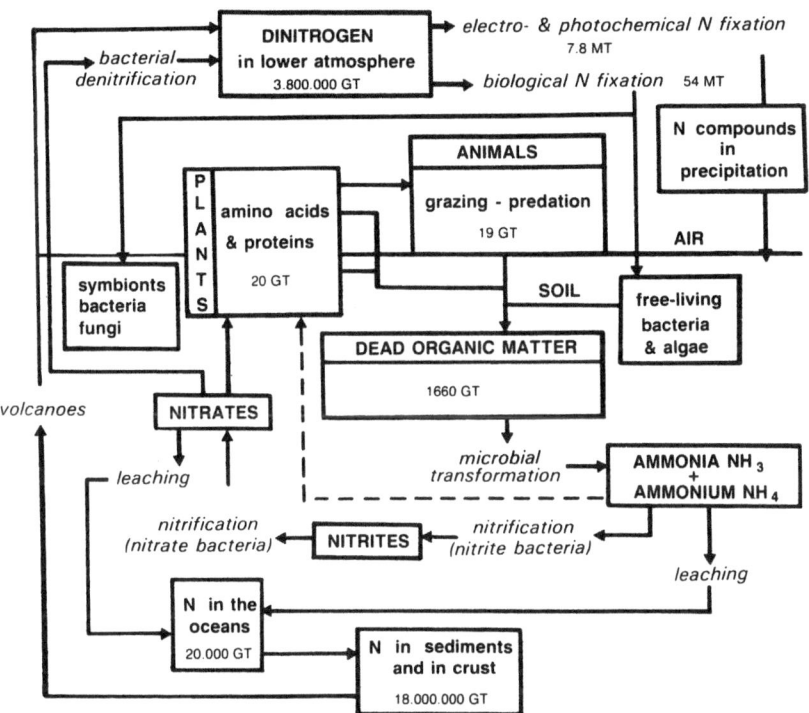

Fig. 3.2. The nitrogen cycle. (After White et al. 1984)

bacteria like *Achromobacter, Bacillus, Micrococcus, Pseudomonas* and *Thiobacillus* convert soil nitrate to N_2 (about 90% on a global scale) and N_2O (about 10% on a global scale). While most of the latter is photodissociated in the upper atmosphere to form N_2 with a resultant life-time of about 50 years, a small fraction is converted to NO and NO_2 in the stratosphere. Here it plays an important role in regulating the ozone concentration. At lower (i.e. tropospheric) altitudes, NO_2 converts into HNO_3, and is then removed by rainout and washout processes (cf. Sect. 3.3.4) and subsequently destroyed by reactions in soils (cf. Bolin and Cook 1983).

Most of the N_2 released to the atmosphere by microbial activity seems to be recycled to the biosphere and related systems by the action of nitrogen-fixing bacteriae and algae which convert N_2 to nitrate ion. Hardy and Havelka (1975) estimate that microbial activity by, for instance, *Azotobacter, Clostridium, Rhizobium* and *Trichodesmium* presently account for about 67% of the total nitrogen removed from the atmosphere, industrial fixation (e.g. Haber–Bosch process) for about 15%, combustion in engines for about 8%, and lightning and other phenomena for the rest. Since the opposing processes of nitrification and denitrification seem to be in a rough balance, an atmospheric N_2 life-time of about 17 million years can be estimated (Burns and Hardy 1975).

Nitrogen is also released in the form of NH_3 during decay of organic matter by bacteria and fungi. This natural process of ammonification seems to add nearly the same quantities of nitrogen to the atmosphere as those produced by denitrifying bacteria. Part of this ammonia is converted in the atmosphere to $(NH_4)_2SO_4$, NH_4Cl and NH_4NO_3 aerosols which are highly water-soluble and consequently quickly rained out in the lower troposphere. After removal from the atmosphere, some ammonia is consumed directly by organisms while some is transformed to nitrite and nitrate by aerobic nitrifying bacteria such as *Nitrobacter*, *Nitrosococcus* and *Nitrosomonas*. As a consequence, the atmospheric life-time of NH_3 of NH_4^- ions is likely to be less than 1 month (Davidson and Wu 1990).

The atmospheric sulphur cycle (Junge 1960, 1963; Ivanov and Freney 1983) as depicted in Fig. 3.3 is of particular interest because sulphur plays an important role in the biosynthesis of proteins and because sulphuric acid became one of the most noxious components of acid precipitation in recent times. Sulphur enters the cells of plants as an inorganic sulphate ion in a way similar to nitrate,

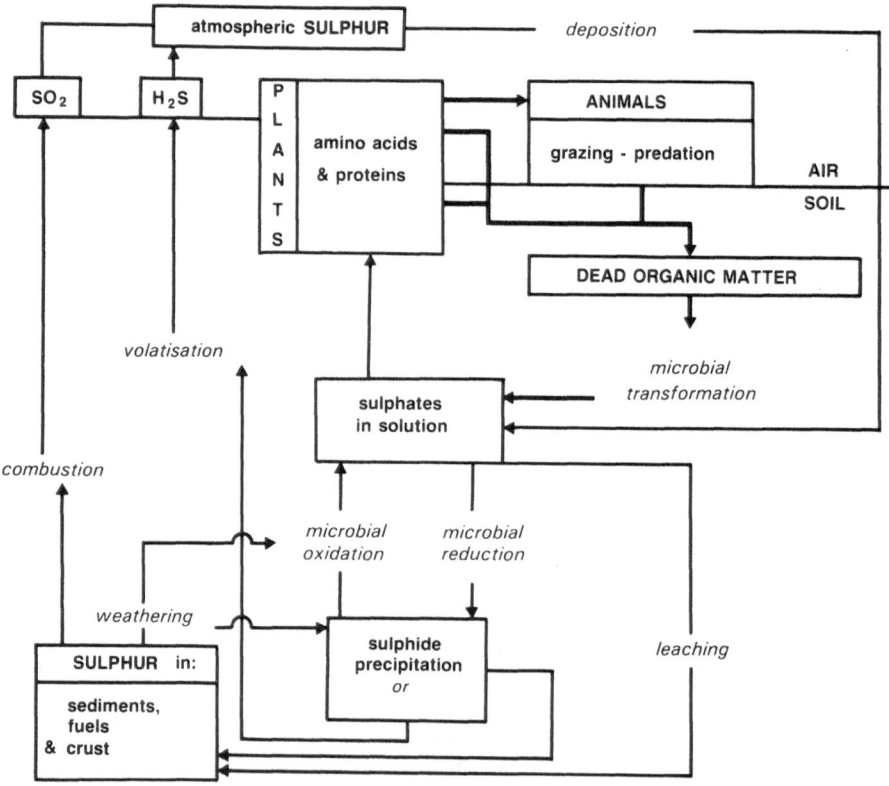

Fig. 3.3. The sulphur cycle. (After White et al. 1984)

and recycling through the atmosphere involves bacterial reduction to sulphides under anaerobic conditions, e.g. by *Desulfovibrio* strains. When neither precipitated in this form nor converted back to sulphates by other sulphur bacteria (e.g. *Thiobacillus*) operating under aerobic conditions, sulphur escapes to the atmosphere as H_2S which is then reoxidized to sulphurous acid or sulphuric acid or sulphate, respectively.

The resultant H_2S lifetime ranges from a few hours in polluted urban air to a few days in clean air far away from conurbations. Biological H_2S is estimated to account for about 60% of the SO_2 entering the atmosphere. $(CH_3)_2S$ may be a further, but as yet unquantified biotic source of atmospheric sulphur contributing to oxidative SO_2 production (Lewis and Prinn 1984). The remaining 40% are due to combustion of fossil fuels (particularly coal), and this source has been steadily increasing with time during the worldwide industrialization process of the last two centuries. In photochemical smogs, the oxidation of SO_2 due to organic peroxy free radicals may be accelerated to such an extent that the resultant SO_2 life-time is reduced to a few minutes (cf. Sect. 3.4.2).

It follows from a comparison of the above partial models that the cycling of sulphur in nature is no less relevant to carboxylation than the cycling of carbon, nitrogen and water. Without downgrading the importance of photosynthesis, it may be said that carbon fixation is only one of at least three critical steps in the global production of proteins. All three are hydrogenations achieved with the aid of enzymes occurring only in the biosphere. Of the three essential reductions mentioned above, however, only that of carbon calls for photosynthesis, while the reduction of nitrogen and sulphur are accomplished anaerobically, by microbes.

Although the partial models deliberately fail to specify fluxes or chemical quantities, they provide a mental framework for the movement of six elements: hydrogen, oxygen, carbon, nitrogen, sulphur and phosphorus, either alone or in combinations or compounds, respectively. The synthetic output is the biosphere, whose central position in these models follows from the fact that for all six elements it is both a natural source and a temporary sink.

The primary aim of these small models is to clarify the essential pathways and processes of natural element transfer between the living systems of the biosphere and their immediate environment, emphasizing the particular role of the atmosphere. By reference to a table of the biospheric composition, it would be instantly clear that phosphorus plays a particular role, along with calcium, potassium, silicon and magnesium, which are all more common in the biosphere than sulphur. P is not a constituent of protein, but no protein can be made without it. The high-energy phosphate bond, reversibly moving between adenosine diphosphate (ADP) and adenosine triphosphate (ATP) is the universal fuel for biochemical work within each living cell. The photosynthetic fixation of carbon, oxygen and hydrogen would be bioenergetically fruitless if it were not followed by the phosphorylation of the sugar produced.

In the present context, the solubility of phosphorus in water can be safely assumed, but the crucial question concerns its volatility. Except as sea spray in coastal zones, or as dust in the vicinity of exposed phosphate rocks (e.g. at the Phalaborwa mines in the Transvaal), phosphorus is unknown in the atmosphere; none of its ordinary compounds has any appreciable vapour pressure. It therefore tracks the hydrologic cycle only part way, namely from the lithosphere to the hydrosphere (cf. Fig. 3.4) and in terms of modelling, this amounts to uncoupling the atmospheric reservoir, except for water.

Another important component of the atmosphere is particulate matter suspended in the air for a reasonable life-time. It includes both solid and liquid particles ranging in size from greater than 100 µm to less than 0.1 µm in diameter. The term aerosol is usually reserved for particulate matter other than water or ice. Over the continents the principal aerosol components are usually insoluble minerals injected into the air by wind erosion, fires and volcanic eruptions. Most commonly, they consist of SiO_2, $CaSiO_3$, $CaCO_3$, oxides and hydroxides of iron, manganese, aluminium, but also include compounds of lead, cadmium, chromium, copper, nickel and asbestos (cf. Chap. 3.5.5). In highly industrialized areas and in regions with relatively frequent forest fires, soot becomes important. Gas reactions, including, for instance, NH_3, NO, NO_2, NO_3^-, N_2O_5, SO_2, SO_3^{2-}, CO_2 and H_2O may lead to ammonium sulphates, nitrates, chlorides, carbonates, sulphurous and sulphuric acids. Pollen, spores, and unsaturated organic compounds are among the commonest aerosols over forests and woodlands. The bulk of aerosols in the atmosphere, however, are marine aerosols ejected from the oceanic surface due to the combined action of bubbles and wind. Consequently, they usually comprise chlorides, carbonates, sulphates, and bromides of the common marine cations Na, K, Ca and Mg.

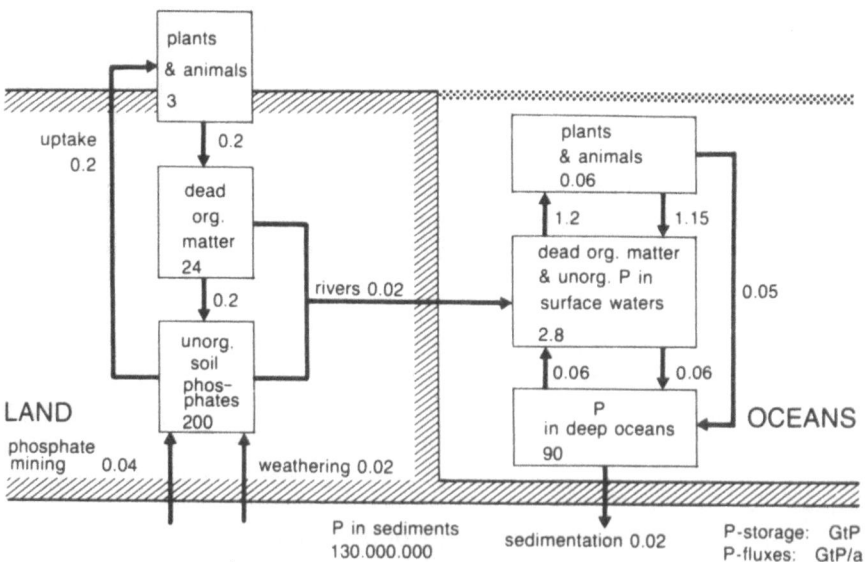

Fig. 3.4. The phosphorus cycle. (After Bossel 1990)

3.1.2 Anthropogenic Air Pollution

The largest sources of air pollution are the *combustion of fuels* such as coal, oil, wood, and gas and the *burning of rubbish* (cf. Table 3.2). During these processes, gases released from fuel combine with oxygen to form CO_2 and H_2O. Since fuel-burning equipment does not always work perfectly, however, some of the CO and carbon formed during the reaction is not converted to CO_2 but appears in flue gases. Impurities in the fuel, such as sulphur, are oxidized and also appear

Table 3.3. Aerosol particle inventory for Los Angeles[a] (basis: 119 μg m^{-3} total mass). (After Hidy and Friedlander 1971)

Source	Amount (wt %)
Natural background	
Primary	
Dust rise by wind (Si)	7–17
Sea salt (Na, Cl)	5
Spores, pollen, etc.	Unknown
	12–22
Secondary	
Vegetation (organic vapours)	2–5
Biological (soil bacterial action, decay of organics)	($NH_4^+ \approx 0.5$)
– NH_3, NO_x, S, etc.	
	2–5
Anthropogenic	
Primary	
Motor vehicles (Pb)	13
Aircraft	3
Chemical	3
Metallurgical	3
Mineral	1
Petroleum	3
Other industry	1
Combustion of fuels	5
	32
Secondary	
Reactive hydrocarbon vapours	9
NO_3^-	12
SO_4^{2-}	11
	32
Total accounted for	78–91

[a] Based on relative humidity less than 40%.

as flue gases. Ore smelters and oil refineries also give off SO_2 in large quantities, and kraft pulp mills and oil refineries emit H_2S. There is always a residue of noncombustible solids that passes to the air unless it is trapped in the combustion system by means of filters or washing devices.

Table 3.4. Aerosol particle inventory for San Francisco[a] (basis: 68 µg m^{-3} total mass). (After Hidy and Mueller 1980)

Sources	Amount (wt %)
Natural background	
Primary	
Dust rise by wind	12
Sea salt	7
Spores, pollen, etc.	Unknown
	19
Secondary	
Vegetation (organic vapours)	2–5
Biological	Unknown ($NH_4^+ \approx 0.5$)
	2–5
Anthropogenic	
Primary	
Motor Vehicles	5
Aircraft	2
Other transport	3
Chemical	3
Metallurgical	4.5
Petroleum	0.5
Organic solvent use	0.7
Combustion of fuels	2
Incineration	0.2
Agricultural burning[b]	7
	38
Secondary	
Reactive organic (from 0.5 × total hydrocarbon vapours)	~ 5
NO_3^-	3
SO_4^{2-}	7
	15
Total accounted for	64–72

[a] Based on relative humidity less than 40%.
[b] Inventory for 1972 and later should exclude this particulate source.

Among the exhaust gases of *internal-combustion engines* are unburned hydrocarbons, oxides of nitrogen and carbon; in diesel-fuel exhaust particularly, excessive quantities of carbon (soot) may occur. NO_x undergo photochemical change in the presence of ozone to produce secondary pollutants, such as peroxacetyl nitrate (cf. Sect. 3.4.4). An aerosol haze made up of such products is characteristic of large urban centers in many parts of the world, depending on the macroclimatic situation. The Los Angeles Basin and San Francisco Bay area of California are classical examples of localities where motor-vehicle exhaust induces noticeable visibility reduction and smog. The preceding two aerosol inventories in Tables 3.3 and 3.4 are illustrative examples of the situation in the late 1960s and early 1970s.

Another combustion process of importance in urban areas is that of *refuse incineration*. Because the composition of refuse, containing garbage, paper, wood, grass, plastic containers, tin cans, and a variety of discarded debris, is highly variable, its burning characteristics change from hour to hour.

An illustrative picture of the regional distribution of refuse incineration plants in a highly industrialized country such as the Federal Republic of Germany is provided by Fig. 3.5.

Combustion products of incineration plants include incompletely converted organic gases, particulates, and oxides of materials in the refuse, such as carbon monoxide and dioxide, NO_x and water. In view of the chemical composition of the residues and the particulate matter resulting from incineration, intricate and complex air-cleaning devices are included in incinerator designs to reduce particulate emissions. In the Federal Republic of Germany, such devices are supposed to be available for more than 80% of the combustion capacity installed in 1985 (Unweltbundesamt 1984).

The character of *industrial wastes* depends on the materials handled, the methods of processing, and the eventual waste-handling procedures. Emissions of sintering plants, coking plants, blast furnaces, slag dumps, and crushers include dusts, SO_2, CO, fluorides, and metal oxides. Rock-crushing operations produce dusts which may, depending on chemical composition and granulometry, have toxic properties resulting in a variety of silicogenic fibroses, for instance.

In the innumerable processes of the extremely diversified chemical industries, small amounts of the products may emerge as mists, vapours, or particulates introducing acids and alkalies and other pollutants into the atmosphere. In general, the emissions of oil refineries include NO_x, SO_2, H_2S, mercaptans, hydrocarbons, other gases, and particulates. Table 3.5 summarizes the temporal changes in the composition of the major emissions in the Federal Republic of Germany in the period 1966–1986.

These changes reflect both variations in the economic situation and changes of the specific emission rates. The latter are customarily defined as emission factors, and are summarized for industrial SO_2, NO_x and hydrocarbons (C_mH_n) in Tables 3.6, 3.7 and 3.8.

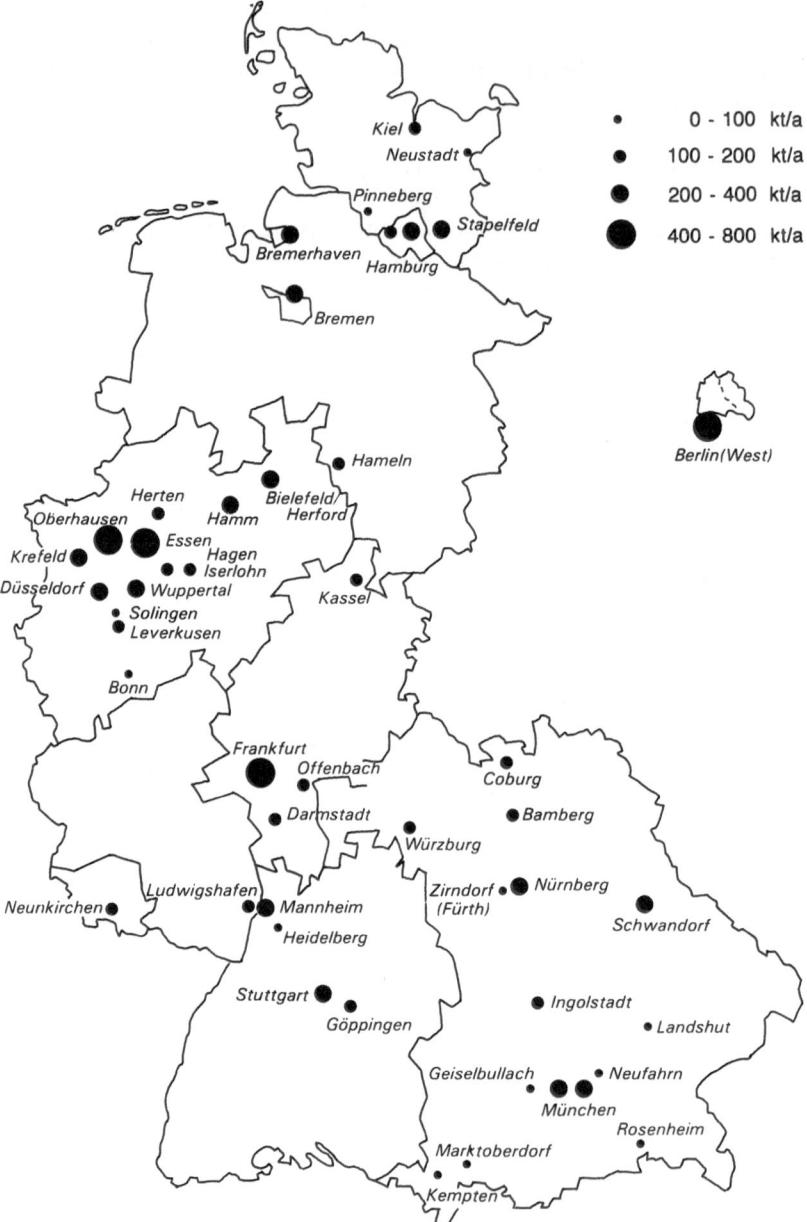

Fig. 3.5. Municipal refuse incineration plants in the Federal Republic of Germany. (After Umweltbundesamt 1984, 1989)

Table 3.5. Composition of emissions in the period 1966–1986 in the Federal Republic of Germany. (Umweltbundesamt 1989)

Pollutant (Mt/a)	1966	1967	1968	1969	1970	1971	1972	1973	1974	1975	1976	1977	1978
NO$_x$ (NO$_2$ equivalents)	1.95	1.95	2.05	2.20	2.35	2.45	2.50	2.65	2.60	2.55	2.70	2.75	2.85
Volatile organic compounds	2.20	2.25	2.35	2.50	2.60	2.65	2.65	2.65	2.55	2.55	2.55	2.50	2.55
CO	12.40	12.40	12.90	13.70	14.00	14.00	14.00	14.10	13.70	13.40	13.00	12.60	12.40
SO$_2$	3.35	3.30	3.40	3.65	3.75	3.70	3.75	3.85	3.65	3.35	3.55	3.40	3.40
Dust	1.75	1.55	1.50	1.45	1.30	1.20	1.10	1.05	0.95	0.81	0.78	0.73	0.70

Pollutant (Mt/a)	1979	1980	1981	1982	1983	1984	1985	1986	Prognoses 1998 Variant I	Variant II [a]
NO$_x$ (NO$_2$ equivalents)	2.95	2.95	2.85	2.85	2.85	2.95	2.95	2.95	2.25	1.85
Volatile organic compounds	2.55	2.50	2.40	2.40	2.40	2.40	2.40	2.45	1.55	1.35
CO	12.30	11.70	10.50	9.80	9.20	9.20	8.80	8.90	6.20	5.10
SO$_2$	3.40	3.20	3.05	2.85	2.70	2.65	2.45	2.30	1.00	0.93
Dust	0.72	0.69	0.65	0.60	0.58	0.59	0.57	0.56	0.47	0.46

[a] Variant I = Current regulations, including increase of road transports.
Variant II = Additional traffic regulations, decrease of sulphur content in diesel and light fuel oil.

Table 3.6. Branch-related SO_2-emission factors. (Nukem 1976)

Industrial branch	Source of emission	Unit	SO_2-emission						
			1960	1965	1970	1975	1975[a]	1980	1980[a]
Power stations	Mineral coal	kg SO_2/MWh	7.14	7.14	7.14	7.14	7.14	7.14	1.44
	Brown coal	kg SO_2/MWh	7.30	7.30	7.30	7.30	7.30	7.30	1.46
	Heavy fuel oil	kg SO_2/MWh	6.50	6.50	6.50	6.50	6.50	6.50	1.30
	Gas	—	—	—	—	—	—	—	—
Industrial firing	Coal	kg SO_2/t MCE[c]	21.00	21.00	21.00	21.00	21.00	21.00	21.00
	Heavy fuel oil	kg SO_2/t MCE	37.30	29.90	22.40	18.70	18.70	18.70	12.40
	Light fuel oil	kg SO_2/t MCE	8.96	8.96	5.60	5.60	5.60	5.60	3.36
	Gas	—	—	—	—	—	—	—	—
Chemical industry Iron industry					≈ 20 process-dependent SO_2-emission factors				
	Sintering plants	kg SO_2/t sinter	9.50	6.90	4.32	4.00	4.00	4.00	2.83
	SM[b]-steel production	kg SO_2/t SM[b]-steel	1.30	1.30	1.30	1.30	1.30	1.30	1.30
Coking plants	Gasworks	kg SO_2/t coke	2.60	1.90	0.90	0.90	0.70	0.90	0.70
	Coking plants	kg SO_2/t coke	2.60	1.90	0.90	0.90	0.70	0.90	0.70
Refuse incineration		kg SO_2/t rubbish	3.30	3.30	3.30	3.30	3.30	3.30	0.66
Mineral oil industry	Refineries and tanks	—	—	—	—	—	—	—	—
Dry cleaning		—	—	—	—	—	—	—	—
Varnishes and paints	Small consumers	—	—	—	—	—	—	—	—
	Large consumers	—	—	—	—	—	—	—	—
Printing-offices		—	—	—	—	—	—	—	—

[a] Date of regulatory measures. [b] Siemens–Martin. [c] Mineral coal equivalents.

Table 3.7. Branch-related NO_x-emission factors. (Nukem 1976)

Industrial branch	Source of emission	Unit	NO_x-emission 1960	1965	1970	1975	1975[a]	1980	1980[a]
Power stations	Mineral coal	kg NO_x/MWh	2.18	2.18	2.18	2.18	2.18	2.18	2.18
	Brown coal	kg NO_x/MWh	3.50	3.50	3.50	3.50	3.50	3.50	3.50
	Heavy fuel oil	kg NO_x/MWh	2.30	2.30	2.30	2.30	2.30	2.30	2.30
	Gas	kg NO_x/MWh	1.70	1.70	1.70	1.70	1.70	1.70	1.70
Industrial firing	Coal	kg NO_x/t MCE[c]	7.20	7.20	7.20	7.20	7.20	7.20	7.20
	Heavy fuel oil	kg NO_x/t MCE	5.20	5.20	5.20	5.20	5.20	5.20	5.20
	Light fuel oil	kg NO_x/t MCE	2.50	2.50	2.50	2.50	2.50	2.50	2.50
	Gas	kg NO_x/t MCE	2.50	2.50	2.50	2.50	2.50	2.50	2.50
Chemical industry			$\approx$ 10 process-dependent NO_x-emission factors						
Iron industry	Sintering plants		—	—	—	—	—	—	—
	SM[b]-steel production		—	—	—	—	—	—	—
Coking plants	Gasworks		—	—	—	—	—	—	—
	Coking plants		—	—	—	—	—	—	—
Refuse incineration		kg NO_x/t rubbish	1.30	1.30	1.30	1.30	1.30	1.30	1.30
Mineral oil industry	Refineries and tanks		—	—	—	—	—	—	—
Dry cleaning									
Varnishes and paints	Small consumers		—	—	—	—	—	—	—
	Large consumers		—	—	—	—	—	—	—
Printing-offices			—	—	—	—	—	—	—

[a] Date of regulatory measures. [b] Siemens–Martin. [c] Mineral coal equivalents.

Table 3.8. Branch-related hydrocarbon-emission factors. (Nukem 1976)

Industrial branch	Source of emission	Unit	C_mH_n-emission						
			1960	1965	1970	1975	1975[a]	1980	1980[a]
Power stations	Mineral coal	kg C_mH_n/MWh	0.0051	0.0051	0.0051	0.0051	0.0051	0.0051	0.0051
	Brown coal	kg C_mH_n/MWh	0.033	0.033	0.033	0.033	0.033	0.033	0.033
	Heavy fuel oil	kg C_mH_n/MWh	0.125	0.125	0.125	0.125	0.125	0.125	0.125
	Gas	kg C_mH_n/MWh	0.378	0.378	0.378	0.378	0.378	0.378	0.378
Industrial firing	Coal	kg C_mH_n/t MCE[c]	0.30	0.30	0.30	0.30	0.30	0.30	0.30
	Heavy fuel oil	kg C_mH_n/t MCE	0.44	0.44	0.44	0.44	0.44	0.44	0.44
	Light fuel oil	kg C_mH_n/t MCE	0.24	0.24	0.24	0.24	0.24	0.24	0.24
	Gas	kg C_mH_n/t MCE	1.10	1.10	1.10	1.10	1.10	1.10	1.10
Chemical industry			≈ 40 process-dependent C_mH_n-emission factors						
Iron industry	Sintering plants		—	—	—	—	—	—	—
	SM[b]-steel production		—	—	—	—	—	—	—
Coking plants	Gasworks		—	—	—	—	—	—	—
	Coking plants		—	—	—	—	—	—	—
Refuse incineration		kg C_mH_n/t rubbish	1.50	1.50	1.50	1.50	1.50	1.50	1.50
Mineral oil industry	Refineries and tanks	kg C_mH_n/t product	0.38	0.38	0.38	0.38	0.30	0.38	0.19
			5 different C_mH_n-emission factors						
Dry cleaning		kg C_mH_n/t material	100	100	20	20	20	20	20
Varnishes and paints	Small consumers	kg C_mH_n/t varnish and diluent	600	600	600	600	600	600	600
	Large consumers		600	600	600	600	543	600	30
Printing-offices		kg C_mH_n/t paint	350	350	350	350	317	350	17.5

[a] Date of regulatory measures. [b] Siemens–Martin. [c] Mineral coal equivalents.

Pollution from *radioactivity* is both natural and man-made, the former being due to cosmic radiation and some minerals, such as uraninite (pitchblende), carnotite, and autunite, the latter resulting from the testing of nuclear weapons. Nuclear reactors do not discharge wastes to the atmosphere, and only in the case of a major accident might there be an emission hazard.

Also *agriculture* contributes to air and water pollution directly by spraying of pesticides by mobilized equipment. The burning of waste plant material, such as stubble and prunings, contributes smoke and by-products. Indirectly the cultivation of land, in particular in semi-arid climates, may leave areas open to deflation; the United States, parts of the Soviet Union, and African states offer conclusive examples of the resulting devastation.

3.2 Transport and Dispersion

Vertical and horizontal movements in the atmospheric boundary layer serve both to transport and to dilute pollutants. While the vertical transport of pollutants is largely controlled by the temperature stratification of the atmosphere or the resulting stability conditions, respectively, wind speed is of prime importance for the amount of forced convection due to internal shearing between air layers, and between the air and the surface roughness elements. As fold-out model I indicates, the regionally differentiating consideration of these facts is of fundamental importance for a genetic understanding of dispersion phenomena. Consequently, the first part of the present section gives a predominantly qualitative description of the relevant boundary conditions in the atmosphere governing pollutant dispersion while the second deals with quantitative and application-oriented aspects in the form of transport models.

3.2.1 General Atmospheric Boundary Conditions of Air Pollution

When a wind is blowing, pollutants are transported and diffused in the along-wind direction and perpendicularly by turbulent eddy diffusion, provided the structure of the atmosphere and the physical-chemical properties of the pollutants exclude an instantaneous deposition. Among these transport-controlling pollutant potentials, the following are of particular importance and consequently figure as regulators in Model I:

- terminal velocity (SG), a function of diameter, configuration and specific gravity of aerosol particles
- density of polluting gas or aerosol in relation to density of ambient cleaner air which controls uplift ($S_T G$)

These gas or aerosol properties may be largely influenced by nucleation and condensation processes on the one hand, chemical transformation processes on the other, which are dealt with in Sections 3.3 and 3.4. Here it may suffice to say that the distinction between "fine particles" (particle diameter $D_p < 2\ \mu m$) and "coarse particles" ($D_p > 2\ \mu m$) is a fundamental one. They constitute two modes in the mass or volume distribution of aerosols usually observed (Lundgren 1973; Graedel and Graney 1974; Mainwairing and Harsha 1975; Kadowski 1976; Sverdrup 1977; Whitby and Sverdrup 1980), and are usually chemically quite different (Appel et al. 1974; Dzubay and Stevens 1975). This separation is due to the fact that condensation produces fine particles which are prevented by specific dynamics from becoming larger than about 1 µm, while coarse ones mostly originate from mechanical processes.

The phenomenon of pollution transport and dilution is directly related to wind speed. The higher it is, the greater is the volume of air passing a point source per unit time, and the smaller the resultant concentration per unit volume. Greater wind speed also means greater turbulence, which involves fluctuations of the local wind vector on various scales, and these eddies act to diffuse a plume. In addition to this "stretching" phenomenon (Oke 1978), wind speed is also important for the distance of transport. In strong winds pollutants may be transported long distances (cf. Chap. 4.2.1) but usually the resultant dilution over 10–50 km is such that most individual plumes have lost their identity, contributing to a more general (background) contamination of the lower atmosphere. Consequently, weaker winds may exhibit a much greater potential for pollution, since both horizontal transport and eddy diffusion are reduced.

This simple fact becomes all the more important since it is under these conditions that local wind systems tend to develop which are difficult to predict with any accuracy (cf. Geiger 1961; Yoshino 1975). As far as they are closed smaller-scale circulation systems (e.g. land and sea breezes, mountain and valley winds, city winds) they not infrequently produce a re-circulation of pollutants coupled with a progressive increase in pollutant loading with time. Particularly dangerous is the city thermal wind system operating mainly in the urban canopy layer which is produced by micro-scale processes in the streets (Sundborg 1951; Chandler 1970; Oke and East 1971; Oke and Maxwell 1975; Oke 1982; Lee 1984; Jauregui 1987). Here flow converges upon the city (or its variable thermal sub-centres) from all directions, rises, diverges, and then moves outward to subside on the urban/rural fringe, where it rejoins the inflow.

The worst conditions for dispersion, however, occur when there is a temperature inversion and consequently the boundary layer stable. With the warmer air layer acting as a lid on the mass of cooler air below, pollutants have no place to go and thus increase in concentration in the confined space. This situation can be brought about by cooling (usually radiative) from below, by warming (usually adiabatic) from above, or by advection of warmer or cooler air.

The driving force of the first type of inversions is quite frequently long-wave radiative cooling and these radiation inversions are based at the ground or another active, i.e. radiating surface (e.g. vegetation cover, cloud tops, or polluted layers) and may extend up to heights of 50–100 m. During the polar night in the winter at high latitudes they may persist for several weeks at a time, but also in mid-latitude depressions their frequency may be such that it significantly lowers the monthly or even seasonal mean temperature (Blüthgen 1966; Kahl 1973). A related type of surface-based inversion is due to evaporation cooling during the day in fine weather. It consequently occurs on lakes or when a summer rain shower cools the ground.

A second and very wide-spread type of inversion is due to adiabatic (subsidence) warming. When unsaturated air sinks within the atmosphere it is compressed and warms. Therefore, subsidence inversions are very common features of anticyclones which are semi-permanent in the sub-tropics and in the mid-latitudes can stagnate over a region for up to 2–3 weeks. In either case stagnation creates a thoroughly murky mixed layer with a sharp upper boundary. In Andalusia, it extends up to 3000 m in summer and has the regional name "calina". Subsidence warming also accounts for the development of almost impenetrable inversions in the lee of mountain ranges (e.g. Alps, Rocky Mountains) in winter. In their forelands, radiative cooling creates cold air masses over which the descending warm air spreads.

Weather fronts are also always characterized by an inversion, since they are sloping planes with the warmer air overlying the colder one. Because of the motion, these frontal inversions are usually short-lived, and pollution problems can only arise with slow-moving warm fronts whose frontal slope (the Margules plane) is usually rather slight. Dispersion conditions become increasingly poor until the front has passed; but even then depressions may trap cold air at their bottom which are overridden by the warm air which acts as a "lid" inhibiting the dilution of pollutants by vertical mixing. Basically analogous advection inversions arise whenever warm air flows across cold surfaces such as cold land surfaces, water bodies, snow or ice cover. The resultant cooling at the underside of the air mass creates a ground-based inversion.

Figure 3.6 gives a synopsis of the seasonal variations in height of inversions in the Ruhr area for the 1966–1976 period (Kuttler 1979). The evaluation is based on 1509 events which display a marked seasonality. Nearly 78% of all inversions occurred during the heating period from October to March while the remaining 22% were registered from April to September.

An even more detailed picture of inversions in the Ruhr area is provided by Table 3.9, which shows that about 70% of the pertinent lapse rates fall into the 0 K to − 1 K range, while about 20% have a lapse rate of − 1 K to − 2 K; 5% range in the − 2 K to − 3 K class, the rest reaches a maximum rate of − 9.2 K.

3.2.2 Atmospheric Lapse Rates and Plume Dilution

Transmission and dispersion of pollutants in the boundary layer are controlled by the lapse rates and these relationships manifest themselves particularly well in the behaviour of individual plumes emitted from elevated continuous point

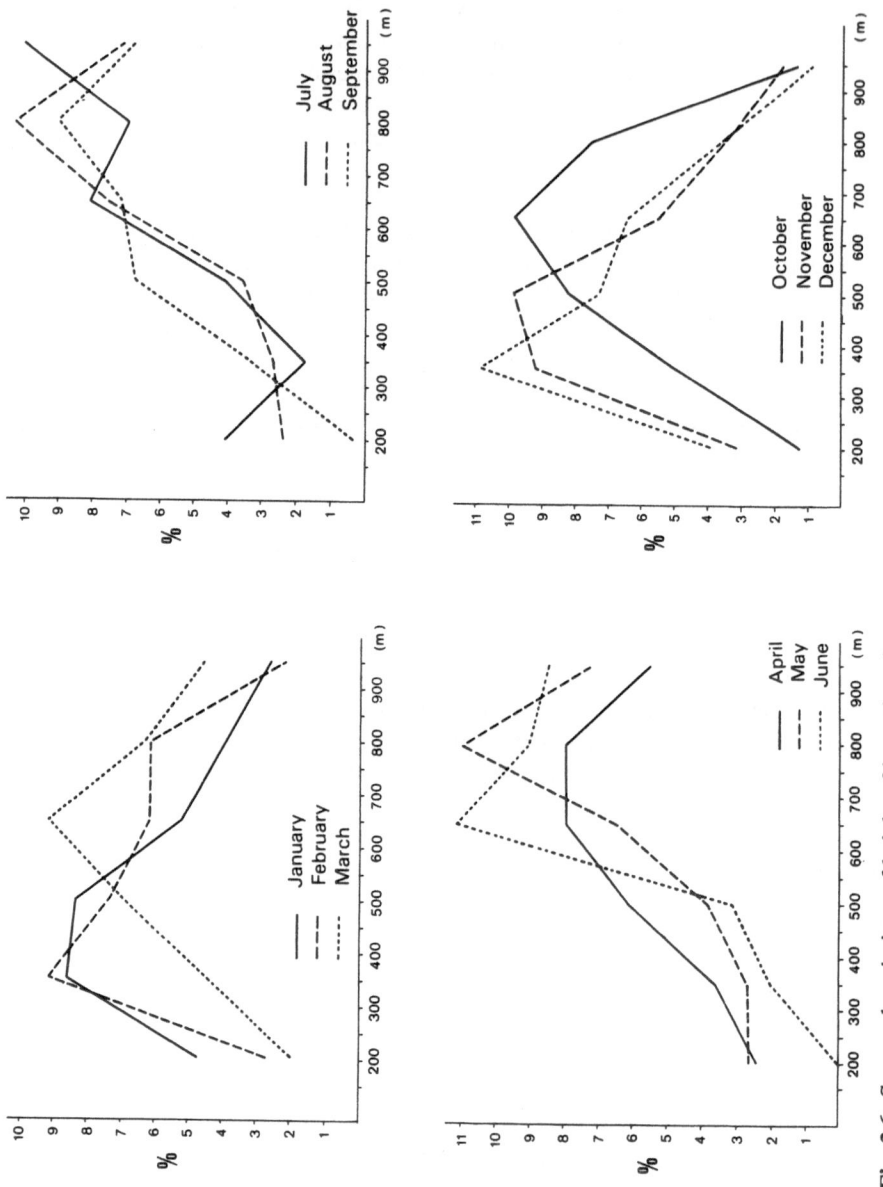

Fig. 3.6. Seasonal variations of height of inversions in the Ruhr area for the 1966–1976 period. (After Kuttler 1979)

Table 3.9. Monthly frequency distribution of inversions in the Ruhr area for the period 1966–1976. (After Kuttler 1979)

Month \ γ	−0.2	−0.4	−0.6	−0.8	−1.0	−1.2	−1.4	−1.6	−1.8	−2.0	−2.2	−2.4	−2.6	−2.8	−3.0
I	2.2	1.6	2.0	1.5	1.6	1.1	1.0	0.9	0.4	0.5	0.3	0.3	0.3	0.2	0.2
II	1.9	1.7	1.9	1.2	0.9	0.8	0.9	0.9	0.1	0.5	0.2	0.1		0.3	0.1
III	2.0	1.5	0.6	1.7	0.6	0.7	0.6	0.4	0.1	0.3			0.1	0.1	
IV	1.1	0.6	0.5	0.5	0.1		0.3	0.1	0.1	0.1		0.1			0.1
V	1.1	0.2	0.6	0.5	0.3	0.2	0.1			0.1	0.1				
VI	0.9	0.7	0.3	0.5	0.1	0.2	0.1	0.1	0.1						
VII	1.2	0.5	0.4	0.5	0.3	0.3	0.2	0.1		0.1	0.1	0.1			0.1
VIII	1.8	0.7	0.5	0.9	0.3	0.1	0.1	0.1	0.1	0.1					0.1
IX	1.8	1.1	0.6	0.9	0.8	0.1	0.1	0.2	0.1						
X	2.4	1.6	1.1	1.3	1.3	0.6	0.7	0.3	0.4	0.3	0.3	0.1	0.1	0.1	0.1
XI	3.0	1.5	1.6	1.5	0.9	0.6	0.5	0.5	0.3	0.4	0.2	0.1	0.1	0.1	
XII	2.8	1.8	2.1	1.7	1.6	1.0	0.5	0.5	0.7	0.4	0.2	0.3		0.2	0.2
S	7.9	3.8	2.9	3.8	1.9	1.0	0.9	0.6	0.9	0.4	0.2	0.2	0.0	0.0	0.3
W	14.3	9.7	9.3	8.9	6.9	4.8	4.2	3.5	2.0	2.4	1.2	0.9	0.6	1.0	0.6
Σ	22.2	13.5	12.2	12.7	8.8	5.8	5.1	4.1	2.4	2.8	1.4	1.1	0.6	1.0	0.9

Table 3.9. (Contd.)

Month \ γ	−3.2	−3.4	−3.6	−3.8	−4.0	−4.2	−4.4	−4.6	−4.8	−5.2	−5.4	−5.6	≥6.0−≤9.2
I	0.1	0.2	0.2			0.1							0.2
II			0.1				0.1	0.1	0.1				0.2
III		0.1		0.1			0.1	0.1	0.1	0.1	0.1		
IV		0.1								0.1	0.1		
V	0.1	0.1											
VI				0.1									
VII													
VIII	0.1							0.1					
IX					0.1								
X	0.1											0.1	
XI			0.1					0.1	0.1				0.1
XII	0.3	0.1	0.1		0.1	0.3	0.1	0.1	0.1		0.1		0.4
S	0.2	0.2	0.0	0.1	0.1	0.0	0.0	0.1	0.0	0.0	0.1	0.0	0.0
W	0.5	0.4	0.5	0.1	0.1	0.4	0.3	0.4	0.4	0.2	0.2	0.1	0.9
Σ	0.7	0.6	0.5	0.2	0.2	0.4	0.3	0.5	0.4	0.2	0.3	0.1	0.9

S = summer; W = winter.

sources like stacks. The following cases can be distinguished following Oke (1978): looping, coning, fanning, lofting, and fumigation. The corresponding plume behaviour is schematically illustrated in Fig. 3.7.

On condition that the environmental lapse rate is smaller than the dry adiabatic lapse rate, a situation typical of sunny days when surface heating is intense, the atmosphere is unstable. Because any vertical displacement of air parcels tends to amplify under these circumstances, plumes transported move up and down in a sinuous track, and these "loops" grow in size as they travel with the wind. *Looping* may bring the relatively undiluted plume in contact with the ground at quite short distances downwind from the stack, resulting in high instantaneous concentrations at these points. Owing to the action of small forced convection eddies, the plume is broken up into increasingly smaller portions, with a Gaussian ground-level pattern of concentration (cf. Sect. 3.2.3).

Coning is characteristic of windy or cloudy conditions with an adiabatic or slightly lapse temperature profile, i.e. with stability close to neutral. After displacement to any level above or below its initial position, the temperature of a parcel of air and its ambient air are the same. Hence there is no relative tendency for the parcel to rise or sink, and if the displacing force is removed, the parcel becomes stationary. In the absence of buoyancy, turbulence is more or less limited to smaller frictionally generated eddies of forced convection. Consequently, the vertical and lateral spreading of the plume tend to be about equal so that it forms a "cone" symmetric about the centre line of the plume, which intersects the ground at a greater distance downwind of the stack than is the case with looping.

Fanning is characteristic in a strongly stable atmosphere, i.e. with an environmental lapse rate greater than the dry adiabatic lapse rate. As described above, the ideal conditions for such inversions are likely to occur with anticyclonic weather, in winter, and especially at night. Stable stratification actively suppresses any buoyant stirring, keeping the plume thin in vertical direction while the erratic behaviour of wind direction viewed in plan may allow a V-shape to form resembling a fan. At other times, the resultant outline may appear as a straight line or meandering ribbon, and plumes of this type can remain essentially unchanged up to 100 km from the source of emission (Oke 1978). Under these circumstances, ground-level pollutant concentrations are insignificant unless the stack is very short, or if there are downwind changes of topography causing the plume to reach the ground. Commonly, this plume type is the precursor of the fumigation type.

Fumigation is due to an inversion "lid" above the plume suppressing upward dispersion while there is a lapse temperature profile beneath so that there is ample buoyant mixing which brings the pollutants to the ground. This unfavourable atmospheric structure occurs, for instance, during the period after sunrise when a nocturnal surface inversion is being eroded by surface heating. When the resultant surficial mixed layer reaches the plume, which may have assumed the above fanning form during the previous night, its pollutants will be carried downwards in the descending portions of convection cells. Temperature

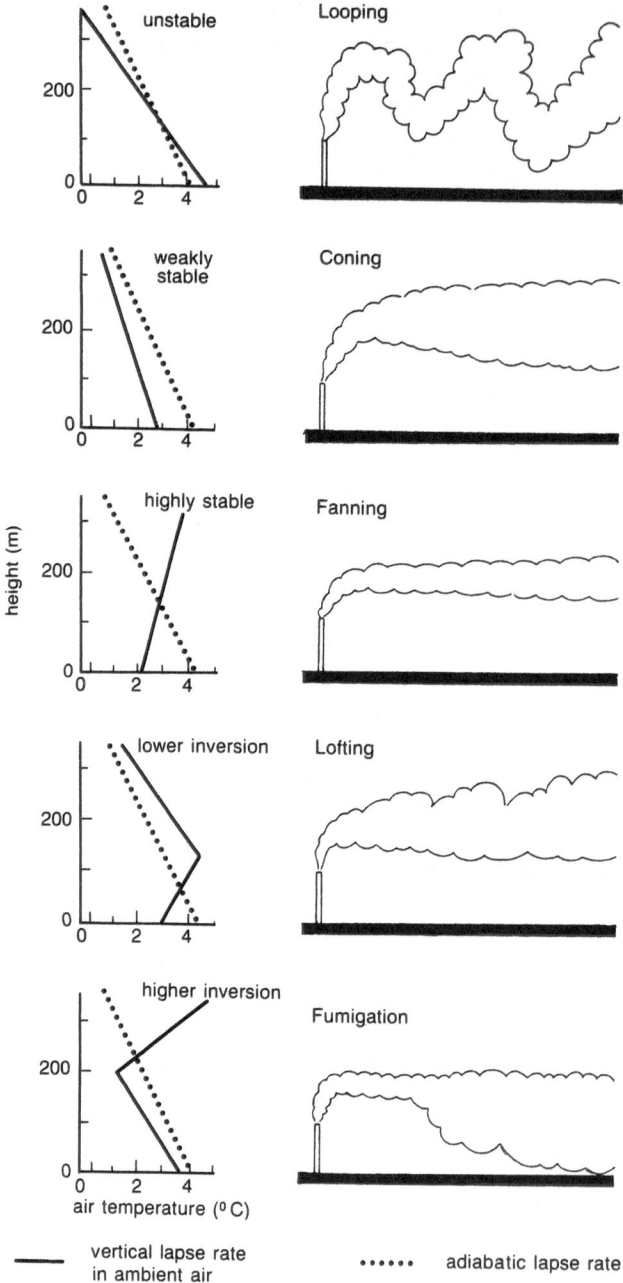

Fig. 3.7. Plume behaviour as related to different atmospheric conditions. (After Rudolf 1986)

profiles conducive to fumigation can occur in valleys, along coastlines, i.e. related to sea breeze, and in cities (Kuttler 1985).

Lofting is the reverse situation to that of fumigation, and as such the most favourable dispersal condition. It occurs in the early evening when a radiation inversion is building up from the surface or a comparable active layer. Then the stable layer beneath the plume prevents pollutant transport downwards but the unstable layer aloft allows the plume to disperse upwards. Unfortunately, it not infrequently happens, depending upon the topographic situation and effective stack height, that the inversion height soon exceeds that of the source of emission; the plume then changes from the lofting to the fanning type.

In view of the importance of atmospheric lapse rates and stability for plume dilution, the reader is finally referred to a pertinent regional analysis by Lautensach and Bögel (1956). It defines the seasonal variations of lapse rates in terms of a macroclimatic typology as listed in Table 3.10.

In view of the importance of a spatially differentiated knowledge of atmospheric lapse rates, in conclusion a new method merits particular attention. It is based on measurements of the angular radiation by means of a mobile microwave radiometer using the emission of the oxygen molecules in the 58 GHz band, which gives a satisfactory temperature resolution up to an altitude of about 500 m (Schönwald 1980 and Blohm and Schönwald 1983).

3.2.3 Selected Dispersion Models

In the course of the last 15–20 years air pollution models have become more and more sophisticated owing to the progressive consideration of pertinent boundary layer effects and concentration or removal processes, respectively (cf. Külske 1975; Hidy et al. 1980; Schneider and Grant 1982; Henderson-Sellers 1987; Pankrath 1987, for instance). This involves not only a considerable increase in computer time but also, and not infrequently more prohibitive, much more detailed input data. In the light of these circumstances, it is a question of practical importance if simpler models are not likely to yield comparable results on condition that their data base can be made proportionally more comprehensive.

Under this perspective the class of box models is particularly interesting and will be dealt with first; thereafter a selected review of the more sophisticated grid and trajectory models will be given.

3.2.3.1 Box Models

Box models, including dry and wet deposition and chemical transformations, are apt to provide calculated concentrations which are in reasonable or good agreement with measured values on a local scale as well as with transport data

Table 3.10. Seasonal variation of atmospheric lapse rates (up to 1 km). (After Lautensach and Bögel 1956)

Type of annual variation (partly following Köppen)	Characteristic distribution	Maximum lapse rate[a]		Minimum lapse rate[a]	
		Season	K/100 m	Season	K/100 m
Af	Af and Aw climates	Months with reduced precipitation	> 0.5	Rainy season	> 0.5
Aw	Aw climates	Dry season	> 0.6	Rainy season	> 0.45
BW, coastal type	Coastal deserts	Mainly summer	< 0.4	Mainly winter	< 0.2
BW, central type	Hot interior deserts	Summer	> 0.8	Winter	> 0.5
Cs	Etesian climates	Winter	> 0.5	Summer	< 0.5
Normal type	Mid-latitude climates	Summer	> 0.6	Winter	0–0.5
Maritime normal type	Coastal areas	Spring	> 0.5	Winter	0–0.5
Cold pole type	Cold poles of northern hemisphere	Summer	> 0.6	Winter	< 0
Polar type	Inland ice caps	Summer	< 0	Winter	< 0

[a] Rate of temperature decrease.

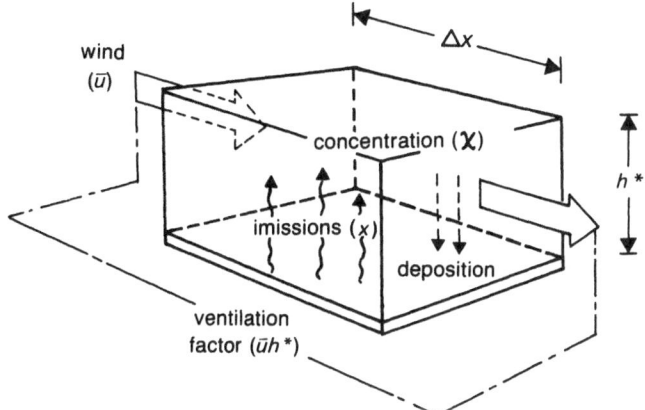

Fig. 3.8. Simplified "box" model of pollution in the boundary layer. (After Klug 1982)

over a distance of 1000 km. An exemplary prototype whose fundamentals have often been used is the following one developed by Klug (1982), describing SO_2 and NO_x concentrations in the industrial areas of Northrhine-Westphalia along the Rhine and Ruhr rivers.

In the Klug model, it is assumed that the box considered has horizontal dimensions (l) in the order of 10–50 km and a vertical dimension (h) corresponding to the height of the boundary layer. Within the box, the concentration s is uniform, while its upwind side receives an advected concentration S_0 with a mean wind $\bar{u}$.

The removal processes taken into account are dry and wet depositions and (first-order) chemical transformations, whence the following rate equation of concentration results:

$$\frac{\delta s}{\delta t} = \frac{Q}{h} - \frac{\bar{u}}{l}(s - s_0) - \left(\frac{v_d}{h} + k_r + k_c\right)s, \tag{3.1}$$

with Q = source strength in g/(m^2 s), v_d = deposition velocity, k_r and k_c = reciprocal time constants for wet deposition and chemical transformation, respectively.

The dimensionless form of the above equation,

$$\frac{1}{u} \cdot \frac{1}{s} \cdot \frac{\delta s}{\delta t} = \frac{Ql}{s \cdot u \cdot h} - \left(1 - \frac{s_0}{s}\right) - \frac{l v_d}{h \cdot \bar{u}} - \frac{k_c + k_r}{\bar{u}} \cdot l, \tag{3.2}$$

$$\qquad\quad \text{I} \qquad\qquad \text{II} \qquad \text{III} \qquad \text{IV} \quad \text{V}$$

provides easier insight into the relative importance of the different terms. With $l = 50$ km, and a Q value averaged from the emission inventory over 50×50 km^2 = 17,000 t SO$_2$/100 km^2 yr, $s = 50\,\mu$g SO$_2$/m^3, $h = 1000$ m,

$\bar{u} = 5$ m/s, $v_d = 0.01$ m/s, $k_c = 0.02/h$, $k_r = 0.1/h$, and an advection concentration $s_0 = 0.2s$, one obtains the following values for the terms:

$$I = 1, \quad II = 0.8, \quad III = 0.1, \quad IV = 0.06, \quad V = 0.5.$$

It should be noted that $I > 0$, $II \lessgtr 0$ depending on the s_0/s ratio, while the last three terms are always negative or zero in case of no precipitation. Neglecting chemical transformation processes stationarity is defined as

$$s = \frac{Ql}{lv_d + \bar{u}h\left(1 - \frac{s_0}{s}\right) + k_r \, lh} \tag{3.3}$$

With the above typical values this yields

$$s_1 - 62 \text{ µg SO}_2/\text{m}^3 \quad \text{for dry conditions}$$
$$s_2 = 48 \text{ µg SO}_2/\text{m}^3 \quad \text{for precipitation.}$$

Integrating Eq. (3.1) yields

$$s = s_0 \, \exp\left[-\left(\frac{v_d}{h} + k_c + k_r\right)t\right]. \tag{3.4}$$

Assuming there are no additional emissions between an areal source and a receptor area 1000 km apart, which corresponds roughly to the distance from the Ruhr area (or the English Midlands) to Norway, Eq. (3.4) yields an s-value of 4 µg SO$_2$ m^{-3} for the boundary conditions inserted into Eq. (3.2). The values for dry and wet depositions are 1 g SO$_2$/(m^2 a^{-1}) or 2 g SO$_2$/(m^2 a^{-1}), respectively, which is in excellent agreement with measurements.

Box models of the type discussed here can still be improved by using emission inventories at continental or national scales and considering wind directions at both the source and receptor areas.

3.2.3.2 Statistical Models

Statistical models define immission concentrations by means of multiple regressions relating immissions to emissions, meteorological parameters, topography or other dispersion-controlling factors. Models of this type vary considerably with respect to both regression analytical approach and selection of data as can be inferred from the exemplary synopsis provided by Table 3.11.

Statistical models are basically characterized by a commendable degree of flexibility. Their application-oriented structure defines kind and number of parameters used, whose factual availability, in turn, constitutes the major limitation of statistical models.

3.2.3.3 Gaussian Dispersion Models

Observations of looping, coning or (to a lesser degree) fanning plumes on an instantaneous basis reveal that the concentration is very peaked with a marked

Table 3.11. Variables considered in three statistical dispersion models. (After Battelle 1977)

Marsh and Withers	Peterson	Dietzer
Wind speed	Temperature	Wind direction
Horizontal and	Precipitation	Wind speed
vertical turbulence	Mixing height	Time of day
Cloud cover	Atmospheric stability	Month
Temperature difference	Wind	Mean lapse rate
($15\,°C - T$)	$-u$ component	(0–200 m height)
Emission data	$-v$ component	Height of lowermost
Background correction	$-$ mean wind speed	inversion
	$-$ resultant vector speed	Thickness of lowermost
	$-$ sine function of wind direction	inversion
	Mixing layer $\times$ mean wind speed	Temperature
	Mixing layer	Wet-bulb temperature
	$\times$ (mean wind speed)$^{-1}$	

maximum in the plume centre. Averaged over growing time intervals, an increasingly smooth and wide plume envelope appears which contains all of the short-period plume fluctuations. The wider spread results in a flatter concentration curve with a correspondingly lower peak centred on the time mean axis of the plume, i.e. the mean wind direction. Analogous bell-shaped concentration profiles also characterize the vertical plane through the plume, although there may be a difference between the width of the horizontal and vertical profiles indicating that turbulence is greater in one or other direction. Figure 3.9 shows that this applies also to the along-wind pattern of ground-level concentrations found downwind from an elevated point source whose plume intersects with the ground.

In statistics, bell-curves of the above type are designated as the normal or Gaussian distribution; they conform to the pattern that a series of completely random errors would assume about the "correct" value of a measurement. With regard to plume dispersion, this is equivalent to the statement that the (almost) random nature of atmospheric turbulence (cf., e.g. Favre et al. 1976; Lesieur 1982), serves to mix pollutants so thoroughly that their concentration is distrib-

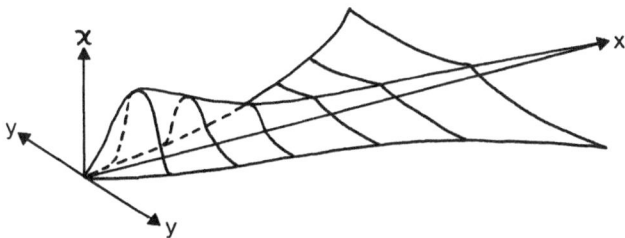

Fig. 3.9. Representative ground-level concentration (χ) distribution downwind of an elevated continuous point source. (After Oke 1978)

uted bi-normally about the plume's central axis. Consequently, the mathematical description of such bi-normal Gaussian curves provides an appropriate means of modelling the dispersion of plumes. The following equation [Eq. (3.5)] is frequently utilized to calculate the concentration of pollution (C) at any point in a plume

$$C(x, y, z) = \frac{Q}{2\pi\sigma_y\sigma_z\bar{u}} \exp\left[-\frac{y^2}{2\sigma_y^2}\right]$$

$$\times \left\{\exp\left[-\frac{(z-h)^2}{2\sigma_z^2}\right] + \exp\left[-\frac{(z+h)^2}{2\sigma_z^2}\right]\right\}, \qquad (3.5)$$

where Q = rate of emission from the source (kg s^{-1}), σ_y, σ_z = horizontal and vertical standard deviations of the pollutant distribution in the geodetic y and z directions, $\bar{u}$ = mean horizontal wind speed through the depth of the plume (m s^{-1}), and h = effective stack height, i.e. stack height plus plume rise (m). Numerically the units of C are therefore kg m^{-3} (more reasonably mg, μg m^{-3}) or units of mass concentration (g s^{-1}) or volumetric rates (m^3 s^{-1}).

While the direct proportionality of C and Q or the inverse one of C and $\bar{u}$ are self-explanatory, it should be noted that the standard deviations are related to the dimensions of the plume as it grows by turbulent diffusion, and hence are functions of downwind distance and atmospheric stability. The concentration is inversely related to σ_y and σ_z because larger values are indicative of better diffusion. The second term of Eq. (3.5) specifies how the pollutant concentration at a given distance downwind is decreased by raising the effective stack height. The last term on the right-hand side is included, by way of contrast, to account for increased concentrations at positions downwind of the point where the plume first intersects the ground. Finally it must be realized that the above formulation implies that all of the pollutant is "reflected" back up into the atmosphere and none is deposited.

Specifications of the fundamental Eq. (3.5) relate to the boundary conditions of plume diffusion considered and the different formulations applied. This is illustrated by the juxtaposition of four dispersion models in Table 3.12 which are frequently used in Germany (cf. Fleischhauer et al. 1983).

For a detailed description of the above diffusion categories, the reader is referred to a comparative analysis by Manier (1975) and Dilger and Nester (1975). The comparison between measured temperature gradients and diffusion categories according to Klug (1969) and Turner (1964), calculated from synoptic weather observations, shows that the variation of temperature gradients in the morning and evening hours is in bad agreement. With a modification of the Klug classification scheme distinctly better results are possible.

3.2.3.4 Grid Models

The Eulerian or grid model is a mathematical representation of the pertinent transport processes and chemical reactions occurring in a grid of cells yielding

Table 3.12. Synopsis of four Gaussian dispersion models

Model	Ra-Ri	TAL-Ent[a]	Klug[b]	Fortak[b]
Effective stack height (h)	$$\Delta h = \frac{1}{\bar{u}_1}\Big\{0.0004[v_s R T_s]^{1/2}$$ $$+0.05[R(T_s - T)]^{1/2}]\Big\}$$ $$\bar{u}_1 = \frac{u(z_0)}{\Delta h}\int_H^h \left(\frac{z}{z_0}\right)^m dz$$	$$h = 109 M^{3/4}\cdot\frac{1}{u_H} + H$$ for $M \le 6$ MW $$h = 143 M^{3/5}\cdot\frac{1}{u_H}$$ for $M > 6$ MW Limits: $$h_{max} = 115\left(\frac{M}{u_H}\right)^{1/3} + H$$ $$u_H = \bar{u}$$	$h = H + \Delta h$ a) Diffusion classes I, II: $$\Delta h = 2.6\left(\frac{F^*}{\bar{u}_1 s(h)}\right)^{1/3}$$ $$F^* = \frac{9.81}{3600}\cdot\frac{R}{}\cdot\frac{(T_s - T)}{273}$$ $$s(h) = \frac{9.81}{273}\frac{\theta}{z}$$ For diffusion class I $$\frac{\partial \theta}{\partial z} = 0.276 z^{-0.59},$$ for diffusion class II	$$\Delta h = \frac{1}{\bar{u}_1} E(x)$$ $$E(x) = 1.6\,F^{*1/3} x^{2/3};$$ $$x < 3.5 x^*$$ $$E(x) = 1.6 F^{*1/3}(3.5 x^*)^{2/3};$$ $$x > 3.5 x^*$$ $$x^* = 14 F^{*5/8};\ F^* \le 55$$ $$x^* = 34 F^{*2/5};\ F^* > 55$$ $$F^* = 3.18\times10^{-6}\,R (T_s - T)$$

Table 3.12. (Contd.)

Model	Ra-Ri	TAL-Ent[a]	Klug[b]	Fortak[b]
Effective stack height (h)		$h = H + \Delta h$		$\bar{u}_1$ cf. Ra-Ri
		Heat emission	$\dfrac{\partial \theta}{\partial z} = 0.17z^{-0.59}$	
		$M = 3.57Rt \times 10^{-7}$ where, R = waste-gas rate under standard conditions ($m^3\ h^{-1}$) t = waste-gas temperature in stack height (°C)	b) Diffusion classes III-1, III-2, IV, V: $\Delta h = \dfrac{1}{\bar{u}_1} Ax^{2/3}$ $A = 1.44F^{*1/3} \cdot f^{*2/3}$ F^* as above $\bar{u}_1$ cf. Fortak $f^* = 0.137 \times 10^{-2}$ $\qquad + 1.0503 \times 10^{-2}\,H$ $\qquad - 3.663 \times 10^{-5}\,H^2$ $\qquad + 0.427 \times 10^{-7}\,H^3$	
Wind direction		36 directions according to 10 degree sectors		

Windspeed	9 classes: 1.0/1.5/2.0/3.0/4.5/6.0/7.5/9.0/12.0 m s^{-1}			
Mean windspeed in an area of diffusion	$\bar{u} = \dfrac{u(z_0)}{m+1}\left(\dfrac{h}{z_0}\right)^m$; $\quad h > 10$ m $\bar{u} = \dfrac{u(z_0)}{m+1}\left(\dfrac{10}{z_0}\right)^m$; $\quad h \le 10$ m	$\bar{u} = \dfrac{u_R}{m+1}\left(\dfrac{2H}{za}\right)^m$	$\bar{u} = \dfrac{u(z_0)}{m+1}\left(\dfrac{2h}{z_0}\right)^m$	$\bar{u} = u(z_0)\left(\dfrac{H}{z_0}\right)^m$
Diffusion parameter	$\sigma_y = (Fx+F_0)^f$; $\sigma_z = (Gx+G_0)^g$		$\sigma_y = Fx^f$; $\sigma_z = Gx^g$	
Diffusion categories	4 categories: 2, 3, 4, 5 (after Turner)	6 categories: I, II, III-1, III-2, IV, V (after Klug 1969)		7 categories: 1, 2, 3, 4, 5, 6, 7 (after Turner 1964)

[a] Ra-Ri and TAL-Ent are acronyms for Raffinerie-Richtlinie (1975) and Entwurf einer Verwaltungsvorschrift zur Änderung der Technischen Anleitung Luft (TA Luft), 10. September 1981

[b] Klug and Fortak relate to the authors of models used in: Gutachten zur Immissionsprognose im Auftrag des Ministers für Arbeit, Gesundheit und Soziales des Landes Nordrhein-Westfalen, 18. August 1978

the spatial and temporal variations of the ground cell concentrations of primary and secondary pollutants. It is characterized by fixed coordinate systems with respect to the ground and consists of a set of non-linear coupled partial differential equations expressing the conservation of mass of each pollutant considered. The inputs to the model are numerical representations of the meteorological and emission conditions as a function of day-time, the initial concentrations and finally the boundary conditions of inflow. Like the trajectory models which are dealt with later, the grid models have their derivation in the atmospheric diffusion equation. Generally speaking, grid models require finite-differencing, whereas trajectory models require neglect of horizontal spatial derivatives and wind shear. In their implementation, grid models usually need considerably higher amounts of meteorological information and computer storage.

For the computation the plan of the area analyzed, e.g. a city, is divided by a square grid of appropriate size. The mixing depth is an input, and the atmosphere may be divided vertically into a (limited) number of layers, giving a grid of cells. Within each cell, conditions are assumed to be uniform. The simulation program predicts the hourly averaged ground-cell concentrations of each pollutant in each square as a function of time. Analogously the vertical concentration profile is computed at each monitor station.

In the urban scale model developed for assessing the pollution situation in European towns by the Committee of Common Market Automobile Constructors with SAI co-operation (Gaddo and Weaving 1982), the simulation package consists of four main programmes:

- the emission data preparation,
- the meteorology data preparation,
- the atmospheric pollution simulation,
- the data plotting programme.

The mathematical representation of the kinetic mechanisms involved, i.e. 15 reactions of 10 chemicals species, is based on four differential equations describing the changes in concentration with time of NO, NO_2, O_3 and hydrocarbons. The basic equation used for describing the conservation of species (i) is:

$$\frac{\partial \langle C_i \rangle}{\partial t} + \frac{\partial (\bar{u} \langle C_i \rangle)}{\partial x} + \frac{\partial (\bar{v} \langle C_i \rangle)}{\partial y} + \frac{\partial (\bar{w} \langle C_i \rangle)}{\partial z} = \frac{\partial}{\partial x} \left(K_h \frac{\partial \langle C_i \rangle}{\partial x} \right) + \frac{\partial}{\partial y} \left(K_h \frac{\partial \langle C_i \rangle}{\partial y} \right)$$
$$+ \frac{\partial}{\partial z} \times \left(K_v \frac{\partial \langle C_i \rangle}{\partial z} \right)$$
$$+ R_i(\langle C_i \rangle, \ldots, \langle C_N \rangle, T)$$
$$+ S_i(x, y, z, t), i = 1, \ldots, N,$$

$$(3.6)$$

where

$\langle C_i \rangle$ is the mean concentration for species i,

$\bar{u}, \bar{v}, \bar{w}$ are the components for the mean wind vector,

R_i is the rate of formation of species i,

S_i is the rate of emission of species i,

x, y, z are the coordinate axes,

t is time,

K_h, K_v are the horizontal and vertical eddy diffusivities, respectively.

The resultant N partial differential equations are solved over the urban area defined by the region:

$$X_w \leq x \leq X_e$$

$$Y_s \leq y \leq Y$$

$$H(x, y) \leq z \leq H(x, y, t),$$

where:

X_w, X_e, Y_s, Y_n are the west, east, south and north boundaries of the airshed $H(x, y)$ is the ground elevation above sea level and $H(x, y, t)$ is the elevation above sea level of the assumed upper limit for vertical mixing or transport.

Elevated point sources are treated individually in the model and appear in the term $S_i(s, y, z, t)$. Near ground sources include domestic heating and automobile emissions which are distributed uniformly within each cell, while the mass flux from the surface is assumed to be proportional to the concentration gradient normal to the surface.

The horizontal wind field is specified by an interpolation procedure defining points throughout the simulation period. The horizontal wind field is assumed constant with height and the vertical wind component is calculated under the assumption the wind field is non-divergent, i.e.

$$\frac{du}{dx} + \frac{dv}{dy} + \frac{dw}{dz} = 0 . \tag{3.7}$$

The vertical turbulent diffusivity K_v is a function of wind speed, height above ground and thermal convection. The horizontal turbulent diffusivity K_u is considered constant. Horizontal fluxes are calculated at the boundaries of the urban area, which requires specification of the concentration field outside the airshed. The turbulent diffusive flux across the elevated inversion which characterizes the upper boundary of the airshed is assumed to be zero. The concentration field above the inversion requires specifications as to advection (cf. Sect. 3.1.2). The initial concentration field is numerically specified at all points in the airshed at the beginning of the simulation by means of interpolation between measuring points.

The value of this and comparable grid models depends on the inherent quality of the underlying physical and chemical assumptions on the one hand, and the spatial validity of the input data and records of monitoring stations on the other. Thus the reference data inserted into the Gaddo and Weaving model as applied to Torino were obviously not specific for photochemical smog formation. Van Egmond and Kesseboom (1982), presenting the results of a multi-layer grid model for mesoscale transport of NO_x for The Netherlands and parts of Germany and Belgium found "reasonable correlations between measured and modelled fields" both in short-term (24 h) and long-term (season-average) applications. They noted, however, that a further improvement of their model, which is based on a comparatively simple NO_x/O_3 chemistry using empirical data for OH production, is complicated by the limited spatial representativity of fixed monitoring measurements.

Hence, and in conclusion, it may be inferred that the validation of grid models by means of more or less extensive measuring networks should imply a thorough spatial analysis of these networks by means of variogram analysis (Matheron 1963; Delfiner 1975; Fränzle 1984a) in order to assess their quality more precisely.

3.2.3.5 Trajectory Models

In the trajectory or Lagrangian model, the time behaviour of the concentration (c) of a pollutant is described by a differential equation of the type

$$\frac{\partial c}{\partial t} = \frac{\partial}{\partial z}\left(\underline{K}\, \frac{\partial C}{\partial z} \right) + S \,, \tag{3.8}$$

where c depends only upon height above ground (z), and the time of travel (t) of an air parcel, while S refers to the net source–sink term representing the combined effects of emissions, chemical processes during transport, deposition and flow across the upper boundary. $\underline{K}$ represents a vertical eddy diffusion coefficient whose values are characteristically dependent on time of day, height and location. In comparison with the above Eulerian models, trajectory models have commendable mathematical advantages resulting from their moving coordinate systems with respect to a fictitious vertical air column moving horizontally with the advecting wind (Eliassen et al. 1982; Pankrath 1983). Considered as the present state of the art in operational long-range transport modelling, the Lagrangian approach was used in numerous studies in the USA and Canada relating to transboundary air pollution. Table 3.13 lists the eight favorite models applied.

Most of these models use sulphur as an appropriate indicator substance instead of nitrogen compounds because its atmospheric behaviour is better known, and measurements of sulphur species in the atmosphere, as deposition, and in ecosystems are not infrequently more reliable.

The difficulties inherent in modelling NO_x chemistry with the same confidence as SO_x chemistry have two basic reasons (cf. Whelpdale and Bottenheim 1982): first, nitrogen reactions proceed quickly in relation to model time steps, and they are not first order with respect to oxidizing species. Second, the data base for model evaluation is usually, i.e. with respect to the pertinent local or regional situation, more sparse than in the case of sulphur. This complication may finally be further enhanced by a lack of reliable information on removal

Table 3.13. Long-range transport models used in North America

Model name (after Hidy et al. 1980)	Acronym
Atmospheric Environment Service Long-Range Transport Model	AES
Advanced Statistical Trajectory Regional Air Pollution Model	ASTRAP
CAPITA-Monte Carlo Model	CAPITA
Eastern North American Model of Air Pollution	ENAMAP-1
Transport of Regional Anthropogenic Nitrogen and Sulphur (TRANS) of Meteorological and Environmental Planning, Ltd.	MEP
Ontario Ministry of Environment Long-Range Transport Model	OME
Regional Climatological Dispersion Model	RCDM
University of Michigan Atmospheric Contributions to Interregional Deposition Model	UMACID

pathways for nitrogen species. The following example of a photochemical column trajectory model is to illustrate this.

The AERE Harwell model (Henderson-Sellers 1987) was employed to investigate the complex relationships between motor vehicle NO_x emissions and the ground level distribution of NO_2 in London. It describes the meteorological situation of the boundary layer based on observations taken at the London Weather Centre. To this end, the hourly windspeeds for the calmest 30 days out of the 77 days during 1973–1975 when ozone exceeded 60 ppb were taken and averaged. They exhibited a characteristic diurnal variation with lowest values in the early morning and maximum values at night.

The diurnal variations in relative humidity and air temperature were included by fitting a three-term Fourier series through the diurnal variations of windspeed, temperature and humidity. Since it is not yet possible to give a formal description of all the processes influencing the diurnal, spatial and height variations of $\underline{K}$ over the complex urban terrain (cf. VDI 1988), a highly parameterized approach is employed, leaving considerable inconsistencies. $\underline{K}$ is thus calculated for each height, location and time of day from the instantaneous values of windspeed, mixing height, flux of sensible heat and temperature for unstable, neutral and stable conditions.

The free radical and other chemical reactions involving the O_x, NO_x and HO_x species define the concentrations of the major free radical species OH and HO_2; they furthermore include some of the processes controlling NO to NO_2 conversion and NO_2 removal. In addition, a simplified scheme of hydrocarbon degradation is included which provides an approximate peroxy radical NO to NO_2 conversion route. Light absorption as driving force of the primary photochemical processes is expressed as a first-order process using a rate coefficient which is time-dependent during the day and zero at night. Deposition is accounted for in the model by removing species from the lowermost layer with definite (measured) deposition flux rates.

Table 3.14. Calculated mean annual SO$_2$ transfer in Europe in the period 1978–1982. (After Umweltbundesamt 1986). Emitting country[a]

	AL	A	B	BG	CS	DK	SF	F	DDR	D	GR	H	IS	IRL	I	L
AL	12	0	0	3	2	0	0	3	2	1	3	3	0	0	12	0
A	0	62	3	0	46	0	0	26	31	40	0	16	0	0	72	1
B	0	0	84	0	2	0	0	34	3	29	0	0	0	0	1	1
BG	2	1	0	189	11	0	0	3	9	6	9	17	0	0	12	0
CS	0	15	5	1	440	1	0	24	131	59	0	61	0	0	24	1
DK	0	0	2	0	4	47	0	5	14	14	0	0	0	0	1	0
SF	0	0	3	0	0	5	92	7	25	20	0	0	0	0	2	0
F	0	2	41	0	10	1	0	760	33	124	0	4	0	0	48	6
DDR	0	2	10	0	15	4	0	25	586	103	0	2	0	0	5	1
D	0	9	45	0	64	6	0	136	149	660	0	5	0	0	31	6
GR	4	1	0	39	5	0	0	4	5	5	111	7	0	0	20	0
H	0	11	1	3	56	0	0	9	25	17	0	227	0	0	36	0
IS	0	0	0	0	0	0	0	0	0	1	0	0	1	0	0	0
IRL	0	0	0	0	0	0	0	2	1	2	0	0	0	22	0	0
I	0	8	3	2	17	0	0	68	17	29	1	14	0	0	948	1
L	0	0	1	0	0	0	0	4	0	2	0	0	0	0	0	4
NL	0	0	21	0	3	0	0	19	7	51	0	0	0	0	1	0
N	0	0	5	0	9	9	2	15	26	25	0	2	0	1	2	0
PL	0	9	11	3	168	10	1	36	270	106	1	48	0	1	26	2
P	0	0	0	0	0	0	0	2	0	2	0	0	0	0	0	0
R	1	5	2	35	49	1	0	11	35	23	5	91	0	0	34	1
E	0	0	3	0	2	0	0	44	6	22	0	0	0	0	4	1
S	0	1	7	1	22	23	12	19	53	45	0	6	0	1	7	1
CH	0	1	2	0	4	0	0	29	6	19	0	1	0	0	45	0
TR	1	1	0	37	8	0	0	5	9	6	25	10	0	0	16	0
SU	4	17	23	80	221	30	60	82	342	233	27	195	0	3	108	5
UK	0	0	9	0	5	1	0	35	12	26	0	1	0	8	1	1
YU	4	19	3	39	52	1	0	32	35	30	6	86	0	0	193	1

	NL	N	PL	P	R	E	S	CH	TR	SU	UK	YU	RE	IND	SUM	Q
AL	0	0	1	0	1	1	0	0	0	0	1	19	1	14	85	50
A	2	0	17	0	1	4	0	4	0	2	11	42	0	36	442	215
B	6	0	1	0	0	1	0	0	0	0	21	0	0	12	198	405
BG	0	0	10	0	23	2	0	0	5	13	2	57	1	35	413	500
CS	4	0	88	0	5	2	0	2	0	7	14	43	0	37	969	1500
DK	2	0	6	0	0	0	2	0	0	1	15	1	0	14	132	228
SF	3	2	22	0	1	0	17	0	0	53	20	4	2	66	363	270
F	17	0	9	3	0	94	1	9	0	1	122	6	0	205	1505	1800
DDR	8	0	29	0	1	1	1	1	0	3	28	7	0	30	918	2000
D	27	0	24	0	1	9	1	8	0	3	88	15	0	96	1388	1815
GR	0	0	4	0	5	3	0	0	5	5	2	37	2	36	305	350
H	1	0	29	0	11	1	0	0	0	5	5	93	0	24	560	750
IS	0	0	0	0	0	0	0	0	0	0	2	0	0	20	26	6
IRL	0	0	0	0	0	1	0	0	0	0	15	0	0	32	86	88
I	2	0	11	0	2	22	0	8	0	2	13	73	6	103	1355	2200
L	0	0	0	0	0	0	0	0	0	0	1	0	0	1	14	21
NL	53	0	1	0	0	1	0	0	0	0	33	1	0	14	210	240
N	5	24	14	0	0	2	11	0	0	9	53	3	0	92	314	76
PL	10	0	776	0	8	3	5	1	1	36	41	45	0	90	1712	2150
P	0	0	0	25	0	16	0	0	0	0	1	0	0	30	80	84
R	2	0	54	0	192	2	1	1	4	53	5	143	1	70	827	100
E	3	0	1	11	0	429	0	1	0	0	17	1	3	113	666	1000
S	7	10	41	0	2	2	100	0	0	27	51	8	0	138	587	275
CH	1	0	2	0	0	4	0	16	0	0	2	3	0	21	165	58
TR	1	0	9	0	8	4	0	0	209	23	2	30	2	93	505	500
SU	29	6	538	0	139	11	45	4	53	4273	126	232	3	1090	7972	8100
UK	7	0	4	0	0	5	0	0	0	1	790	1	0	87	996	2560
YU	3	0	30	0	17	10	0	2	1	10	13	678	3	106	1377	1475

[a] Labels corresponding to international registration plates.

EMEP sulphur balances as of October 1978 to September 1982, based on 1978 emission factors Unit 1000 t S/a.

IND = wet deposition, SUM = approximate total deposition in the respective country, Q = calculated annual sulphur emission per country,

RE – deposition onto the remaining part of study area

The emissions from NO_x, CO, SO_2 and 16 hydrocarbon species are represented on a $10 \times 10\,km^2$ grid by hour of the day. In greater detail, these emissions comprise contributions from three specific classes of processes, namely: exhaust emissions from petrol-engined vehicles, emissions from stationary combustion and emissions from other sources. Since these data are derivations from primary statistical data and emission factors (cf. Sect. 3.1.2, Tables 3.6, 3.7, 3.8), they can be considered accurate only to one order of magnitude, and the geographical and diurnal variations as mere indications of possible systematic differences between areas and times of day.

The AERE model was run in a complete and a simplified version which required much reduced computer time. During daytime the simple model comprising only three photostationary state reactions tended to overestimate the NO_2 concentrations, most of all close to the ground. The percentage error increased steadily as the air column passed over London. At the downwind model boundary the overestimation in the NO_2 concentrations amounted to 3–5%, which is due to an underestimation of O_3. The overestimation seems to be due to the neglect of the $OH + NO_2$ reaction and the corresponding underestimation of ozone to the neglect of the $RO_2 + NO$ reactions.

During nighttime, the simple model tends to overestimate both the O_3 and NO_2 concentrations and underestimate the NO values by 6–7%, which seems to be caused in part by the omission of the nighttime $NO_2 + O_3$ chemistry. With regard to the impact of NO emissions in the urban areas on NO_2 concentrations in rural areas further downwind, however, the neglect of the above $OH + NO_2$, $RO_2 + NO$ and $NO_2 + O_3$ reactions will become progressively more significant. It ensues from the foregoing that the primary products of trajectory models are medium to large-scale (monthly, seasonal or annual) concentration and deposition fields and transfer matrices. Table 3.14 is an illustrative example of such a matrix.

On condition that appropriate allowance is made for their inherent limitations resulting from the assumption of a simple linear superposition of source contributions, matrices have the particular advantage of an easy manipulation of the relationships for various emission scenarios and optimization procedures. For instance, they can be used to examine the relative importance of different source regions to given receptor areas. Furthermore, using coefficients in the matrix which are normalized against emissions in a particular source area permits identification of the relative effectiveness of controlling the various source regions for a given receptor. This, in turn, permits an identification of likely emission reduction configurations (cf. OECD 1979; Battelle 1982; Deutscher Wetterdienst 1982).

In conclusion, it should be noted that there are only few comparisons between Gaussian and Lagrangian models (cf. Umweltbundesamt 1977; Bundesminister des Innern 1981a; Pankrath 1983). According to Schorling (1990), a Lagrangian calculation showed much better fit to measurements of a tracer experiment than the Gaussian model, which described neither the correct magnitude nor the trend of the concentration pattern. An additional advantage of Lagrangian models seems to be the possibility of a relatively easy extension for application in complex terrain. This requires an additional term to cover the convective accelerations of the wind and, consequently, a sufficiently detailed knowledge of the spatially and temporally varying windfield in the area under consideration.

3.3 Physical Transformation Processes

The fate of pollutants in the troposphere is largely influenced by the various processes of condensation, coagulation and nucleation which occur prior to or in conjunction with rainout and washout processes.

3.3.1 Nucleation

Nucleation is the generic term describing the whole of the processes that lead to aerosol or nuclei formation in the atmosphere. They are probably very complex, with no single mechanism dominating, since the lower atmosphere contains substantial amounts of airborne particles which are continuously interacting with trace gases and especially with water as vapour and in the condensed phase. In elucidating atmospheric nucleation processes, both homogeneous processes of precursor formation and heterogeneous interactions with primary aerosols must be accounted for. Regardless of the chemistry, which is dealt with in greater detail in the following chapter, certain physical constraints control aerosol–gas interactions. They relate mainly to the thermodynamic stability of condensed particles of small diameter, particle nucleation and condensation processes, and finally to gas-phase diffusion-limited rates combined with absorption or adsorption on particles (Hidy and Burton 1980).

The partial pressure of condensed species on particles must be less than or equal to the saturation vapour pressure at atmospheric temperature, which is particularly important to the stability of organic aerosols. Most organic vapours identified in the atmosphere are in the range of carbon number less than four, and only material with a carbon number much greater than C_6 would be thermodynamically stable in the condensed phase at the concentration of ~ 1 ppb. In contrast, the requirements of low vapour pressure present no great problem for most inorganic salts or for sulphuric acid. It places, however, a severe constraint on HNO_3 to exist as a pure compound or as an acid diluted in water, since it has a partial pressure over aqueous solutions more than 100 times higher than that of concentrated H_2SO_4 (cf. Sect. 3.4.3).

Accumulation of condensable material as aerosols may take place by two basically different processes: (1) by condensation of supersaturated vapour or by chemical reaction leading to spontaneous formation of new particles (homo-geneous nucleation); and (2) by condensation, absorption, adsorption, or reaction on or in existing particles (heterogeneous nucleation). It is generally accepted that the latter set of processes are most likely in the atmosphere because of the large number of existing nuclei (Husar et al. 1976). This assumption is based on the general decrease in nuclei concentration with large-particle concentration, suggesting a distinctly lesser new particle formation rate.

Also the thermodynamic is of importance for the growth of nuclei. If the radius of the particles is too small, the partial pressure of condensable species increases according to Kelvin's relation significantly with diminishing radius of curvature such that effective growth is limited to particles greater than 0.1 μm.

3.3.2 Coagulation

A further important factor controlling nuclei size distributions is *coagulation*, which describes the adhesion and fusion processes following the collision of two or more particles. As such they may be of thermal, electric, magnetic, hydro-dynamic or molecular nature, to cite only the apparently more important phenomena. Under the assumption of exclusively effective Brownian or thermal collisions of globular particles, the decrease in total number concentration of an aerosol can be described by the following relation (Billings and Gussmann 1976):

$$-\frac{dn}{dt} = Kn^2 Cc , \qquad (3.9)$$

with n = number of particles, K = coefficient of coagulation and Cc = Cunningham's correction factor.

$$K = \frac{2kTS}{3\mu f} , \qquad (3.10)$$

with k = Boltzmann constant, T = absolute temperature, S = number of interacting particles, and μf = coefficient of viscosity of gas or liquid.

If aerosols are electrically charged, their coagulation rate is increased or decreased in relation to the number of positive and negative charges. Turbulence also increases coagulation rates. According to Fuchs (1964), the ratio of turbulent to thermal coagulation coefficients is

$$k_T = \frac{\alpha}{16\pi} \left(\frac{\varepsilon}{v}\right)^{1/2} \cdot \frac{r^2}{D_B} , \qquad (3.11)$$

with α varying between 4 and 25, ε = energy distribution per $g\,s^{-1}$, v = kinematic viscosity of gas (air), and D_B = coefficient of diffusion.

The resultant mechanism of kinematic coagulation is due to the different relative velocities of particles of different size induced by gravity or other external forces. The rate of effective collisions Φ then is

$$\Phi = \pi\eta\, r_1^2 \sqrt{s}\,, \tag{3.12}$$

with η = efficiency of impact, r_1 = radius of larger particle and s = speed of larger particles in relation to the (smaller) background aerosol. If larger and smaller particles carry electrical charges

$$\eta = \frac{3|q_1 q_2|}{\pi\rho r_2 r_1^2 (r_1^2 - r_2^2)g}\,, \tag{3.13}$$

with q_1 = charge of larger particle, q_2 = charge of smaller particle, ρ = density of particles, r_2 = radius of smaller particle, and g = gravity acceleration.

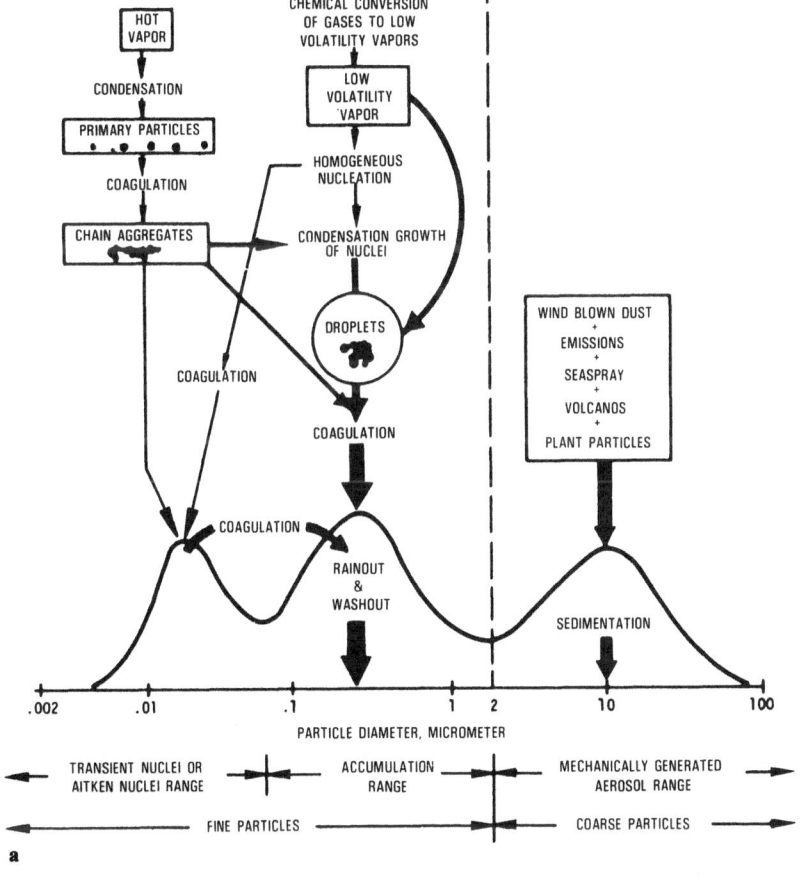

Fig. 3.10. a Trimodal size distribution model of atmospheric aerosol and the principle processes involved in mode formation. (Whitby 1977).

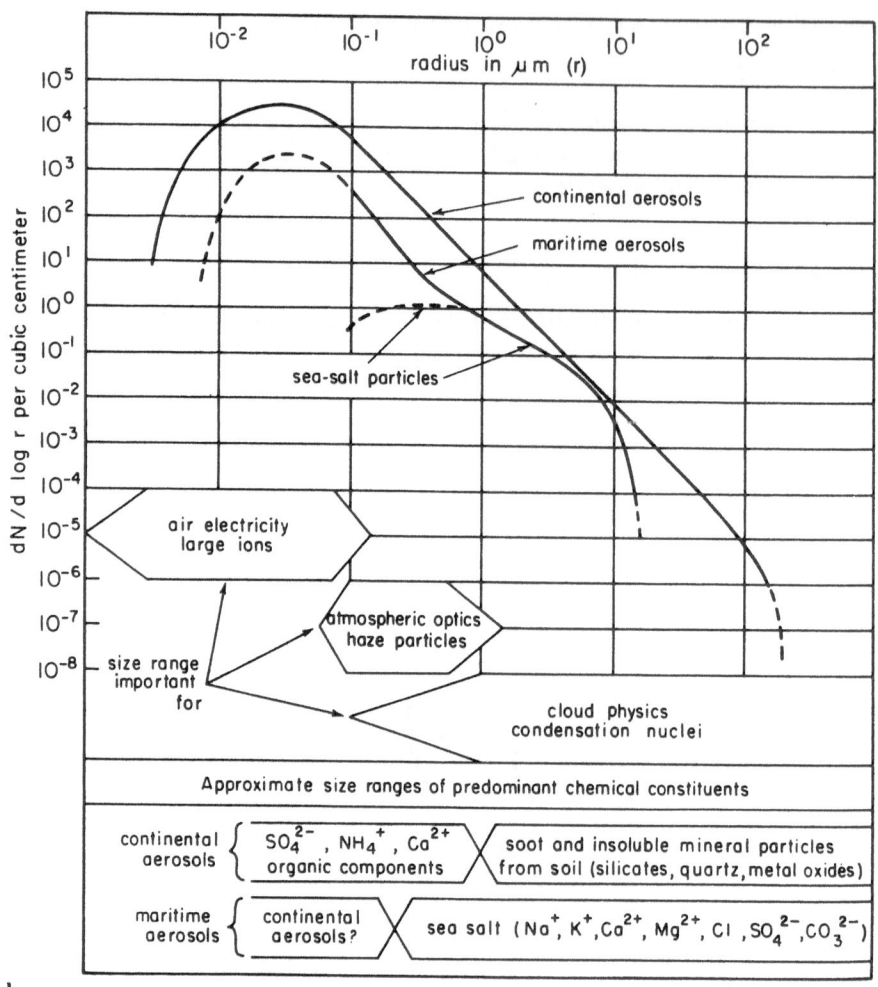

b

Fig. 3.10. b Size ranges, abundances, compositions, and origins of the major classes of tropospheric aerosol. (Lewis and Prinn 1984)

Figure 3.10 (after Whitby 1977) summarizes the observed influence of nucleation and coagulation processes on atmospheric aerosol size distribution.

This partial model, which should be considered in conjunction with the corresponding components of foldout model I, shows a basic trimodal distribution described in terms of three additive log-normal size distributions. Reasonably well established to interpret both physical and chemical size distributions which are of particular importance for the discussion of water vapour condensation, it must be noticed, however, that for most atmospheric aerosols no more than two distinct modes are observed (Whitby and Sverdrup 1980).

3.3.3 Condensation

Condensation takes place whenever the atmosphere is sufficiently cooled in the presence of receiving terrestrial surfaces or suspended nuclei. The relative humidity at which this occurs is largely dependent upon the nature and number of these nuclei whose radii range in size from less than $10^{-2}\,\mu m$ to almost 100 μm. As Fig. 3.10 shows, the most effective particles are in the range $10^{-1}\,\mu m$ to 2 μm, while either extreme in the range of sizes is least effective in initiating condensation. The transient nuclei produce unstable droplets readily evaporated back into the atmosphere, while the mechanically generated particles exceeding 2 μm fall quickly to the ground under the influence of gravity. Hygroscopic nuclei such as sodium chloride or ammonium sulphate may initiate condensation well before the air reaches its saturation vapour pressure, i.e. at relative humidities as low as 80%. Recent work on the chemistry of aerosols, particularly in urban air, has furthermore suggested that water plays an important part in particle behaviour even at a relative humidity well below 70% (Ho et al. 1980). Indirectly, a liquid water content of at least 10% is supposed to account for the total mass composition of filter-collected aerosol equilibrated with air of a relative humidity < 50% (Miller et al. 1972; Meyer et al. 1973).

A comprehensive classification of atmospheric and terrestrial condensation and associated precipitation forms is due to Berg (1948), who distinguished no less than 23. For the purposes of this presentation and with regard to Section 3.4, it may suffice to classify according to the nature of the underlying cooling processes, the most important of which are radiative cooling, advection cooling and dynamic cooling.

Radiative cooling affects both terrestrial surfaces and the atmosphere, and is largely dependent upon the marked differences in thermal conductivity of air and solids. While soils or rocks lose heat rapidly by radiation and may consequently reach dew-point temperature quickly, radiative cooling of the atmosphere proceeds at much slower rates only. Hence it is rarely the sole cause of condensation but can enhance the cooling of the air already in contact with cold ground surfaces.

Advection cooling occurs when two streams of air mix together provided there is a relatively large temperature difference between them. If, for example, warm moist air is transported over a cool moist surface, it mixes with the shallow layer of cooled air overlying the surface, and the mixture may attain saturation vapour pressure or even be supersaturated. This excess is then condensed in the form of advection fog which may be associated with relatively turbulent airflow and may consequently be fairly deep. A further form of advection fog occurs where cold air flows over warm water. Water evaporating from the comparatively warm surface then condenses in the cold air above, producing a fog which resembles smoke or steam rising (sea or river smoke).

In rising air *dynamic or adiabatic cooling* takes place. The form of the resultant condensation depends upon both the extent and velocity of the enforced rise and the stability of the air (cf. Sect. 3.2.1). A gradual rise caused by slow-moving warm fronts will produce less spectacular condensation processes than violent uplift due to localized convection of unstable air or the passage of air over mountain ranges (foehn). In all cases, the rising air first cools at the dry adiabatic lapse rate until it reaches saturation. Owing to condensation, the further cooling continues at the saturated adiabatic lapse rate. The shape and depth of the clouds originating in the course of these various condensation processes is indicative of the relative amount of atmospheric stability. It consequently varies from shallow layered forms (stratus) to clouds with considerable vertical extent (cumulus) which are associated more with unstable than stable air. According to the nature of the dynamic condensation process involved, the resultant precipitation may be broadly classified as orographic, convectional or cyclonic.

3.3.4 Rainout and Washout

While the international cloud atlas (WMO 1956) provides a comprehensive description of cloud characteristics, quantitative synoptics (Hantel et al. 1984; Matveev 1984; Emeis 1985) and satellite imagery (cf. Flohn 1968; Weischet 1979, 1980) constitute the basis for statistical analyses of the regionally varying types and frequencies of condensation processes.

In polluted air they are associated with scavenging processes which are customarily distinguished as rainout and washout and figure in Model I under the label "AUS". *Rainout* comprises all processes of incorporation or accretion of gases or aerosols to cloud droplets whose radii normally range from 1 μm to 20 μm. In still air they would consequently have terminal velocities between 0.1 and 50 mm s^{-1}; normally, however, these velocities are so exceedingly small in relation to those of the cellular updraughts that the droplets are held within the cloud mass and are not precipitated.

As a rule, precipitation is only initiated when drops have reached a radius of approximately 1000 μm, which means a millionfold increase in volume with corresponding scavenging effects, and can be brought about by a number of processes. Among these, the ice crystal process, collision and coalescence seem to be the most important. As a result of the freezing of supercooled water, ice crystals may grow further by sublimation and coalescence with other crystals, which increases their earthward velocity. If the droplet melts in lower and warmer parts of the cloud, it continues to fall as a water droplet and keeps growing as a result of coalescence with much smaller droplets in its path. In warm clouds with temperatures well above the freezing point, the only movement of air produces water droplets of different sizes through chance collisions and coalescence. Their further growth eventually leads to droplets large enough to fall to the ground. On their way through the atmosphere below the base level of condensation they may incorporate gases and aerosols which process is called *washout*. It is either defined as a "washout coefficient", i.e. ratio of particle loss rate per unit volume to initial particle concentration or as a "washout factor". The latter is the proportionality factor of the concentration of a trace substance in rain water and the concentration in the ambient air. Although terminologically well distinct, the relative importance of rainout and washout processes is difficult to assess in practice. Georgii (1984) considers washout to account for 20–60% of the total pollutant concentration in rain water.

The essential meteorological variables controlling the efficiency of rainout and washout processes are:

– drop size spectra in clouds and precipitation,
– intensity of precipitation,
– pH and chemical composition of droplets,
– temperature.

In addition to the meteorological boundary conditions, clearly a set of chemical and physical pollutant properties and the eventual resultant reactions are of importance (Dana and Hales 1976; Pruppacher et al. 1983):

– the initial concentration and distribution of pollutants,
– changes in concentration due to scavenging by droplets,
– the size and reactivity of aerosol particles,
– the solubility and reactivity of gases.

Other things being equal, i.e. irrespective of the above-mentioned physical and chemical reactions, the control of scavenging processes by the *size distribution of cloud droplets*, i.e. the efficiency of rainout, can be described by Henry's law:

$$\varepsilon = \frac{N_L}{N_0} = \left(1 + \frac{KM_{H_2O}}{RTL} \right)^{-1}, \tag{3.14}$$

with ε = scavenging rate, N_0 and N_L = concentration before and after condensation, K = equilibrium constant, M_{H_2O} = molecular mass of water, and L = mass of condensed water in a volume of air (cf. Rasool 1973).

Based on laboratory experiments, Beilke (1969) showed for constant precipitation intensities that the washout rate of SO_2 is directly proportional to the radius of droplets. If the radius of droplets remains constant, washout increases with the *intensity of precipitation*. Experiments with NO_2 yielded similar results but saturation was achieved more quickly. Under natural atmospheric conditions, however, highest concentrations are coupled with low-intensity precipitation. This is due to the fact that normally an increase in intensity is brought about by a distinctly higher percentage of larger drops which attain higher terminal velocities. Thus reaction times are reduced considerably, as Georgii (1982) substantiated by comparative measurements of SO_2 concentrations in convectional and cyclonic precipitation. In the latter, the rainout/washout ratio is lower than in showers, where rainout processes are relatively more important. In view of these apparently trivial relationships it was thought that the changes in concentration during precipitation events as determined by sequential rain sampling could yield reliable information on scavenging processes; but the results of comparative measurements (Asman et al. 1982; Kins 1982) fell short of such expectations, since concentrations in precipitation are the result of the highly variable influence of the variables enumerated above. Thus each precipitation event seems to have its own character and is consequently difficult to compare with others.

Yet for elementary chemical reasons the *pH control* of scavenging is particularly marked. SO_2 absorption, for instance, increases with pH and may be further enhanced by catalysts like $MnSO_4$ or $FeCl_2$ which introduce a tendency to accelerated formation of SO_4^{2-} ions (Johnstone and Conghanour 1958). Aqueous oxidation processes including O_2 and O_3 are also highly pH-dependent; the reactions are suppressed in an acidic medium. In contrast, the H_2O_2 reaction is much less acidity-dependent, especially in the pH range below 4.5, where cloud and wet aerosol particle processes are likely to occur (Husar et al. 1976).

One of the major sources of the pH increase of droplets is the incorporation of alkaline pollutants such as ashes or NH_3, which frequently account for the marked seasonality of pH values in precipitation (cf. Georgii 1982; Horváth 1982; Müller 1982; Perseke 1982). At a coastal site near Den Helder (Netherlands), the relative importance of NH_3 as a neutralizing agent ranged between 53% for continental weather situations and 29% for maritime ones (Asman et al. 1982). This implies that the amount of acid after neutralization is almost equal in either weather situation, although the amount of acid generated by SO_2 and NO_x is much higher for continental weather. Ca and other components seem to be much less important as neutralizing agents than NH_3. With respect to the effects of precipitation it must be kept in mind, however, that the NH_4^+ ion can undergo rapid oxidation to HNO_3 in aerated soils (cf. Sect. 4.4). Consequently, there will be a considerable difference in impact of precipitation associated with continental and maritime air masses with respect to ultimate acidification.

According to Le Chatelier's principle solution processes in cloud droplets or precipitation are highly temperature-dependent. Hence, and other things being equal, rising *temperatures* are normally linked with increased solution rates of aerosol particles in droplets while the reverse holds for gases.

Figures 3.11 and 3.12 define washout efficiency in relation to particle diameter distinguishing the three most important scavenging mechanisms operative. Table 3.15 provides some complementary information on washout factors of metals which result from deposition measurements in the German deposition network during the 1979–1981 period (Rohbock 1982). The specific enrichment mechanisms in the table need further clarification. From a comparison of the factors and the corresponding deposition velocities it ensues, however, that high washout factors correlate with high deposition velocities.

The correlation is in agreement with the findings of Figure 3.11 and indicates that wet removal and dry deposition are more effective for metals attached to or incorporated into larger particles. In contrast, wet removal plays the dominant role in removing submicron particles from the atmosphere whereas dry depos-

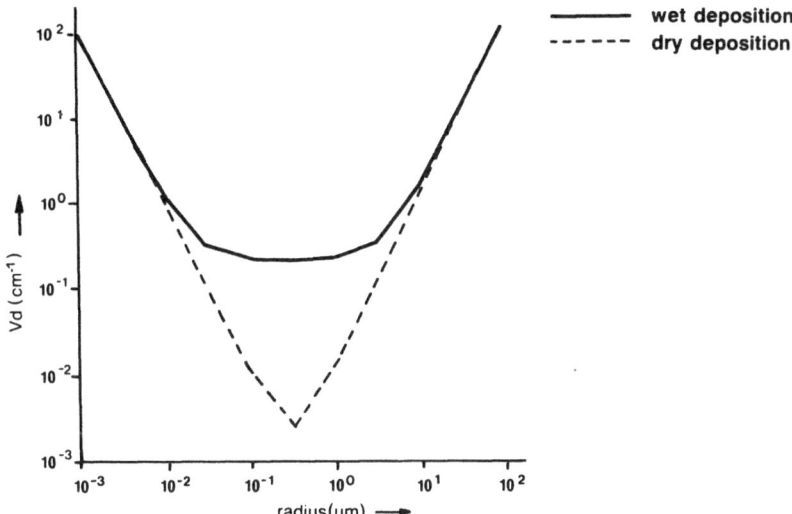

Fig. 3.11. Correlation of deposition velocity with particle size. (Jaenicke, VDI-Berichte Nr. 500, 1983)

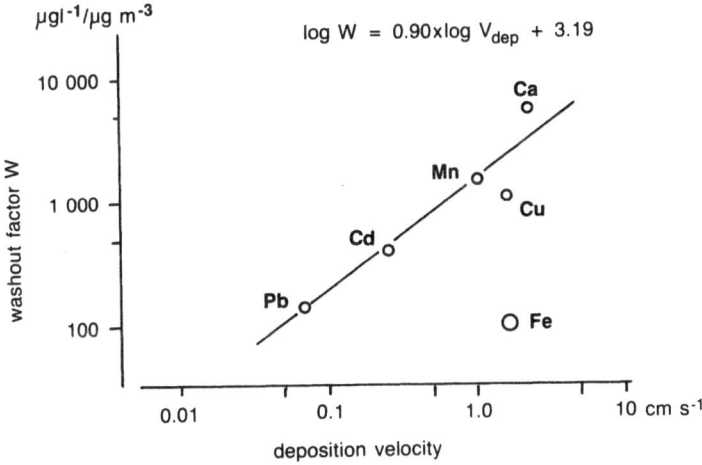

Fig. 3.12. Comparison of washout factors with deposition velocities. (After Rohbock 1982)

ition is negligible (cf. Sects. 3.5.5 and 3.5.7). Owing to their considerably enhanced specific surface (cf. Nakaya 1954), the correspondingly much reduced terminal velocity and, finally, the normally much longer way through the atmosphere snow crystals have distinctly better scavenging properties than water droplets. Gosch (1989), for instance, compared trace element concentrations in snow with those in rainwater at the Feldberg observatory (Taunus, Germany) and found always significant differences. Nguyen et al. (1979) provided comparable data for toxic metals.

Table 3.15. Washout factors of various metals. (After Rohbock 1982)

Element	Washout factor[a]
Pb	140
Cd	400
Mn	1500
Fe	70
Cu	1200
Ca	1800

[a] Ratio of concentration in rain ($\mu g \, l^{-1}$) to concentration in air ($\mu g \, m^{-3}$)

3.4 Chemical Transformation Processes

The terrestrial atmosphere, and in particular the troposphere, provide an excellent matrix for photochemical or oxidative alterations of pollutants which essentially contribute to enhance the effectiveness of physical removal processes. Solar radiation (cf. the pathways of ESI in Model I) is able to break C–C and C–H bonds, cause the photodissociation (PHOTO) of NO_2 to NO and atomic oxygen, and produce photolytically considerable amounts of hydroxyl radicals. The high concentration of oxygen (Table 3.1) makes this element one of the most important participants in various reactions with contaminants since the rates are concentration-dependent. A similar reasoning applies to reactions with water and carbon dioxide (summarily labelled Z in foldout model I).

The main transformations leading to the removal of chemicals from the atmosphere involve photolysis (direct gas phase and adsorbed on dispersed particulate matter), homogeneous (gas phase) and heterogeneous reactions (on aerosol surfaces). Direct phototransformation, i.e. transformation resulting from direct photo-excitation of the molecule, is a possible removal pathway only for those chemicals which absorb in the region of solar radiation ($\lambda \geq 290$ nm). It is therefore limited to a few pollutants (NO_2, O_3, SO_2, HNO_2, HNO_3, CH_2O, alkylnitrites, alkylnitrates) but hitherto data have been insufficient to allow an assessment of its relative importance compared to the removal by photochemically produced radicals (Zellner 1980; ECETOC 1983).

Chemicals may be sorbed onto aerosol particles from the vapour phase (Sects. 3.3.1 and 3.3.3), and thence removed from the atmosphere with the aerosol. Phototransformation of chemicals in an aerosol is different from that in air, since its adsorbed or absorbed state and the physicochemical properties of the aerosol substrate will influence the heterogeneous reactions (Jaeschke 1986). Therefore it is still questionable whether, in practice, the phototransformation of

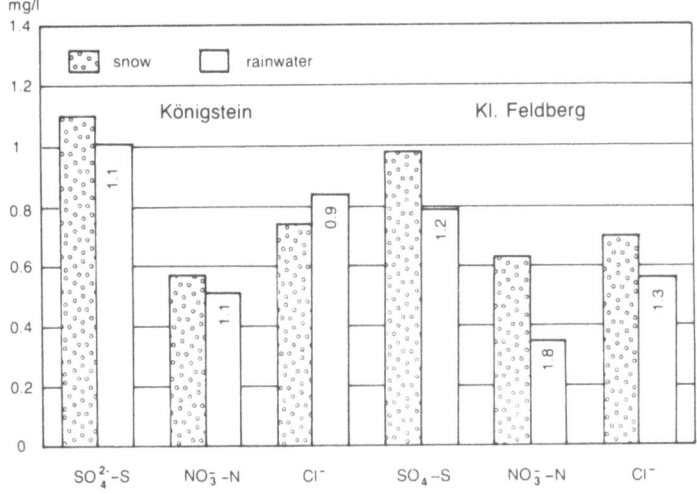

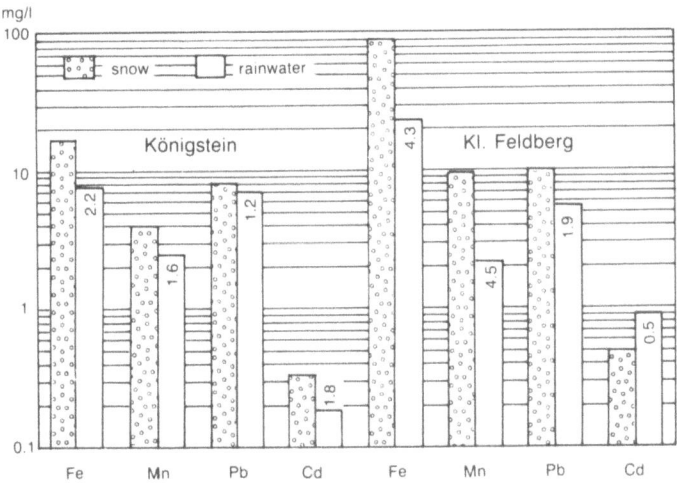

Fig. 3.13. Comparison of trace element concentrations in snow and rainwater at two Taunus stations during the November 1986–April 1987 period. Figures in *open columns* indicate concentration ratios snow/rainwater. (After Gosch 1989)

chemicals in aerosols contributes to the overall removal of substances of low volatility (Cupitt 1980; Heicklen 1981a).

Undoubtedly photochemical activation plays the major role among the homogeneous gas reactions, and consequently the present section is deliberately focussed on their description. In this connection it should be kept in mind that the probability that a compound occurs in the gas phase depends not only on its vapour pressure but also on its water solubility and adsorption/desorption behaviour (cf. the regulators A E, KoN and LO in Model I). Therefore even substances of a relatively low vapour pressure, i.e. down to 10^{-3} Pa, can be found in the atmosphere in measurable quantities (Giam et al. 1980; Eisenreich et al. 1981).

3.4.1 Formation of Free Radicals and Reactive Trace Constituents

Direct or indirect photochemical activation is the basis for all chemical reactions in the atmosphere. The quantum yield of direct phototransformation of a molecule in the gas phase depends essentially on the excitation wavelength. When a polyatomic molecule absorbs quasi-monochromatic radiation, a vibrational excited level of an electronically excited state, e.g. the lowest excited singlet state is populated. From this specific level many events can occur, for instance: a very fast monomolecular transformation into fragments, radicals or an isomer; a vibrational intra- or intermolecular relaxation or a bimolecular process with a reactive molecule like oxygen (photooxidation) or with the primary photoproducts (ECETOC 1983). Provided they are fast enough, all these processes compete on each vibrational level of the excited state, and the quantum yield of phototransformation is a function of:

- the wavelength of the excitation light absorbed,
- the total pressure of the gas phase,
- the temperature,
- the nature of the inert gas, and
- the presence of reactive molecules.

In the light of these interrelationships, direct phototransformation of pollutants is likely to be highly variable in space and time, but data are very scarce as yet. Pitts et al. (1981) have suggested a calculation procedure for determining maximum photolysis rates which is based on the Beer–Lambert law.

The most significant phototransformations for the removal of chemicals in the atmosphere are, in order of importance, reaction with OH and O_3 (cf. Table 3.16). Other conversions are initiated by free radicals such as H, O, $O(^1D)$, $H\dot{O}_2$ and reactive trace constituents like $O_2(^1\Delta_g)$, $N\dot{O}_3$ which derive from photochemical precursors.

In view of these reactions, which will be described in greater detail with respect to sulphate, nitrate and selected organics, the removal rate R_A of a pollutant A in the homogeneous gas phase may be defined by

$$R_A = \frac{d \ln [A]}{dt} = \sum_{i=1}^{n} k_i [X_i] , \qquad (3.15)$$

with X_i = potential reactants (i.e. OH, O, HO_2, O_3, ...) and k_i = corresponding rate coefficients. Provided that concentrations of X_i are time-independent

Table 3.16. Formation processes and concentration ranges of free radicals and reactive trace compounds in the troposphere. (After Zellner 1980)

Species	Formation process	Concentration [molecules/cm³] max.[a]	yearly average
H	$CH_2O + hv \rightarrow CHO + H$ $OH + CO \rightarrow CO_2 + H$		$< 10^{-1}$
O	$NO_2 + hv \rightarrow NO + O$	10^5	$\ll 10^5$
$O(^1D)$	$O_3 + hv \rightarrow O_2^* + O(^1D)$		$< 10^{-1}$
OH	$O(^1D) + H_2O \rightarrow 2OH$ $HONO + hv \rightarrow OH + NO$ $HO_2 + NO \rightarrow OH + NO_2$	5×10^6	5×10^5
HO_2	$H + O_2 + M \rightarrow HO_2 + M$	5×10^8	10^8
O_3	$O + O_2 + M \rightarrow O_3 + M$	5×10^{12}	10^{12}
$O_2(^1\Delta_g)$	$O_3 + hv \rightarrow O + O_2(^1\Delta_g)$ $NO_2^* + O_2 \rightarrow O_2^*(^1\Delta_g) + NO_2$		$< 10^7$

[a] The maximum concentration refers to the conditions of photochemical smog.

partial (τ_{Ai}) and total life-times (τ_A) may be defined as:

$$\tau_{Ai} = (k_i[X_i])^{-1} \tag{3.16}$$

or

$$\tau_A = \sum \tau_{Ai}, \text{ respectively.} \tag{3.17}$$

It must be kept in mind, however, that the underlying assumption of a constancy of X_i concentration hardly applies in reality. In addition there may be emissions of A, and in this case the concentration of the pollutant A enters the steady-state equation for X_i:

$$[X_i]_{ss} = p_i \Big/ \sum_{i=1}^{n} k_i[M_i] . \tag{3.18}$$

As a consequence $[X_i]$ may be drastically reduced and with it also the rate of active removal of the pollutant. Furthermore also the rate coefficients k_i are, as a rule, temperature-dependent. Due to the diurnal and seasonal changes of the tropospheric lapse rates (cf. Sect. 3.2.2) k_i and therefore τ_i change considerably with altitude, and the net tropospheric life-time thus becomes (Zellner 1980):

$$\tau_i^{-1} = [X_i] \cdot k_i^0 \left[1 - \exp\left(-\frac{E_i}{R} \alpha Z_T \right) \right] \Big/ \left(\frac{E_i}{R} \alpha Z_T \right), \tag{3.19}$$

with k_i^0 = rate coefficients at $T_0 = 296$ K, E_i = activation energy of the respective reactions, and $\alpha = l/T_0^2$ where l = atmospheric lapse rate, and Z_T = average height of the troposphere.

3.4.2 Sulphate Reactions

There are more than a dozen sulphate-forming processes that are relevant to atmospheric pollution. They have been reviewed by several investigators, including Hidy and Burton (1980), and can be grouped in terms of homogeneous gas-phase reactions and heterogeneous reactions involving suspended particles.

Table 3.17 summarizes the existing knowledge breaking down the whole set of exothermic homogeneous reactions into three subcategories whose end products are SO_3, RO_2SO_2, and $ROSO_2$.

The first five reactions listed have been considered severely rate-limited on the basis of available data, but the remaining ones appear to have rates which are sufficiently rapid to be of importance in polluted air with active photochemical processes. Reactions (7) and (8) indicate that SO_2 may be oxidized at

Table 3.17. Estimated rates of theoretically possible homogeneous removal paths for SO_2 in a simulated polluted atmosphere. (After Hidy and Burton 1980)

Reaction	ΔH_{298} (kcal/mol)	Approximate rate (%/h)
A Inorganic reactions forming SO_3		
1. $SO_2 + (1/2)O_2 + sunlight \rightarrow SO_3$	-24	< 0.021
2. $O(3P) + SO_2 + \underline{M} \rightarrow SO_3 + \underline{M}$	-83	0.014
3. $O_3 + SO_2 \rightarrow SO_3 + O_2$	-58	0.00
4. $NO_2 + SO_2 \rightarrow SO_3 + NO$	-10	0.00
5. $NO_3 + SO_2 \rightarrow SO_3 + NO_2$	-33	0.00
6. $N_2O_5 + SO_2 \rightarrow SO_3 + N_2O_4$	-24	0.00
B Organic reactions forming SO_3		
7. $CH_2\overset{O_3}{\overset{/\backslash}{—}}CH_2 + SO_2 \rightarrow SO_3 + 2CH_2O$	-81	$< 0.4–3.0$
8. $\cdot CH_2OO\cdot + SO_2 \rightarrow SO_3 + CH_2O$	~ -117	$< 0.4–3.0$
$\quad CH_2{=}O \rightarrow O + SO_2 \rightarrow SO_3 + CH_2O$	~ -85	
9. $HO_2 + SO_2 \rightarrow HO + SO_3$ (a)	-19	0.85
$\quad \rightarrow HO_2SO_2'$ (b)	< -25	?
10. $CH_3O_2 + SO_2 \rightarrow CH_3O + SO_3$ (a)	-30	~ 0.16
$\quad \rightarrow CH_3O_2SO_2'$ (b)	< -25	?
C Reactions forming $HOSO_2$ or $ROSO_2$ radical		
11. $HO + SO_2 \rightarrow HOSO_2'$	~ -82	$\sim 0.23–1.4$
12. $CH_3O + SO_2 \rightarrow CH_3OSO_2'$	~ -73	~ 0.48

Total potential rate of conversion of SO_2 to SO_3 (or sulphates) in moderate smog $\cong 1.7$ to $5.5\%/h$.

appreciable rates in the dark in ozone-olefin-air mixtures. Radical addition reactions are exemplified by the series (9) to (12). Their products, such as $HOSO_2$, should react rapidly with other species to generate sulphuric acid, peroxysulphuric acid, alkylsulphates, and mixed intermediates, e.g. $HOSO_2ONO_2$. Any of these eventually should lead to sulphate in the presence of water. The theoretical rate of sulphate production as rationalized in the range 1.7–5.5%/h, however, requires unstable intermediates at relatively high concentrations. In non-urban air or in cities with minimal photochemical activity, this condition is certainly not always met.

In the cases where photooxidative homogeneous reactions cannot be important, the hetereogeneous processes must be considered. They are listed in Table 3.18 and categorized as aqueous and nonaqueous. The class of reactions most frequently invoked to explain high SO_2 rates in the presence of liquid water containing aerosols is that of series (13). The catalytic activity of heavy metal salts such as Mn^{2+} ion can realize oxidation rates in excess of 1%/h in clean water solutions.

The absorption of SO_2 can be considerably promoted by the buffering effect of the simultaneous extremely rapid absorption of ammonia. Since the ammonium ion also enhances the aqueous SO_2 oxidation, including the intermediates $SO_2NH_2^-$ and $SO_3NH_2^-$, rates approaching 10%/h are achieved in fog. This process can attain particular importance in The Netherlands, Belgium and Northern Germany and parts of the USA, where highly intensified stockbreeding yields excessive amounts of manure which in turn release huge quantities of ammonia. The resultant wet and dry deposition of large amounts of ammonium sulphate on vegetation and soil may cause severe forest damage, while subsequent nitrification of the ammonium leads to a marked pH decrease in part of the soils affected (Tjepkema and Cartica 1981; Adema and van Ham 1984; Roelofs et al. 1985). While these and related effects are described in greater detail

Table 3.18. Heterogeneous sulphate-forming reactions. (After Hidy and Burton 1980)

A Aqueous

13. $SO_2 + H_2O(\ell) \rightleftarrows H_2SO_3$
$$H_2SO_3 \rightleftarrows H^+ + HSO_3^-$$
$$HSO_3^- \rightleftarrows H^+ + SO_3^=$$
13A. $2SO_3^= + O_2\,(aq) \rightarrow 2SO_4^=$
13B. $SO_3^= + O_3\,(aq) \rightarrow SO_4^= + O_2$

B Non-aqueous

14. $\left.\begin{array}{l} SO_2\ (ads) \\ H_2O\ (ads) \\ O_2\ (ads) \end{array}\right\} + carbon(s) \rightarrow H_2SO_4\ (ads)$

Table 3.19. Ammonia emissions in selected parts of Europe
(t/year and percent of European total)

Federal Republic of Germany	371,000	5.8
DDR	207,000	3.2
Denmark	117,000	1.7
Netherlands	150,000	2.3
USSR (European part)	1,256,000	19.5

in Chaps. 4.2.1, 4.2.2 and 4.4.1, a tabular synopsis (Table 3.19) may indicate the order of magnitude of ammonia emissions which are at least to a considerable extent, if not mostly, due to massive stockbreeding.

While ammonia has proved to be an outstanding and widespread promoter of SO_2 oxidation, organic contaminants have been shown to significantly suppress such aqueous reactions. For instance organic acids or alcohols may well reduce the aqueous absorption of SO_2 and its subsequent oxidation by as much as an order of magnitude. Hence it may be surmised that the aqueous oxidation will probably be most efficient in clean clouds.

In contrast to the homogeneous gas-phase reactions and the above heterogeneous mechanisms of SO_2 oxidation those occurring in the absence of liquid water are only poorly understood as yet. However, Novakov et al. (1974) have shown that significant amounts of SO_4^{2-} can be found on carbon particles generated by combustion of hydrocarbons in ppm-level SO_2-enriched air. The experiments of Yamamoto et al. (1973, cited in Hidy and Burton 1980) indicate that SO_2 oxidation can be as high as 30%/h on activated charcoal particles ~ 5 mm in diameter. It must be emphasized, however, that the mechanism involved obviously depends on a variety of factors, ranging from grain size of the carbon to temperature and to concentrations of SO_2, water vapour, and oxygen, as well as to the micropore structure of the particle surface, i.e. their specific surface. On the one hand, these findings may well account for a significant enhancement of the SO_4^{2-}-concentration in snow, which is frequently enriched in organic matter. On the other hand, it would seem that oily, gummy, wet particles collected from the atmosphere are only poorly suited for the non-aqueous reactions to form sulphate since their specific surface is but minimal.

3.4.3 Nitrate-Forming Reactions

Like sulphates, also nitrates in atmospheric aerosols can be formed by a variety of homogeneous and heterogeneous reactions. As ensues from Table 3.20, mixed intermediates involving sulphuric acid occur which are of particular environmental interest since they link the SO_x and NO_x chemistry.

Since nitrous acid and nitric acid are much more volatile than sulphuric acid, it does not appear possible that particulate NO_x species can exist in the

Table 3.20. Reactions potentially involved in nitrate formation. (After Hidy and Burton 1980)

Species	$d[NO_2]/dt$, or rate constant (ppm/min)[a]
A Nitrogen oxides	
15. $O_3 + NO \rightarrow NO_2 + O_2$	2.7×10^{-2}
16. $O + M + NO \rightarrow NO_2 + M$	—
17. $RO_2 + NO \rightarrow NO_2 + RO$	2.5×10^{-3}
18. $O_3 + NO_2 \rightarrow NO_3 + O_2$	-4.0×10^{-4}
19. $NO_3 + NO_2 \rightleftarrows N_2O_5$	-1.0 to -23×10^{-4}
B Volatile acids	
20. $N_2O_5 + H_2O \rightarrow 2HONO_2$	2×10^{-5}
21. $HO + NO_2 + M \rightarrow HONO_2 + M$	-10^{-5} ($M = 1$ atm N_2)
22. $NO + NO_2 + H_2O \rightarrow 2HONO$	—
23. $HOSO_2O + NO \rightarrow HOSO_2\,ONO$ $+ H_2O \rightarrow H_2SO_4 + HONO$	—
24. $HOSO_2O + NO_2 \rightarrow HOSO_2\,ONO_2$ $+ H_2O \rightarrow H_2SO_4 + HONO_2$	—
C Gaseous nitrates	
25. $NH_3 + HONO_2 \rightarrow NH_4NO_3$	$\sim 10^{-6}$
26. $RO_2 + N_2O_5 \rightarrow R^1C\overset{O}{\underset{ONO_2}{\diagdown}}$	-10^{-3}
$NO_2 + R^1C\overset{O}{\underset{ONO}{\diagdown}} + \ldots$	

[a] Typical for smog reactant concentrations in the first hour of reaction.

atmosphere in pure form as acids. Nitrate formation consequently involves production of condensable species such as NH_4NO_3 in the gas phase or heterogeneous processes in aqueous or non-aqueous medium, e.g. absorption of a nitrogen oxide followed by stabilization through chemical reaction.

Classical in terms of (A) important NO_x species, (B) volatile acids HONO and $HONO_2$, and (C) gaseous nitrates, the precursors for particulate nitrate formation are summarized in Table 3.20.

Reactions (20) and (21) are currently believed to be most important. The rate of conversion of NO_2 and $HONO_2$ in the latter reaction should be in the range of 2 to 8%/h, and with absorption in wet aerosols and neutralization with ammonium ion, this value may well account for the morning peak condition of

Table 3.21. Aqueous reactions of nitrogen oxides.
(After Hidy and Burton 1980)

27. $N_2O_5 + H_2O(\ell) \rightarrow 2H^+ + NO_3^-$
28. $NO + NO_2 + H_2O(\ell) \rightarrow H^+ + NO_2^-$
29. $2NO_2 + H_2O(\ell) \rightleftarrows H^+ + NO_3^- + HONO$
    $HONO + OH^- \rightarrow H_2O + NO_2^-$
29A. $2NO_2^- + O_2 \text{ (aq)} \rightarrow 2NO_3^-$
29B. $NO_2^- + O_3 \text{ (aq)} \rightarrow NO_3^- + O_2$
30. $2NO_2 + H_2SO_4 \rightleftarrows HONOSO_4 + HNO_3$
    $HONOSO_4 + H_2O(\ell) \rightleftarrows HNO_2 + H_2SO_4$
    $3HNO_2 \rightleftarrows HNO_3 + 2NO + H_2O$
31. $RONO_2 + H_2O(\ell) \rightarrow H^+NO_2^- + R^1OH$

Los Angeles smog (Hidy and Burton 1980). In contrast, reaction (26) has been hypothesized only on the basis of an analogy to the NH_3–HCl reaction.

The complexities of the aqueous reactions of NO_x become apparent in Table 3.21. It shows that all of the nitrogen oxides contained in the atmosphere will react with liquid water, forming traces of nitrite and nitrate. Since these reactions are generally reversible, however, they must be stabilized by bases such as NH_4^+.

Reaction (30) describing nitrate formation via absorption of NO_2 to form nitrosylsulphuric acid, may be of interest if significant quantities of concentrated sulphuric acid are formed in atmospheric aerosols. Although the rate of NO_2 absorption in sulphuric acid is fast, it appears that a higher efficiency is linked to the absorption of ammonia or another base, which drives the equilibria to nitrate formation. Consequently, in this, as in all other of these aqueous reactions, nitrate production may be much more limited by the concentration of ammonia than by any of the NO_x species (cf. in particular Schneider and Grant 1982) .

3.4.4 Oxidative Chemistry of Organic Pollutants

During the past decades, certain aspects of the reaction mechanisms and products of OH radicals and O_3 with simpler organics have become better known from a combination of laboratory, environmental chamber and computer modelling studies. Under atmospheric conditions, i.e. in the framework of homogeneous gas-phase removal processes, the OH radical reacts with all organics except for perhalogenated compounds, while O_3 reacts primarily with alkenes and substituted alkenes (Herron et al. 1979; Atkinson and Pitts 1980).

In contrast to these major oxidants and NO_3^-, other reactive species like O (^3P), O (^1D) and O_2 ($^1\Delta_2$) are of comparatively little importance. A comparison of the apparent first-order rate constants for their reactions with propene, butane and toluene shows that the O (^3P) atom reaction rate is approximately 100 times slower than the OH and ozone reaction rates, while reaction rates for O (^1D) and O_2 ($^1\Delta_2$) are even 10^6–10^7 times slower (Hendry and Mill 1980; Mill et al. 1981).

Therefore the following description focuses on OH and O_3 processes and the elucidation of the subsequent reaction pathways. It is important to recognize the products and mechanisms of the atmospheric degradation, since it is obviously possible that the degradation products (NS in foldout Model I) may be more toxic, carcinogenic or mutagenic than the intially emitted parent compound (i.e. I, χ_1).

3.4.4.1 Atmospheric Reaction Pathways for Alkanes

Under ambient air conditions, the sole chemical removal process of alkanes is via reaction with the OH radical, i.e. hydrogen abstraction from a carbon–hydrogen bond, which leads to the formation of substituted carbon radical and water:

$$OH + R \cdot CH_2 \cdot R_1 \;\rightarrow\; R\dot{C}H \cdot R_1 + H_2O \; . \tag{3.20}$$

For the simple ($< C_3$) alkanes a series of reactions follows:

$$R\dot{C}HR_1 + O_2 \;\rightarrow\; RR_1CHOO\cdot$$

$$RR_1CHOO\cdot + NO \;\rightarrow\; RR_1CHO\cdot + NO_2$$

$$RR_1CHOO\cdot + NO_2 \;\rightleftarrows\; RR_1CHO_2 + NO_2$$

$$RR_1CHO\cdot + O_2 \;\rightarrow\; RCOR_1 + HO_2 \;\text{(reaction with } O_2\text{)}$$

$$RR_1CHO\cdot \begin{cases} \rightarrow RCHO + R_1\cdot \\ \rightarrow R_1CHO + R\cdot \end{cases} \quad \text{(decomposition)}$$

$$RR_1CHO\cdot + NO_2 \;\rightarrow\; RR_1CHONO_2 \; .$$

Finally the chain carrier, i.e. the OH radical, is regenerated:

$$HO_2 + NO \;\rightarrow\; OH + NO_2 \; . \tag{3.21}$$

This reaction scheme with intermediate formation of carbonyl compounds which in turn react with OH radicals or photodecompose further is only applicable to methane, ethane and propane. Two other processes involving RO_2 and RO reactions can occur for the larger alkane systems: alkyl nitrate formation, and alkoxy radical isomerization (Atkinson and Pitts 1980).

With regard to alkyl nitrate formation, the following addition reaction pathway was suggested (Darnall et al. 1976a):

$$RO_2 + NO \longrightarrow \begin{cases} RO + NO_2 \\ RO_2NO* \end{cases}$$

$$RO_2NO* \rightleftarrows RONO_2^*$$

$$RONO_2^* \rightarrow RO + NO_2 \text{ decomposition}$$

$$RONO_2^* \xrightarrow{M} RONO_2 \text{ stabilization.}$$

Since this reaction sequence is a substantial sink for NO_x and also a radical termination process, its relative importance must be assessed experimentally.

Thermochemical calculations by Carter et al. (1976) and Baldwin et al. (1977) have shown that under ambient air conditions the isomerization of alkoxy radicals, i.e. 1.4- and 1.5-H shifts via five- and six-membered ring transition states, can be of importance. For $>C_4$ alkanes, this means that alkoxy radicals formed during their NO_x photooxidation can react via the following four routes:

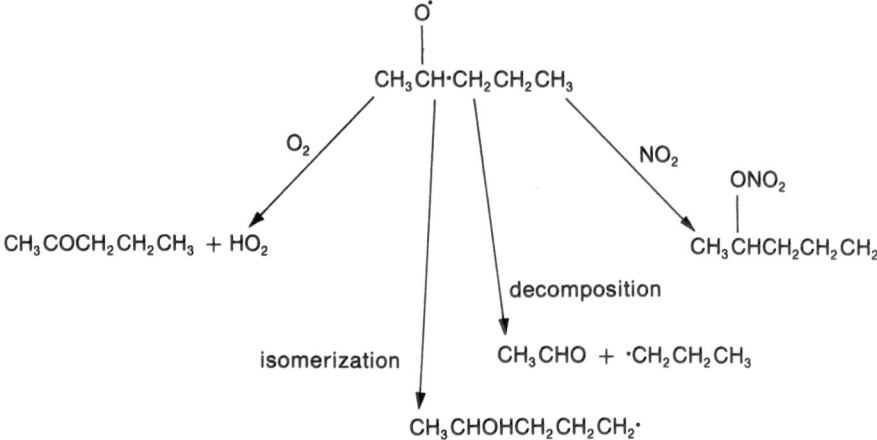

Thus, in addition to isomerization, combination with NO_2 to form the above-mentioned alkyl nitrates, reaction with O_2, and unimolecular decomposition may occur (Atkinson and Pitts 1980). It was found that for the higher alkanes such as n-pentane, 70% of the total number of alkoxy radicals formed can undergo isomerization, and this process becomes increasingly important for the higher straight chain alkanes (Carter et al. 1976).

3.4.4.2 Atmospheric Reaction Pathways for Alkenes

In the atmosphere, alkenes react predominantly with ozone and OH radicals. Ethenes and methyl-substituted ethenes have as sole initial reaction pathway the OH radical addition to the double bond. Alkenes with longer side chains ($>C_2$) also undergo H atom abstraction from the weak allylic C–H bonds (Hendry and

Mill 1980). According to smog chamber studies and experiments with irradiated HONO–alkene–NO–air mixtures (Niki et al. 1978), the β-hydroxyalkoxy radicals formed from the OH–alkene adducts seem to decompose rather than react with O_2.

Ozone cleaves the double bond via the Criegee route forming a carbonyl and an initially energy-rich oxygen bi-radical ($>C\dot{O}\dot{O}$). A significant fraction, i.e. $\sim$ 20–40% of the latter species are thermalized under atmospheric conditions, with rearrangement of decomposition accounting for the other reactions of the biradicals (Atkinson and Pitts 1980). Among these, reactions with aldehydes to form secondary ozonides, with SO_2 to form SO_3 and the corresponding aldehydes, and finally with H_2O producing organic acids merit special attention.

$$R\dot{C}HO\dot{O} + R_1CHO \rightarrow$$

$$R\dot{C}HO\dot{O} + SO_2 \rightarrow products$$

$$R\dot{C}HO\dot{O} + H_2O \rightarrow RCOOH + H_2O$$

Reactions involving high-molecular-mass alkenes are known to produce polymerized oxygenated species condensing out at ppb vapour concentrations (Hidy and Burton 1980). Such reactions are likely to be promoted on the surfaces of larger particles which act as preconcentrators for the alkenes or air-intermediate species (Guderian 1985). Thus organic aerosols can also be formed readily in nonurban air from reactions between "background" ozone and hydrocarbon vapours emanated in significant quantities from the vegetation (cf. Table 3.2).

3.4.4.3 Atmospheric Chemistry of Aromatics

Aromatic rings react largely by addition of OH although abstraction of benzylic hydrogens will normally compete. Atkinson et al. (1979) provided data for both the overall OH radical rate constants and the relative amounts of the two reaction pathways indicated and labelled (a) and (b) in the following examplary formula:

In the atmosphere the benzyl radical will undergo further reactions leading to benzaldehyde and benzyl nitrate, which reactions are analogous to the relatively well-known alkyl radical reactions.

The aromatic OH adducts can in turn react with O_2 via H atom abstraction to form cresols or via reversible O_2 addition to the adduct.

The prompt formation of biacetyl and other α-dicarbonyls in the NO_x photooxidation of o-xylene is indicative of the formation of bicyclic radicals of the following type:

Form (C) can add O_2 and react with NO

and this reaction is followed by a spontaneous decomposition of the bicyclic oxy radicals

$$CHOCH=CHCHCHOHCOCH_3$$

$$\downarrow$$

$$CHOCH=CHCHO + CH_3CO\dot{C}HOH$$

$$\downarrow O_2$$

$$HO_2 + CH_3COCHO$$

In conclusion, Table 3.22 may provide a summary review of photo-transformation rates of selected and frequently chlorinated hydrocarbons determined under simulated atmospheric conditions in the presence of NO and NO_2. In addition to the preceding compilation, Table 3.23 provides a comparative reactivity classification of hydrocarbons. It should be pointed out that this table, like several others (e.g. Altshuller and Cohen 1963 or Kopczinski 1964), have only inherent consistence while, in comparison, multivariate classifications of chemicals based on PC potentials may prove to be quite different (Fränzle 1982).

Table 3.22. Photodegradation of selected hydrocarbons under simulated atmospheric conditions. (After Dilling et al. 1975)

Compound[a]	Time (h) for 50% disappearance of compound	
	With NO[b]	With NO_2[c]
CH_3CCl_3	d	a
CH_2Cl_2	f	g
t-Butyl alcohol	34.5[h]	
Epichlorohydrin	16.0	
1,1,2-Trichloroethane	15.9[i]	
1-Butene oxide (c-CHEtCH$_2$O)	15.9[j]	15.2[k]
Ethyl acetate (EtOAc)	14.6[l]	
$CCl_2=CCl_2$	14.2[m]	8.3[n]
Methyl ethyl ketone (MeCOEt)	9.8[o]	
Nitromethane	9.2	
Chlorobenzene	8.7[p]	
Cyclohexane (c-C_6H_{12})	6.9	6.5, 7.5[k]
Toluene (PhMe)	6.8	
n-Butyl alcohol (n-BuOH)	6.5	
Ethyl benzene	5.0	
Trioxane	4.7	
Vinyl chloride ($CH_2=CHCl$)	4.3	
s-Butyl alcohol	4.0	
$CHCl=CCl_2$	3.5	2.9

Table 3.22. (Contd.)

Compound[a]	Time (h) for 50% disappearance of compound	
	With NO[b]	With NO$_2$[c]
Methyl isobutyl ketone (MeCOi-Bu)	3.5	
Isobutyl alcohol (i-BuOH)	3.5	
Dioxane [c-O(C$_2$H$_4$)$_2$O]	3.4	
1-Methoxy-2-propanol (MeOCH$_2$CHOHMe)	3.1	
p-Xylene[q] (p-Me$_2$C$_6$H$_4$)	3.1	
m-Xylene[q] (m-Me$_2$C$_6$H$_4$)	2.9	
cis-Dichloroethylene (cis-CHCl=CHCl)	3.0	3.0
$trans$-Dichloroethylene	2.9	2.8
Ethylene (CH$_2$=CH$_2$)	2.9	2.5
Vinylidene chloride (CH$_2$ = Cl$_2$)	2.1	
Cyclohexene (c-C$_6$H$_{10}$)	0.87	0.19[k]
2,4,4-Trimethyl-1-pentene	0.60	
$trans$-2-Butene ($trans$-MeCH=CHMe)	0.30	0.17[k]
N-Methylpyrrole	0.16	

[a] All compounds at 10 ppm in air initially, unless specified otherwise. [b] All initial concentrations, 5 ppm. [c] All initial concentrations, 16.8 ppm, except as noted. [d] Reaction, < 5%, in 23.5 h. [e] Reaction, < 5%, in 8.0 h. With 50-ppm CH$_3$CCl$_3$, and 10-ppm NO$_2$: < 5% reaction in 28 days. [f] Reaction, < 5%, in 21.0 h. [g] Reaction, < 5%, 7.5 h. With 50-ppm CH$_2$Cl$_2$ and 10-ppm NO$_2$: < 5% reaction in 4 days.
Anomalous reactions were observed at longer times; these reactions are currently under study. [h] Extrapolated from 22% reaction in 15.2 h. [i] Extrapolated from 22% reaction in 7.0 h. [j] Extrapolated from 47% reaction in 14.9 h. [k] Initial NO$_2$ concentration, 5 ppm. [l] Extrapolated from 24% reaction in 7.0 h. [m] Extrapolated from 44% reaction in 12.5 h. [n] Extrapolated from 42% reaction in 7.0 h. [o] Extrapolated from 33% reaction in 6.5 h. [p] Extrapolated from 43% reaction in 7.5 h. [q] p-Me$_2$C$_6$H$_4$ and m-Me$_2$C$_6$H$_4$ measured simultaneously at initial concentrations of 2 and 8 ppm respectively.

Conclusions

It is broadly agreed that reactions with OḢ are the dominant and generally the fastest photo-induced hydrocarbon reactions in the atmosphere. The only class of compounds which does not react with OḢ are the fully halogenated alkanes (see, e.g. Mill et al. 1981; Wuebbles and Cornell 1981). Only for some alkenes, alkadienes or terpenes does the rate of removal by O$_3$ exceed that by OḢ, the resultant life-time as derived from $k_{OḢ}$ is very short, i.e. less than a day. OḢ and O$_3$ molecules both add to the double bond of alkenes, forming an addition complex which afterwards disintegrates to the reaction products. If abstraction of an H atom by OḢ is the dominant transformation process, the competitive reaction with O$_3$ is expected to be much slower.

Table 3.23. Proposed reactivity classification of hydrocarbons and CO based reaction with hydroxyl carbons. (After Darnall et al. 1976b)

Compound	K_{OH} + Cpd $(l\,mol^{-1}\,s^{-1})$ $\times 10^{-9}$	Reactivity rel. to methane	Proposed class
Methane	0.0048	1	I
CO	0.084	18	II
Acetylene	0.11	23	II
Ethane	0.16	33	II
Benzene	0.85	180	III
Propane	1.3	270	III
n-Butane	1.8	375	III
Isopentane	2.0	420	III
Methyl ethyl ketone	2.1	440	III
2-Methylpentane	3.2	670	III
Toluene	3.6	750	III
n-Propylbenzene	3.7	770	III
Isopropylbenzene	3.7	770	III
Ethene	3.8	790	III
n-Hexane	3.8	790	III
3-Methylpentane	4.3	900	III
Ethylbenzene	4.8	1000	III–IV
p-Xylene	7.45	1530	IV
p-Ethyltoluene	7.8	1625	IV
o-Ethyltoluene	8.2	1710	IV
o-Xylene	8.4	1750	IV
Methyl isobutyl ketone	9.2	1920	IV
m-Ethyltoluene	11.7	2420	IV
m-Xylene	14.1	2920	IV
1,2,3-Trimethylbenzene	14.9	3100	IV
Propene	15.1	3150	IV
1,2,4-Trimethylbenzene	20	4170	IV
1,3,5-Trimethylbenzene	29.7	6190	IV
cis-2-Butene	32.3	6730	IV
β-Pinene	42	8750	IV
1,3-Butadiene	46.4	9670	IV–V
2-Methyl-2-butene	48	10000	V
2,3-Dimethyl-2-butene	67	14000	V
d-Limonene	90	18800	V

Reaction with ozone is, however, of much greater interest when processes related to smog formation with higher O_3 levels are involved. On the basis of a limited number of results, it may be inferred that reactions with HO_2 are of only secondary importance. There is some evidence that alkanes, phenols and cresols react with the NO_3 radical formed in photochemical smog (Carter et al.

1981). From these exploratory studies it may be concluded, however, that these NO$_3^-$ reactions apply only to special environmental stiutations, e.g. at night in a moderately polluted atmosphere (ECETOC 1983). The rate constants for reactions with singlet oxygen seem to be very low, and hence this reaction can be neglected as a removal pathway for organics (Heicklen 1976). In contrast, available data are as yet insufficient to allow an assessment of the relative importance of direct phototransformation as a possible elimination pathway of atmospheric chemicals.

3.5 Deposition

Processes of pollutant removal from the atmosphere may be conveniently classified under the headings of gravitational settling, dry deposition and wet deposition.

Gravitational settling is the most effective removal process for particulate matter greater than 1 µm in diameter. Even strong turbulence is unable to hold particles greater than 10 µm in suspension for long, and their deposition rate is a relatively simple function of their size and density on the one hand (Stokes' law), and the strength and turbulence of the wind on the other (cf. regulators SG, S$_T$G, $_T$A, $_T$AB and u in foldout Model I). Particles less than 1 µm in diameter are under the influence of turbulent diffusion which slows their descent such that they are in practice mostly removed by dry and wet deposition, or by settling after their aggregation into larger particles (cf. Fig. 3.10). Dry deposition is a turbulent transfer process comparable to that of heat, water vapour and momentum, and can appropriately be described by flux-gradient expressions (Schmidt 1925). Clearly the deposition process involves a downward flux to the underlying surface, which eventually acts as a pollutant sink. Thus deposition is strongly governed by the degree of forced and free convection, and highly sensitive to surface roughness and a number of further physical and chemical surface attributes. For instance, over vegetation, the actual adsorption rate is affected by stomatal aperture, over soil by moisture and bacterial or fungal activities, over water by the surface tension, and in all cases uptake may be further influenced by electrostatic attraction and a sometimes bewildering variety of chemical processes (cf. Sects. 4.2.1 and 4.2.2).

Wet deposition due to rainout or snowout and washout processes as described in Sect. 3.3.4 is the most effective cleansing process for gaseous and small particulate pollutants. The efficiency and rate of washout scavenging depends upon the rainfall or snowfall intensity, the sizes and electrical charges of the droplets or flakes and, of course, the physical and chemical characteristics of the pollutants. During rain events of longer duration, such as warm front precipitations, concentrations are frequently high at the beginning, then decrease rapidly to remain fairly constant thereafter. It should be noticed, however, that quite a few exceptions to the rule may occur in relation to specific local or regional synoptic conditions (Asman et al. 1982; Kins 1982). For convective showers, in contrast, rainfall intensity and pollutant concentrations show strong inverse relationships.

3.5.1 Deposition Networks

An assessment of deposition rates on a regional scale must be based on deposition networks which provide valid data for the identification of unequivocal spatial structures related to emission, transport and sedimentation. This normally involves the construction of isopleth maps which, in turn are usually constructed by means of metric interpolation. The underlying assumption is that of an approximately continuous distribution of the basic punctiform data, which, however, requires statistical verification. Unless the observed realm of influence of spatial data – in more technical terms, their spatial autocorrelation – is, in fact, sufficient to permit interpolation isopleth maps are not only of inferior quality but simply erroneous.

Variogram Analysis

A statistical method which permits testing the validity of interpolation procedures is variogram analysis, which is normally applied in conjunction with kriging methods (Matheron 1963; Journel and Huijbregts 1978; Verly et al. 1984). Although the latter set of procedures is most useful for the construction of unequivocal isopleth maps by means of weighted (moving average) interpolation, it need not be discussed in greater detail here. Variogram analysis is based on the observation that the variabilities of all regionalized variables have a particular structure. When considering two numerical values $z(x)$ and $z(x + h)$ at two points x and $x + h$ separated by the vector h, the variability between these two quantities is characterized by the variogram function $2\gamma(x, h)$ which is defined as the expectation of the random variable $[z(x) - z(x + h)]^2$, i.e.

$$2\gamma(x, h) = E\{[z(x) - z(x + h)]^2\} . \qquad (3.22)$$

In general, this variogram is a function of both the point x and the vector h, and basically the estimation of such a variogram would require several realizations of the pair of random variables. Normally, in practice, only one such realization $z(x)$, $z(x + h)$ is available, and this is the actual measured couple of values at points x and $x + h$. Introducing the intrinsic hypothesis that function (3.22) depends only on the separation factor h and not on the location x, it is possible to estimate the variogram $2\gamma(h)$ from the available data. The estimator $2\gamma^*(h)$ is the arithmetic mean of the squared differences between two experimental measures $[z(x_i), z(x_i + h)]$ at any two points separated by the vector h; i.e.

$$2\gamma^*(h) = \frac{1}{N(h)} \sum_{i=1}^{N(h)} [z(x_i) - z(x_i + h)]^2, \qquad (3.23)$$

where $N(h)$ is the number of experimental pairs $[z(x_i), z(x_i + h)]$ of data separated by the vector h.

In a given direction, the variogram – or the normally computed semi-variogram $\gamma(h)$ – may become stable beyond some distance, $|h| = a$, called the range, or it may attain a maximum. Beyond this critical distance a, the mean square deviation between two quantities $z(x)$ and $z(x + h)$ no longer depends on the distance $|h|$ between them, i.e. the two quantities are no longer correlated. Thus the range a gives a precise mathematical and physical meaning to the intuitive concept of the zone of influence of a sample $z(x)$. In terms of interpolation, this is equivalent to the statement that a mathematically and factually meaningful interpolation must be limited to quantities of measuring points whose distance is inferior to range a (Figs. 3.14, 3.15).

There is no reason for the range to be the same in all directions of space since the observed variability of a phenomenon is most often due to several, or many, causes ranging over various scales. The resultant directional variograms exhibit a characteristic anisotropy. In addition to an anisotropy of the structural

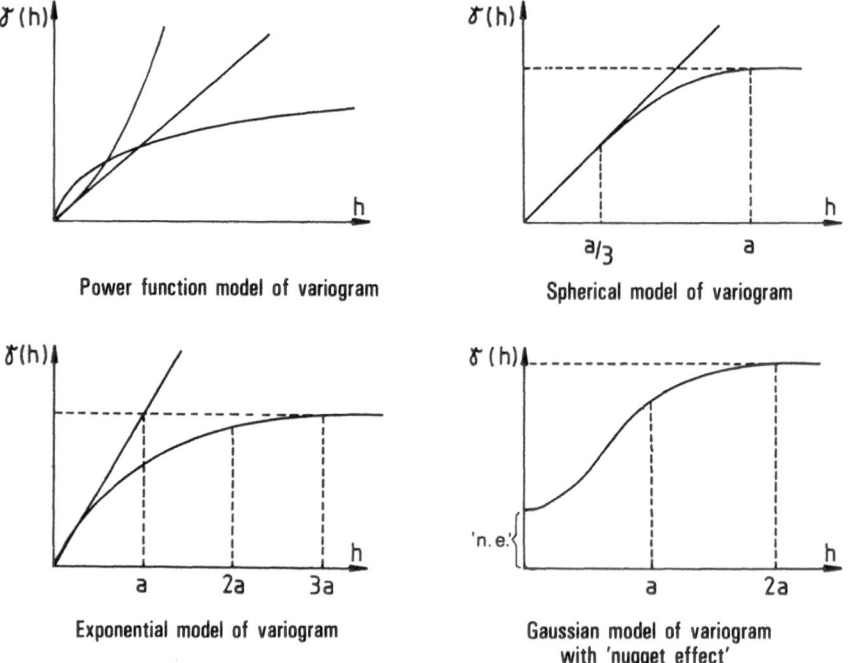

Fig. 3.14. Variogram models

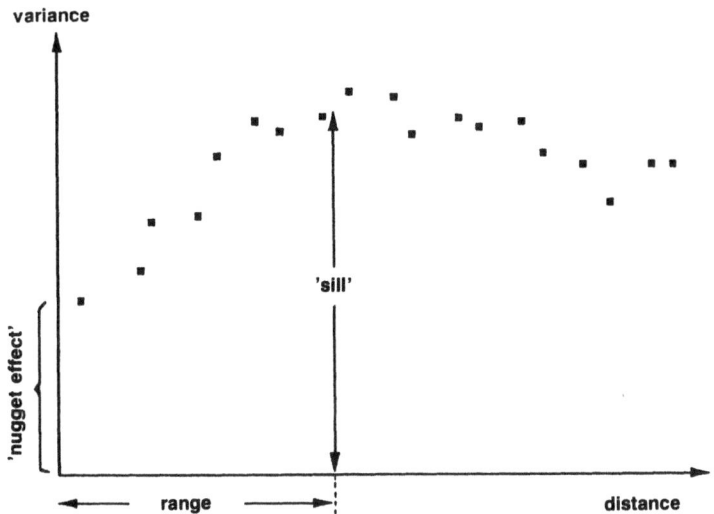

Fig. 3.15. Model of empirical variogram

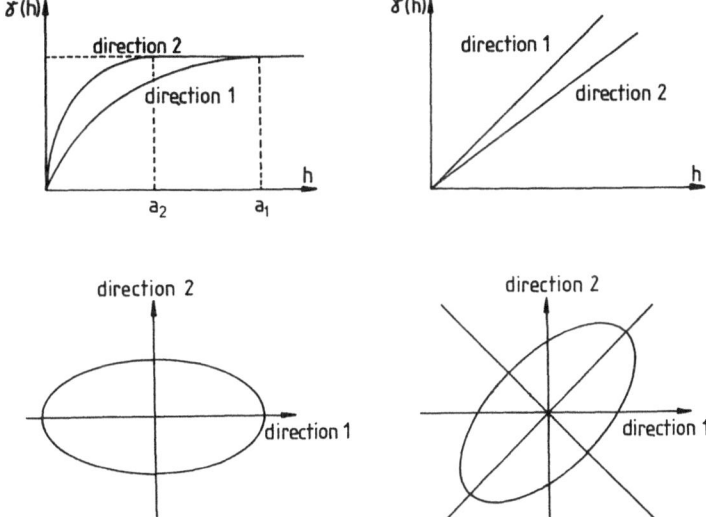

Fig. 3.16. Models of anisotropic variograms

function $\gamma(h)$ also nested structures may occur. Such a nested model $\gamma(h)$ then consists of the sum of a microstructure $\gamma_1(h)$ and a (or several) macrostructure(s) $\gamma_{2+n}(h)$. Again, there is no reason for these two or multi-component structures to have the same directions of anisotropy (cf. Figs. 3.16 and 3.17).

In practice, only a minor part of the sources or structures of variability which come into play simultaneously and for all distances, h, are reflected in nested

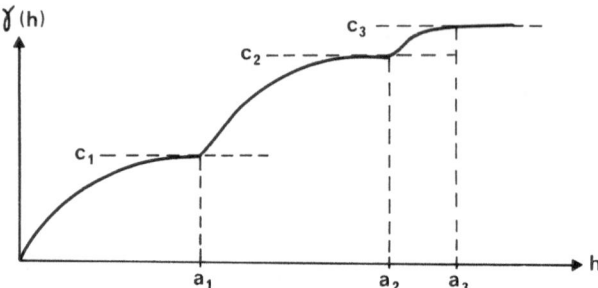

Fig. 3.17. Model of nested variogram

variograms. Their comprehensive observation would require an enormous, and normally prohibitive, amount of data of quasi-point support covering the entire field of variability from 1 μm to 100 (or 1000) km. At each scale of observation or evaluation, the support of the samller-scale measurements integrates the larger-scale variabilities into one undifferentiated variability called "nugget effect".

As a generalization from its specific meaning in gold-bearing deposits, the term is used in geostatistics to characterize the residual influence of all variabilities which have ranges much smaller than the available distances of observation, i.e. $h \gg a$. Consequently, a nugget effect will appear on the variogram as an apparent discontinuity at the origin. As such, it is closely related to the scale of observation. The same structure $\gamma_1(h)$ with range $a_1 = 10$ km would be quite evident on a sampling grid of 1 to 5 km, but would appear as no more than a nugget effect on a grid of 10–50 or, more conspicuously, 100–500 km as the following examples are to show.

The EMEP Monitoring Network

One of the essential aims of the EMEP monitoring progamme is to identify the spatial structure of air pollution in Europe, which is, as a rule, plotted in isopleth form (cf. Schang et al. 1984). The maps are based on 43 to 55 stations providing data from at least 6 months. In the following exemplary analysis the variogram procedure is applied to data of wet sulphate deposition from 40 stations providing complete date for at least 10 months (Zölitz 1985). Three variables were selected:

- mean sulphate deposition in January 1981,
- mean sulphate deposition in June 1981,
- annual mean sulphate deposition in 1981.

The mean (non-directional) variogram of the annual mean sulphate deposition has a marked slope between steps 2 and 7 (cf. Fig. 3.18). Beyond step 7, where a mean distance of approximately 2000 km between pairs of points is exceeded,

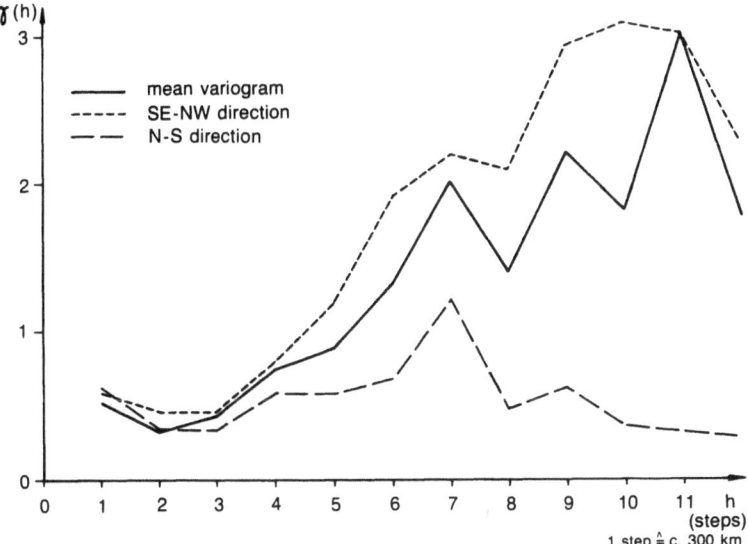

Fig. 3.18. Variograms of mean annual sulphate deposition in Europe. (After Zölitz 1985)

the statistical validity decreases. Thus the shape of the empirical variogram indicates that metric interpolation is possible upto this range. There is, however, also evidence of a marked nugget effect which amounts to about 25% of the semivariance at step 7. It clearly indicates a discontinuous distribution of values in the close proximity of the stations, if observational errors can be excluded. This implies that isoline maps based on the data under consideration must not be interpreted in detail or with respect to singular smaller area. They are only suitable to provide some spatial information about the overall, supraregional trend of sulphate deposition in Europe.

The directional variograms are of even lower statistical validity because they are based on a distinctly lesser number of points coupled. Nevertheless, they provide some evidence for an anisotropy in NW–SE direction, which stands for a NW–SE-oriented distribution pattern of values. Hence the range is greater in this direction than in others.

In addition to the mean variogram of annual mean sulphate deposition discussed, Fig. 3.19 shows non-directional variograms of January and June. The latter is similar to the variogram of the mean annual deposition. In contrast to it, the January variogram hardly increases, and at step 1 has nearly twice as much semivariance. This indicates a significantly higher level of background noise, and consequently any metric interpolation would be basically misleading.

By means of an analogous sequential spatial analysis of the larger-scale deposition networks of the Federal Republic of Germany or parts thereof, the residual influence of smaller-range variabilities can be detected and the nugget effect reduced. Yet it remains considerable with regard to sulphate even in

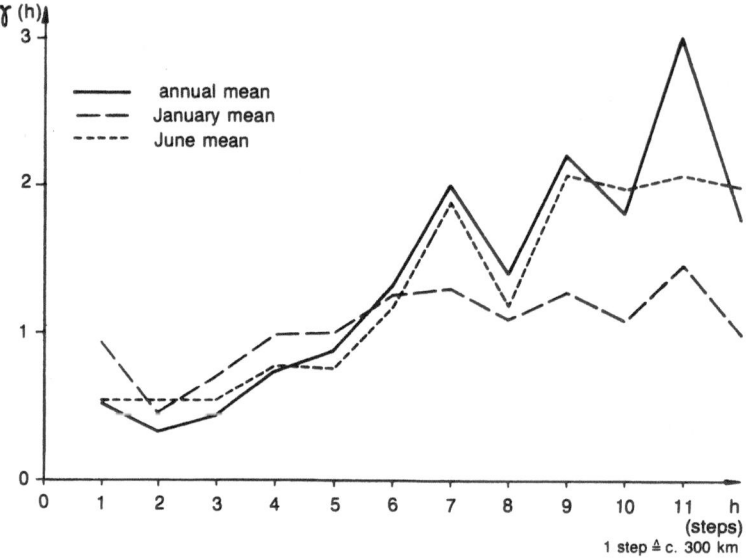

Fig. 3.19. Non-directional variogram of sulphate deposition in Europe in January and June 1981. (After Zölitz 1985)

temporary networks based on a 10 km grid, while the situation is still worse in regard to NO_3^- and Cl^- (Fränzle 1984 unpub.). In view of the complex physical and chemical transformation processes involved in pollutant transport and deposition (cf. Sects. 3.3 and 3.4), this is far from astonishing, and can already be amply demonstrated by recourse to the spatial structure of the underlying SO_2 distribution as determined in relatively large-scale local or regional networks (Fränzle et al. 1980) (Fig. 3.20).

Kriging

In view of the obvious shortcomings related to a great many measuring networks, the problem of valid local estimations is of particular importance. In geostatistics the available information used to this end is generally made up of a set of data and structural information, e.g. the variogram model characterizing the spatial variability in the area studied. The optimal technique to find the best unbiased estimator of the mean value of a regionalized variable over a limited domain is kriging. In the sense of minimum estimation variance, it is unequalled by the more classical linear estimation methods such as polygons of influence, inverse distances, inverse square distances and least-square polynomials, to cite only the commonest methods (cf. Munn 1981). Global estimates, e.g. of deposited pollutants, may be obtained by combining the various local kriged es-

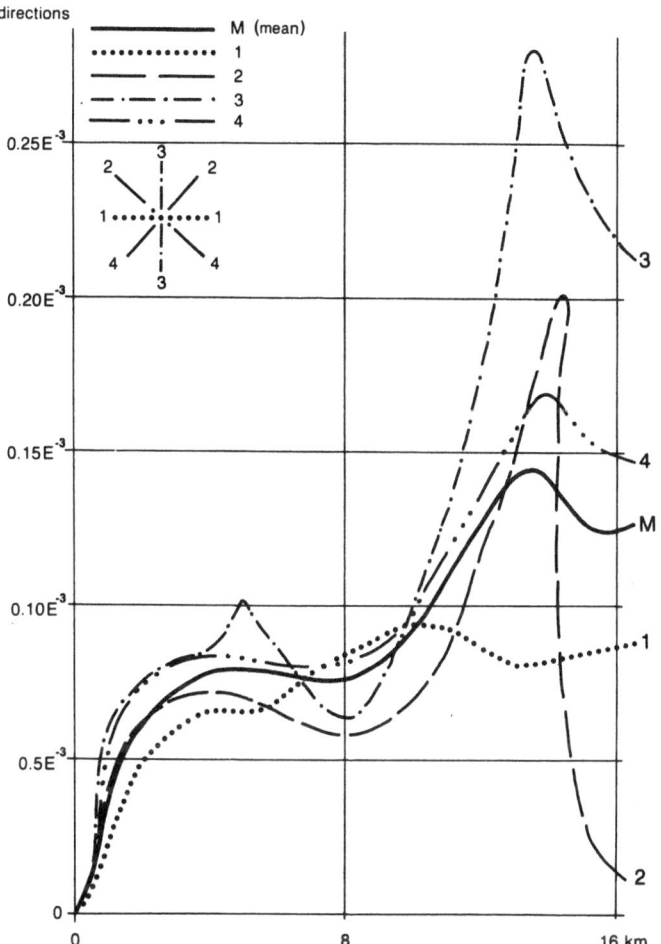

Fig. 3.20. Mean and directional variograms of SO_2 measurements in the Cologne area

timators of elementary units which make up a higher-order space (cf. Journel and Huijbregts 1978).

3.5.2 Sampling Procedures

In addition to a critical analysis of the geometry of regional networks, an assessment of their quality also implies a consideration of the punctiform sampling procedures adopted, the analytical methodology and quality control. While analysis no longer poses particular problems for many components of precipitation, given the right methodology such as chromatography, atomic absorption spectrometry, neutron activation analysis, polarographic techniques etc., it is much more difficult to maintain good quality control. In round robins organized in The Netherlands, for instance, differences of 10% or more in the results of synthetic rain samples were found for bulk elements, even in a case when only four experienced laboratories were involved (Slanina 1983).

Sampling is normally rendered difficult by contamination or adsorption of compounds in the samplers. Provided there are no financial or other technical limitations, these problems can be tackled by the use of identical samplers, placed closely together and operated in the same way. In general, the standard deviation calculated from the results of eight identical rain samplers stays constant or goes down at increasing concentrations. If the contrary is the case, contamination is likely to have occurred during sampling, sample pretreatment or subsequent analysis. Comparative determinations in The Netherlands showed (Slanina 1983) that two samplers are sufficient to obtain mean values within 5% of the value found for eight samplers for parameters such as rainfall, sulphate, nitrate, fluoride and lead. The relative standard deviation is 5% or less.

The results for H^+, NH_4^+, Ca^{2+}, Mg^{2+}, K^+, Na^+ and Cl^- indicate that a minimum of three samplers is necessary to guarantee that the mean value is within 5% of the value for eight samplers, while heavy metals present greater problems. For instance, at least four or five samplers are needed to reach the 10% accuracy level for Zn and Cd.

Even more illustrative examples are provided by the following scatter diagrams (Fig. 3.21) showing the seasonal variation of throughfall under an oak forest canopy (Fränzle 1986). Each point represents the mean variability of throughfall as a function of above-canopy precipitation. Throughfall was determined by means of 49 rain gauges in a 7.25 m^2 grid.

Closely related to the problem of deposition measurements at defined levels of accuracy by means of pooled samplers is the determination of deposition rates. They are commonly expressed by the deposition velocity v_g, i.e. the quotient of: flux density to the surface/concentration at height z. It ensues from the foregoing that the locations for concentration measurements must be such that the values obtained are representative of the concentration the surface experiences. Thus it is most suitable for measurements involving a vegetation canopy, to measure the relevant concentration at a reference height, r_z, above the canopy but within the "constant flux layer". Since concentration frequently varies with height (cf. Sect. 4.2.1) r_z should be specified.

A "constant flux layer" is normally found over extensive flat areas of snow cover, fallow, cropland or forest. To ensure that measurements are made in this layer, the uniform area should extend upwind for a fetch of at least 100 times the height of observation, or preferably even two or three times as much. Only under very specific circumstances can useful observations be made without meeting this criterion (Garland 1983; Heidland 1986).

While variogram analysis and kriging procedures or, for other purposes, the sophisticated grouping algorithms of numerical taxonomy, yield excellent results for discrete independent and unambiguously identifiable objects or entities,

a) **summer 1983**

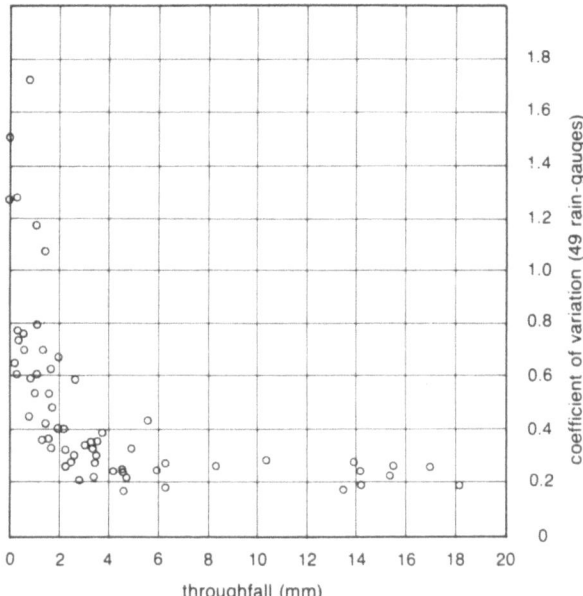

throughfall (mm)

b) **winter 1983/84**

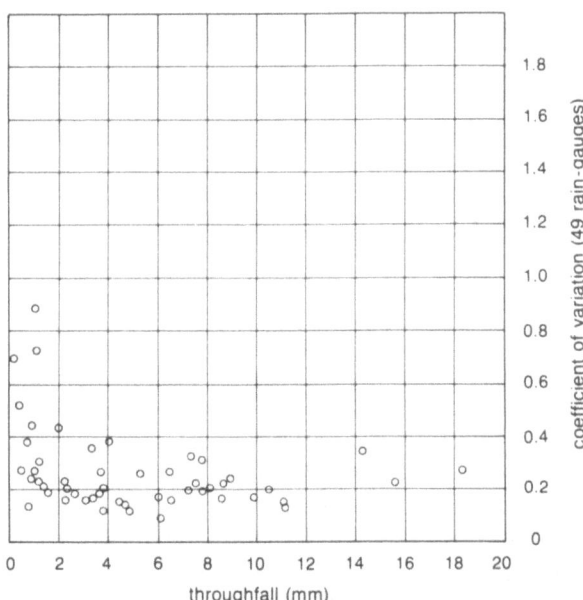

throughfall (mm)

Fig. 3.21a, b. Variability of throughfall in an oak forest near Kiel.

problems arise when these techniques are applied to areal data. As Mather pointed out "these problems concern (1) the arbitrariness involved in defining a geographical individual; (2) the effects of variation in size and shape of the individual areal units; (3) the nature and measurement of location" (cf. Berry and Marble 1968). Difficulties encountered in separating individual areal units from a continuum are most frequently, and at least partially, overcome by the selection of grid squares as the basic units, geographical characteristics (topography, climate, hydrology or soils) being averaged out for each grid square. Since grid squares are all of the same shape and size, their use eliminates variability in these properties and thus solves the second problem. The most common solution of the third problem, which is peculiar to geography, is to make relative location, as measured by spatial contiguity, the dominant variable of analysis. It can be accomplished by means of special diversity analyses or regionalization procedures (Renk 1977; Fränzle 1978b; Fränzle and Kuhnt 1983; Fränzle et al. 1986), which are based on comprehensive geographical data matrices. These methods are essential for selecting spatially representative sites for measurement purposes out of a continuum of complicated areal structures. As such, they are of particular importance in the framework of determining interactions of environmental chemicals with ecosystem compartments and are consequently described in detail in Chap. 4 of the present book.

In the light of the preceding considerations on representative measurements or reliable networks, respectively, the following last sections of this chapter deliberately give preference to punctiform data in the place of biased areal data deduced from individual samplers not infrequently placed at distances exceeding the critical maximum as defined by variogram analysis.

3.5.3 Principles of Deposition

3.5.3.1 Dry Deposition of Gases

Dry deposition transfers gases and particulate matter from the boundary layer to the earth's surface without the interaction of precipitation. The exploration of the phenomena involved, ranging from Brownian motion to gravitational settling, began in the 1940s, starting with the demonstration of sulphur accumulation by SO_2-exposed plants (Thomas et al. 1943). Meetham (1950) studied the fate of SO_2 in the lower atmosphere over Great Britain and showed the importance of dry deposition as a sink. In the nuclear industry, the necessity to understand the environmental behaviour of fission products prompted early investigations of the deposition of iodine (Chamberlain and Chadwick 1953; Parker 1956), and the study of the deposition of particles started at about the same time (Gregory 1945; Chamberlain 1953). Comprehensive surveys of the voluminous literature that has developed subsequently are provided by

McMohan and Densison (1979). Sehmel (1980), Straughan et al. (1981), Garland (1983) and Lindberg et al. (1990).

Deposition involves transport to a surface and subsequent capture, and many field and laboratory investigations whose methodology is summarized in Table 3.24 have shown that both processes may impede deposition and either may be limiting. All these methods, whose basic principles are also suited to investigate particle deposition, have limitations due to the availability of instruments and inherent specific requirements of the methods (Naujokat 1991; Spranger 1992).

With regard to SO_2 deposition , isotopes make it possible to safely distinguish the small amount of sulphur deposited during an experiment from the substantial quantities present in all plants and soils (cf. Chap. 4.2.1.1). The method is not applicable for ozone, however (Garland 1983). Stable ^{15}N has been used to trace the fate of atmospheric ammonia, but does not seem to have been applied to nitrogen oxides.

Surface mass balance studies are normally based on the determination of trace substances in throughfall and stemflow below forest canopies to deduce dry deposition. This involves corrections for the changes in storage in the

Table 3.24. Methods for measuring dry deposition. (Garland 1983)

Method	Spaces and time scales	Notes
Tracers: stable or radio-active isotopes, fluoroscent particles etc.	Leaf to field Minutes to hours	Unambiguous measurements of deposited material, examples: $^{35}SO_2$, $^{34}SO_2$, $^{14}CO_2$, $^{3}H_2$
Surface mass balance	Leaf or plant to river catchment. Weeks to years	Changes in storage usually uncertain
Mass balance in air	Box or wind tunnel: $= 1$ min upwards Plume: 1 h to 1 day Region: 1 year	Artifical conditions in enclosure Dual tracer method Requires extensive data collection
Micrometeorological methods		
Eddy correlation	Uniform fetch of > 100 m > 10 min	Fast-response detectors needed, e.g. SO_2-flame photometry, O_3, NO_x-chemiluminescence
Gradient or profile methods	Uniform fetch of > 100m > 10 min	High resolution in mean concentration measurements

canopy, as well as the leaching of compounds derived from the soil via root uptake (Gravenhorst et al. 1982; Mayer and Ulrich 1982; Jensen 1985; Müller 1987), but the success of these corrections is still uncertain. The same applies to the interpretation of river catchment and soil drainage data (Fränzle 1982). Attempts to replace the complexity of natural canopy elements by surrogate surfaces of simpler chemical behaviour are only useful if similarity between surrogate and natural surfaces in all significant aspects of the complicated deposition process can be ensured.

Mass balance studies in air, including box, bag and cuvette experiments in field and laboratory, have been widely applied to gases. As in studies in the free atmosphere, difficulties arise, however, due to variations and uncertainties in the extent of upward mixing in the boundary layer (Garland and Branson 1976; Slinn 1982).

The micrometeorological methods which require extensive, uniform areas both measure the flux through a horizontal plane above the surface, and are based on the assumption that the measured and surface fluxes are equal. This, however, requires careful evaluation, since horizontal or vertical concentration gradients, due to nearby sources or advection of different air masses, can give rise to spurious vertical fluxes.

In regard to the gradient method, requirements for concentration measurements are particularly strict, since they must be carried out at two or more heights above the canopy but within the constant flux layer. As in practice the concentration difference is usually less than 10%, methods of measurement with a resolution of order 1% or less are required (Garland 1977, 1983). The eddy correlation method, finally, depends on detecting fluctuations in concentration of a few percent or less, which usually occur with frequencies in the range of cycles per minute to a few cycles per second. Consequently, the response speed of the detector is the critical feature, but suitable instruments are available for SO_2 (Galbally et al. 1979; Hicks et al. 1983), O_3 (Wesely et al. 1978), NO_2 (Wesely et al. 1982) and particulate matter (Wesely et al. 1977; Katen and Hubbe 1983).

3.5.3.2 Dry Deposition of Particles

Particles, like gases, are transported by a turbulent transfer process to the neighbourhood of the surface elements, so that the methods outlined above are basically applicable. Depending on the size of the particles and the surface geometry, however, several processes have to be distinguished. The smallest particles with diameters below 0.1 µm diffuse to the canopy or soil by Brownian motion. On contact with the surface they adhere, i.e. they behave like gases of low diffusion coefficient but zero surface resistance. Particles ranging in diameter between 1 and 20 µm are deposited chiefly by impaction (cf. Fig. 3.10). As particle size and speed increase, so does the ratio of momentum to drag, so the rate of impaction increases with size and wind speed. In laminar or turbulent flow, particles may be impacted onto obstacles if the Stokes number, i.e. the ratio of the particles stopping distance to the diameter of the obstacle is greater than about 0.2 (Fuchs 1964). Thus only the smallest objects such as twigs, pine needles etc., would be effective for particles as small as a few micrometers.

In turbulent flow, particles may be further impacted if the turbulent eddies import sufficient momentum to carry them to the surface. Larger particles whose diameter exceeds 20 μm also deposit by gravitational settling. Deposition of aerosol particles about 5 μm in diameter and larger is frequently complicated by bounce-off (Chamberlain 1975), whereas smaller particles appear to stick on contact at normal wind speeds, regardless of the nature of the surface. Bouncing is indicative of the energy of rebound exceeding the forces of molecular attraction, and is often observed as a decrease of deposition velocity with increasing wind speed or particle size. Since wet surfaces or particles usually absorb the momentum of impact, rebound is eliminated or at least reduced, which is of particular importance for hygroscopic aerosols.

In the size range around 0.1–1 μm, neither Brownian motion nor impaction are efficient removal processes. They are supplemented by (aerodynamic) interception, i.e. the collection of particles which are carried so close to a surface by a streamline that they touch. Since this is a fairly slow process, however, it should result in a marked minimum in deposition. In fact, wind tunnel and some field experiments have indicated deposition velocities below $0.1 \, \mathrm{cm \, s^{-1}}$ in this region. Other field experiments, however, yielded deposition velocities of about $1 \, \mathrm{cm \, s^{-1}}$ for aerosol components in this particle size range. In some cases, this discrepancy may be partly explained by the wide size spectrum of the aerosols detected, or by effects of humidity, but as a whole the experimental issue and results of theoretical studies remain controversial.

3.5.3.3 Model Concept of Deposition

In the light of the above experimental findings and theoretical considerations, it may be concluded that both transport to and capture at a surface are the limiting processes of deposition. In algebraical terms, these two sets of processes act like resistances in series in an electric circuit (Garland 1983, Wesely 1989):

$$r = \frac{1}{v_g} = r_{air} + r_s , \tag{3.24}$$

where r denotes the total resistance to deposition, r_{air} the resistancce to transport through the air, and r_s the resistance to uptake at the surface.

This comprehensive formulation allows an analysis of the highly variable influence of aerodynamic effects and surface effects on deposition rates. It has the further commendable advantage of the above-mentioned electrical analogy, since the two processes act like resistances in series in an electric circuit, current representing flux F, and voltage, concentration $\hat{\chi}$. The concentration decreases as the sink is approached, and in contact with the leaf surface $\hat{\chi}_s$ should be

$$\hat{\chi}_s = r_s F \ . \tag{3.25}$$

Since the surface concentration in reality is far from being the same for all leaves and surfaces in a canopy, Eqs. (3.24) and (3.25) are simplifications. Yet they give useful indications of both the effects of surface and atmospheric properties on deposition rates. More elaborate resistance models (cf. Wesely 1989) would, in general, be much less useful because of the increasing difficulties of evaluating their various components (Naujokat 1991).

The transport of pollution, heat and water vapour all resemble that of momentum, since turbulence conveys them all at similar rates (Schmidt 1925). The "concentration" of momentum in air is $\rho u(z)$, where ρ is the density of air and $u(z)$ the mean horizontal wind speed at height z. Drag at the surface removes momentum from the air, and consequently the downward flux density is equal to the drag per unit area of the surface, τ:

$$\tau = \rho \cdot u_*^2 \ , \tag{3.26}$$

where u_* is the friction velocity.

Under particularly simple circumstances, i.e. in the absence of significant heat transfer at the surface, u_* can be determined by measuring wind speeds at several heights in the constant flux layer:

$$u(z) = \frac{u_x}{k} \ln\left(\frac{z-d}{z_0}\right) , \tag{3.27}$$

where d and z_0 denote the zero plane displacement height, and roughness length, respectively, while k is the Karman constant.

When heat exchange between surface and atmosphere cannot be neglected, for instance, when the surface is heated strongly by the sun or cooled by intense radiation at night, corrections are necessary. Garland (1977) showed how the correction can be applied for treating gaseous deposition, and Fig. 3.22 illustrates the effects of wind speed, surface roughness and atmospheric stability on atmospheric resistance.

The surface resistance, r_s, is dependent on the rate of sorption or reaction of the gas at the surface. The complexity of the latter, e.g. leaves, bark and other plant surfaces and the highly variable nanno- and microrelief of soils, makes theoretical predictions almost impossible, and experiments are therefore essential. The effect of wind speed and turbulence must be borne in mind, and under this perspective the frequently critical value of $u = 2 \ \mathrm{m \ s}^{-1}$ has been explicitly introduced into foldout Model I as a specific boundary condition.

Under certain circumstances the surface resistance proves to be very small, and deposition is critically dependent on turbulence. Then experiments in the undisturbed field stituation have clear advantages (cf. Geiger 1961). When, however, r_s is large so that v_g is correspondingly small, experiments are rather insensitive to turbulent mixing, and laboratory results may be directly applicable to field conditions.

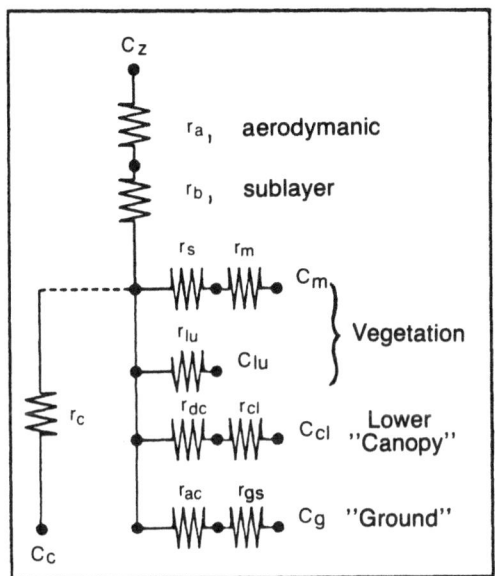

C_c concentration at the surface
C_{cl} concentration representative of substrates in the lower canopy
C_g concentration representative of substrates at the ground surface
C_{lu} concentration representative of substrates in the upper canopy
C_m concentration representative of the plant mesophyll
C_z concentration in air at height z
r_a aerodynamic resistance
r_{ac} resistance for transfer that depends only on canopy height and density
r_b sublayer resistance
r_c surface resistance
r_{cl} resistance for leaves, twig, bark or other exposed surfaces in the lower canopy
r_{dc} gas-phase transfer resistance affected by buoyant convection in canopies
r_{gs} resistance for soil, leaf litter, etc., at the ground surface
r_{lu} resistance for leaf cuticles in healthy vegetation and otherwise the outer surfaces in the upper canopy
r_m leaf mesophyll resistance
r_s surface bulk resistance for leaf stomata

Fig. 3.22. 1 Resistance model by Wesely (1989). 2 (p. 142) Variation of atmospheric resistance r_{air} with (a) wind speed, surface roughness and (b) surface heat flux. (After Garland 1983)

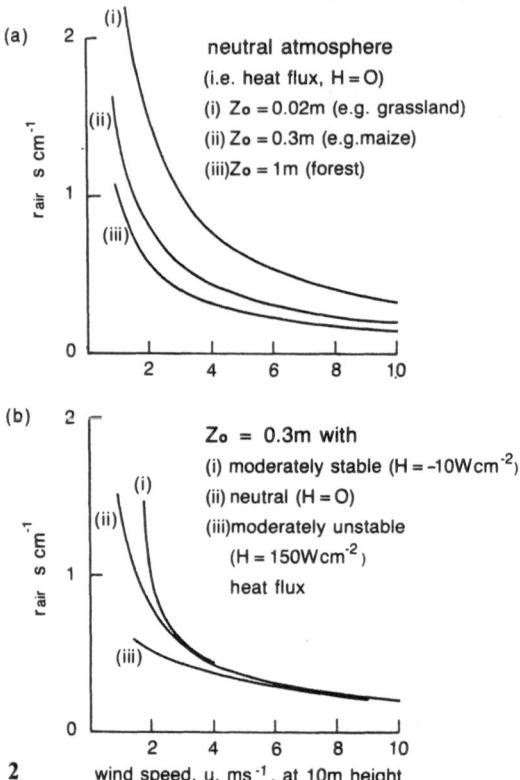

(a)

r air s cm⁻¹

neutral atmosphere
(i.e. heat flux, H = O)
(i) $Z_0 = 0.02$m (e.g. grassland)
(ii) $Z_0 = 0.3$m (e.g.maize)
(iii)$Z_0 = 1$m (forest)

(b)

r air s cm⁻¹

$Z_0 = 0.3$m with
(i) moderately stable (H = –10Wcm⁻²)
(ii) neutral (H = O)
(iii)moderately unstable
     (H = 150Wcm⁻²)
heat flux

wind speed, u, ms⁻¹, at 10m height

Fig. 3.22. (Contd.)

3.5.4 Deposition of Gases

Experimental studies and theoretical considerations show a wide range of behaviour for gases, and emphasize the need to thoroughly investigate deposition for each gas and a variety of surfaces and atmospheric conditions. In the light of results obtained hitherto it appears possible, however, to make some generalizations.

3.5.4.1 General Features of Deposition

According to present knowledge, some gases, such as HF (Israel 1977) and HNO_3 (Huebert 1983), are so reactive that they penetrate most natural surfaces, including the waxy cuticle of leaves. Surface resistance is correspondingly small, and consequently deposition is likely to be determined by the resistance to transport through the air. The uptake of other gases, such as SO_2, O_3 or NO_2

with subsequent rapid interior absorption, is controlled by stomatal mechanisms which allow gas exchange through the cuticle. The ratio of surface resistance for uptake of a gas to that for water vapour is then equal to the inverse ratio of the molecular diffusion coefficients in air for the two molecular species. Gases with a marked affinity to water, e.g. SO_2 and NH_3 in particular, are preferably deposited to water surfaces and to moist soils, while ozone is much more retained by dry soil (Turner et al. 1973).

Gases such as CO, CO_2, H_2 are consumed in metabolic processes of plants or soil microbes (cf. Pochon et al. 1969). Consequently, the rate of uptake is on the one hand limited by the demand of the relevant physiological process, and physical parameters such as water solubility and diffusion coefficients, on the other (Garland and Cox 1980).

3.5.4.2 Deposition of Acidic Species and Ozone

For ecologists and soil scientists, deposition rates of potentially toxic substances are normally of higher interest than the concentrations of these substances in the air. Since precipitation is usually limited to a comparatively short time in relation to the total length of precipitation-free intervals, dry deposition can for many substances contribute essentially to total deposition. In the case of acidic species, however, precipitation accounts for the highest contributions (Jensen-Huß 1990; Fränzle et al. 1992).

Hence deposition can be approximately assessed by multiplying the concentrations with the precipitation amounts. In view of the considerable number of homogeneous and heterogeneous sulphate and nitrate-forming processes (cf. Sects. 3.4.2 and 3.4.3), it is understandable that also the dry deposition SO_2 and NO_x has received most attention while knowledge of other gases is far less adequate. Selected field and laboratory investigations of SO_2 to various surfaces are summarized in Table 3.25, which shows that in the succession from grass via crops to forest, the deposition velocity decreases slightly as crop height increases, due to an increase in stomatal resistance which is greater than the reduction in aerodynamic resistance. Particularly noteworthy are the diurnal and seasonal variations, and the marked increase of v_g due to the melting of snow.

Nitric acid seems to deposit more rapidly than SO_2 throughout the year (Huebert 1983) while the deposition velocity of NO_2 is rather smaller, with comparable diurnal and seasonal fluctuations. Hill (1971) states a significant return flux of NO to the atmosphere. Ammonia is readily absorbed by soil and vegetation (Malo and Purvis 1964; Hutchinson et al. 1972), but may also evaporate, for example from pastures (Denmead et al. 1974; Spranger 1992). Since also PAN may make a contribution to nitrogen exchange the magnitude of the net gaseous nitrogen flux is still problematical, and further systematic research required. This applies in particular to the small-scale spatial deposition patterns and local exchange rates involved. The other gases specified in Table 3.25 probably make minor contributions to the general flux of acidity, but may be of high local importance in the neighbourhood of specific source areas.

Table 3.25. Dry deposition of selected gases related to acidic deposition. (After Garland 1983)

Gas	Surface	Conditions	v_g cm s^{-1}
SO$_2$	Soil	Calcareous, field	1.1
	Soil (rendsina)	Laboratory, pH 7.6, dry	0.55
		moist	0.6
	(fen)	pH 4.5, dry	0.19
		moist	0.38
	Short grass	Field, winter and summer, wet and dry	0.18 to 2.37 mean $\simeq$ 0.8
		Dry summer mean	0.6
		Wet	1.14
		Dry, winter	0.48
		Wet	0.68
	Wheat	Growing season, day	1.2
		night	0.3
		night with dew	0.5
		Senescent	0.3
	Forest,	day	0.6
	Scots pine	night	0.1
		day	0.2
		Dry, day	0.5
		night	0.1
		Wet, day	0.3
		night	0.3
	Forest	Winter	< 0.09
	Snow	$T < -3\,°C$	0.1
		melting	high
NO$_x$	Short grass	Day, spring	0.1 to 0.6
	Soybean field	Day	up to 0.6
		Night	0.05
	Forest,	Summer day	0.4 to 0.8
	Scots pine	night	small
		Winter	< 0.09
HNO$_3$	Tall grass		2.9
HF	Alfalfa		1.9
H$_2$S	Grass	Day	comparable to SO$_2$
	Pine needles	Day	
	Soil	Laboratory	0.016
Peroxyacetyl	Grass	Wind tunnel	0.25
nitrate	Soil		0.25
NH$_3$	Vegetation	Laboratory	comparable to SO$_2$
	Soil	Field	0.45

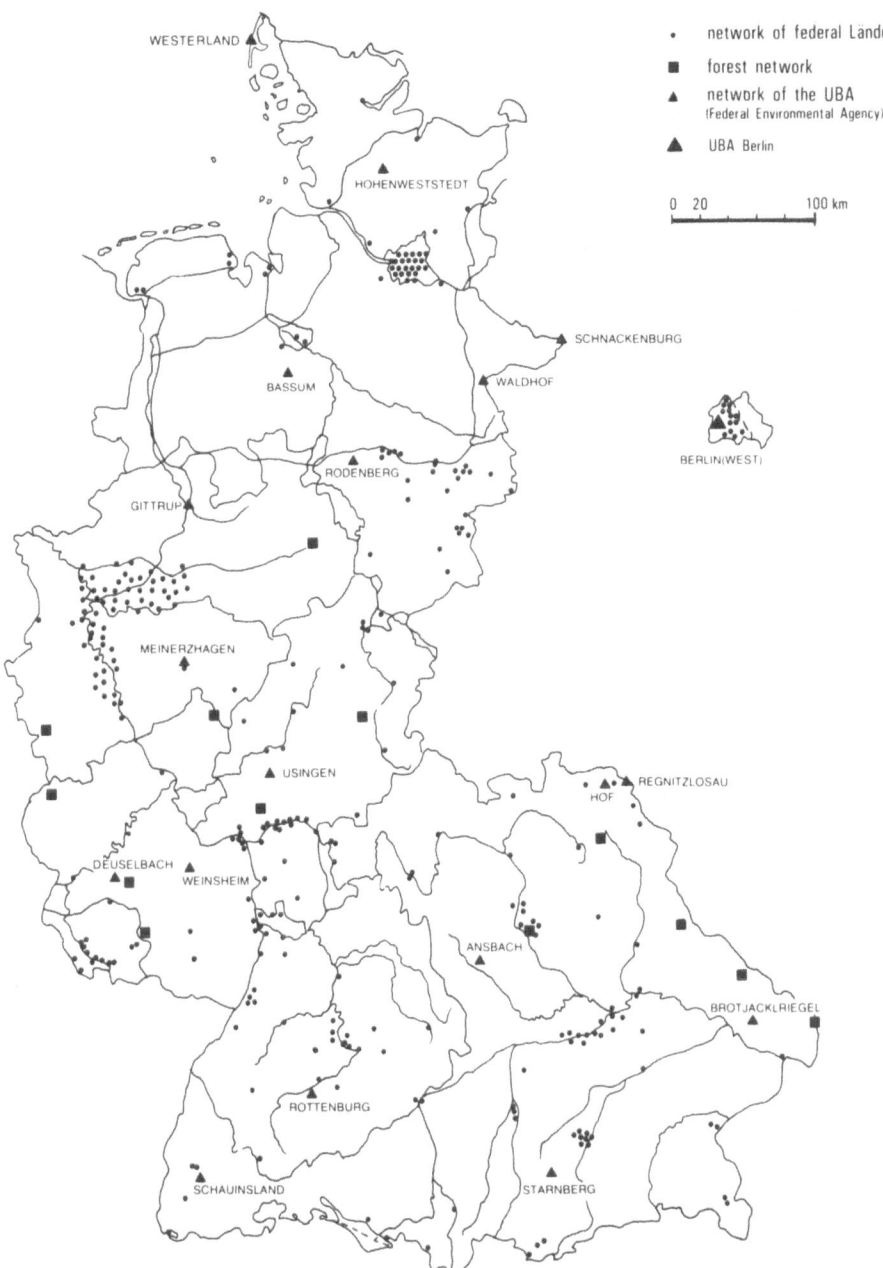

Fig. 3.23. German deposition network. (After Umweltbundesamt 1988/89)

The stations of a special German deposition network as depicted in Fig. 3.23 are located in polluted areas like the Ruhr and Rhein-Main areas or Hamburg, while the stations Deuselbach (Hunsrück), Kleiner Feldberg (Taunus), Hoher Peißenberg (Bavaria) and Schauinsland (Black Forest) represent less polluted areas.

The range of pH values encountered is illustrated by the following cumulative frequency distributions which corresponds to a range in $H^\pm$ ions by three orders of magnitude. The differences in pH between Frankfurt and Hof are particularly interesting because comparable differences are found for the sulphate concentrations, which is indicative of relationships. Klockow et al. (1978) established a good correlation between H^+-concentrations and the excess sulphate concentration, i.e. (SO_4^{2-})–(Ca^{2+}).

These findings are further corroborated by investigations in the framework of the OECD project *Long-range transport of air pollutants* and measurements in the USA which all demonstrate that sulphate, nitrate and chloride are the main

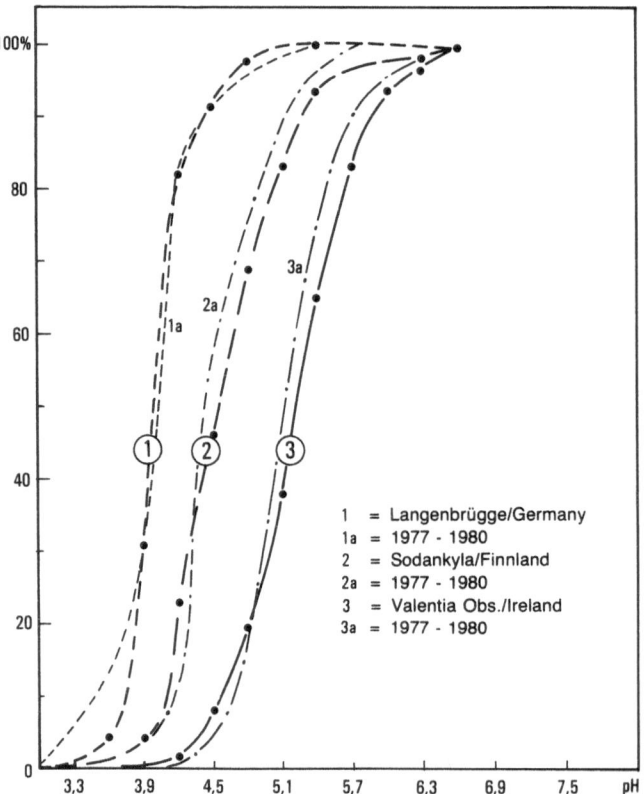

Fig. 3.24. Cumulative frequency distribution of pH in precipitation. (After Georgii 1982)

acidic components in precipitation. An analysis of the German deposition network yielded the following relative amounts of these acidifying species, assuming for each mole of sulphate an equivalent of two moles H^+ or an equivalent of one mole H^+ for each mole of nitrate and chloride, respectively.

On the basis of these concentration measurements and precipitation amounts wet deposition of acidifying species can be assessed within the limits imposed by the temporal and spatial resolution of the underlying network. With these restrictions, Fig. 3.26 may illustrate the regionally different variability of sulphate deposition in the western Federal Republic of Germany.

Beilke (1983) attempted to calculate emission, transformation and deposition averages for SO_2 and NO_x in Central Europe. Figure 3.27 illustrates the situation.

The percentages deduced from various sources correspond to an estimate by Garland (1978) based on mass balance studies for NW Europe. Twenty to 55% of SO_2 are here supposed to be removed by dry deposition, hence the mean atmospheric residence time of sulphur is limited by dry and wet deposition to about 2–4 days.

In view of the faster conversion of NO_x to nitric acid and nitrates than the comparable conversion of SO_2 to sulphuric acid and sulphates (cf. Sect. 3.4.3) and the higher deposition velocity of nitric acid, it may be surmised that NO_x contributes less to long-range acidification. This is corroborated by gradient

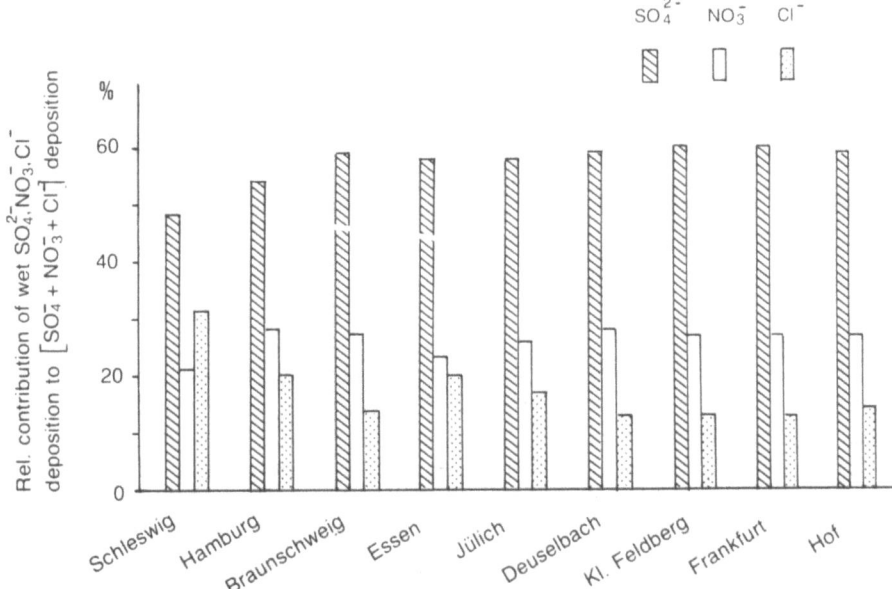

Fig. 3.25. Relative contribution of wet SO_4^{2-}, NO_3^- and Cl^- deposition to total wet deposition in val %. (After Perseke 1982)

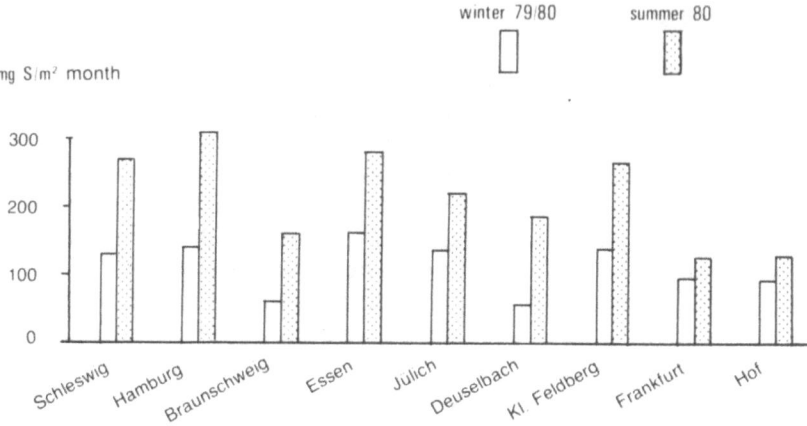

Fig. 3.26. Wet SO_4–S deposition during winter and summer at selected German stations. (After Perseke 1982)

analyses of rain samples. In the north of Scandinavia, i.e. far from the large European source areas, the sulphate/nitrate ratio of rain is greater than 5; in the south of Scandinavia, nearer to the source areas, this ratio is about 2. In The Netherlands and surrounding highly industrialized countries the ratio is about 1–1.5 (Rodhe et al. 1981; Alcamo et al. 1987).

The implication is that NO_x contributes, in relation to SO_2, more to the acidification of rain nearby the large source areas than at large distances. For The Netherlands, for instance, this seems to be supported by correlation studies. They reveal that a higher correlation exists between the nitrate and H_3O^+ content of rain than for the sulphate and H_3O^+ content (Slanina and Asman 1981). On the basis of these observations Zwerver (1982) concludes that NO_x contributes for at least one third to possibly more than one half to acidification of Dutch rains. As a consequence of the time-consuming conversion of NO_2 to HNO_3 and NO_3^- aerosol, the Dutch NO_x emissions mainly affect the acid rain abroad, while NO_x emissions from the neighbouring countries dominate, probably for more than 70%, the composition of Dutch rain with respect to the nitrogen compounds.

In the light of these observations it appears particularly interesting to study the temporal trend of the ammonium ion in precipitation since it partly neutralizes the atmospheric nitric acid. Brimblecombe and Stedman (1982) collected sets of rainfall analyses from non-urban sites in North America and western Europe, which start from the last century. The reliable data suggest an increase in nitrate ion deposition over eastern North America of ≈ 0.06 val m^{-2} a^{-1} since 1880. In regions receiving a 1000 mm annual precipitation, this would imply a precipitation-weighted mean pH shift from 5.6 to about 4.2 in CO_2 equilibrated water. The present-day deposition of sulphate (0.09 val m^{-2} a^{-1}; 1.2 g S m^{-2} a^{-1} wet deposition and 0.25 g S m^{-2} a^{-1} dry deposition) is not very

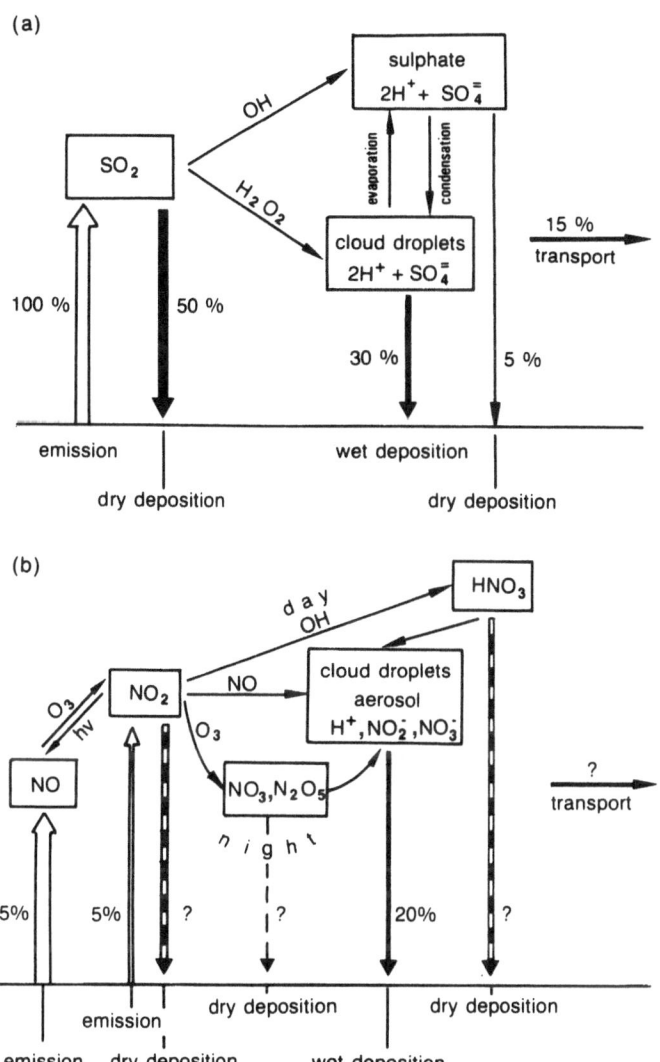

Fig. 3.27. Emission transformation and deposition of (**a**) SO_2 and (**b**) NO_x in Central Europe. (After Beilke 1983)

much larger than the current deposit of nitrate (0.065 val m^{-2} a^{-1}), reaffirming that the nitrate component of rain makes an important contribution to its acidity. Agricultural records are indicative of an increase in sulphur deposition, but problems experienced in early analyses make these changes less clear than those for the nitrate and ammonium ion. In the English East Midlands, the relative contribution of the nitrate and sulphate components of precipitation are different: NO_3^- 0.03; SO_4^{2-} 0.055 val m^{-2} a^{-1}. This difference might result from differing patterns of fuel usage, as the number of equivalents provided by the NO_x component of the English fuel emission is less than one-quarter of that derived from sulphur while in the US it is about one-half. In comparison with the marked increase in the annual deposit of nitrate ion the ammonium ion level seems to have remained relatively stable at the sub-continental scale over the past hundred years, independent of the amount deposited (0.1 to 1.8 $g\,NH_3$–$N\,m^2\,a^{-1}$). Thus the pre-industrial value for the fraction of inorganic nitrogen present as the ammonium ion, $NH_4^+/(NH_4^+ + NO_3^-)$, of 0.7 seems much reduced in present-day North America and Europe, where values as low as half this are found. Such a change in the ratio should be expected if the NO_3^- concentration of precipitation were increasing and that if the NH_4^- ion remaining fairly constant. This implies that ammonia emanates largely from moderately manured or natural soil and industrial inputs are relatively small. At the local or sub-regional scales, however, where highly intensified stockbreeding yields excessive amounts of manure which in turn release major quantities of ammonia (cf. Table 3.19), emanation intensities may be such that the above pre-industrial ratio may well be exceeded.

Maximum wet sulphur deposition occurs in highly industrialized areas and in regions with high precipitation. In Germany it is coupled with advection from SW to W as a result of the prevailing winds during frontal rainfall. It is worth noting that, owing to this mechanism, also in less polluted areas comparatively far downwind the emission areas, the concentrations of acidic species in rain are reduced by a factor of 2–3 only. A critical comparison of earlier pH measurements with present-day ones shows that during the last five decades precipitation acidity has not changed essentially, although SO_2 emissions accounting mostly for acidification have doubled (Winkler 1982, 1983).

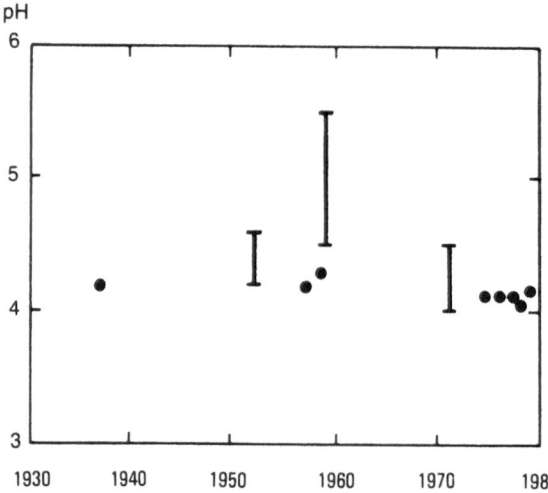

Fig. 3.28. pH of rainwater in Germany during the last five decades. (After Winkler 1983 after different sources)

This means that an increase of acidifying emissions does not lead to a correspondingly lower local or sub-regional pH in rainwater but rather induces a proportional extension of acid rain events.

Georgii (1982), who examined the temporal trend of acidity in rainwater in North America and Western Europe, came to similar conclusions: "It appears that the tendency of increasing resp. decreasing acidity and its variation from year to year is more a regional problem depending largely on the changing production, transport and scavenging of pollutants but also on the year-to-year variation of the rainfall rate". This statement is further supported by an evaluation of the pH in precipitation at five WMO-regional stations in the USA over the period 1972–1976 which showed no definite trend. It must be taken into account, however, that the 8-year period is too short for valid extrapolations in temporal respect.

The temporal variation of SO_2 in Germany is summarized in Table 3.26. The regional differentiation of NO_x concentrations in Table 3.27 does not show a uniform trend either, although there appears to be a tendency to an increase in the overcrowded, heavily industrialized areas, which is in agreement with the overall increase of emissions (Schmölling and Jörß 1983). The corresponding measurements in the Ruhr area, as reviewed from 1975 by Buck et al. (1982), also lack a definite temporal trend.

The same applies to the concentrations of SO_4^{2-} and NO_3^- resulting from the oxidation of SO_2 and NO_x emissions. Kallend (1983) examined 120 stations of the EACN network in Europe and found in only 23 cases a statistically significant increase in SO_4^{2-} concentrations for the 1956–1976 period, while one station proved to have an opposite trend (cf. Field 1976; Bettleheim and Littler 1979; Sørensen 1983). By contrast, NO_3^- concentration had significantly risen in no less than 55 cases, the mean increase being 6%. Large deposition rates are possible when low cloud or fog, driven by stronger wind, encounter vegetation or the ground surface (Lovett et al. 1982; Dollard et al. 1983). Under these circumstances the concentrations of dissolved species and H_3O^+ ions may be an order of magnitude higher in the deposited water than in rainwater, and consequently the small volumes of water deposited under such conditions may make a disproportionately large contribution to the total deposition of acidity.

Table 3.26. Temporal variation of SO_2 concentrations at selected stations of the German deposition network (after Jost and Beilke 1983); 98% values based on daily means ($\mu g\ SO_2/m^3$)

Location	1973	1974	1975	1976	1977	1978	1979	1980	1981	1982
Waldhof	90	87	62	105	112	96	106	97	101	99
Deuselbach	50	47	44	67	45	85	89	83	93	74
Schauinsland	33	31	36	33	26	37	35	32	30	35
Brotjacklriegel				42	51	58	58	42	48	45
Westerland					50	43	45	37	32	37

Table 3.27. Temporal variation of NO$_2$ concentrations at selected stations of the German deposition network (after Jost and Beilke 1983); 98% values based on daily means (μg NO$_2$/m^3)

Location	1969	1970	1971	1972	1973	1974	1975	1976	1977	1978	1979	1980	1981	1982
Waldhof			39	35	36	31	36	36	36	34	37	34	43	41
Deuselbach	20	25	27	32	32	28	28	32	30	34	38	36	49	34
Schauinsland	15	10	13	12	13	14	13	13	11	15	17	15	18	13
Brotjacklriegel	17	18	16	22	18	17	17	15	16	17	29	19	30	16
Westerland	39	33	39	37	32	24	35	26	32	36	30	31	29	45

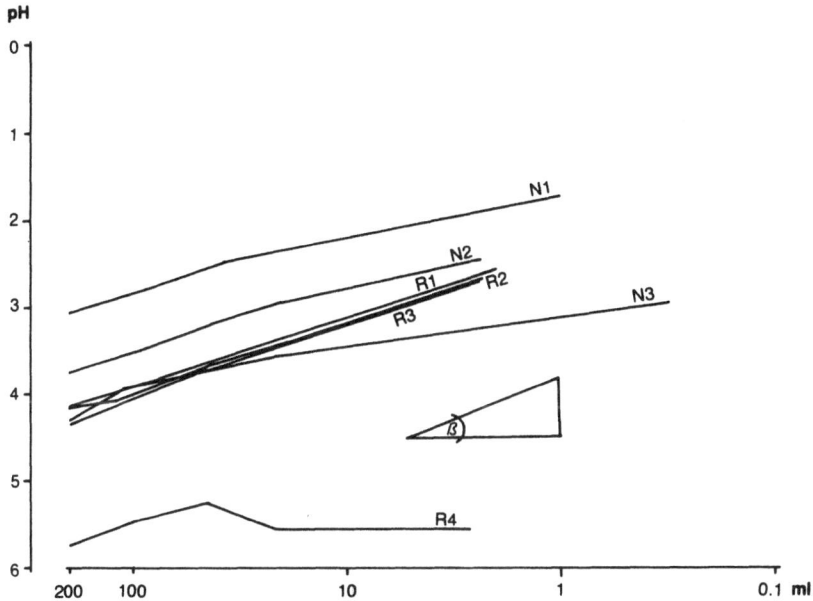

Fig. 3.29. pH changes due to evaporation of rainwater (*R1–R4*) and fog water (*N1–N3*) in samples. β indicates the theoretical slope if an acid of low vapour pressure were present. R4 contained predominantly CO_2 which was lost by degassing. (After Frevert and Klemm 1984)

This is due to selective retention of acidic species by vegetation cover and soil, which, in detail, proves to be the result of two competing processes. The first one is the well-known fog-catching effect which is, however, not yet quantifiable for inherent difficulties in measurement. The other is the result of the evaporation of intercepted rainwater or fog, and its importance can be derived from Fig. 3.29. The influence of these processes is revealed by the sulphur fluxes in a beech stand, investigated by Mayer and Ulrich (1982) in the framework of the interdisciplinary Solling project. The assessment is based on the mass balance approach

$$P_c = D_d + D_w - R_c + L_c , \tag{3.28}$$

where P_c is the element flux below the canopy, D_d is dry deposition, D_w is wet deposition to the canopy, R_c is retention of air-borne substances in the canopy, and L_c is canopy leaching.

It ensues from Table 3.28 that the major portion of total deposition was measured directly, namely wet and dry depositions during the winter months. Dry deposition estimates for the vegetation period are considered as quite reliable, since total uptake into the aboveground biomass was found to be but small. It must be noted, however, that the spatial validity of the measurements as expressed in terms of quantities per hectare is left open (cf. Sect. 3.5.1).

Table 3.28. Sulphur fluxes in a beech stand (Solling, Federal Republic of Germany) 1968–1976 average. (After Mayer and Ulrich 1982)

1. Measured rates	
– Wet deposition (soluble fraction found in bulk samplers)	23.8 kg S ha^{-1}a^{-1}
– Wet deposition under canopy (stemflow included)	53.1 kg S ha^{-1}a^{-1}
– Total uptake of S into above ground biomass per year	10.1 kg S ha^{-1} a^{-1}
– Dry deposition during the dormant season (Nov–April)	
assuming leaching and retention to be zero	13.9 kg S ha^{-1}
2. Estimates	
– Dry deposition during vegetation period (May–October)	12.0 kg S ha^{-1}
3. Total deposition calculated from flux balance (1)	49.7 kg S ha^{-1}a^{-1}

A comparison of sulphur flux by throughfall exhibits a close relation to composition and structure of a stand. Figure 3.30 shows deposition rates of spruce and beech sites in the Solling area (Federal Republic of Germany), clearly underlining the much greater effect of the spruce stand.

Table 3.28 shows that the total deposition of acidic species to a forested area may considerably exceed wet deposition rates collected in a bulk rain gauge. It further ensues from the Solling measurements that dry deposition to a spruce canopy exceeds that to a beech canopy as a consequence of different surface resistance. This effect is strongest in winter when beech has lost its leaves and therefore has a reduced surface area as compared to the evergreen spruce canopy.

It should be borne in mind, however, that these results reflect complicated interactions between rain and foliage which will be dealt with in greater detail in Sect. 4.2.1. Here it may suffice in order to illustrate the necessity of comparative quantitative studies in the framework of comprehensive ecosystem research programmes (Ellenberg et al. 1978; Fränzle et al. 1986), to refer to investigations in North American chestnut oak (*Quercus prinus*) and white oak (*Q. alba*) stands (Lindberg et al. 1983). They clearly show that the influence of a deciduous canopy on the nature and concentration of acidity in intercepted rain may be quite different from the above findings in the Solling Mountains.

The fully developed oak canopy decreased the strong acid concentration reaching the forest floor by 20–40% and altered the relative contribution of weak acids to total acidity substantially. Weak acids accounted for an average of only 30% of the total acidity in rain above the canopy, but increased to nearly 60% of the acidity below the chestnut oak canopy. Because of these changes, the total acidity reaching the forest floor remained practically unaffected, i.e. total acidity averaged 140 µval/l in both rain and throughfall. Although the pH values indicated "acid" rain above the canopy (pH ranging from 3.5–4.5), on several occasions nearly "normal" rain occurred below (pH from 5.0–6.0). The white oak canopy generally removed a substantially higher amount of the incoming strong acidity, in some events up to 100%, than did the chestnut oak

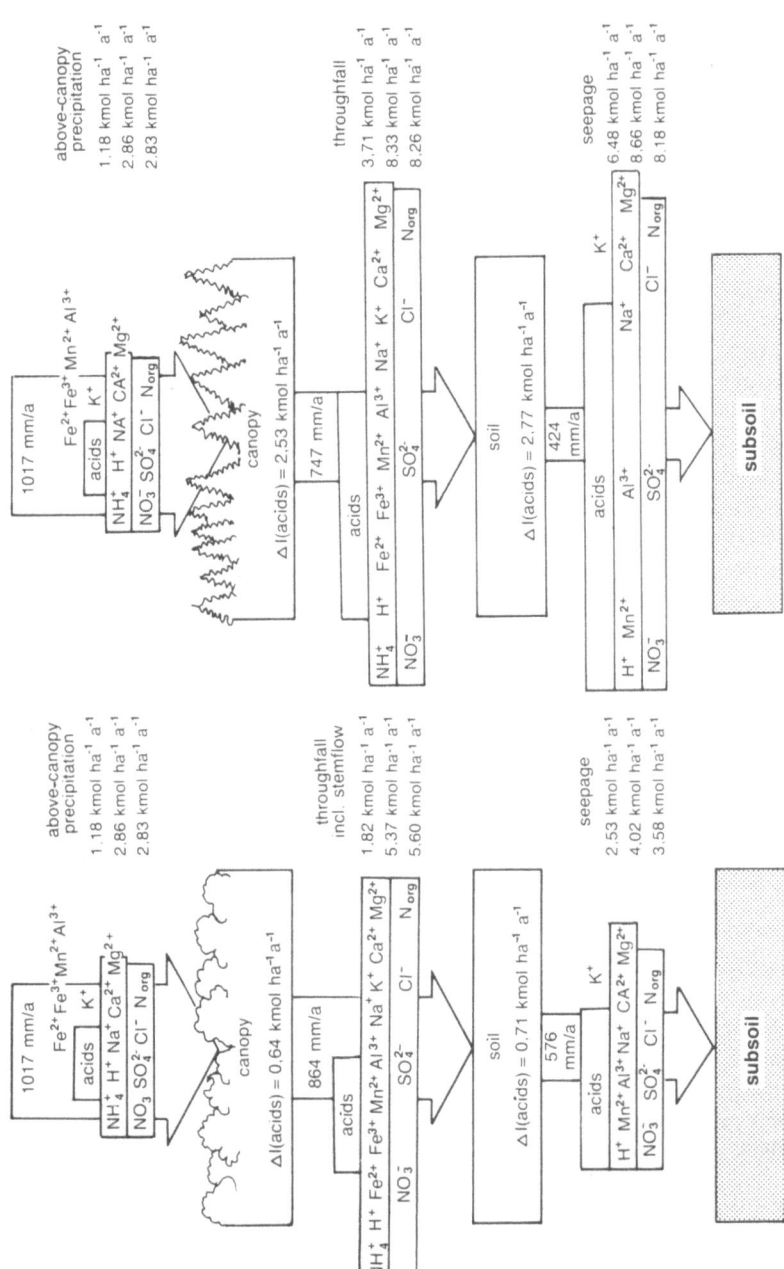

Fig. 3.30. Fluxes and balances of water and ions in a beech (*left*) and spruce stand of the Solling Mts. Average ion equivalents (kmol ha^{-1} a^{-1}) for the 1969–1983 period. (After Benecke 1987)

canopy. In both canopies, the more leaf layers with which the rain interacted, the greater the degree to which strong acids were scavenged. Again the primary physical factor influencing acid uptake was rain intensity; when it decreased, and rain residence time consequently increased, strong acid scavenging rose (cf. Spranger 1992).

In comparison to the studies on sulphate and nitrate deposition in rain there are but few longer-term investigations of NO_3^- and SO_4^{2-} deposition in atmospheric aerosols. Cox and Penkett (1983) report on a feeble SO_4^{2-} increase of about 20% in 18 years (1957–1974) at Harwell (UK) which is exclusively due to an increase of the summer amounts. For NO_3^-, however, the increase amounts to more than 100%. Either trend is attributed to enhanced photochemical processes during the summer months (cf. Sects. 3.4.2 and 3.4.3), but the decisive role is played by the exhaust components NO_x and reactive hydrocarbons of motor vehicles. Table 3.29, summarizing the aerosol sulphate measurements in Germany, does not indicate a temporal trend. The yearly averages have a range from 4 to 10 $\mu g\,m^{-3}$ while the daily means attain maxima of about 70 μg $SO_4^{2-}\,m^{-3}$.

Several recent studies of the deposition of ozone to various surfaces have been summarized by Wesely (1983). Table 3.30 gives typical deposition rates to land and water surfaces. A comparison with Table 3.25 shows that the deposition rate to land surfaces is similar for O_3 and SO_2, but the deposition of ozone at sea and over lakes is much reduced. Total deposition rates appear not to be a good indicator of the potential for damage to plants (Turner et al. 1974;

Table 3.29. Sulphate concentrations in aerosols at selected stations of the German deposition network (after Jost and Beilke 1983). Yearly averages in $\mu g\,SO_4^{2-}\,m^{-3}$, determined as sulphur by means of X-ray fluorescence analysis

	1974	1975	1976	1977	1978	1979	1980	1981	1982
Waldhof	4.5	4.5	6.2	7.2	7.5	6.6	6.9	5.4	9.8
Deuselbach	5.4	5.7	7.2	5.4	6.2	4.8	6.6	5.4	6.1
Schauinsland	3.9	2.4	4.2	3.9	3.9	3.6	4.8	3.9	3.5
Brotjacklriegel	3.0	2.4	6.0	6.3	5.1	5.1	5.1	4.2	5.1
Westerland	5.1	4.5	6.9	6.6	6.9	5.7	6.0	4.2	5.7
Hohenwestedt		4.2	7.2	7.8	9.9	6.9	6.9	5.4	7.4
Barsum		5.4	7.8	7.8	9.9	9.9	7.8	6.6	8.6
Rodenberg		7.2	6.3	6.0	6.3	6.3	7.5	6.6	6.8
Meinerzhagen		5.7	7.5	7.8	8.1	7.8	6.9	6.0	7.1
Usingen		6.9	9.9	8.1	9.3	7.5	8.1	6.9	8.0
Weinsheim		4.8	7.2	6.6	8.1	7.2	6.3	6.0	6.2
Rottenberg		6.9	8.7	6.3	7.5	7.2	6.9	5.1	5.0
Ansbach		6.3	8.4	7.5	9.0	8.1	7.2	5.7	6.7
Starnberg		3.9	5.7	5.7	6.0	5.7	6.3	4.5	5.2
Hof							6.0	6.3	7.3

Table 3.30. Deposition of ozone to land and water surfaces. (After Garland 1983)

Surface	Conditions	v_g cm s^{-1}
Short grass	Summer day	0.58
	night	0.29
Grassland	Wide range of surface and wind speed	0.1–1
Soybean	Good growing conditions with incomplete canopy, complete	1.3
	canopy, day	0.85
	night	0.3
Maize	Growing season, day	0.5
	night	0.15
	Senescent, day	0.4
Forest	Mixed, summer morning	1.0
	Loblolly pine, day	0.5
	night	0.08
	Deciduous, winter, day	0.19–0.37
	night	0.05
Snow		0.06
Water	Lake Michigan	0.01
	Seawater	0.03–0.05
		0.05
Soil	Field of bare soil, Summer	0.25–0.6
	Waterlogged	0.1

Wesely 1983). The chemical impact of O_3 on soils is still unknown, but presumably only that part of the ozone flux absorbed by leaves is relevant to crop damage. Large diurnal variation in O_3 deposition rates are due to stomatal closure, but variations in air resistance due to ground level inversions at night are also of importance (Garland and Derwent 1979). Variations in concentration near the surface can also be considerable, with minima at night and maxima during the day, and they may also be induced by changes in the stability of the boundary layer. The combined effect of these diurnal changes is that the deposition rate of O_3 is many times smaller on stable nights than typical day-time rates.

Conclusions

Apart from the problems resulting from a possibly incoherent geometry of sampling networks, the very measurement of precipitation chemistry is complicated, in particular with respect to determining realistic areal deposition rates. It seems to be fairly fallacious to suppose that the "total deposit" in an open collector actually represents the sum of wet and dry deposition to any natural surface (cf. Cox et al. 1976; Garland 1983; Jensen-Huß 1990; Spranger 1992). A collector, in almost any case, does not resemble the natural surface, either in physical-chemical or geometric characteristics, so that the deposition rates of both gases and particles to the collector and to the natural surface are most likely to be substantially different.

Not infrequently, also dry deposition to rain collectors introduces a decisive element of uncertainty to the measurement of precipitation chemistry. Open rain collectors necessarily combine the deposition of particles and gases. A comparative analysis in Northern Britain has shown that 13–35% of the non-marine sulphur collected in open rain collectors is due to dry deposition (Fowler and Cape 1984). The corresponding figures determined in the framework of ecosystem research at forested and non-forested sites near Kiel (Germany) are 10–33% dry deposition. Only the use of apparatus which automatically covers the collecting tunnel or vessel when rain is not falling reduces this spurious flux, probably to insignificant proportions. Further difficulties in obtaining reliable areal deposition rates arise from the wide variation, in both deposition velocity and concentration, with season, time of day and weather. In principle, the mean should be weighted by concentration, but normally the variations are not sufficiently well known to allow this to be done. As a result, the estimates of the mean fluxes for SO_2 and NO_x may well be subject to considerable uncertainties, probably in the order of 50%, as air pollution measurements at Harwell (Oxfordshire, UK) show (Cox et al. 1976). For nitric acid, fewer measurements of either concentration or deposition rates are available, and consequently the uncertainty in this, potentially important, flux and its spatial and temporal variation must be larger than for SO_2. In like manner, the possibility that the particle flux is underestimated by a large factor must also be admitted (see also Sect. 3.5.5).

Lastly, where fog or low clouds occur frequently, they may make a dominant contribution to deposition whose assessment can be considerably supported by meteorological modelling approaches (Hacker-Thomae 1985). In areas more remote from sources, the pollutants may be mixed into deeper atmospheric layers, and a larger fraction may be in aerosol rather than gaseous form owing to transformation processes during transport. Both changes are liable to favour wet deposition relative to dry deposition, so that in remote areas the relative importance of the two removal modes may be different.

3.5.5 Removal of Airborne Metals by Wet and Dry Deposition

By contrast to C, N, O, S and P, the biogeochemical cycles of metals are considerably less known. In many cases it is difficult, therefore, to decide whether a metal or its compound has been anthropogenically channelled into a natural cycle or whether a novel route has been started. Garrels et al. (1975), for instance, have defined an "interference index" (i.e. ratio between estimated anthropogenic emission into the boundary layer and the estimated total amount removed by precipitation) and concluded that anthropogenic emissions were comparable in importance to natural ones. Jenkins (1976), however, estimated the amount of anthropogenic emissions to be considerably larger than that of natural ones. In view of the great number of metals existing in nature and used in industry, however, the question appears to be open still (Duce et al. 1975), and much regionalized research is needed to obtain a better insight into the situation on both the regional and global scales (Munger and Eisenreich 1983). Yet at the regional level, and considering selected metals, there is ample evidence that a strong increase in deposition rates is synchronous with the beginning of industrialization or the introduction of petrol engines (cf. e.g. Murozumi et al. 1969; Tyler 1972; Wilkniss 1973; Bruland et al. 1974; Erlenkeuser et al. 1974; Förstner and Müller 1974; Suess and Erlenkeuser 1975; Gydesen et al. 1981; Galloway et al. 1982; Mart et al. 1982; Hutton 1983; Kremling 1983; Ouellet and Jones 1983).

Following Wood (1976), metals may be divided into non-critical ones, highly toxic and relatively accessible ones, and toxic but hardly soluble or rare ones. Members of the first group are Na, K, Mg, Ca, Fe, Li, Rb, Sr, Al, Si, while the second group comprises: Be, Co, Ni, Cu, Zn, Sn, As, Se, Te, Pd, Ag, Cd, Pt, Au, Hg, Te, Pb, Sb, Bi. Toxic but of very low solubility or very rare are: Ti, Hf, Zr, W, Nb, Ta, Re, Ga, La, Os, Rh, Ir, Ru, Ba.

Metals can be transported in the atmosphere in gaseous form or as aerosols, so that volatility is a characteristic of prime importance. According to Garrels et al. (1975) and Heit (1977), the following volatility sequences can be defined:

- As, Hg > Cd, Pb, Bi, Tl > In, Ag, Zn > Cu, Ga > Rb as oxides, sulphates, carbonates, silicates and phosphates,
- Hg > As > Cd > Zn, Sb > Bi > Tl > Mn > Ag, Sn, Cu > In, Ga as elements, and
- Ag, Hg > Sn, Ge > Cd > Sb, Pb > Zn, Tl > In, Cu > Co, Ni, Mn as sulphides.

On the basis of these sequences and the geogenic occurrence in coal, Heit (1977) concluded that As, Cd, Hg, Pb and Tl had the highest emission rates as a result of coal combustion. Comparative studies have shown that As, Cd, Hg, Pb, Sb, Se, Te and Tl are emitted to 50% and more in gaseous state while the others were chiefly in the ash (cf. Sect. 3.1.2). It should be noted, however, that irrespective of the manifold possibilities of an atmospheric transport, the main environmental route of metals is via aquatic systems.

3.5.5.1 Factors Influencing Total Atmospheric Deposition of Metals

Wet and dry atmospheric deposition of soluble metals results in both chronic and episodic exposure of terrestrial surfaces to atmospheric pollutants. Light rain, fog and dew dissolve previously dry-deposited metals on leaves, enhancing the potential for interaction with internal tissues of plants. In quite a few areas, such as the subalpine forests of the New and Old World or the Mist Belt of East Africa, cloud-water deposition can also contribute substantially to the flux of atmospheric pollutants to vegetation and soil. Rainfall removes some fraction of both soluble and insoluble particulate metals via stemflow and throughfall from the canopy and produces an episodic flux to forest soils. The concentrations and speciation of metals deposited are modified by manifold physical and chemical interactions in the canopy and on twigs, branches and stems among rain, particles, and dissolved organic material leached from plant tissues.

The investigation of the water-soluble fractions of aerosol-bound heavy metals is essential for the assessment of their bio-availability and a pre-requisite for the interpretation of the data of authors who analyzed only the water-soluble species, e.g. Nürnberg et al. (1983). Figure 3.31 based on data of the German deposition network, shows the distribution of the water-soluble fractions of Cd, Pb, Mn and Fe in rainwater in form of cumulative frequencies.

The figure shows that Cd, Pb and Mn occur predominantly in soluble form while the iron distribution shows much higher proportions of insoluble compounds. Some smaller differences exist in the soluble/insoluble ratio in samples from stations of different air quality. For instance, in polluted and urban areas, the insoluble fractions or Pb and Fe are higher than in clean-air regions. This trend may be attributable to intensified washout in conurbations, which predominantly removes coarser particles from the boundary layer. For elementary physical reasons their solubility is less than that of aerosol particles of the sub-micron range. In addition to Fig. 3.12, Fig. 3.32 shows the bi- and trimodal distributions of various metals in aerosols as determined at the Frankfurt pilot station of the German deposition network.

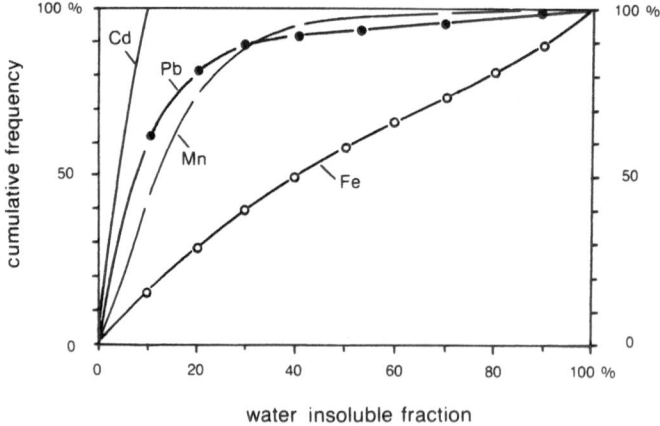

Fig. 3.31. Cumulative frequency distribution of metals in rainwater. (After Rohbock 1982)

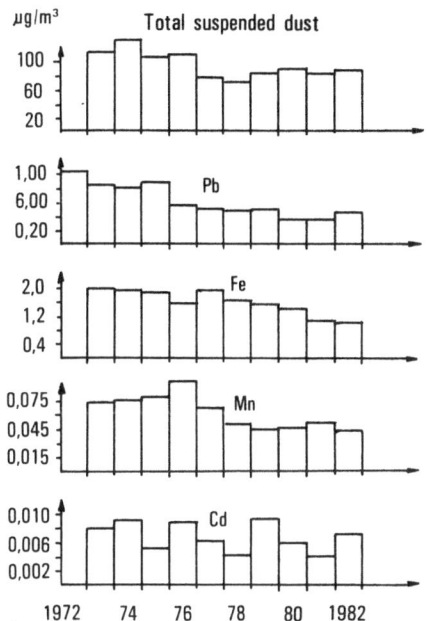

μg/m³   Total suspended dust

Fig. 3.32. Bi- and trimodal distributions of various metals in aerosols as determined by means of an Anderson impactor at the Franckfurt pilot station of the German deposition network. (After Umweltbundesamt 1983)

The absolute metal concentrations in rainwater are the result of the highly variable influence of drop size spectra, intensity of precipitation, temperature, pH, chemical composition of droplets, concentration and direction of advection. As substantiated in Sect. 3.3.4, this complexity of boundary conditions involves the specific character of each precipitation event, which renders comparisons with others difficult.

Therefore Fig. 3.33 should be considered rather as an illustration of specific differences in the short-term deposition behaviour of common heavy metals than an indication of an average decrease in metal concentration with time due to washout processes.

In the light of current knowledge and assessment techniques, the elements in deposition can be grouped into two classes:

– The first class comprises S, Na, K, Mg, Ca, Mn, Fe and Zn which predominantly display high solubility on the forest canopy, at least under acid-rain conditions (cf. Ulrich et al. 1979; Frevert and Klemm 1984). Therefore retention due to physico-chemical processes is of comparatively little importance, and also uptake due to metabolic processes should play only a lesser role, since the elements contribute relatively little to the element content of organic matter (S, Na, Ca, Zn), or are leached from the canopy in considerable quantities (K, Mg, Zn, Mn) during the vegetation period. During the dormant season, when deciduous trees have lost their foliage, leaching of metabolites from the canopy becomes negligible. This permits

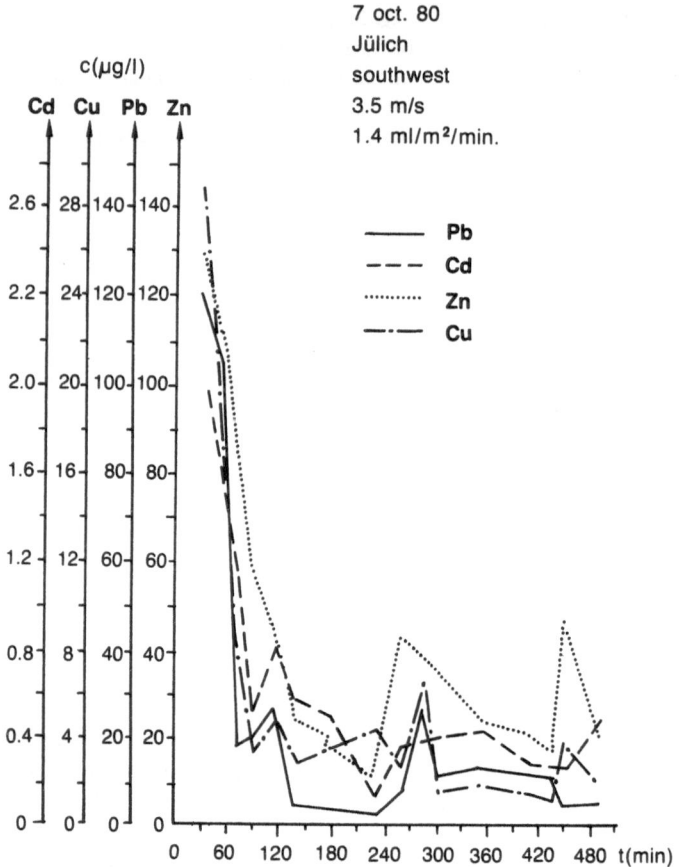

Fig. 3.33. Decrease in metal concentration due to washout processes in the boundary layer. (After Schladot and Nürnberg 1982)

to deduce dry deposition from the balance equation [Eq. (3.28)] by means of estimating dry deposition during the vegetation period on the basis of the winter wet/dry deposition ratio (Mayer and Ulrich 1982).

- The second class of elements includes Al and the heavy metals Cr, Co, Ni, Cu, Cd and Pb which may be considered as relatively immobile. Numerous experiments indicate that uptake via roots is also restrained in comparison to nutrients (Mayer and Heinrichs 1981). Under these circumstances, a rough estimate of the maximum uptake rate is possible. It shows that total uptake into the above-ground biomass is predominantly supplied from the atmosphere. Since total annual uptake can be determined experimentally, a rough estimate of root uptake permits to calculate the rate of retention in a balance equation, e.g. Eq. (3.28).

3.5.5.2 Deposition Rates and Mechanisms in Deciduous and Coniferous Forests

In the course of a 1-year project, the role of forest vegetation as a particularly important sink for atmospheric emissions was studied at Walker Branch Watershed. The watershed is an oak-hickory-dominated catchment in eastern Tennessee situated within 20 km of two coal-fired electric generating stations of about 2500 MW capacity. The methods involved collection of precipitation as wetfall-only on an event basis above and below the canopy and collection of dry deposition using inert Petri-dish collectors located in the canopy (Lindberg and Harriss 1981; Lindberg 1982).

The estimated annual fluxes of heavy metals to Walker Branch Watershed are summarized in Table 3.31. The figures indicate that both dry and wet processes are significant in the atmospheric deposition of Cd, Mn, Pb and Zn to the forest; the ratio of annual dry/wet deposition varies from 0.2 (Cd and Zn) to 1 (Pb) to 8 (Mn). About 20% of the total annual atmospheric deposition of Pb to the canopy may be taken up by plant tissue. Net removal from the canopy, i.e. the difference between deposition above and below the canopy may be attributed to both wash-off of dry deposition and to foliar leaching. Dry deposition of Pb to the canopy exceeds net removal, suggesting canopy absorption of ~ 3 mg m^{-2} a^{-1} of dry-deposited lead under the assumption of no foliar leaching. The anual atmospheric Pb input to the forest floor exceeds Pb flux in leaf-fall by a factor of approximately 100. The large difference between the apparent rate of Pb uptake by the canopy and the transfer of Pb in leaf-fall to the forest floor is indicative of translocation of absorbed Pb from the foliage as reported by Krause and Kaiser (1977) or Lalubie (1991). By contrast, the atmosphere contributes only $\sim 50\%$ of the Cd and Zn and $\sim 10\%$ of the Mn flux to the forest floor. These elements are not retained in the forest canopy and net removal from the canopy exceeds estimated dry deposition by factors of 5 to 10.

Variations in meteorology and canopy characteristics necessarily result in a wide range of heavy metal deposition rates and mechanisms. For low-volume rain events of short duration, wet deposition represents a more intense flux of trace metals to the vegetation than does dry deposition during the intervening dry periods (cf. Sect. 3.5.3).

Wet-deposition rates normalized to the event duration exceed dry-deposition rates by factors ranging from 40 to 11,000. The ratio of wet/dry deposition inputs during these periods ranges from 0.06 to 24, which indicates that total input may be dominated, by either dry or wet deposition. Lovett et al. (1982) found that cloud-water deposition rates can exceed typical dry-deposition rates by two orders of magnitude, and that also total deposition of most ions in the subalpine ecosystems studied is dominated by cloud-water input. The pronounced differences in the temporal structure of wet, dry, and cloud-water deposition is likely to have important implications for ecological assessment; duration and intensity of exposure may exert a strong influence on pollutant effects on plants (cf. Chap. 4.2.1.6).

Table 3.31. Annual atmospheric deposition and internal flux of heavy metals to a chestnut oak stand and single-event deposition rates to upper-canopy surfaces in Walker Branch Watershed (Tennessee). (After Lindberg et al. 1983)

Constituent	Annual deposition and internal fluxes in a chestnut oak stand (mg m^{-2} year^{-1})								Normalized short-term deposition rates to upper-canopy receptors (pg cm^{-2} h^{-1})[a]					
	Dry deposition to canopy and soils	Wet deposition		Net removal from canopy[b]	Foliar leaching[c]	Leaf fall	Total flux to forest floor		Dry			Wet		
		Above canopy	Through-fall				Internal[d]	External[e]	W1[f]	W2	W3	W1	W2	W3
Cd	0.09	0.43	1.2	0.77	0.69	0.04	0.73	0.52	0.14	0.75	0.03	0	270	270
Mn	30.9	4.0	154	150	123	91	214	35	62	120	190	0	19,000	7100
Pb	8.0	7.3	11	3.7	(−3.3)[g]	0.16	0.16	12	15	62	12	0	2400	4600
Zn	1.73	7.6	16	8.4	6.9	5.0	12	9.3	8.3	9.6	3.2	0	2300	1100
Rainfall amount (cm)	—	142.4	—	—	—	—	—	—	—	—	—	0	0.13	0.30

[a] Rates are normalized to the duration of either the wet or dry event. [b] Wet deposition below canopy minus wet deposition above canopy. [c] Net removal minus dry deposition to canopy. [d] Leaf fall plus canopy leaching, not including possible translocation to soils of foliar-absorbed material. [e] Total dry deposition plus above-canopy wet deposition minus in-canopy uptake (if appropriate). [f] W1 = May 9–16, 1977 (total duration 167 h), W2 = May 16–20, 1977 (wet event duration 0.17 h, total duration 101 h), W3 = May 30–June 6, 1977 (wet event duration 0.48 h, total duration 168 h). [g] A negative foliar leaching term suggests in-canopy uptake of some fraction of dry deposition.

Assuming spatial validity of the underlying data metal inputs to other forested sites in the United States and Europe provide an interesting exemplary comparison with inputs to Walker Branch. Particularly interesting are the findings of Lindberg et al. (1983) relating to four watersheds in the southeastern United States for the period March–June 1981, i.e. synchronous to the above measurements (Table 3.32).

In comparison with Table 3.1, it ensues from the figures in Table 3.32 that Cd input by dry deposition was three to five times lower at the Cross Creek and Coweeta sites than at the Walker Branch and Camp Branch sites, whereas Cd input by wet deposition was considerably higher at Camp Branch and Cross Creek than at Walker Branch and Coweeta. Zn input by wetfall was three to seven times higher at Coweeta than at other sites. Dry deposition contributed smaller proportions of the total Cd input, and perhaps Mn, at Camp Branch and Cross Creek than at Walker Branch and Coweeta.

These results indicate differences among the four sites either in local or regional metal emissions or in the underlying meteorological situations. Sampling locations at Walker Branch and Coweeta were in forest clearings and consequently somewhat protected from the wind, while collectors at Camp Branch and Cross Creek were positioned immediately above the forest canopy. Assuming equality of other factors, the differences in exposure should favour a higher relative proportion of dry deposition at the sites in forest clearings. Tables 3.33 and 3.34 which should also be considered in the framework of the German deposition network provide comparative data.

Particularly interesting are the measurements in a spruce and beech stand of the Solling area. These neighbouring sites are characterized by high precipitation, frequent fog with predominant western and southwestern winds transporting immissions from the large industrial complexes of Kassel (50 km) and the Ruhr (130 km). Höfken and Gravenhorst (1982) analyzed aerosol and precipitation samples collected in 1980 in 20 precipitation funnels placed above and outside the beech and spruce forests. To assess leaching of ions out of the plant substance by rain, beech and spruce twigs were washed with rain water of known chemical composition. Table 3.34 shows that the concentration of compounds and elements in throughfall was significantly (i.e. 1.5 to 41.2 times) higher than in above-canopy precipitation.

A comparison with Table 3.36 shows that the values are in good agreement with the findings of Ulrich et al. (1979), Mayer et al. (1980) and Mayer (1981) based on another methodology and longer periods of measurement. The remarkable increase in concentration of throughfall is due to the incorporation of aerosol particles deposited onto the canopy during dry intervals. Only in the case of Mn was a considerable leaching from the very foliage observed so that the exceptional enrichment by factors 26.5 or even 41.2, respectively, are due to both dry deposition and leaching.

Including winter deposition rates the following Table 3.35 summarizes the data on deposition fluxes to the forest canopy. By means of the balance approach as summarized in Eq. (3.28) Mayer and Ulrich (1982) tried to assess long-term averages of deposition in the same stands.

Table 3.32. Wet and dry deposition (g ha^{-1}) of metals to four watersheds in the southeastern United States for the period March–June 1981. (After Lindberg et al. 1983)

	Dry deposition				Wet deposition				100 dry/(dry + wet)			
	WB[a]	CB	CC	CW	WB	CB	CC	CW	WB	CB	CC	CW
Cd	0.32	0.33	0.06	0.10	0.34	1.7	1.0	0.14	48	17	6	42
Mn	16	11	5.3	11	13	16	14	8.5	56	41	27	57
Pb	5.2	3.6	2.7	2.0	13	28	13	14	28	12	17	12
Zn	4.2	4.2	2.8	6.8	16	9.9	21	73	21	30	12	8

[a] Abbreviations and rainfall amounts at each site for the sampled period area as follows: WB = Walker Branch (37.9 cm rain), CB = Camp Branch (32.7 cm), CC = Cross Creek (29.9 cm), CW = Coweeta (46.2 cm).

Table 3.33. Heavy metal deposition in European forest ecosystems. (After Höfken et al. 1981)

Location	Above-canopy Pb ppb	Pb $g\,ha^{-1}a^{-1}$	Cu ppb	Cu $g\,ha^{-1}a^{-1}$	Cd ppb	Cd $g\,ha^{-1}a^{-1}$	Throughfall Pb ppb	Pb $g\,ha^{-1}a^{-1}$	Cu ppb	Cu $g\,ha^{-1}a^{-1}$	Cd ppb	Cd $g\,ha^{-1}a^{-1}$	Species	Period
Solling	27	285	23	236	1.5	15.9	31	188	19	78	1.5	8.6	Beech	Nov. 74
Solling	27	285	23	236	1.5	15.9	64	467	30	227	2.7	20.1	Spruce	to Apr. 79
Bärhalde		110		18				66		22		13	Spruce	Nov. 80
Lüneburger	24	214			1.6	5	16	140			0.9	5	Oak	to
Heide						5.6	21	170			1.3	6.7	Fir	Oct. 81
Thetford (GB)		470		180		2.6		240		90		2.6	Fir	Aug. 77 to May 78
Ascot (GB)		190		120		2.6		270		130		2.6	Fir	May 76 to May 78
Upsala (S)		29.2				0.7								
Kohlbrunnen Schwäb. Alb		84		22		3.9								
Spundgraben Schwäb. Alb		43		23		2.0								

Table 3.34. Absolute and relative concentration in precipitation May–October 1980. (After Höfken and Gravenhorst 1982)

	NH_4	Cl	Pb	Fe	Cd	NO_3	SO_4	M_n
Above forest (µg/l)	1010	970	15	115	3.6	2830	4100	22
Beneath beech +	1.7	1.5	1.6	2.5	3.8	1.8	2.9	26.5
Beneath spruce +	3.3	3.9	3.9	6.7	6.7	7.4	13.4	41.2

+ = Ratio of the concentration in the throughfall to the above-forest concentration.

Table 3.35. Deposition rates in $mg/(m^2\ month^{-1})$. (After Höfken and Gravenhorst 1982)

Components	Wet deposition		Dry deposition rate			
			Beech		Spruce	
	W	S	W	S	W	S
Mn	2	2	< 8	< 6	< 9	< 18
Fe	13	10	11	10	< 26	23
Cd	0.04	0.03	0.09	0.06	0.12	0.08
Pb	1.3	1.2	1.1	2.3	4.3	3.3
NH_4^+	130	84	61	28	120	61
Cl^-	160	80	6	13	90	67
NO_3^-	360	230	97	90	1200	670
SO_4^{-2}	660	340	350	440	2600	1400

(W: Feb.–Apr. 1980, S: May.–Oct. 1980)

It should be pointed out that this method cannot substitute for carefully performed meteorology-based measurements. While it yields only very limited information on the deposition process, it can lead to long-term averages of deposition rates which are urgently needed by ecologists, soil scientists, geochemists and conservationists. For the environmental meteorologist the balance methodology, provided it is based on spatially valid data, may provide a very useful temporal and spatial framework for process analyses.

3.5.5.3 Microscale Variability of Deposition in Forests

The comparative evaluation of deposition networks in Sect. 3.5.1 has shown that only a minor part of the sources or structures of variability which come into play simultaneously and for all distances are reflected in nested variograms. At each scale of observation the support of the smaller-scale measurements integrates

Table 3.36. Annual deposition rates of a beech (B) and a spruce
(S) stand on the Solling plateau. (After Mayer and Ulrich 1982)

Element		Dry deposition kg ha^{-1}a^{-1}	Total deposition kg ha^{-1}a^{-1}
S	B	25.9	49.7
	S	49.8	73.6
Cl	B	13.2	30.1
	S	19.0	35.9
Na	B	4.7	12.4
	S.	8.3	16.0
K	B	8.8	12.1
	S	18.2	22.5
Mg	B	2.0	3.9
	S	2.9	4.8
Ca	B	10.8	21.2
	S	16.9	27.3
Al	B	1.1	2.2
	S	1.8	2.9
Cr	B	0.135	0.149
	S	0.152	0.166
Mn	B	1.5	1.76
	S	4.9	5.2
Fe	B	0.89	1.77
	S	1.24	2.12
Co	B	0.001	0.015
	S	0.002	0.017
Ni	B	0.096	0.123
	S	0.113	0.140
Cu	B	0.234	0.470
	S	0.420	0.659
Zn	B	0.255	1.632
	S	0.355	1.732
Cd	B	0.0	0.016
	S	0.004	0.020
Pb	B	0.152	0.437
	S	0.448	0.732

1968–1976 average for S, Cl, Na, K, Mg, Ca, Al, Fe, Mn
1974–1979 average for Cr. Co, Ni, Cu, Zn, Cd, Pb

the larger-scale variabilities into one undifferentiated variability called the
"nugget effect". This term characterizes the residual influence of all variabilities
with ranges much smaller than the distances of observation in the wider-
spaced grids of higher-order networks. Its importance for appropriately assessing ecological situations in general and site structures in particular can be easily

demonstrated by a consideration of the microscale differentiation of heavy metal deposition as related to stemflow and throughfall in forests.

Normally stemflow exhibits a strong increase in acidic species which is particularly marked in fog or when precipitation intensity is low, or after longer dry periods. This is due to bark leaching on the one hand and the filter mechanisms described on the other. Thus the sulphate and chloride concentrations in stemflow of beech and oak stands in the Haard Forest and the Bergisches Land (Federal Republic of Germany) exceeded above-canopy values by a factor 14 (Block and Bartels 1983) while nitrate and phosphate concentrations were distinctly less increased and partly even lower. The H-ion concentrations of beech stemflow amounted to 1000 mg/l (Haard) or 253 mg/l (Bergisches Land). In oak and fir stands of the Lüneburger Heide Mayer (1983) found a doubled Pb concentration is stemflow but a tenfold reduction of Cd concentration as compared to above-canopy precipitation.

The essential hydrological boundary conditions operative in these chemical cascading systems are exemplarily illustrated in Fig. 3.34. In the above figure, which also shows the marked differences in throughfall and stemflow as related to stand structure, the upper interrupted line indicates zero-interception. The intersection of the lower interrupted line with the abscissa defines the average interception capacity of the respective stand, while the orthogonal projection a of the tangential point onto the abscissa indicates the minimum amount of precipitation necessary to bring about maximum interception. Thus in the Solling beech stand a 2.6 mm interception is reached with a 9 mm above-canopy precipitation; the corresponding figures for the spruce stand which has practically no stemflow are 4.7 mm and 13 mm.

Under these circumstances, it is easily understandable that the small-scale variability of stemflow and throughfall chemistry poses particular problems in the framework of geostatistically based measurement programmes. Table 3.37 summarizes the results of comparative measurements in forest sites SE of Kiel (Jensen 1985; Fränzle 1986) and demonstrates that spatial validity of deposition chemistry as related to throughfall can only be achieved on the basis of a 5 m network. A comparison with the spatial structure of heavy metal concentrations in soil solutions (Chap. 4.4.1.3) exhibits a marked influence of stemflow, indicating the quasi-punctiform character of the latter even at such a scale. Its spatial differentiation could be appropriately assessed only by means of decimeter sampling grids.

3.5.5.4 Large-Scale Spatial Variability of Atmospheric Removal Processes of Air-Borne Metals

An analysis of national deposition networks always reveals strong local influences as a result of spatially and temporally varying pollution strains. In view of the geometrical limitations outlined in Sect. 3.5.1, normally only a punctiform comparison of deposition rates is appropriate. The following consideration,

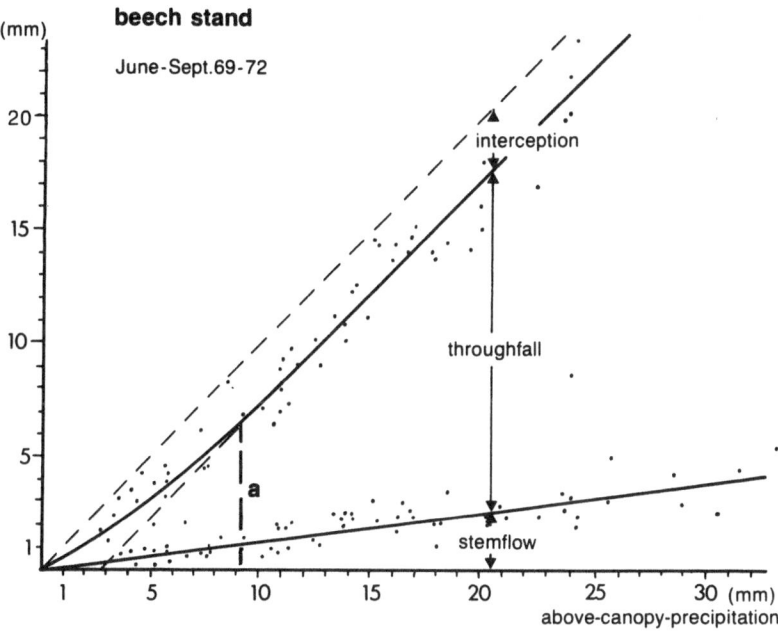

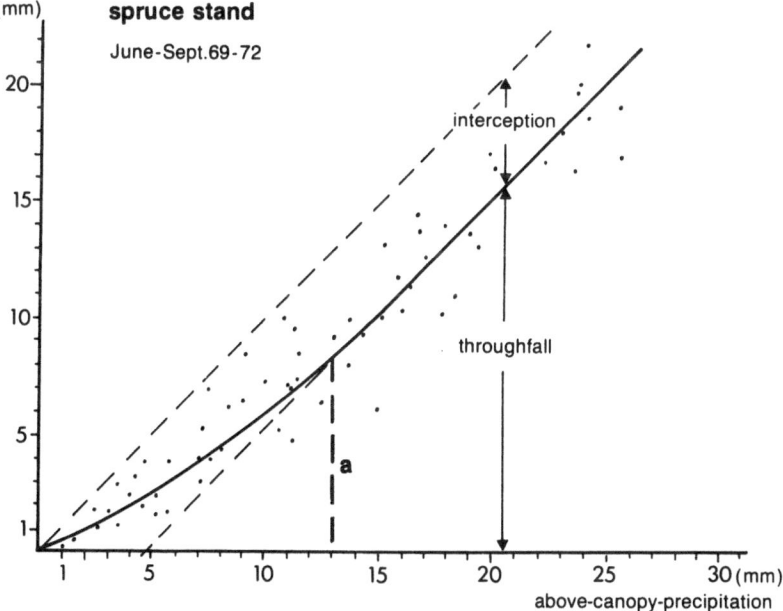

Fig. 3.34. Above-canopy precipitation and resulting interception, throughfall and stem-flow in a beech and spruce stand for the June to September 1969–1972 period. *a* above-canopy precipitation as related to maximum interception. (After Benecke 1984)

therefore, will be based on selected stations of the German network, and from this perspective Fig. 3.35 provides an illustrative example.

Essen represents a polluted area near the southern fringe of the heavily industrialized Ruhr district while Deuselbach reflects the situation in a (relatively) clean-air part of the Hunsrück Mts. The deltoids indicate the range of concentrations found during the 1979–1981 period while the peaks represent maximum or minimum values, respectively, and cross-beams mark 50% values.

Table 3.37. Stemflow chemistry of neighbouring oak and beech trees in a 90-year-old stand SE of Kiel. (After Jensen 1985; Fränzle 1986)

			Cd	Pb	Cu
			Cumulative monthly values in µg		
March[a] 1984	S2	Oak	8.37	14.35	48.35
	S4	Beech	0.61	12.50	12.99
	S6	Oak	0.77	5.75	4.98
	S8	Oak	1.66	14.19	13.41
	S10	Oak	2.90	17.74	14.10
	S12	Oak	0.68	0.69	5.39
April 1984	S2		3.33	17.71	139.05
	S4		0.17	14.74	43.69
	S6		1.45	11.93	31.92
	S8		1.50	3.90	45.99
	S10		5.91	21.07	186.88
	S12		3.18	5.51	101.30
May 1984	S2		4.36	83.00	160.07
	S4		0.04	0.68	2.70
	S6		0.42	9.37	16.60
	S8		1.53	23.26	46.93
	S10		2.40	36.12	87.81
	S12		1.06	50.75	107.34
June 1984	S2		11.51	83.52	141.10
	S4		4.12	19.32	29.70
	S6		0.35	1.53	15.72
	S8		0.23	23.88	32.75
	S10		1.65	19.87	35.95
	S12		3.88	21.16	139.46
July 1984	S2		1.83	45.35	109.91
	S4		0.40	2.89	4.54
	S6		0.38	5.68	17.28
	S8		0.65	2.85	23.65
	S10		4.02	17.57	164.70
	S12		1.36	51.91	146.80

Table 3.37. (Contd.)

		Cd	Pb	Cu
		Cumulative monthly values in µg		
August[b] 1984	S2	0.27	2.43	19.86
	S4	0.09	1.09	3.22
	S6	0.05	0.39	1.84
	S8	0.05	0.69	3.48
	S10	0.01	0.06	0.29
	S12	0.05	0.34	1.76
Sept. 1984	S2	5.00	79.00	377.04
	S4	1.03	11.93	62.28
	S6	0.16	4.03	14.79
	S8	0.22	7.63	31.66
	S10	0.69	31.22	139.78
	S12	1.97	40.41	171.49
Oct. 1984	S2	1.16	40.63	200.10
	S4	0.15	8.01	43.59
	S6	0.16	8.96	42.13
	S8	0.03	1.11	4.98
	S10	0.85	50.11	236.10
	S12	0.62	51.33	248.61

[a] March to July: totals of concentration measurements relating to individual stemflow events.
[b] August to October: totals derived from the product of average concentrations and sum of individual stemflow rates.

At Deuselbach, total Pb deposition is in the range of $30\ \mu g\,m^{-2}\,d^{-1}$ over extended periods, but contrary to current assumptions (Rohbock 1982), this rate cannot be regarded as a representative value for backround deposition since geostatistical validity is not ensured (see Sect. 3.5.1). In comparison to wet deposition, dry deposition ranges only from 5 to $10\ \mu g\ Pb\ m^{-2}\,d^{-1}$. In accordance with measurements of total dust deposition (Lahmann and Fett 1980), also dry lead deposition slightly increased. In industrialized urban areas such as Essen the total lead deposition, normalized to a 1-year period, amounts to $160\ \mu g\,m^{-2}\,d^{-1}$, and during 2-week periods even $300\ \mu g\,m^{-2}\,d^{-1}$ occur. In urban areas the fraction of dry-deposited lead is about 17–22%, whereas in remote rural areas it sinks below 10% of the total lead deposition.

Cadmium shows a generally similar deposition pattern but the fraction of dry deposition is slightly increased as compared to lead, i.e. in urban areas it may amount to 20–36% of the total deposition of 1–$4\ \mu g\ Cd\ m^{-2}\,d^{-1}$. This increase may be attributed to gaseous cadmium.

Manganese and iron are bound onto coarser aerosol particles, and consequently they are removed by dry deposition to a larger extent. Total manganese

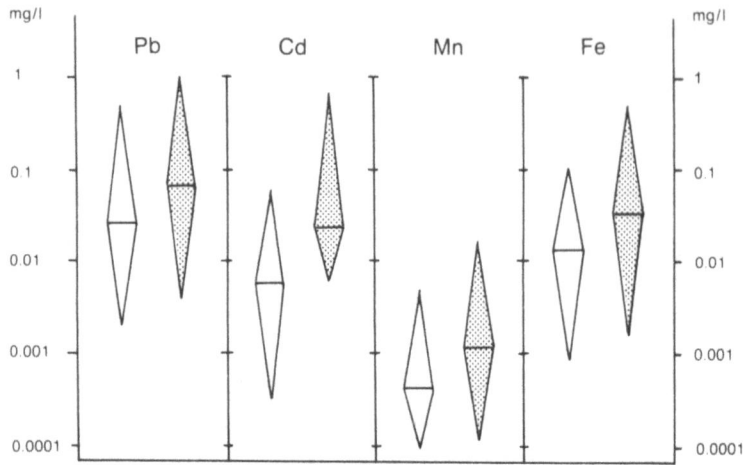

Fig. 3.35. Metal concentrations in rainwater at a clean-air station (Deuselbach, Hunsrück *white deltoids*) and in a polluted area (Essen, Ruhr district *dotted deltoids*). (After Rohbock 1982)

depositions amount to 24–70 µg Mn m^{-2}d^{-1}, 50–80% of which are dry-deposited. Iron depositions range from 320 to 1530 µg Fe m^{-2}d^{-1}, the maxima being found in the Ruhr district with its iron and steel producing industry and at Jülich in the neighbourhood of large open-cast brown coal mining. Dry deposition amounts to no less than 70% of total deposition on an annual average.

Table 3.38 reflects both the spatial and temporal variability of heavy metal deposition in Germany in absolute figures. Only the sampling locations in the Ruhr district, Essen and Dortmund in particular, have significantly higher, although ecotoxicologically sub-critical, Cu deposition values as compared to the other stations. In the case of Dortmund, the influence of one of the biggest copper plants in Europe situated at a windward distance of about 50 km is quite distinct.

Within a factor 2 the deposition of Pb is similar for most locations with the exception of urban agglomerations like Essen which is under the combined influence of dense automobile traffic, and the emissions of numerous industrial sources; but even here the average daily wet Pb deposition did not exceed the cut-off value of 250 µg m^{-2}d^{-1} defined as tolerable for orchards, vegetables or fish farming (Davids and Lange 1986). Clearly the situation may be much worse in closer neighbourhood to efficient emission sources and under unfavourable meteorological conditions. In 200 m distance from the Stolberg-Binsfeldhammer lead smelter, for instance, an average daily deposition of 23.3 µg Cd m^{-2}d^{-1} and 650 µg Pb m^{-2}d^{-1} was registered. This exceeds the tolerable threshold value by the factors 4.7 or 2.6, respectively (Nürnberg et al. 1983).

In this connection it should be noted that normally a substantial amount of toxic metals emitted from a source is deposited in its vicinity. Typical distances

Table 3.38. Average daily heavy metal deposition in the Federal Republic of Germany in $\mu g \, m^{-2} \, d^{-1}$. (After Nürnberg et al. 1983)

Site	Precipitation ratio 1981/80	Cu 1980	Cu 1981	Pb 1980	Pb 1981	Cd 1980	Cd 1981	Zn 1980	Zn 1981	Ni[a] Oct. 80–May 81
Rural areas	1.13	5–12	3–15	20–40	20–40	0.4–1.0	0.5–0.9	25–70	20–70	2.4–4.2
Hamburg	0.90	19	13	48	38	1.0	0.7	81	57	5.9
Frankfurt area	1.35	8–16	12.5	50–60	50–80	1.2	1.4	80–90	45–60	4.8–6.5
Essen	1.19	25	27	116	158	2.3	2.3	225	300	8.5
Dortmund	1.22	37	24	78	96	2.2	1.8	210	190	—
Stolberg-Werth	1.09	15	27	90	196	1.65	4.0	73	122	—
Goslar	1.40	15	30	88	152	7.0	9.0	520	1535	22

[a] Preliminary measurements.

under the macroclimatic conditions of Central Europe are in the range of 10–20 km (Kloke 1972; Nürnberg et al. 1983). The remaining proportion, often small but by no means negligible, may undergo long-distance transport, and be deposited in the Arctic or even Antarctic regions.

For Cd, a steady increase is observed from coastal over rural sites to urban and industrial agglomerations. The peak value found at Goslar is due to the influence of zinc smelters at 5 km distance. Consequently the threshold value of 2.5 µg Cd m^{-2} d^{-1} as defined for agricultural areas, is grossly exceeded by a factor of 1.4–1.8. For Essen and Dortmund the mean wet Cd deposition remains just below the tolerance level.

Zn deposition varies for the majority of rural and urban stations between 50 and 90 µg m^{-2} d^{-1}, but the stations in the Ruhr district, as well as Goslar, have exceedingly high deposition values. The unusually high deposition rates of the Hohenpeissenberg station are due to a nearby zinking plant which does not contribute to increase Cd deposition, however, as a smelter would do.

In addition to these determinations in permanent deposition networks, measurements related to specific weather situations (cf. Sect. 3.2.2) and based on random sampling approaches or cumulative assessment techniques in the framework of biomonitoring programmes are of particular methodological interest. Herrmann (1978, 1984), Schrimpf and Herrmann (1978), Schrimpf et al. (1979) studied the spatial distribution of Cd, Cu, Pb and Zn in random snow samples in northeastern Bavaria. Thomas et al. (1983) analyzed the accumulation of these heavy metals in the epiphytic moss *Hypnum cupressiforme* L. ssp. *filiforme* Brid. It ensues from these findings that the predominantly meridional orientation of mountain barriers exposed to mainly westerly winds causes an increase of heavy metal deposition with increasing height and intensified luff–lee contrasts. Long-distance transport contributes considerably to overall pollution, while turbulent transport near the ground brings about a decrease of immissions within short distances from the sources. Therefore high concentrations in heavy metal deposition occur in densely populated areas and increasing concentrations along mountain ridges.

Seasonal variations of weather conditions (cf. Sect. 3.2.2) and emissions are reflected in the immission pattern exhibiting a distinct maximum in winter. This regionalized picture can be complemented by a historical analysis of heavy metal deposition based on the microchemical study of firtree rings (Herrmann et al. 1978; Schrimpf 1980). It shows, under the assumption of representative sampling, that trace metal pollution in NE-Bavaria has increased considerably during the last 100 years. The amount of zinc incorporated has doubled, and that of cadmium and lead is even four times as high.

3.5.5.5 Deposition Velocities of Metals

Deposition velocity, defined as the proportionality factor of material deposited and ambient air concentration, describes the efficiency of transport mechanisms

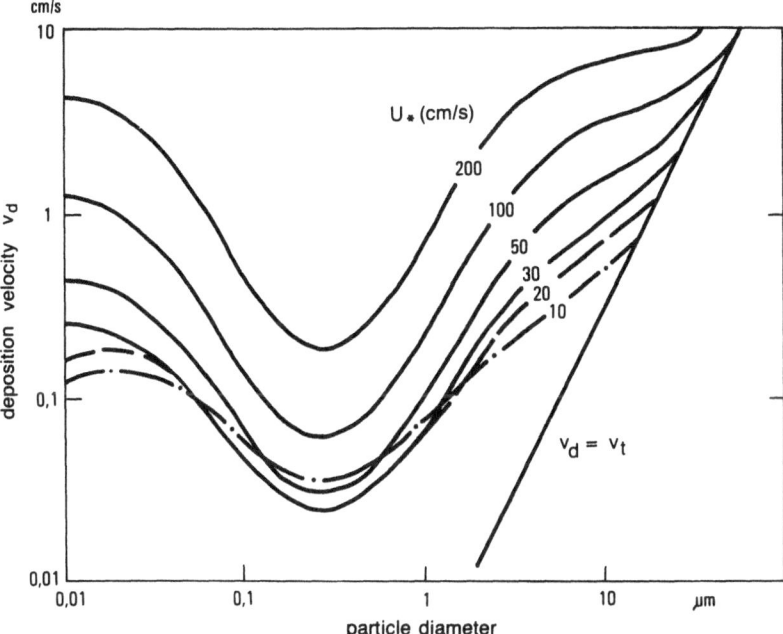

Fig. 3.36. Deposition velocity as a function of particle diameter and friction velocity u_*. Reference height $z = 3$ cm, neutral stability. (After Marggrander and Flothmann 1982)

onto surfaces. As such, this summary parameter combines the whole set of pertinent physical and meteorological conditions operative, namely atmospheric stability conditions, microphysical behaviour of single suspended particles and surface conditions (Sehmel 1980; Rohbock 1982).

Figure 3.36 gives characteristic (average) deposition velocities of metal components in aerosols plotted versus median particle diameter. The data show that the characteristic deposition velocities increase with particle size, which indicates that sedimentation is the prevalent mechanism in dry deposition over long periods.

During shorter periods, typically two-weeks intervals, changes in the meteorological boundary conditions may strongly influence the deposition velocity. Figure 3.37 provides an illustrative example of the highly variable deposition velocity of lead. The marked increase of deposition velocities from April to June 1980 as observed at the three northern stations is attributed to a prolonged high pressure situation over the North Sea. It resulted in pronounced daily variations of both temperature and relative humidity which induced efficient coagulation processes. Complementary data on deposition velocities are provided by trace element measurements in an industrial area in an inland part of England (Table 3.39), where a number of ferrous and non-

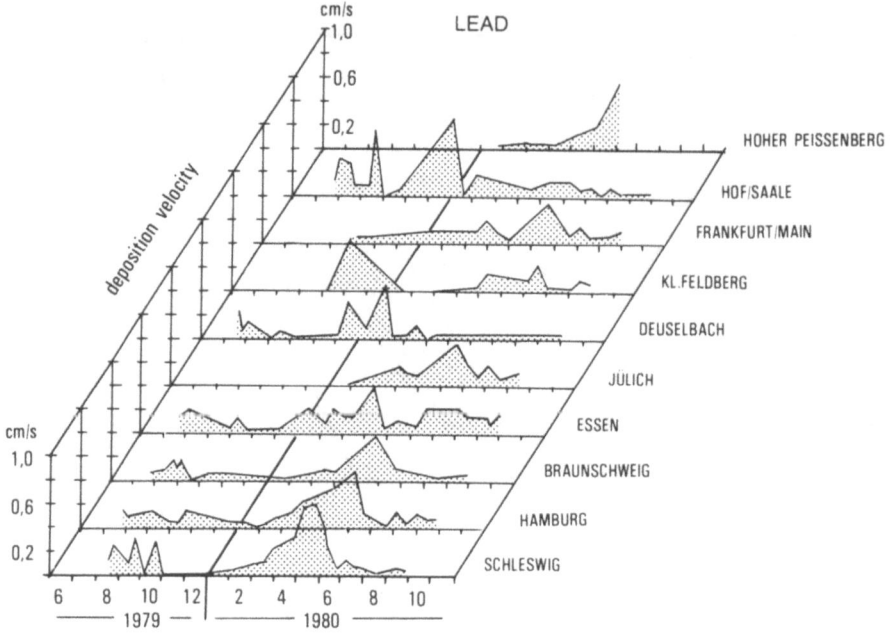

Fig. 3.37. Temporal trends of the deposition velocity of lead measured at selected stations of the German deposition network. (After Rohbock 1982)

ferrous smelting and manufacturing works are concentrated (Pattenden et al. 1982).

It is interesting to note, in comparison to the findings of the German deposition network, that the dry-to-wet ratios are roughly similar in magnitude in this urban-polluted area. The chief exception is Na which comes mostly from sea spray, with a ratio of only 0.043, while Co, Cr, Fe and Sc have ratios > 2. There is a striking difference between the exceptional ratio for Na and for other elements with high solubility, such as Cu, Pb and Zn with ratios of 0.44 to 1.6. This indicates that these, like the other elements studied, are likely to come from much closer sources, and for these the dry deposition process is more favoured, even for very soluble elements. Other correlations are summarized in Table 3.40.

The concentrations in air correlate with both wet and dry deposition, but the correlation with the wet is better. This indicates, in corroboration of German results, that the dry deposition process tends to apply to larger particles than the wet (Fränzle et al. 1992).

The values of dry deposition velocity show a moderate correlation with the dry-to-wet ratios, and an equally moderate anti-correlation with the rain-soluble fraction. Explained in terms of particle size, this suggests that elements associated with larger particles tend to have a lower rain-soluble fraction than those associated with smaller particles (cf. Galloway et al. 1982).

Table 3.39. Deposition velocities in an industrial area of England, August 1980–January 1981. (After Pattenden et al. 1982)

Element	Deposition velocity (mm s^{-1})		Ratio Dry/wet
	Dry (v_d)[a]	Wet (v_w)[b]	
Antimony	4.7	3.3	1.4
Arsenic	1.9	1.3	1.5
Cadmium	—	4.3	—
Chromium	15	4.8	3.1
Cobalt	14	5.7	2.5
Copper	14	19	0.73
Iron	16	5.6	2.8
Lead	4.6	2.9	1.6
Manganese	4.3	3.6	1.2
Nickel	10	10	1.0
Selenium	1.0	1.2	0.86
Silver	2.3	2.1	1.1
Vanadium	4.6	5.3	0.86
Zinc	2.0	4.6	0.44
Aluminium	20	18	1.1
Indium	7.9	6.3	1.3
Scandium	11	3.8	2.5
Sodium	1.6	37	0.043

[a] v_d (mm s^{-1}) $= \dfrac{\text{dry deposition rate } (\mu g\, m^{-2} s^{-1})}{\text{concentration in air } (\mu g\, m^{-3})} \times 10^3$.

[b] v_w (mm s^{-1}) $= \dfrac{\text{wet deposition rate } (\mu g\, m^{-2} s^{-1})}{\text{concentration in air } (\mu g\, m^{-3})} \times 10^3$.

Table 3.40. Correlations of deposition behaviour. (After Pattenden et al. 1982)

Pairs of parameters	Correlation coefficient	Degrees of freedom	Significance level
Wet and dry deposition[a]	0.70	14	> 99%
Air conc. and wet deposition[a]	0.76	15	> 99.9%
Air conc. and dry deposition[a]	0.57	14	> 95%
Dry/wet ratio and solubility	− 0.74	15	> 99%
Dry dep. vel. and solubility	− 0.48	15	> 90%
Dry dep. vel. and dry/wet ratio	0.45	15	> 90%

[a] Excluding Na results.

3.5.6 Deposition Patterns of Hydrocarbons

Hydrocarbons form a large group of substances which are used worldwide, yet only limited information is available about their environmental behaviour. By far the majority of publications deals with the different possibilities and routes of entry into the environment while systematic studies about distribution and fate are much lesser in number (cf. Bidleman and Olney 1975; McClure 1976; Benarie and Detrie 1978; Chang and Penner 1978; Dobbins 1979; Thibodeaux 1979; Atlas and Giam 1981; Hanna et al. 1982; Schmitt 1982; Umweltbundesamt 1982; Versino and Ott 1982; Goldstein 1983; Kirschmer et al. 1983; Fränzle and Zilling 1984; Haase and Fränzle 1984; Figge et al. 1985; BUA-Stoffberichte 1985–1989).

In the light of the specific emission factors given in Tables 3.6–3.8, estimates of the average annual hydrocarbon release are in the order magnitude of 90×10^6t, the major part of which is likely to enter into the atmosphere. According to Garrels et al. (1975), more than 2×10^6t of mineral oil hydrocarbons are released into the oceans as a result of oil transports, and their fate can be assessed in qualitative terms only.

Because of the cancerogenic potentials of a number of polycylic aromatic hydrocarbons (PAH) (Schoental 1964), these compounds constitute a particularly important sub-group of hydrocarbons, and consequently the following presentation is deliberately focused on them. PAHs are widely distributed in the environment; they are found in plant and animal tissues, in sediments, soils, air and water. Although partly of natural origin, the present concentrations in the environment are largely due to human activity. This is indicated by the fact that the detectable PAH concentrations in towns are 10 to 100 times higher than in rural areas (Table 3.41).

Table 3.41. Concentration of PAH components in urban and rural air from February 3–6, 1981 (ng m^{-3}). (After Umweltbundesamt 1983)

	PHE	FLU	PYR	CHR	BaP	DbA	BghiP	COR
Schleswig	0.4	—	—	0.9	—	—	—	—
Hamburg	0.6	1.6	2.1	3.7	0.9	0.5	1.7	0.6
Braunschweig	0.6	1.1	1.3	2.7	0.3	0.9	1.9	0.8
Essen	1.4	2.2	2.5	10.5	1.4	0.7	1.7	1.0
Jülich	1.1	1.6	1.8	6.0	1.1	0.6	1.3	0.7
Deuselbach	0.4	—	—	2.1	0.5	0.4	0.6	0.2
Frankfurt/M.	1.5	3.1	2.3	7.9	1.0	0.6	0.8	1.4
Hof	0.8	1.4	1.5	2.3	0.5	0.3	1.0	0.3
Hohenpeissenberg	—	—	—	0.9	—	—	—	—

PHE phenanthrene, FLU fluoranthene, PYR pyrene, CHR chrysene + benzo-(a)anthracene + triphenylene, BaP benzo-(a)pyrene, DbA dibenzo-(ah)anthracene + dibenzo-(ac)anthracene, BghiP benzo-(ghi)perylene, COR coronene.

The major anthropogenic sources are traffic, petrochemical industries and combustion of coal and fuel in power plants (Youngblood and Blumer 1975; Blumer 1976; Heinrich and Güsten 1978; Laflamme and Hites 1978). The normal way by which PAHs find their entrance into the environment is by transport in adsorbed form, since many have high boiling and melting points, coupled with very high distribution coefficients and poor water solubility (Heinrich and Güsten 1978). Thus the concentrations of adsorbed PAH in urban air amount to some ng m^{-3}.

The concentrations in ambient air depend largely on temperature, intensity of irradiation, traffic density and industrial activity. With half-life periods of some hours, PAHs can relatively quickly be degraded photolytically by oxidation reactions with singlet oxygen, ozone, alkylperoxy and hydroxy radicals, chlorine etc. (cf. Sect. 3.4.4). Consequently, higher PAH concentrations normally occur in winter, since then photolytic degradation is reduced on the one hand, and emission due to increased combustion of fossil fuels in higher on the other.

Deposition of PAHs seems to occur to a great extent in the vicinity of the various sources as systematic analyses have shown in soils (Grimmer et al. 1972; Blumer et al. 1977), rivers (Althaus and Sörensen 1971) and marine mussels (Dunn and Stich 1976; Dunn and Young 1976).

Particularly illuminating are random sampling studies of the deposition of PAHs in newly fallen snow in NE Bavaria during the period of January 18–21, 1978 (Herrmann 1978). Here the most important and densely populated area is the Erlangen–Nürnberg–Fürth conurbation with a predominance of electrical industries, engineering and metal works. The scarplands between the rivers Pegnitz and Naab with cuesta scarps facing W or SW are sparsely populated, with little industry. North and west of the Fichtelgebirge food and textile industries prevail, the population varying from thin to dense. In the mountains and the adjacent eastern areas an important pottery and glass industry is found.

The following regional means ($\bar{x}$) and standard deviations (s) for the four PAH examined were found:

- 3,4-benzopyrene: $\bar{x} = 29 \text{ ng} \, l^{-1}, s = \pm 28 \text{ ng} \, l^{-1}$
- fluoranthene: $\bar{x} = 200 \text{ ng} \, l^{-1}, s = \pm 140 \text{ ng} \, l^{-1}$
- 1,12-benzoperylene: $\bar{x} = 50 \text{ ng} \, l^{-1}, s = \pm 40 \text{ ng} \, l^{-1}$
- 3,4-(o-phenylene)-pyrene: $\bar{x} = 62 \text{ ng} \, l^{-1}, s = \pm 46 \text{ ng} \, l^{-1}$

For the sum of these PAHs together with 3,4-benzofluoranthene and 11,12-benzofluoranthene and World Health Organization (1971) has proposed an upper concentration limit of 200 ng l^{-1} in drinking water.

The regional distribution of PAH concentrations during the period of observation with an average snowfall of 7 l m^{-2}, and ranging from 1.5 to 22 l m^{-2}, clearly reflects two influences:

- The high concentrations at and near the conurbation of Nürnberg are due to heating and traffic. The slight shift of the maximum (> 400 ng l^{-1} in the case of fluoranthene, and > 100 ng l^{-1} for 3,4-benzopyrene) to the NE may be explained by southwesternly winds during deposition.

– Rather high concentrations, i.e. > 50% of the above maximum values, along
  the mountain barriers, on the one hand, and rather low concentrations in
  leeward situations on the other, even irrespective of higher human activities,
  may be due to intensified scavenging processes in rising air, while adiabatic
  descent in the lee exerts an opposite influence.

3.5.7 Deposition of Air-Borne Dust

The distinction between "fine particles" with a diameter < 2 μm and "coarse
particles" (> 2 μm) is a fundamental one. The physical separation of the two
modes originates because nucleation, coagulation and condensation produce
fine particles while mechanical processes produce mostly coarse ones (cf.
Fig. 3.10). Consequently, as a first approximation, the two modes are also
transformed separately, are removed separately, and are usually chemically
different. Figure 3.10 shows in addition that the fine particles comprise two
generic modes, namely the nuclei and accumulation modes while coarse particles comprise wind blown dust, emissions, sea spray, volcanic ash, and plant
tissues. Table 3.42 indicates settling velocities of particulate matter as a function
of diameter and specific density. In the light of this fundamental distinction,
Whitby and Sverdrup (1980) hypothesize that the following categories are
universally applicable for describing the physical properties of particle suspen-
sions in air:

– Marine surface background. Measured a few meters above the remote ocean
  it is characterized by a low Aitken nuclei concentration (ANC) of about
  400 cm^{-3}, by low nuclei and accumulation mode concentrations, and by
  steady concentrations observed over large areas.
– Clean continental background with negligible anthropogenic contributions.
  For these conditions the ANC is lower than 1000 cm^{-3}, the fine-particle
  volume is less than $2 \text{ μm}^3 \text{ cm}^{-3}$, and the coarse-particle volume is of the order
  of $5 \text{ μm}^3 \text{ cm}^{-3}$.

Table 3.42. Settling velocity (cm s^{-1}) of coarse
particles in air. (After Moll 1978)

Diameter (μm)	Specific density (g cm⁻³)	
	2.0	3.0
100	40	54
50	14	20
20	2.4	3.6
10	0.6	0.9
5	0.16	0.23

- Average continental background. It is a mixture of clean background plus various kinds of anthropogenic aerosols. Consequently, typical ANCs range from 4000 to 6000 cm^{-3}, and coarse-particle volume concentrations are around 25 μm^3 cm^{-3}.
- Background plus aged urban plumes. Because of the frequently considerable transport distances of urban plumes (cf. Sect. 3.1.2.2), the nuclei mode is about the same as that for average background. The volume of aerosols in the accumulation mode averages 45 μm^3 cm^{-3}, however, i.e. almost ten times the concentrations in average continental background.
- Background plus local sources. Among these motor vehicles are most common, tending to raise the ANC into the 10^4–10^5 cm^{-3} range; the accumulation mode is but little affected, however. Vehicles may stir up considerable quantities of dust, resulting in coarse-particle concentrations about one-third above the average background.
- Urban average aerosol is characterized by an ANC of about 10^5 cm^{-3}, accumulation mode volumes ranging from 30–40 μm^3 cm^{-3}, while coarse-particle concentrations tend to be the same as those for average background.
- Urban plus motorway. Near motorways or other large roadways the average nuclei mode volumes are of the order of 5–10 μm^3 cm^{-3} while maxima attain 30 μm^3 cm^{-3}. Since coarse particles are frequently resuspended, their concentration is normally elevated in relation to that of the average urban aerosol.

4 Fluxes of Chemicals in the Soil–Vegetation Complex

Environmental effects of chemicals can only be fully understood and influenced in a lasting manner if studied in their synergetic and systematic relationships. Basically, this requires a long-term ecosystem research providing a comprehensive insight into structures, functions and absorptive capacities or stability and resilience of ecosystem compartments, respectively. In the nearer future, however, other ecologically oriented approaches appear necessary in order to tackle these problems in compliance with political and commercial exigencies. In order to avoid unrealistic predictions of chemicals distribution, this involves an appropriately detailed consideration of the ecological characteristics of target areas in connection with the anticipated use pattern. Ecological and chemical data will become meaningful only if they can be judged in relation to the type of habitat where the compound will eventually be present. A chemical which is unacceptable in a certain application in one place is not necessarily unacceptable in another.

In order to enable realistic hazard predictions to be made, the basic chemical and toxicological data have consequently to be matched to additional data on the properties of the most important types of environments where a substance may ultimately occur. This can be accomplished by defining the fluxes or distribution of chemicals by means of specific transformation operators or regulatory mechanisms, respectively. Summarized in a synthetic model, they describe the input and state variables of terrestrial ecosystems and their linkage with the adjacent atmospheric and aquatic systems, sorption processes in soil as controlled by moisture and microbial activity, and the interactions and cascading of matter and energy in air, water, vegetation and soil in their capacity as essential ecosystem compartments.

Thus the structure of the cascading sub-systems depicted in foldout model II "Soil–water balance and fluxes of potentially toxic substances" are largely controlled by regulators which can be divided into two groups. First, there are "threshold regulators" such as the cationic or anionic exchange capacities (KAK or AAK), interception (IU, IL) or albedo (αIS) which control storage decisions concerning the energy, water or toxic substances entering a sub-system. Second, there are "dispositional regulators" which, although not presenting such obvious thresholds, control the disposition of energy or mass: for example, will one particular chemical be subject to solution in water, or not (LO)? After passing the regulators the fluxes of energy and mass (double arrows) may be diverted into stores of spatially and temporarily variable capacity (S_{BO}, OS, SSA, . . .).

Both regulators and stores are influenced by sets of boundary conditions such as temperature (T), pH value (PH), redox potential (RED), which are depicted by means of dotted lines. It should be emphasized that part of these regulators – particulárly the threshold regulators – and of the boundary conditions are susceptible to human intervention, and it is one of the most important problems to reliably determine extent and consequences of such manipulations.

4.1 Energy and Water Balances as Boundary Conditions of Chemical Fluxes

Soil, as the result of physical, chemical and biotic modifications of weathering mantles (regolith) is a complicated three-phase system characterized by a distinct horizonation. The solid phase has two components, the first of which is made up of inorganic material ranging from fragments of more or less unaltered rock, through weathering disaggregated mineral grains to secondary clay minerals of highly variable composition. The second component of the solid phase consists of decaying organic matter and a range of humic substances resynthesized from the products of decay. While the inorganic fraction has a determinable granulometry defining soil texture, the solid phase does not usually consist of discrete particles. Frequently, organic or inorganic particles are intimately associated to form soil aggregates or peds.

The forces of attraction leading to peds involve the formation of hydrogen and ionic bonds, and are frequently promoted by wetting and drying processes, root pressure and by polysaccharide gums secreted by the soil microflora. Between these soil aggregates are voids of various kinds, customarily classified into micro-, meso-, macropores and cracks. Together, the kind, size and distribution of soil aggregates and voids define the structure or fabric of the soil.

The liquid and gas phases are represented by soil water and air occupying the pores and voids. Water entering the soil profile by infiltration and percolating its horizons is subject to forces which, together, determine the soil water potential. Soil water normally contains a wide variety of substances in solution which are derived from a number of sources. Some are already present in precipitation, some originate in the vegetation canopy (cf. Chap. 3.5.5), while others result from solution processes in the very soil which are due to weathering, decomposition and ion-exchange. In addition, clay particles, humic polymers and the hydrated amorphous oxides and hydroxides of iron, manganese, aluminium and titanium may occur as a colloidal suspension. Both the solutes and the colloidal fraction form the background load of soil water to which anthropogenic pollutants or contaminants have to be matched.

The composition of the soil atmosphere partly filling the pores and cracks of soil may differ considerably from the free atmosphere above (cf. Table 3.1). The main difference is that O_2 concentrations are somewhat lower and CO_2 concentrations significantly higher than in the atmosphere. Furthermore, these differences may increase with depth and fluctuate with time, depending on the respiratory demand for oxygen and the rate of carbon dioxide evolution by roots and soil organisms. The relationship is quite complicated by the very different diffusion coefficients of the two gases in air and water and by their different solubilities in water or soil solution. With these reservations, Table 4.1 provides some comparative information of the O_2 and CO_2 composition of the pedospheric air of well aerated soils and dry atmospheric air.

The properties of the three-phase system soil vary both vertically and laterally. The vertical organization is referred to as the soil profile, and the sequence of layers constituting it are called horizons. They differ in the relative proportions of the solid-phase constituents and a number of other characteristics, so that each horizon presents a different set of physical, chemical and biological attributes. These specific combinations of attributes reflect the pedogenic processes operating, or which have operated, in the particular soil horizon concerned, but they also form the boundary conditions under which contemporary processes take place (cf. foldout model II).

Laterally, the characteristics of the horizons vary in a largely predictable manner in response to the soil's position in a landscape, particularly in relation to slope.

The formal processes of pedogenesis are lucidly represented in Fig. 4.1. Detailed descriptions of pedogenesis and soil classification are given in textbooks of pedology. The following are recommended for further reading: Bunting (1965), Duchaufour (1960), Knapp (1979), Millot (1964), Mückenhausen (1985), Schachtschabel et al. (1989).

4.1.1 Energy Balance at the Earth's Surface

The disposition of radiant energy at the surface of the earth is of prime importance for the understanding of soil–water balances and the related chemical transport and transformation processes. A formalized transcription of the

Table 4.1. The oxygen and carbon dioxide composition as percentage by volume of the gas phase of well aerated soils, compared with that of dry tropospheric air. (After White et al. 1984)

		O_2 (%)	CO_2 (%)
Dry air		20.95	0.03
Arable soil	Fallow	20.7	0.1
	Unmanured	20.4	0.2
	Manured	20.3	0.4
Manured sandy soil cropped with potatoes		20.3	0.6
Pasture		18–20	0.5–1.5

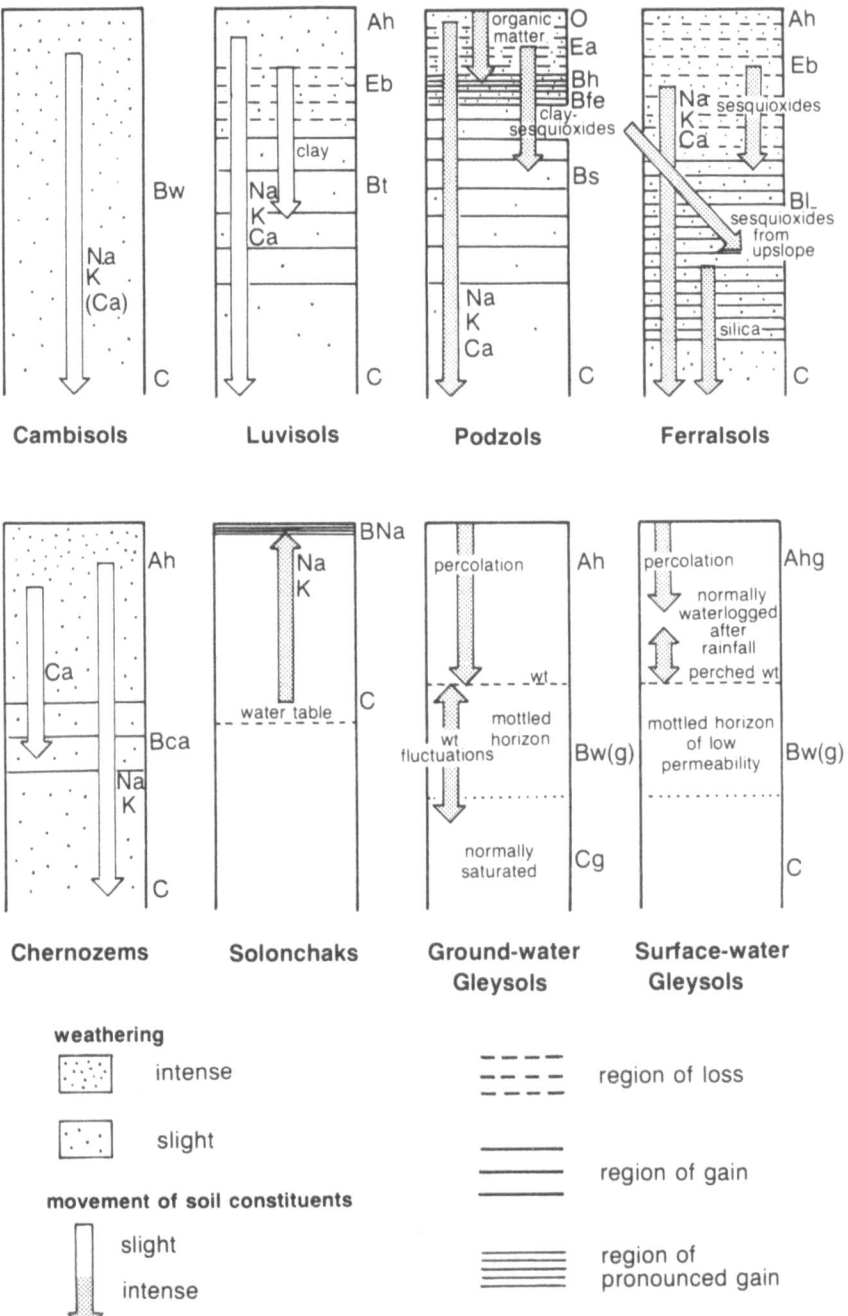

Fig. 4.1. Processes of pedogenesis. (After White et al. 1984)

energetic relationships depicted in foldout model II yields the following form of the energy balance equation:

$$R_n + W_B + W_L + r_{ET} = 0 \, . \tag{4.1}$$

In this equation, which summarizes the solar energy cascade with its numerous regulation and transformation components, any form of energy that is flowing toward the surface is considered positive and any form that is moving away from it is considered negative. R_n is the *net radiant flux*, W_B is the *heat flow* in the soil at the surface, W_L is the *sensible heat* transfer through the air at the surface, r_{ET} is the transfer rate of *latent heat* due to evapotranspiration. Conventionally, all terms are expressed in joules per minute and cm^2 ($J\,cm^{-2}\,min^{-1}$). $4.19\,J\,cm^{-2}\,min^{-1}$ equal $69.8\,mV\,cm^{-2}$ $W\,cm^{-2}$ and equal approximately 1 mm of water depth evaporated per hour.

4.1.1.1 Net Radiation

The earth's surface receives short-wave and long-wave radiation from the sun, some of which is lost through reflection, and long-wave counter radiation from the atmosphere, while itself emitting long-wave radiation. While photochemical reactions are mostly controlled by short-wave radiation, the diurnal variability of surface temperatures and the related energy fluxes are very much under the influence of long-wave radiation. Net radiant flux (R_n) may be derived from the short- and long-wave radiation balances:

$$R_n = ES_{II} \, (1 - \alpha) + R_c - {}_DA \, , \tag{4.2}$$

where ES_{II} is the sum of direct beam radiation ES_s and diffuse radiation ES_D, each term comprising short-wave and solar long-wave components, R_c the long-wave counter radiation of the lower atmosphere, ${}_DA$ long-wave radiation from the earth's surface (Stefan–Boltzmann radiation) and α is the reflectivity (albedo). To relieve foldout model II, R_c has been included into the energy input ES_{II}.

Regionally the net radiant flux of the earth's surface varies considerably with a general trend towards a positive balance or surplus in low latitudes and a negative balance or deficit in high latitudes (for details cf. for instance, Geiger 1961; Kessler 1968; Kondratyev 1972). Factors such as cloudiness and atmospheric humidity, which in conjunction with aerosols effect the transmission of both solar and terrestrial radiation, and the radiative properties of the soil and vegetation canopy, create the spatial variation in radiation balance over the earth's surface. It is not entirely latitudinal because the net radiation balance of ocean surfaces is generally greater than that of land surfaces at the same latitude, the continental maximum net radiant flux being $3600\,MJ\,m^{-2}\,a^{-1}$ in South America, while the oceanic maximum with $5880\,MJ\,m^{-2}\,a^{-1}$ is located in the Timor Sea (Budyko 1968). The low reflection and high absorption of solar radiation at ocean surfaces largely accounts for this disparity, which is further enhanced by the considerably higher amount of atmospheric counter radiation above the sea.

ES_s is measured by pyrheliometers, ES_D by pyranometers, $_DA$ by pyrgeometers. ES_{II} and $_DA$ may be determined simultaneously by one instrument, the pyrradiometer, and the radiation balance R_n of any point can be estimated directly by the net radiometer (Hofmann 1961). The pyranometer is perhaps the most versatile of these instruments, as it may be used, with corresponding modifications, to measure the components of ES_{II} separately, together with solar intensity and albedo. For many purposes, the net radiometer has been the most efficient instrument, as it provides an immediate measurement of the change in storage at any point.

Net radiometers may be of ventilated or unventilated type. In both cases, the difference in radiation flux on either side is measured. To achieve this in all wavelengths with equal intensity, major attempts are made to make the surfaces behave as nearly as possible as perfect black paint. Since the temperature of either side of the net radiation transducer depends not only upon the radiation balance, but also upon the intensity of the exchange of sensible heat with the surrounding air, the influence of convection must be eliminated. This can be achieved by means of ventilation with a sufficiently strong stream of ambient air or by covering the net radiometer with a suitable shield of a transparent material. Although no material is perfectly translucent to all wavelengths, including the long-wave radiation originating at ambient temperature, which makes the choice of the radiometer cover always somewhat of a compromise, radiometers with convection shields have proved advantageous. They can be made quite small and thus permit measurements close to the surface where their own shade does not interfere with the measurement, and they can be used within a vegetative cover without disturbing effects.

4.1.1.2 Soil Heat Flow

The rate of which heat is transferred downwards into the soil and subsurface substrate (W_B) is directly related to the nature and efficiency of the distribution mechanisms. In solids, heat is redistributed by conduction, and the flow rate is dependent on *thermal conductivity*. It is defined as the rate of flow of heat through a unit area or plate of unit thickness when the temperature difference between the faces is unity. Units are Watts per metre per degree Celsius ($W\,m^{-2}\,°C^{-1}$) or joules per centimetre per second per degree Celsius ($J\,cm^{-1}\,s^{-1}\,°C^{-1}$). Under steady-state conditions, flow of heat ($\partial Q/\partial t$) is related to the temperature gradient ($\partial \Theta/\partial x$) and thermal conductivity (λ) by

$$W_B = \frac{\partial Q}{\partial t} = \lambda \frac{\partial \Theta}{\partial x}.$$

(4.3)

A major problem in soils is that steady-state conditions are only rarely achieved and thermal conductivity is a complicated function of granulometry, mineralogical composition, compaction and water content. Its conventional determination is consequently fairly difficult. An alternative parameter, *thermal diffusivity*, is used, which is given by $\lambda/(c\rho)$, with $c =$ specific heat and $\rho =$ density. Units are metres squared per second ($m^2\,s^{-1}$). Thermal diffusivity, for a homogeneous medium, defines the rate at which temperature changes ($\partial\Theta/\partial t$) take place:

$$\frac{\partial\Theta}{\partial t} = \frac{\lambda}{c\rho}\frac{\partial^2\Theta}{\partial x^2} \, . \tag{4.4}$$

The above product, $c\rho$, is called *thermal capacity*. It defines the amount of heat required to raise the temperature of a unit volume of a substance by one degree. Units are joules per cubic centimeter per degree Celsius ($J\,cm^{-3}\,°C^{-1}$). Table 4.2 provides figures of the above thermal properties of selected materials. It ensues from Eq. (4.4) that a harmonic oscillation of temperature due to the diurnal or annual energy regime at the surface ($z = 0$) of a homogeneous medium induces damped oscillations o_1 and o_2 at the depths z_1 and z_2 (Geiger 1961):

$$o_2 = o_1 \exp\left[(z_1 - z_2)\frac{c\pi\rho}{\lambda T}\right] (°C) \, , \tag{4.5}$$

where T is the diurnal 86,400-s period or the annual 365.25 multiple of it.

Another solution of Eq. (4.4) describes the delay a temperature oscillation experiences during its downward propagation:

$$t_2 - t_1 = (z_2 - z_1)\cdot\frac{1}{2}\frac{c\rho T}{\lambda\pi} (s) \, , \tag{4.6}$$

Table 4.2. Thermal properties of selected materials. (After Schachtschabel et al. 1982; White et al. 1984)

Material	Thermal conductivity λ ($J\,cm^{-1}\,s^{-1}\,°C^{-1}$)	Thermal capacity ($J\,cm^{-3}\,°C^{-1}$)
Quartz	8.8×10^{-2}	2.1
Clay minerals	2.9×10^{-2}	2.1
Still clear water	5.7×10^{-3}	4.18
Pure ice	2.24×10^{-2}	1.93
Still air	2.5×10^{-4}	1.3×10^{-3}
Fresh snow	8×10^{-4}	0.21
Moist sand	2.2×10^{-2}	2.96
Dry sand	3×10^{-3}	1.08
Moist peat	5×10^{-3}	4.02
Dry peat	6×10^{-4}	0.58
Granite	4.61×10^{-2}	2.18
Iron	8.79×10^{-1}	3.47

Table 4.3. Maximum penetration of diurnal and annual variations in temperature. (After Geiger 1961)

Substrate	Rock	Moist sand	Snow	Dry sand
Diurnal regime (cm)	108	76	64	24
Annual regime (m)	20.6	14.5	12.2	4.6

where $t_{1,2}$ are the times a maximum or minimum temperature occurs at the depths $z_{1,2}$. Applied to several substrates characterized in Table 4.2, Eq. (4.5) may be used to determine the depth at which the amplitude of the diurnal of annual temperature variation has been damped to one per cent of the surface value. For the diurnal regime this yields 764 $\lambda/(c\rho)$ (cm) under the assumption of homogeneity, for the annual regime 19.1 times as much. In natural, inhomogeneous substrate the annual temperature variation exceeds the penetration of the diurnal one by a factor of about 10 (Richter 1986). Table 4.3 summarizes some illustrative values.

The models formulated in Eqs. (4.3) and (4.4) or their solutions (4.5) or (4.6), respectively, are generalizations from the real-world situation implying homogeneous substrate and sinusoidal temperature regime at the surface. With these reservations they may nevertheless provide useful estimates of the energy distribution in soils, which is an essential boundary condition for the soil moisture regime and a whole set of related physical and chemical processes. If higher accuracy is required, more sophisticated models are available (Lettau 1954; Richter 1986).

Heat flow in the soil is measured with sets of thermometers or thermotransducers placed at different depths. Ideally, a soil heat flow transducer, i.e. an embedded thermopile, should be located at the surface, or, at the most, covered with a thin layer of soil material. Even so, it is often difficult to exclude radiation sufficiently. This applies in particular if the soil is subject to cracking, for instance in the case of vertisols and vertic cambisols. When buried transducers are used, the heat flow at the very surface is not measured and corrections must be attempted. This can be accomplished by estimating the heat capacity of the uppermost soil layer and its change in temperature from time to time by means of thermocouples. This procedure is necessarily approximate but the inherent inaccuracies are tolerable even with bare surfaces in arid climates since, at any given time, the heat flow in the soil at the surface seldom exceeds 20% of the entire energy balance (van Bavel and Fritschen 1965).

4.1.1.3 Sensible Heat Flow in the Lower Atmosphere

The magnitude of the heat flow (W_L) into the air at the surface can be obtained by a difference when the other components of the energy balance in Eq. (4.1) are measured or it can be determined by direct measurements. Among the latter approaches two methods merit particular attention: (1) aerodynamic method, (2) Sverdrup–Albrecht method.

The aerodynamic method is based on very precise determinations of the vertical temperature and wind profiles and involves a horizontal homogeneity of the surface over considerable distance in the luff of the measuring station. It is therefore important to emphasize, in comparison to the energy balance approach, that the aerodynamic method does not provide an estimate of the heat flux at the very surface but at some height, customarily taken to be the average of the two elevations at which temperatures and wind speeds were measured. The formula utilized is of the following form (van Bavel and Fritschen 1965):

$$A_{1,2} = \frac{60\rho C_p k^2 (U_2 - U_1)(\Theta_2 - \Theta_1)}{[\ln(z_2/z_1)]^2},$$
(4.7)

in which:

$A_{1,2}$ is sensible heat flow between the elevations 1 and 2 (ly min^{-1}),
ρ is density of air (g cm^{-3}),
C_p is specific heat of air at constant pressure (cal g^{-1} °C^{-1}),
k is von Karman's constant (= 0.4),
U is wind speed at level (cm s^{-1}),
Θ is air temperature at level (°C),
z is elevation above surface at level (cm).

The application of this method is not infrequently fraught with difficulty, both experimental and theoretical, particularly on areas of limited extent such as those on which the experiments described below were carried out. They call attention to the possible existence of vertical divergence of sensible heat flux and, by inference, of vapour flux. Discrepancies between measurements of sensible heat fluxes at the very surface and at levels above it thus seem to permit an assessment of the degree of non-uniformity in terms of horizontal temperature gradients.

The Sverdrup–Albrecht method determines either W_L or ET as components of the energy balance in Eq. (4.1):

$$W_L = \frac{W_L}{ET}(R_n - W_B) \cdot \left(1 + \frac{W_L}{ET}\right)^{-1}.$$
(4.8)

The ratio W_L/ET (Bowen 1926) can be estimated from measurements of the vertical gradients of temperature ($\Delta\Theta/\Delta z$) and vapour pressure ($\Delta e/\Delta z$) above the surface:

$$\frac{W_L}{ET} = \frac{bc_p}{0.623r} \cdot \frac{\Delta\Theta/\Delta z}{\Delta e/\Delta z} \sim 0.49\frac{\Delta\Theta/\Delta z}{\Delta e/\Delta z},$$
(4.9)

where b is the atmospheric pressure, c_p is the specific heat of air, and r is the latent heat of water vapour. Substituting W_L/ET in Eq. (4.8) yields:

$$W_L = 0.49\frac{\Delta\Theta/\Delta z}{\Delta e/\Delta z}(R_n - W_B) \cdot \left(1 + 0.49\frac{\Delta\Theta/\Delta z}{\Delta e/\Delta z}\right)^{-1}.$$
(4.10)

The above formulation of Bowen's ratio is valid provided two requirements are fulfilled: firstly W_L and the rate of evaporation (or evapotranspiration) at the surface must be constant with height, and secondly the transfer coefficients for heat and water vapour must be equal (Swinbank 1958).

It is possible to ensure with some confidence that the first condition is met by working on terrain of sufficiently extended horizontal uniformity. With regard to the postulated equality of transfer coefficients, however, there is both theoretical reason and empirical support by observation for eddy conductivity to exceed eddy diffusivity in unstable conditions, increasingly with instability and with height. An estimate of evaporation by this method under unstable conditions, as will be the case when the heating is very strong, must therefore be excessive (cf. Baader 1978). Furthermore, it should be emphasized that this method, as the preceding one, demands high accuracy in measurement, in particular with regard to specific humidity. If this is done by means of the wet-bulb hygrometer, the precision demanded for ordinary climatological purposes is insufficient for the present method.

4.1.1.4 Latent Heat Flux

The latent heat flux can be estimated as a result of the vaporization or condensation of water. Consequently, five general methods are used to evaluate the water vapour flux caused by evapotranspiration (cf. Milthorpe 1962, Schrödter 1985).

1. Analogous to Eq. (4.8) the Sverdrup–Albrecht method may be used to determine ET:

$$ET = (R_n - W_B)\left(1 + \frac{W_L}{ET}\right). \tag{4.11}$$

2. Analogous to flux Eq. (4.7) Thornthwaite and Holzman (1939) suggested:

$$ET = \frac{\rho k (U_2 - U_1)(q_1 - q_2)}{[\ln(z_1/z_2)]^2} \tag{4.12}$$

with q = water vapour content at level, ρ = air density and the other terms being those of Eq. (4.7).

3. A continuous record of water vapour convection by means of the eddy-correlation technique as described by Swinbank (1958), Taylor (1958), Frankenberger (1958) and Dyer (1961) yields estimates of ET by

$$ET = \frac{1}{t}\int_0^t \rho w q \, dt , \tag{4.13}$$

where ρ = air density, w = momentary vertical wind speed, and q = instantaneous value of water vapour concentration. The problem of measuring ET as the time average of the vertical flux of water vapour thus becomes one of designing instrumentation capable of measuring the turbulent air motion and structure of

Table 4.4a. Latent heat of vaporization as a function of temperature

	0	10	20	30	40	50	(°C)
r	2499	2478	2452	2428	2407	2382	($J\,g^{-1}$)

q. This involves an equipment whose response time is sufficiently short to take account of all frequencies in the turbulent spectrum contributing significantly to the flux. It is found that this requirement depends both upon the height of measurement and the stability of the air, which means, in general terms, that the sensing elements should respond adequately to signals of a period of one second.

4. If all terms of the energy budget are known except the flux due to evaporation (or evapotranspiration), the latter can be obtained by a difference.

5. Finally the latent heat flux can be found by direct measurement, i.e. by means of weighing lysimeters on the local scale or, on the regional scale, as the difference term of the water budget of a drainage system.

The latter determination is only an approximation since the regionally differing temperature of the soil surface is not known with the adequate amount of accuracy required to use the correct value for the latent heat of vaporization (Table 4.4a). In lysimeters these measurements do only pose the problems described in Section 4.1.1.2. Thus the water vapour flux as determined by any of the above methods provides information on the flux of latent heat $r \cdot ET$.

4.1.1.5 Efficiency of Photosynthesis

In comparison to the above components of the energy balance summarized in Eq. (4.1), the absorption of the photosynthetically active radiation in the plant cover plays a minor role only. Yet it merits some interest because it furnishes useful information of the amounts of solar energy available for photochemical reactions at the surface of soil and vegetation under natural conditions.

A quantitative description of the absorption process in terms of foldout model II is provided by the equation

$$dP = \gamma(ES)_v\, s\, dz \,, \tag{4.14}$$

where γ is the coefficient of absorption of photosynthetically active radiation (i.e. wavelength ranges from 400–480 nm and 640–690 nm), $(ES)_v$ is the incident direct beam and diffuse short-wave radiation, s is the specific leaf surface (i.e. the surface of leaves in a unit volume) and z is the vertical coordinate counted upward (Budyko 1968). When solving Eq. (4.14), value $(ES)_v$ is the radiation at the upper boundary of the vegetation cover, and the distribution of the leaves' specific surface by height, $s = s(z)$, can be considered to be known.

According to Monteith and Szeicz (1960) the daily energy rates required for gross assimilation are about 28.5, for respiration 6.7, and for net assimilation 21.3 $J\,cm^{-2}\,d^{-1}$. This means that normally the photosynthetic efficiency of the vegetation cover is in the order of magnitude of 1–2% of the total incident radiation. These findings were corroborated by numerous authors. Exemplary studies from different parts of Central Europe are, for instance: Nanson (1962), Baumgartner (1967), Ruetz (1973), Runge (1973), Duvigneaud and Kestemont (1977), Ellenberg et al. (1986) and v. Stamm (1992).

Recent accounts for maize with full cover and under ideal conditions which defined an efficiency of about 10% are based on the visible light spectrum only. Since the visible spectrum is approximately 40% of the total global radiation, the efficiency on the basis of the energy balance is only 4%. If this value is further corrected for respiration, the latter being 10–20% of photosynthesis, it decreases to about 3% and therefore is not important from the purely quantitative budgetary point of view.

With regard to photochemical transformation reactions, however, the situation may be different (cf. Sect. 2.1.3.2 and Sect. 4.2.1.4). A number of compounds, e.g. various chlorinated aromatics, show distinctly higher conversion rates if adsorbed on particulate matter than if deposited as solids or thin films (Gäb et al. 1974). This can be attributed to bathochromic shift, changes in the relative extinction, or appearance of new absorption bands as a consequence of adsorption, and finally to fixation of the dispersed molecules in the adsorbed phase resulting in an intensified compound–oxygen contact (Korte 1978).

4.1.1.6 Energy Balance of Wet, Dry and Vegetated Surfaces

The distribution of energy along the pathways summarized in Eq. (4.1) varies considerably according to the type of surface. For land surfaces, lateral heat transfer beneath more extensive, level surfaces can be ignored for the sake of simplicity. Rock and soil are not only poor conductors of heat, but temperature gradients within them are relatively modest. This does not apply, however, to isolated rocks of smaller dimensions (e.g. large blocks) or near-vertical slope facets where considerable horizontal temperature gradients may develop (Friedel 1976; Jäkel and Dronia 1976; Smith 1977). They then have important consequences for the near-surface movement of weathering solutions, which in turn gives rise to specific nano-relief features (Mosebach 1974; Haberland and Fränzle 1975; Benet and Jouanna 1983). Under a fluid surface and in the atmosphere, however, a lateral transport of heat takes place, and the basic energy balance Eq. (4.1) must consequently include a lateral movement of latent and sensible heat energy.

For the sake of illustrative simplicity, the energy balance of bare surfaces is discussed first, since they display the interesting relationships particularly well. The data are derived from measurements in Arizona and pertain to one day on which the surface was very wet and another on which the surface was relatively dry. Both days had weather conditions typical for an arid climate, i.e. cloudless sky, high radiation level, high temperatures and low moisture content of the air.

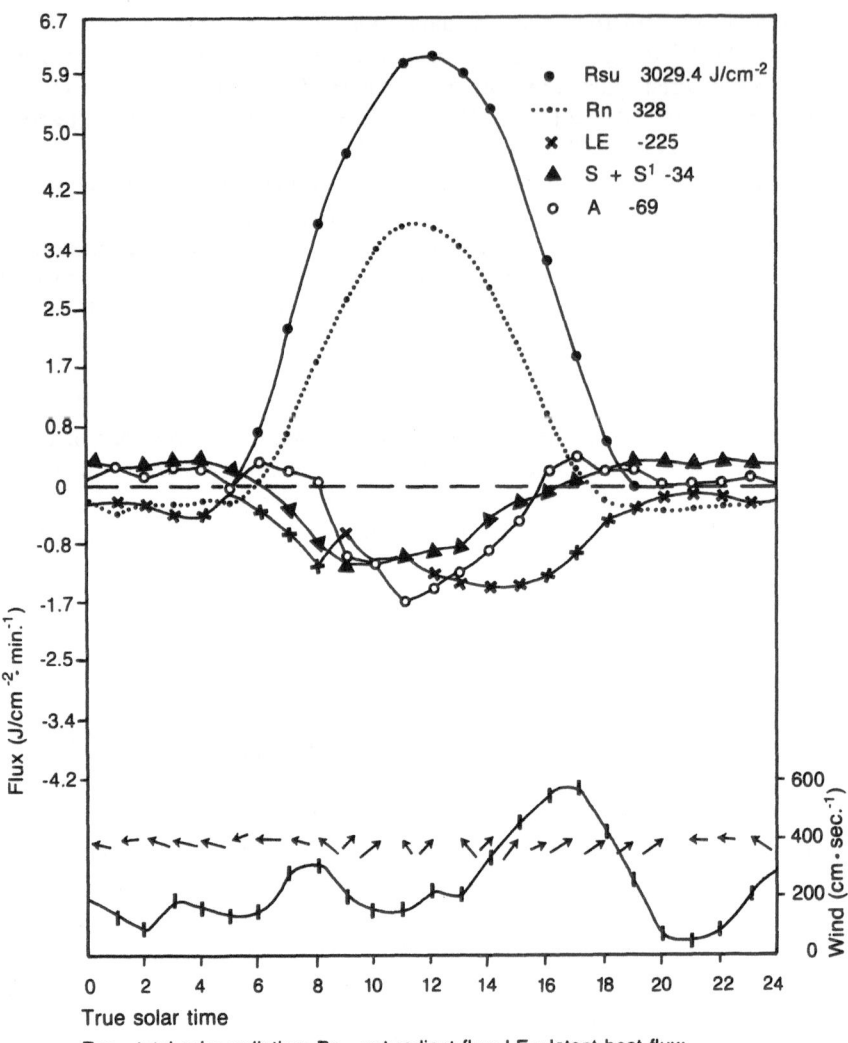

Fig. 4.2. Hourly energy balance for a bare, dry surface on May 2, 1961 at Tempe, Arizona. (After Van Bavel and Fritschen 1965)

The wet-day measurements on a typical day in an arid climate, with high radiation intensity and air temperatures, and extremely low humidity of the air indicate that the distribution of energy over the various terms of the budget is not radically different from what has been measured under more temperate and humid conditions (cf. Albrecht 1941; Lettau and Davidson 1957; Tanner 1960; Frankenberger 1961; Baumgartner 1966). Considering the whole of the day,

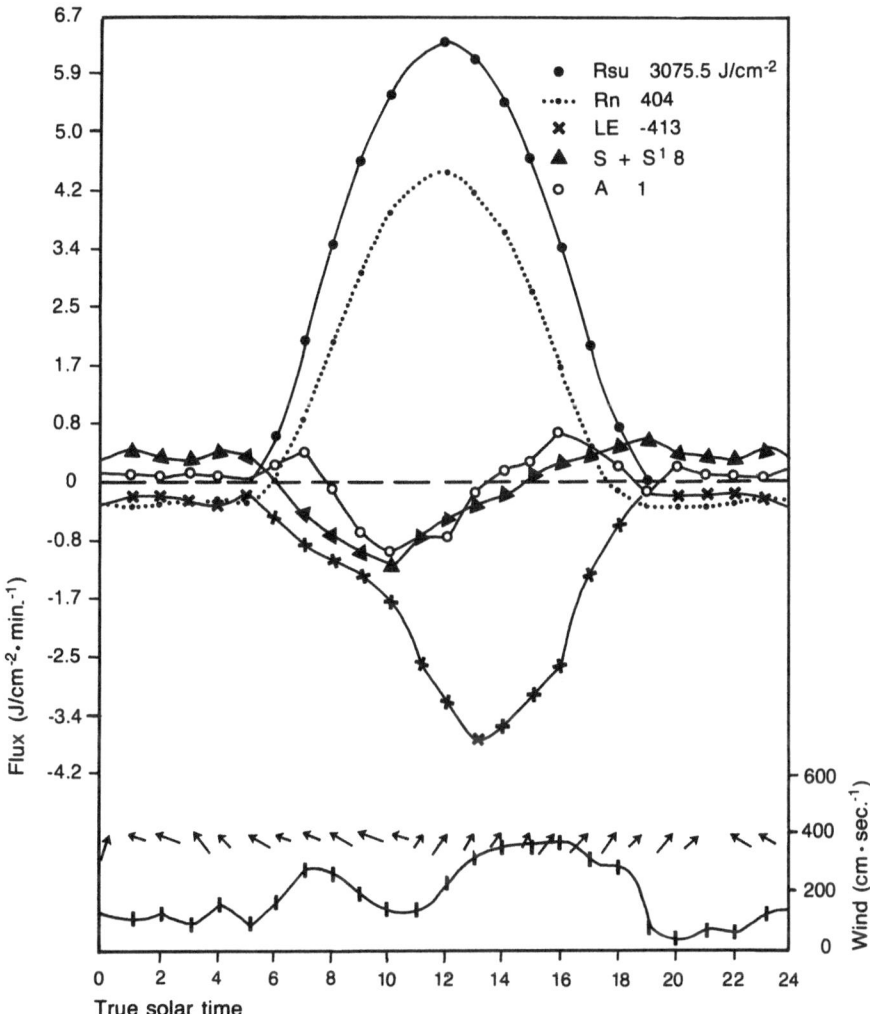

Fig. 4.3. Hourly energy balance for a bare, wet surface on April 29, 1961 at Tempe, Arizona. (After Van Bavel and Fritschen 1965)

both soil and air show no appreciable net gains or losses in energy; i.e. evaporation can be almost entirely equated to net radiation, while advection at the surface plays a certain role only during the later part of the afternoon.

In the case of the dry soil, solar radiation is practically identical, but net radiation is diminished by about 15% throughout the day as a consequence of the higher albedo of the dry surface and its higher temperature. Heating of both

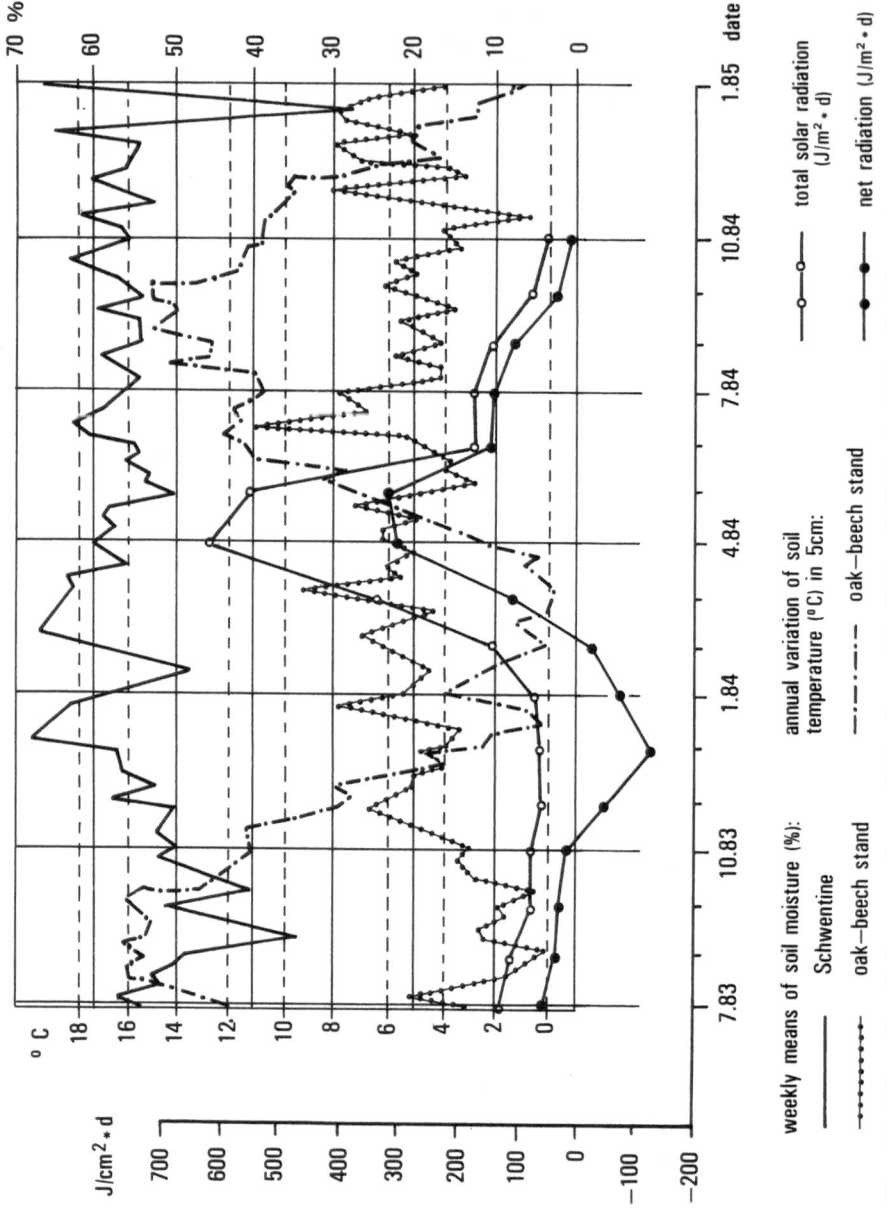

Fig. 4.4. Microclimatic features of an oak-beech stand in Schleswig-Holstein. (After Grünberg 1987)

air and soil continues for the greater part of the day. Evaporation takes place at a distinctly lower rate, and is obviously governed by wind movement which was slight but showed two distinctive maxima associated with a noticeable increase in the evaporation rate.

Under these circumstances, the soil surface maximum temperature can exceed 70°C, which may cause a substantial increase in physical desorption rates (cf. Sect. 4.1.2). From energy and microclimate measurements in a natural mulga (*Acacia aneura* F. Muell.) plant community in central Australia, Turner (1965) derived the following multiple regression:

$$y = -30.370 + 0.711x_1 + 1.809x_2 - 0.011x_2^2 , \qquad (4.15)$$

where y = soil surface maximum temperature °C, x_1 = screen maximum temperature °C, x_2 = solar radiation in 0.1 langleys per day.

The 95% confidence limits of the estimation at the means are $\pm 3.5°C$ and they increase by only one-tenth of this amount at the extremes of the range.

On the basis of comparative microclimatic investigations on three neighbouring sites (oak-beech stand, meadow on meliorated low-fen, and uncultivated garden) in Schleswig–Holstein, partly summarized in Figs. 4.4 and 4.5, the correlations of Table 4.4b were determined (Grünberg 1987; Fränzle et al. 1989).

4.1.2 Water Balance of Bare Soil and Vegetation Stands

A number of models for plant–environment interactions, and particularly for the utilization of water and energy by plants, have been developed in the course of

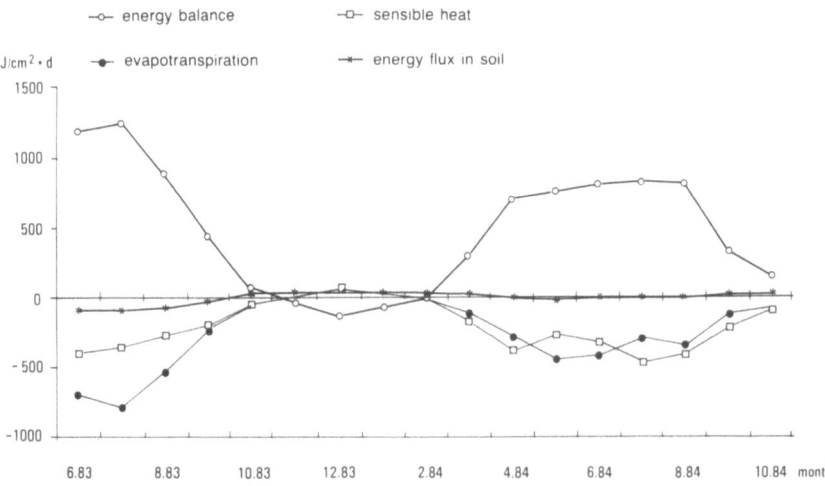

Fig. 4.5. Microclimatic features of a meadow on meliorated low fen in Schleswig-Holstein. (After Grünberg 1987)

the last four decades. Several of these attempt to understand plant growth and water use as related to specific physiological and environmental parameters (cf. Cowan 1965; Monteith 1965; de Wit 1978). These models, however, cannot be applied to situations for which only few data are available unless a number of simplifying assumptions are used.

On the regional and geographical levels, other models of a predominantly qualitative character have been suggested (Thornthwaite 1948; Prescott and Thomas 1949; Meigs 1953; Turc 1954; Dzerdzeevskii 1958). As a consequence of this dichotomous development, there have been attempts to bring these two approaches together with a view to simplifying the comprehensive models, so as to make them applicable to regional use in areas with limited data, without introducing misleading oversimplifications.

Examples of soil–water balance studies which appear, in some ways, to partly fill in this middle ground are those of Butler and Prescott (1955), Arkley and Ulrich (1962), Fuchs and Stanhill (1963), Slatyer (1968), Benecke and van der Ploeg (1978).

The soil–water balance equation as depicted in foldout model II is generally written in the form:

$$N - A_0 - (GWA + TF) - ET + \Delta(BF + SW + GW) = 0 , \qquad (4.16)$$

where N = precipitation, A_0 = surface run-off, GWA = ground water run-off,

Table 4.4b. Correlation matrix of essential climatic elements of three neighbouring sites in Schleswig-Holstein

1. Macroclimate solar radiation	All correlation
2. Macroclimate net radiation	coefficients are
3. Solar radiation (forest)	significant on
4. Net radiation (forest)	the 99% level
5. Macroclimate PAR	
6. PAR (forest)	
7. Soil heat flow	
8. Soil heat flow (meadow)	
9. Soil heat flow (garden)	
10. Screen temperature (meadow)	
11. Screen temperature (forest)	
12. Screen temperature (garden)	
13. 5 cm soil temperature (meadow)	
14. 5 cm soil temperature (forest)	
15. 5 cm soil temperature (garden)	
16. 20 cm soil temperature (meadow)	
17. 20 cm soil temperature (forest)	
18. 20 cm soil temperature (garden)	
19. Evapotranspiration (meadow)	
20. Evapotranspiration (garden)	

Table 4.4b. (Contd.)

	1.	2.	3.	4.	5.	6.	7.	8.	9.	10.	11.	12.	13.	14.	15.	16.	17.	18.	19.	20.
1.	■	0.98	0.46	0.69	0.96	0.57	0.70	0.68	0.62	0.66	0.76	0.82	0.74	0.70	0.73	0.70	0.70	0.67	0.95	0.89
2.		■	0.42	0.72	0.95	0.46	0.69	0.75	0.70	0.74	0.80	0.84	0.78	0.74	0.77	0.73	0.70	0.70	0.95	0.92
3.			■	0.86	0.37	0.98		0.50												
4.				■	0.61	0.82	0.33	0.79	0.66	0.50	0.40	0.53	0.76	0.72	0.76	0.71	0.72	0.73	0.61	0.47
5.					■	0.98	0.79	0.44	0.45	0.33	0.75	0.77							0.91	0.94
6.						■	0.33	0.46	0.37										0.46	
7.							■	0.54	0.52	0.73	0.72	0.73	0.65	0.64	0.67	0.61	0.63	0.61	0.73	0.67
8.								■	0.93	0.68	0.59	0.63	0.49	0.47	0.53	0.40	0.46	0.43	0.72	0.57
9.									■	0.73	0.58	0.60	0.47	0.45	0.49	0.40	0.44	0.39	0.61	0.56
10.										■	0.99	0.99	0.97	0.97	0.96	0.95	0.96	0.93	0.70	0.78
11.											■	0.99	0.97	0.97	0.96	0.96	0.97	0.95	0.75	0.82
12.												■	0.97	0.97	0.97	0.96	0.97	0.95	0.77	0.82
13.													■	0.99	0.99	0.99	0.99	0.99	0.73	0.82
14.														■	0.99	0.99	0.99	0.99	0.69	0.77
15.															■	0.99	0.99	0.99	0.72	0.80
16.																■	0.98	0.99	0.69	0.78
17.																	■	0.99	0.68	0.78
18.																		■	0.31	0.34
19.																			■	0.34
20.																				■

TF = throughflow, ET = evapotranspiration, BF = soil moisture, SW = perched water (above a compact horizon), GW = ground water.

For the sake of greater convenience the sum of (GWA + TF) is termed deep drainage and Δ(BF + SW + GW) change in soil water storage (initial minus final) during the period, and for the depth of measurement. For a number of cases, deep drainage may be adequately defined as the amount of water passing beyond the root zone, or, for experimental purposes, as the amount passing below the lowest point of measurement. All symbols have dimensions of length, e.g. mm (Benecke and van der Ploeg 1978; Benecke 1984). Equation (4.16) can be more specifically written as

$$\int_{t_1}^{t_2} [(N - A_0) - (ET) - v_z]\, dt = \int_{t_1}^{t_2} \int_0^z \frac{\partial (BF)}{\partial t}\, dz\, dt, \tag{4.17}$$

where $(t_2 - t_1)$ is the time interval over which the measurements are made (s), z is the depth to the lowest point of measurement (cm), v_z is the net downward flux of water at depth z (cm s^{-1}) and BF is the volumetric soil water content (cm^3 water cm^{-3} soil). N, A_0 and ET are in units of mm s^{-1} or g cm^{-2} s^{-1}. In the absence of a water table near the surface, v_z is generally positive.

In comparison with aboveground measurements or estimates of the water vapour flux (cf. Sect. 4.1.1.4) the advantages of the soil–water balance approach are those of ease of data processing and integration, since the soil–water reservoir (BF) automatically integrates extraction rates between observations. The disadvantages are largely associated with a somewhat lower level of measurement accuracy and the difficulty of adequately assessing evapotranspiration during periods of rainy weather. Therefore its applicability is, to a considerable degree, restricted to regions of relatively high potential evaporation rates and sufficiently well-defined alternations of rainy and dry weather.

4.1.2.1 Precipitation and Interception

The measurement of precipitation (N) at a site or in a region is generally considered as simpler and more straightforward than that of the other terms in the water balance equation. The technical requirements are comprehensively described by Gilman (1964), WMO (1965), v. Hoyningen-Huene and Nasdalack (1985) and Sevruk (1981). The identification of unequivocal spatial structures should be controlled by means of variogram analysis (cf. Chap 3.5.1), which is particularly important in vegetation stands.

Here, marked differences in the pattern of precipitation actually reaching the ground normally develop in many plant communities because of the gross interception of precipitation by the vegetation. Subsequently, it is partly transferred to the soil by channelling down the main stems ("stemflow"), partly by dripping from branches, twigs and foliage ("canopy leaching") or it may be lost by evaporation from the wet surfaces. This latter proportion constitutes the term "net interception" (I). Further differences in the amount of precipitation reaching the ground between plants ("throughfall") are particularly due to the disturbed wind structure, and are most noticeable in the case of snow.

In view of the complicated physical nature of net interception, attempts to assess it by means of measurement and by indirect estimations are quite numerous. Critical reviews of relevant older literature are provided by Kittredge (1948), Delfs (1955), Penman (1963), Lull (1964) and Fränzle (1976b, c). In the framework of a comprehensive agroclimatological model, Braden (1985) developed the following interception estimate:

$$N_i(R) = a \cdot \text{LAI} \cdot \left(1 - \frac{1}{1 + N_0 b / a \, \text{LAI}}\right), \tag{4.18}$$

where (in the author's notation) $N_i(R)$ = net interception, $a \cdot \text{LAI}$ = saturation parameter dependent on leaf area index (LAI) and a species-specific maximum interception a, N_0 = above-canopy precipitation, b = density of vegetation cover.

The Braden approach has the double advantage of mathematical simplicity and physical foundation, and its validity has been widely tested (Löpmeier 1987).

The amounts of precipitation retained on the surfaces and the proportion of stem flow vary considerably with the morphological characteristics of the species concerned and the structure of the precipitation events (Delfs 1955; v. Hoyningen-Huene 1983; v. Hoyningen-Huene and Nasdalack 1985). With rain, it is usual to observe drip or stemflow after an area rain total of about 2 mm has been received but, with freezing rain or snow, under conditions favouring retention on the leaves, twigs branches and stems (i.e. low wind, temperatures a few degrees below freezing) several times this amount may be accumulated.

Stemflow is enhanced by a smooth bark and by branches and leaves which are inclined upwards. Figure 3.34, together with Tables 4.5 and 4.6, show that in deciduous (beech) and evergreen (spruce) forests the amounts vary considerably,

Table 4.5. Interception of beech and spruce stands of the Rothaar Mts., North Rhine-Westphalia, in relation to precipitation. (After Eidmann 1960)

Precipitation (mm)	Spruce (%)	Beech (%)	Difference
0–1.0	81.7	71.9	9.8
1.1–2.0	63.2	55.2	8.0
2.1–3.0	54.8	24.8	30.0
3.1–5.0	46.9	18.4	28.5
5.1–10.0	32.8	16.5	16.3
10.1–15.0	30.3	16.2	14.1
15.1–20.0	25.5	17.2	8.3
> 20.0	24.1	17.7	6.4

Table 4.6. Precipitation, throughfall, stemflow, and infiltration in spruce stands of the Schönbuch. (After Bücking and Krebs 1986)

	(1) Precipitation mm	(2) Throughfall mm	% of (1)	(3) Stemflow mm	% of (1)	(4) Infiltration (lysimeter) mm	% of (1)	% of (2)
Site 1								
S 79	271.9	109.2	40	0.12		161.3	59	148
W 80	338.6	220.0	65	1.01		212.4	63	97
S 80	447.9	236.8	53	0.56		261.7	58	111
1980	786.5	456.8	58	1.57	< 1	474.1	60	104
W 81	252.2	114.3	45	0.29		110.7	44	97
S 81	464.7	244.5	53	0.74		275.1	59	113
1981	716.9	358.8	50	1.03	< 1	385.8	54	108
W 82	390.9	263.1	67	0.66		189.2	48	72
S 82	611.9	382.4	62	0.81		393.6	64	103
1982	1002.8	645.5	64	1.47	< 1	582.8	58	90
W 83	403.8	244.8	61	0.79		250.9	62	102
Average	795.5	454.3	57			463.7	58	102
W	346.4	211.1	61			190.8	55	90
S	449.1	243.2	54			272.9	62	112
Site 3								
S 79	287.4	188.5	66	—	—	149.9	52	80
W 80	356.5	150.6	42	—	—	174.7	49	116
S 80	443.8	243.3	55	—	—	235.7	53	97
1980	800.3	393.9	49	—	—	410.4	51	104
W 81	208.4	102.2	49	—	—	125.7	60	123

S 81	402.2	219.4	55	—	—	282.9	70	129
1981	610.6	321.6	53	—	—	408.6	67	127
W 82	350.4	226.9	65	—	—	163.6	47	72
S 82	509.1	304.8	60	—	—	376.0	74	123
1982	859.5	531.7	62	—	—	539.6	63	101
W 83	377.1	212.6	56	—	—	277.9	74	131
Average								
W	733.7	412.1	56	—	—	446.6	61	108
S	323.1	173.1	54	—	—	185.5	57	107
	410.6	239.0	58	—	—	261.1	64	109
Site 5								
S 79	279.1	174.9	63	—		144.4	52	83
W 80	333.4	212.7	64	0.36		220.8	66	104
S 80	406.9	264.0	65	0.26		249.6	61	95
1980	740.3	476.7	64	0.62	<1	470.4	64	99
W 81	200.9	123.0	61	0.23		124.1	62	101
S 81	415.6	275.6	66	0.38		259.9	63	94
1981	616.5	398.6	65	0.61	<1	384.0	62	96
W 92	345.3	250.4	73	0.46		151.7	44	61
S 82	516.7	352.9	68	0.45		330.8	64	94
1982	862.0	603.3	70	0.91	<1	482.5	56	80
W 83	377.8	246.2	65	0.52		208.9	55	85
Average								
W	719.0	474.9	66			422.6	59	89
S	314.4	208.1	66			176.4	56	85
	404.6	266.8	66			246.2	61	92

S = Summer, W = Winter

Table 4.7. Precipitation, throughfall, stemflow, and infiltration in beech stands of the Schönbuch. (After Bücking and Krebs 1986)

	(1) Precipitation mm	(2) Throughfall		(3) Stemflow		(4) Infiltration (lysimeter)		
		mm	% of (1)	mm	% of (1)	mm	% of (1)	% of (2)
Site 2								
S 79	273.7	151.3	55	—	—	110.4	40	73
W 80	351.4	273.8	78	~ 18.0	5.1	281.9	80	103
S 80	460.2	291.7	63	23.2	5.0	278.1	60	95
1980	811.6	565.5	70	41.2	5.0	560.0	69	99
W 81	221.0	165.5	75	5.6	2.5	165.9	75	100
S 81	432.2	265.1	61	29.7	6.9	268.0	62	101
1981	653.2	430.6	66	35.3	5.0	433.9	66	100
W 82	363.5	300.1	83	15.5	4.3	207.4	56	69
S 82	599.8	402.9	67	35.4	5.9	391.9	65	97
1982	963.3	703.0	73	50.9	5.0	599.3	62	85
W 83	414.9	328.5	79	23.2	6.0	288.6	70	88
Average	779.2	544.7	70	45.0	5.0	498.1	64	91
W	337.7	267.0	79	15.6	5.0	236.0	70	88
S	441.5	277.7	63	29.4	5.9	262.1	59	94
Site 4								
W 80	310.2	236.9	76	11.7	4.0	—	—	—
S 80	370.0	253.7	69	14.5	4.0	236.2	64	93
1980	680.2	490.6	72	26.2	4.0	—	—	—
W 81	193.3	138.4	72	~ 3.2	2.0	145.1	75	105
S 81	429.0	293.9	69	28.3	7.0	270.3	63	92

1981	622.3	432.3	69	31.5	5.0	415.4	67	96
W 82	340.1	266.0	78	9.7	3.0	201.8	59	76
S 82	524.3	366.2	70	35.2	7.0	369.2	70	101
1982	864.4	632.2	73	44.9	5.0	571.0	66	90
W 83	388.3	304.0	78	21.8	6.0	266.0	69	88
Average	749.1	540.9	73	37.6	5.0	496.2	66	92
W	308.0	236.3	77	11.6	4.0	204.3	66	86
S	441.1	304.6	69	26.0	6.0	291.9	66	96
Site 6								
S 79	279.1	195.4	70	—	—	95.0	34	49
W 80	333.4	255.4	77	—	—	243.6	73	95
S 80	406.9	280.2	69	—	—	231.6	57	83
1980	740.3	535.6	72	—	—	475.2	64	89
W 81	200.9	140.0	70	1.57	1	124.1	62	89
S 81	415.6	296.4	71	6.98	2	258.0	62	87
1981	616.5	436.4	71	8.55	1	382.1	62	88
W 82	345.3	262.5	76	6.03	2	159.2	46	61
S 82	516.7	373.8	72	8.82	2	328.1	63	88
1982	862.0	636.3	74	14.85	2	487.3	57	77
W 83	377.8	302.6	80	8.6	2	271.8	72	90
Average	719.0	526.6	73	13.3	2	427.9	60	81
W	314.4	240.1	76	5.4	2	199.7	64	83
S	404.6	286.5	71	7.9	(2)	228.2	56	80

S = Summer; W = Winter

beech providing much more stem flow and much less interception loss while oak would have an intermediate position.

4.1.2.2 Infiltration and Surface Runoff

Surficial runoff (A_0) occurs whenever the rate of effective precipitation (i.e. $N - I$) exceeds the rate of infiltration ($F*$) and the resultant accumulation of surface water exceeds the pondage capacity ($S*$) at the point of measurement. The most important regulator being $F*$, it is useful in applied hydrology and in relation to pollutant transport to characterize the dynamics of infiltration by a small number of parameters. Various forms of algebraic infiltration equation have been used in the 1930s and 40s (Philip 1964) but neither of these equations can be related directly to the physical processes involved in infiltration, and neither gives a good fit to the dynamics of infiltration even under the simplifying assumption of a uniform soil. Philip (1957) developed a simple physical model of infiltration which is, however, closely related to more precise diffusion descriptions of infiltration (cf. Philip 1964):

$$i = St^{0.5} + At \, , \tag{4.19}$$

where i is the cumulative infiltration at time t, and the constants S ("sorptivity") and A have a physical meaning related to the diffusion analysis of infiltration. The first term of the right-hand side of Eq. (4.19) describes the contribution to infiltration due to capillarity, whilst the second term mainly represents the contribution due to gravity. The differential form of Eq. (4.19) is

$$v = \tfrac{1}{2}St^{-1/2} + A \, ,$$

where v = rate of infiltration (cm s^{-1}).

Table 4.8 gives some values of the infiltration capacity for particular soils in specified conditions. These values are much lower than those determined by means of a field rain simulator in the Schönbuch Nature Reserve (Baden-Württemberg), where the use of buffered cylinder infiltrometers proved imposs-

Table 4.8. Minimum infiltration rates of different soils and rocks. (After Musgrave 1955)

Soils, rocks	Rate of infiltration (mm h^{-1})
Sand, loess, silt	11–7
Sandy loam	7–4
Clayey loam, soils poor in organic matter	4–1
Clays, alkaline soils (solonets)	< 1

ible because of the extremely high spatial variability of the infiltration-relevant soil characteristics. The experimental conditions were such that a constant surface runoff rate was brought about by a constant amount of rainfall when the soil had reached its saturation point. Since evaporation is negligible during the short duration of the experiment in comparison to the rainfall applied ($100-250$ mm h^{-1}), and the increment in soil water storage is assumed to be zero, the difference between precipitation and surface runoff may be equated to infiltration (Schwarz 1986).

For two clay soils tested the minimum infiltration capacity amounted to 58 and 60 mm h^{-1}, for sandy soils 79 mm h^{-1} were measured while the infiltration rates of loamy soils varied between 63 and 76 mm h^{-1}. These results are, on the one hand, indicative of a marked enhancement of infiltration due to organic matter and, in particular, desiccation cracks and root voids (cf. in this respect also Klaer and Krieter 1982); on the other, they point to high throughflow rates close to the surface, e.g. piping. As a consequence, no differences in runoff characteristics could generally be found on soils in coniferous, broadleaf or mixed forests. Variation in the runoff rates was high when the soils were initially dry but low when the soils were wet.

A comparison of these minimum infiltration rates with the maximum net precipitation rates recorded within the last hundred years leads to the conclusion that surface runoff is an exceedingly rare phenomenon in temperate forests. In relatively uniform soils three moisture zones can be distinguished during the infiltration process:

1. the thin saturated zone at the very surface whose water content decreases rapidly downwards to pass into
2. the transmission zone whose water content diminishes more gradually to
3. the basal wetting zone and wetting front, where soil moisture decreases rapidly once more.

On condition infiltration events from the surface are repeated several times at relatively short intervals, air is trapped between successive transmission zones. Thus water movement in the soil, which is dealt with in greater detail from the generalized viewpoint of potential theory in Sect. 4.1.2.5, assumes the character of a pressure-wave-induced process. It has not only far-reaching consequences for the profile differentiation of luvisols (Bartelli and Odell 1960a, b) and rhythmical clay translocation in sands (Fränzle 1971a), but also for water-borne pollutant transport (cf. Flügel 1981; Duysings et al. 1983; Einsele et al. 1986). In comparison to one-dimensional infiltration, it is a more difficult problem to study quantitatively infiltration from furrows or small punctiform sources, since this demands two- and three-dimensional treatment on the basis of a generalized diffusion analysis. As far as the redistribution of small quantities of water supplied at a point on the surface of a soil is concerned, it predicts, data conforming to the result, that the mean moisture content was a unique function of $t \cdot V^{-2/3}$, whatever the value of V, the volume of water applied (Philip 1964).

Most of this discussion has dealt with the situation that net precipitation has exceeded the infiltration capacity of the soil, even when the latter has not been saturated. The resultant type of overland flow is customarily called *Horton overland flow* (A_0 in foldout model II) and appears to be a common process in semi-arid and arid regions, where precipitation intensities are high and infiltration capacity of the sparsely vegetated soil is low (Fränzle 1976c). It is further caused or intensified by the development of a crust on the soil as the surface layer becomes compacted and the pores blocked as a result of the redistribution of soil particles following raindrop impact (de Ploey 1983). Crust formation due to lateral iron translocation is a particularly widespread phenomenon in ferric luvisols of the semi-humid tropics, where it largely contributes to enhance pediplanation processes (Maignien 1966; Fränzle 1977b, 1978b).

In temperate environments with normally modest precipitation rates and well-structured soils, except under certain conditions of cultivation and when the ground is frozen, Horton overland flow is the exception rather than the rule. Here all the pore spaces may become filled with water after a period of prolonged rainfall, thus saturating the soil. At this point the water table has risen to the surface and the effective infiltration capacity is consequently reduced to zero, and subsequent rainfall runs off directly across the surface of the slope as *saturated* (or *saturation*) *overland flow*. This situation is likely to come about towards the base of a slope or in microtopographical depressions on a slope where both local infiltration and throughflow received from higher up the slope contribute to soil moisture.

Smith and Parlange (1978) describe simple relationships which enable saturating or ponding times to be estimated from values of saturated hydraulic conductivity (k_s) and sorptivity (s). In Morocco it was found by Imeson (1983) that the amount of rain required to pond the soil (p_r) could be estimated reasonably well with one of these equations, namely

$$\int_0^{t_p} p_r \, dt = \frac{A}{k_s} \ln \frac{r_p}{r_p - k_s} , \qquad\qquad (4.20)$$

where $A = 0.5s^2$ and r_p = rainfall intensity.

It ensues from the number of boundary conditions operative in runoff that the latter varies considerably with amount, intensity and duration of precipitation, furthermore with slope configuration and soil fabric, which determine the degree and extent to which pondage can take place. (cf. Einsele et al. 1986; Schwarz 1986; Drescher et al. 1988). In natural situations, slope is rarely constant and, while runoff tends to reduce soil water recharge at the top of a slope and increases it at the bottom, minor changes of slope generally modify the slope–runoff interrelation.

Because of these factors, many of which interact to produce complex relationships (Chow 1964; Barsch and Flügel 1978; Richter 1978; Jung 1980; Dikau 1983; Leser 1983) and because it is difficult to measure runoff directly, without affecting the pattern of runoff over an experimental site (cf. Bork 1980, 1983; Jung 1980; Scholles 1985) runoff is frequently determined indirectly by

Table 4.9. Hydrometeorological characteristics of two ploughland sites in the northern DDR for the period 1961–1965. (After Flegel 1970)

	Müncheberg	Bochow
Number of erosion-inducing showers	30	39
Precipitation (mm)		
Median value	14.8	11.8
Mean extreme value	23.6	18.4
Absolute extreme value	34.8	25.2
Runoff (mm)		
Median value	1.2	2.1
Mean extreme value	3.3	5.1
Absolute extreme value	10.2	8.1

means of the water balance. This implies that evapotranspiration is either measured or assumed to be zero. Estimates of soil water storage are required before and after runoff-inducing rainfalls. If deep drainage, i.e. throughflow and groundwater runoff can also be neglected, as is the case in many brief showers, surface runoff is given as the simple difference between precipitation and the observed increment in soil water storage of the uppermost soil horizon.

Table 4.9 provides for some exemplary runoff figures from two sites in the FRG. Bochow is representative of loamy luvisols developed from Weichselian till while Müncheberg is characterized by a podzolized cambisol on coversand overlying Saalian till in a humid environment.

4.1.2.3 Deep Drainage

The term deep drainage in balance Eq. (4.16) comprises throughflow and groundwater flow and can be equated to a vertical flow which, in turn, may be calculated from hydraulic conductivity and soil water potential data (Rose and Stern 1965). The normal equation for vertical flow of water, v_z, is

$$v_z = K + K\frac{\partial h}{\partial z},\tag{4.21}$$

where K is the hydraulic conductivity (cm s^{-1}) and $(\partial h/\partial z)$ is the rate of change of soil water suction (h in cm) with depth (z in cm). Soil water suction is derived from soil water potential, in dyn cm^{-2} by the relationship $h = -\psi/\rho_w g$, where ρ_w is the density of water and g the vertical acceleration due to gravity (cf. Sect. 4.1.2.5). Unless h is very small, $(\partial h/\partial z)$ is usually much greater than unity, so that the K term in Eq. (4.21) is often negligible (Slatyer 1968). Under these circumstances, the deep drainage term of the water balance equation is given by

$$GWA + TF = \int_{t_1}^{t_2} v_z dt,\tag{4.22}$$

where $(t_2 - t_1)$ is the time period between observations (cf. Blume et al. 1978; Dörrhofer and Josopait 1980; Petzold 1984; Renzer and Strebel 1980).

In other situations there can be a net upward flux of soil water into the root zones from wetter underlying soil horizons or, in particular, from a water table close to the surface (Schroeder 1983; Wohlrab 1983; Thöle and Schreiber 1985; Wessolek et al. 1985). For comprehensive reviews of methods available for the determination of deep drainage with a particular emphasis on groundwater recharge the reader in referred to Arbeitskreis Grundwasserneubildung (1977), Fränzle et al. (1987), Freeze and Cherry (1979), Heckmann et al. (1985), Petzold (1981), Sager (1983) and Wonderen and Sage (1985).

Under certain soil conditions the diffuse water movement through the intergranular pore spaces and voids may be supplemented by concentrated turbulent throughflow in networks of pipes. These result from large voids which exist in many soils and are enlarged further by soil fauna (e.g. mice, rats, hamsters, moles, weasels, ground squirrels) and the growth and decay of roots. Frequently soil pipes develop at the interface between organic soil and the underlying mineral soil (cf. Klaer and Krieter 1982; White et al. 1984).

Discharge in completely filled pipes varies in dependence on pressure and gravity potentials (cf. Sect. 4.1.2.5), in partly filled pipes in response to the gradient of the water surface. Usually therefore, pipeflow velocity is much more rapid than matrix flow. Table 4.10 quotes estimates of flow velocities. It ensues from these figures that pipeflow may attain a considerable importance for chemical transport although the distances covered are normally small in comparison to channel flow.

Piping is perhaps more strongly associated with semiarid areas than with humid regions (Jones 1987). Drainage and slope development in badlands all over the world are frequently dominated by piping (Bryan and Yair 1982). The most spectacular enhancement of the phenomenon seems to be reported from the varved late-glacial South Thompson Silt at Bornhart Vale in interior British Columbia, where chambers up to 6 m long and 9 m wide have been formed (Evans and Buchanan 1976; Clague et al. 1987).

Table 4.10. Flow velocities along different routes in the catchment. (After Weyman 1975, from various sources)

	Flow routes	Velocities (m h^{-1})
Surface	Channel flow	300–10 000
	Overland flow	50–500
Soil flow	Pipeflow	50–500
	Matrix throughflow	0.005–0.3
Groundwater flow	Limestone (jointed)	10–500
	Sandstone	0.001–10
	Shale	10^{-8}–1

4.1.2.4 Change in Soil Water Storage

Measurement of change in soil water storage is most accurately conducted by the use of weighing or hydraulic lysimeters provided they are properly designed and sited (Bloemen 1964; Klausing 1970; Liebscher 1970; Schroeder 1976). Lysimeters cannot be used, however, when the nature of the species composition, the spatial structure of the vegetation cover, the depth and ramification of the root system or other factors make it impossible to simulate the natural environment inside the lysimeter itself. In such cases, determinations of changes in soil water storage at different points in the plant community provide the only technique for evaluating $\Delta(BF + SW + GW)$. As for precipitation or infiltration measurements, the location of sampling points requires careful planning and the application of variogram analysis to ensure spatial validity and representativeness of the entire sampling network. Rambal et al. (1984) have shown that principal component analysis is a suitable means to indentify the factors which influence the spatial variability of soil water storage and to select, from 13 primary measurement points, two representative points which are indicative of the hydrodynamic functioning of a *Pinus pinea* stand in southern France. The first principal component explains 82% of the variability which is highly correlated with the bulk density of the soil, while the next component is associated with multidimensional flow.

Soil water sensing equipment may still fall short of operator requirements, although marked advances have been made in recent years. Probably the most commonly used techniques, at the present time, are those of neutron moderation (Trost 1966; Wendling 1967; Klausing and Salay 1976a, b; Mosimann 1980), tensiometer measurements (Richards 1942; Klute and Peters 1962; Strebel et al. 1970; Bläsig et al. 1984; Eichinger et al. 1984; Merkel and Grimmeisen 1985), and direct auger sampling for gravimetric moisture determination (Diez 1965; Czeratzki 1968; Klausing and Salay 1976a, b). The accuracy obtainable with these procedures is generally insufficient to reveal diurnal trends in evapotranspiration, but soil water changes equivalent to differences of the order of 1–2 mm surface depth of water can be detected with care. Thus, for actively growing vegetation, under moderate evaporation conditions, soil water determinations at 2- to 3-day intervals provide a quantitative picture of crop evapotranspiration (Slatyer 1968; Sommer 1980/81, 1988) or soil water movement in its quality as transporting vehicle of a great number of chemicals.

The relationship between soil water content and deep drainage rate is illustrated in Fig. 4.6. The field capacities of the soils are:

1. Loessic Cambisol-Luvisol intergrade   32–34%
3. Loamy Cambisol   27–29%
5. Sandy Calcaric Regosol   20–22%
6. Loessic Calcaric Regosol   24–26%

It must be emphasized, however, that the traditional concept of "field capacity" is no more than a crude index of the drainage and water-holding

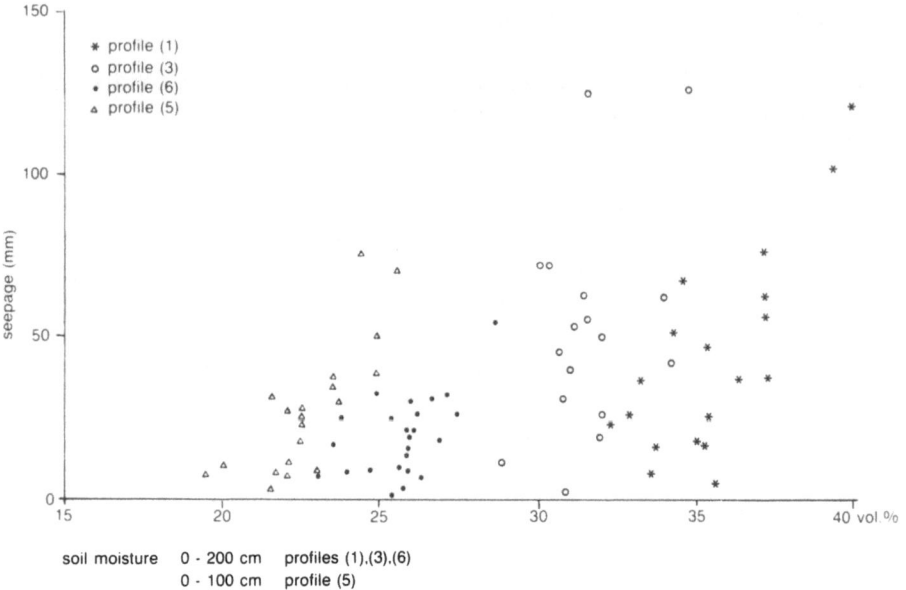

Fig. 4.6. Relationship between soil moisture content and deep drainage; see text for further information. (After Mosimann 1980)

properties of a soil, and the identification of any such quantity with a point on the moisture potential curve may be rather misleading. Any quasi-equilibrium moisture distribution which may be found under particular ecological or environmental–chemical conditions arises from highly complex interactions between the hydraulic conductivity, the moisture potential with the hysteresis properties, and gravity (cf. Sect. 4.4.3.4).

4.1.2.5 Soil Water Potential and Matrix Flow

Movement of water in the soil and bedrock complex depends on the difference in potential between any two points (Buckingham 1907 cited by Tschapek 1959; Gardner et al. 1922), which in turn is a function of soil moisture conditions on a slope. Soil water potential (Ψ_s) is a composite quantity and may be defined as follows:

$$\Psi_s = \Psi_m + \Psi_g + \Psi_p + \Psi_o , \tag{4.23}$$

where Ψ_m = matric (or matrix) potential, Ψ_g = gravitational potential, Ψ_o = osmotic potential, and Ψ_p = pressure potential (Tschapek 1959; White et al. 1984; Schachtschabel et al 1989).

Matric potential is the surface tension force in the water menisci of the pore spaces; in addition there are also forces of adsorption between water molecules and charged particles, particularly in the organic and inorganic colloidal fraction of the soil. Which force predominates, depends on both the textural properties and water content of the soil, but usually adsorption forces increase as soil water content decreases.

The gravity potential (Ψ_g) is relatively insignificant in dry soil, but has some importance in saturated soils on slopes and below the water table. It is proportional to the density of the water, the height of a point above its base level and to the accelaration due to gravity. A related component is the pressure potential (Ψ_p), which is directly proportional to the excess hydrostatic pressure exerted over atmospheric pressure by the column of soil water above a point. Therefore it can be of some importance at depth, but only below a water table in saturated soils.

The last component contributing to the water potential in soils is the osmotic potential (Ψ_o) due to the concentration of solutes in the soil solution. In moist soils, when the matric potential approaches zero as saturation is approached, the osmotic potential may largely predominate in Ψ_s.

Thus it can be said that differences in Ψ_m and Ψ_o determine the direction and rate of water movement in unsaturated conditions, while Ψ_p and Ψ_g predominate under saturated conditions of slope soils. Since the determination of Ψ_s is complicated, interest is frequently in the hydraulic potential (Ψ_H) alone, i.e. the sum of the following components:

$$\Psi_H = \Psi_m \; (\text{or } \Psi_h) + \Psi_g \,, \tag{4.24}$$

where Ψ_m refers to points above the groundwater table, Ψ_h to points below.

Figure 4.7 illustrates the equilibrium situation $\Psi_H = \Psi_m + \Psi_g = 0$ and positive and negative deviations from it in the unsaturated zone above the water table. Whenever $\Psi_H < 0$ vertical water movement upwards is induced, while for $\Psi_H > 0$ compensating seepage results.

Unsaturated Flow

The unsaturated flow of water in soil may be described by Darcy's law:

$$V = - K_{(\Theta)} \cdot \frac{\delta \Psi_s}{\delta z} \,, \tag{4.25}$$

where V is the volumetric flow velocity, $K_{(\Theta)}$ the capillary conductivity, $(\delta \Psi_s / \delta z)$ the potential gradient, and z the height above the groundwater level.

Furthermore the flow has to satisfy the law of conservation of matter which is given in an equation of continuity, stating that the change in moisture content in a volume element of the soil equals the difference between in- and out-flow:

$$\frac{\delta V}{\delta z} = - \frac{\delta \Theta}{\delta t} \,, \tag{4.26}$$

where Θ is the volume fraction occupied by the soil moisture. Combining the

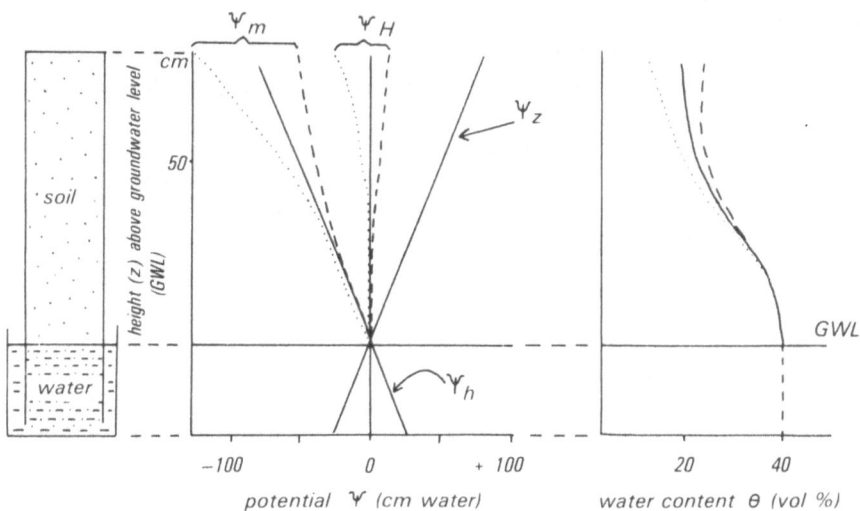

Fig. 4.7. Hydraulic potential, matric potential, gravitational potential and water content of a soil monolith under conditions of equilibrium (—), seepage (----) and capillary rise. (After Schachtschabel et al. 1989)

above Eqs. (4.25) and (4.26) yields:

$$\frac{\delta \Theta}{\delta t} = \frac{\delta}{\delta z}\left(K_{(\Theta)} \cdot \frac{\delta \Psi_s}{\delta z}\right). \tag{4.27}$$

Capillary conductivity $K_{(\Theta)}$ is maximum if the soil is fully saturated, because in this case all pores are filled with water and take part in water flow. When soil moisture content decreases, capillary conductivity decreases very rapidly because the largest pores are emptied first by withdrawal of water and become filled with air. They no longer participate in the flow of water, so that the cross-sectional area available for the movement of water is reduced. At a moisture content of approximately 10–15% of saturation capillary conductivity becomes zero, because the water remaining in the soil no longer forms a closed system of flow channels. Hence a further flow of water occurs only in comparatively very small quantities by means of vapour transport (cf. subsection "Water transfer under temperature gradients", below).

Since there is a functional relation between moisture content and suction (the inverse of the matrix potential), capillary conductivity has also to be a function of the soil moisture tension. Gardner (1958) described the approximate relation between capillary conductivity and suction by the general equation

$$K = \frac{a}{(\Psi_m)^n + b}, \tag{4.28}$$

where a is a constant related to the saturated hydraulic conductivity of the soil, while n has a value 2 in clay soils and higher values in sandy and peaty soils (Rijtema 1962). This equation is such that the saturated conductivity equals a zero tension.

Gardner (1956) developed a method for the calculation of capillary conductivity from pressure-plate outflow data, Butijn and Wesseling an analogous approach on the basis of suction-plate measurements which were modified by Rijtema (1959), Kretschmar (1979) and Opara-Nadi (1979), so as to take the flow impedance of the membrane into account. The advantage of either method is that they can be combined with determinations of the soil moisture characteristic. It describes the relationships between soil moisture content and soil moisture tension or suction, which are of fundamental importance in the study of hydrological and plantphysiological factors dependent upon soil moisture. In considering the wide range of suction values occurring in natural soils, Schofield (1935) introduced the pF scale, expressing suction as the decadic logarithm of the height in centimetres of the water column which the suction could support, irrespective of the forms of energy involved. This has the appreciable advantage that the suction can be used as a negative hydraulic head, which facilitates a unified approach to the examination of moisture conditions both above and below the water table.

The relation between moisture content and suction can be determined experimentally on undisturbed samples by means of a suction-plate apparatus (Hartge 1966; Wolkewitz 1959/60), or a pressure-membrane apparatus as introduced by Richards (1941). Figure 4.8 represents the soil moisture characteristics of four Dutch soils of which the mechanical analysis is given in Table 4.10.

In addition to the relation of soil moisture characteristics to granulometry, there is also a marked functional dependence upon humus content, which is illustrated in Fig. 4.9. A closer study of these soil constants, and some other relevant soil properties, will provide a reliable estimate of the soil moisture characteristic, which is of major importance for hydrological or environmental chemical soil mapping (cf. Beven and Germann 1982; Klaer and Krieter 1982; Schwarz 1986).

Provided soil moisture content and moisture tension can be measured periodically at various depths below the surface by means of tensiometers and neutron or X-ray probes, the depth function of flow velocity or capillary conductivity can be determined (Wind 1955; Krikham and Powers 1972; Hofius 1977; Hartge 1978; Kreutzer et al. 1978; Opara-Nadi 1979; Hillel 1980).

Water Transfer Under Temperature Gradients

Under certain circumstances the temperature gradients in the soil influence the unsaturated soil–water transfer associated with evaporation. Quantitative determinations of this process (e.g. Gurr et al. 1952; Taylor and Cavazza 1954; Philip 1964) have revealed that the transfer rates are of the order of five to ten times above the rates estimated by assuming that transfer is due to molecular

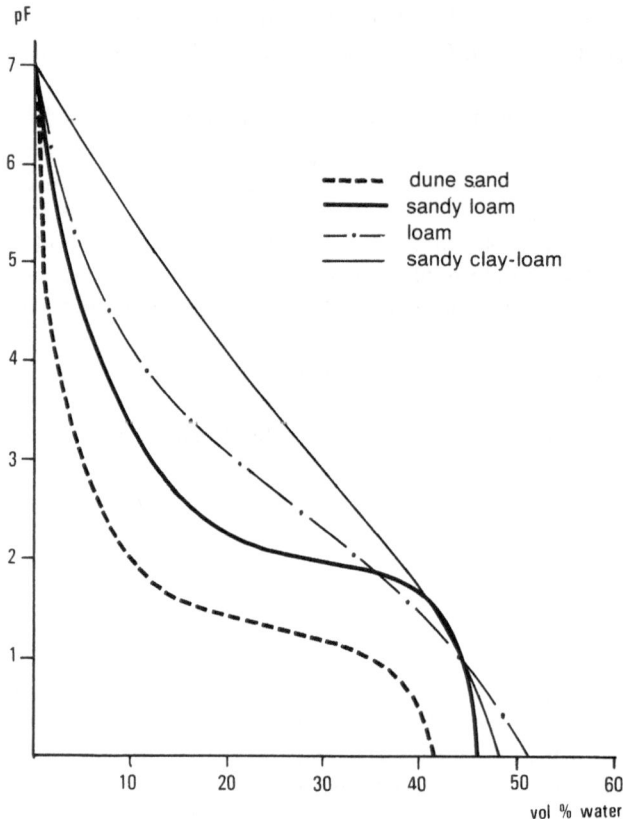

Fig. 4.8. Soil moisture characteristics of four Dutch soils. (After Rijtema 1961)

Granulometry and lime content of the Dutch soils represented in Fig. 4.8. (After Rijtema 1961)

No.	Soil type	Organic matter	Granulometry < 16	16–90	(μm) > 90	CaCO$_3$
1.	Dune sand	—	0.5	3	94	2.5
2.	Sandy loam	0.4	9	46	28	16.5
3.	Loam	0.8	18	50	17	14.1
4.	Sandy clay loam	0.8	29	54	1	15.2

diffusion in the air-filled voids down a vapour pressure gradient. In addition, the moisture content at which maximum transfer occurs is greater than would be predicted from this assumption.

Philip and de Vries (1957) showed that this simple hypothesis neglects the interaction of the water vapour with the liquid and solid phases in the soil, and takes no appropriate account of the difference between temperature gradients in

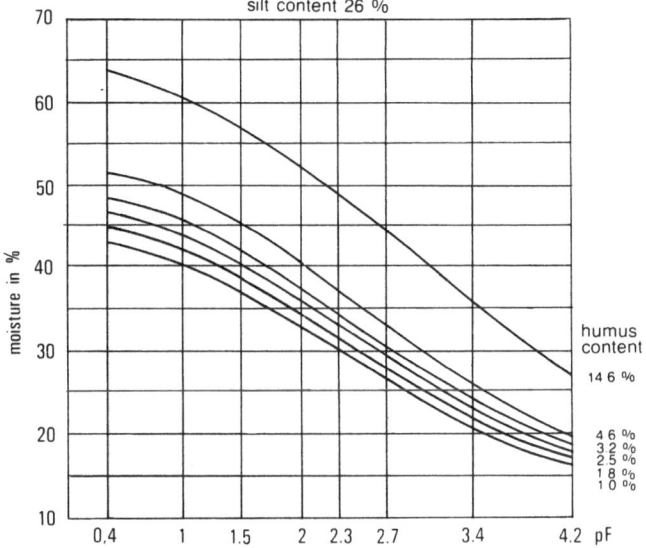

Fig. 4.9. Soil moisture characteristics of soils with a silt content of 26% but various humus contents. (After Visser 1958)

the air-filled voids and of the medium as a whole. They consequently put forward the concept that, in a soil in which liquid continuity has failed, the transfer occurs as a series–parallel flow process through regions of vapour and liquid. Contrary to the simple hypothesis, treating the liquid "islands" as obstacles blocking the passage of diffusing vapour, these are, in fact, regions of very rapid transfer of water. Condensation at the upstream end of the island and evaporation at the opposite (downstream) end produce changes in the curvature of the menisci at each end, which result in a virtually zero resistance to liquid flow through the island. Furthermore, when the heat transfer through the solid, liquid and gaseous phases of a soil was examined in detail, it was found that the temperature gradients across air-filled voids could be of the order of twice the mean temperature gradient in the soil as a whole (Philip 1964). This implies that the liquid phase transfer in response to thermal gradients, when no counter water potential is developed, is not small with respect to the vapour transport (Taylor and Cary 1960; Taylor 1962). Thus, when speaking of flow, all the forces and all the fluxes must be considered. If there is a flow of heat, electricity or any other form of matter in a direction opposite to the water potential, the water may be induced to move in a direction opposite to its own potential gradient (Taylor 1962; Taylor and Cary 1965). Such phenomena, which may exhibit a specific importance in the framework of weathering or chemical transport processes, are commonly considered in the basic literature on thermodynamics of irreversible processes (e.g. de Groot and Mazur 1962).

Saturated Flow

For the purposes of an environmental chemical discussion "saturated flow" may be generally defined as movement of water in the interstices of essentially saturated soils or rocks. It thus comprises groundwater flow and a variable part of interflow. In the light of the general statements of Sect. 4.1.2.4, a significant quality of groundwater in the hydrological cycle is the storage life or stability, which can be most readily expressed by the storage/flow ratio, i.e. as the ratio of the storage upstream of any cross-section to the flow past the section (Chapman 1964). At any location, it will be a constant only if the flow is linearly related to storage, and in general the ratio will tend to decrease as the storage increases, but such variations do not affect the utility of the ratio for characterization of the flow regime in relation to chemical persistence, for instance.

If the storage/flow ratio is sufficiently large to damp out fluctuations in recharge, groundwater hydrology becomes a relatively simple problem of steady flow analysis. A ratio of at least 50 years in the arid zones and rather less in more humid areas is generally adequate for this purpose (Langbein 1960; Chapman 1964). The situation is slightly more complex when the groundwater regime is markedly seasonal with storage/flow ratios in the range 1–5 years, but the most difficult problems occur where the ratio is inadequate to provide a sufficient stabilizing influence. The ratio in an aquifer normally increases in the down-stream direction, and there is also a general tendency for this increasing stability of the flow regime to be associated with higher salt contents. Probably the most spectacular man-made example of this type is provided by the groundwater salinity in The Netherlands.

A further distinctive feature in both the hydrological cycle and transport of persistent chemicals is the lag effect. It is partly due to storage, and insofar analogous to surface storage (S) increasing the lag of a flood crest, but is more directly the result of the low or very low velocities involved, i.e. 10–1000 cm a^{-1}, and hydrodynamic dispersion phenomena (Kinzelbach 1983). Owing to tortu-ous or smaller passages, some of the water accomplishes less direct line movement than a solution taking a more direct path. In addition, a chemical transported in solution may also diffuse into stagnant pores to be released slowly when the main body of chemical has passed (Schmidt et al. 1983).

Also groundwater hydraulics are based on Darcy's law [Eq. (4.25)] which originally resulted from experiments in a sand-bearing water under a uniform hydraulic gradient (Darcy 1856). Philip (1957) has shown that it is a consequence of the Navier–Stokes equation for an incompressible fluid, with the inherent restriction that the inertia terms are negligible. This involves an upper limit of the velocities for which Darcy's law is valid, and experiments show that this limit is equivalent to Reynolds numbers of the order 1–10, based on a mean particle diameter. While this limit is normally exceeded only near artificially created discharge surfaces, such as wells, there is evidence that deviations from Darcy's law occur in soils containing appreciable quantities of clay (Swartzendruber 1962), which induces a non-linear behaviour.

Groundwater can be appropriately classified according to the lithology of the top boundary. In confined flow it is the limit of a stratum of highly reduced permeability, while in unconfined flow the boundary is not fixed geometrically but is a surface of constant pressure corresponding to the upper limit of the essentially saturated zone. This type of boundary increases the complexity of the mathematics implied in solving the general unsteady flow Eq. (4.27), especially in the normal case of heterogeneous aquifers, and consequently the majority of analyses of unconfined flow problems rely on approximations (cf. Chapman 1957; Mattheß and Ubell 1983). For confined steady flow in homogeneous and isotropic media, the familiar Laplace equation may be used, for which Scheidegger (1957b) has given a comprehensive list of solutions.

Numerous experimental methods are available to give estimates of the near-saturated or saturated hydraulic conductivity K_s of soils or sediments (cf. Talsma 1960; Schröter 1983; Matthess et al. 1985). Among these, the use of ring infiltrometers (Talsma 1969; Dunin 1976) and well permeameters (Talsma and Hallam 1980) is particularly widespread. The ring infiltrometer favours primarily the vertical as against both horizontal and vertical components of hydraulic conductivity combined in the well permeameter method (Bonell et al. 1983). Bouwer and Jackson (1974) noted, however, that the latter technique measured $K_{(\Theta)}$ mostly in horizontal direction. The well permeameter K values are normally only 50–85% of the true saturated hydraulic conductivity (K_s) as measured by the auger hole method (Talsma 1960; Winger 1960). The under-estimation is due to clogging of the pores of the sidewalls of the hole.

Using Darcy's law [Eq. (4.25)] the estimation of hydraulic conductivity $K_{(\Theta)}$ from the ring infiltrometer is based on a consant head of 4 cm of water over a sample 10 cm deep.

$$K_{(\Theta)} = Q\frac{dz}{d\phi}, \tag{4.29}$$

where $K_{(\Theta)}$ is measured in cm min^{-1}, and

$$\frac{dz}{d\phi} = \frac{\text{sampling length}}{\text{total head}}.$$

Q = flow rate through the sample (cm min^{-1}) which is calculated from:

$$\frac{\Delta\chi \cdot \text{scale factor}}{\text{area of ring}},$$

where $\Delta\chi$ is the rate of fall in head in the perspex permeameter and the scale factor is the volume of water (cm^3) between the inner and outer perspex tubes for each cm head of water.

The corresponding formula for calculating hydraulic conductivity from well permeameter measurements was developed by Zanger (1953):

$$K = \frac{Q[\sin\,h^{-1}(Hr^{-1} - 1)]}{2\pi H^2}, \tag{4.30}$$

where Q = rate of water flow into auger hole (cm^3 min^{-1}) which is calculated from:

$$\frac{\Delta\chi \cdot \text{scale factor}}{\text{time}},$$

with $\Delta\chi$ and the scale factor determined in the same way as above, H = constant water depth in the auger hole, and r = radius of hole.

Valid use of Eq. (4.30) requires that $Hr^{-1} > 10$ (Winger 1960) and the depth to the water table or impermeable layer below the hole exceed $2H$ (Zanger 1953).

Bonell et al. expanded Zanger's formula to the following, which is slightly different from that used by Boersma (1965) or Bouwer and Jackson (1974):

$$K = \frac{Q}{2\pi H^2}\left[\ln\left(\frac{H}{r} + \sqrt{\frac{H^2}{r} + 1}\right) - 1\right]. \tag{4.31}$$

Comparative date analyses by Tukey (1977) and comparable field investigations by Rogowski (1972), Nielsen et al. (1973) and Baker (1978) showed that the K data frequency distributions are closely approximated by the log-normal function (cf. Lévy 1954; Cramér 1957). Consequently, the use of log means for interlayer comparison of K appears more appropriate than the use of arithmetic means.

In addition to these direct methods for evaluating hydraulic conductivity, two indirect ones are worthy of note because of their simplicity. Hazen derived conductivity from the d_{10} value of a granulometric analysis of the sediment concerned; Beyer (1964) attained an even better estimate by additionally introducing the d_{60} value. Figure 4.10 illustrates the wide variability of K estimates (cf. W. Müller 1975) as dependent upon the experimental method applied.

4.2 Impact of Environmental Chemicals on Vegetation

In agronomy and forestry the traditional concern about productivity losses has been associated with soil fertility and with insect and disease damages. Losses affect the production, availability and cost of food and lumber and therefore have important economic, social and political implications, especially when the worldwide economy is depressed. In particular, concern for accelerated forest dieback in Germany, Sweden and the United States of America has caused an increased awareness of the potential long-term impact that deposition of acidic and acidifying substances may have. These concerns have stimulated a renewed interest in forest ecosystem research, and increased the awareness of the many variables which control the performance and productivity of plants in the natural setting (Ulrich and Sumner 1991). Soils, soil water, nutrient cycling, air quality, drought, disease, insects and human activity are the most important factors which contribute significantly to the ultimate productivity of a forest

cm sec^{-1}

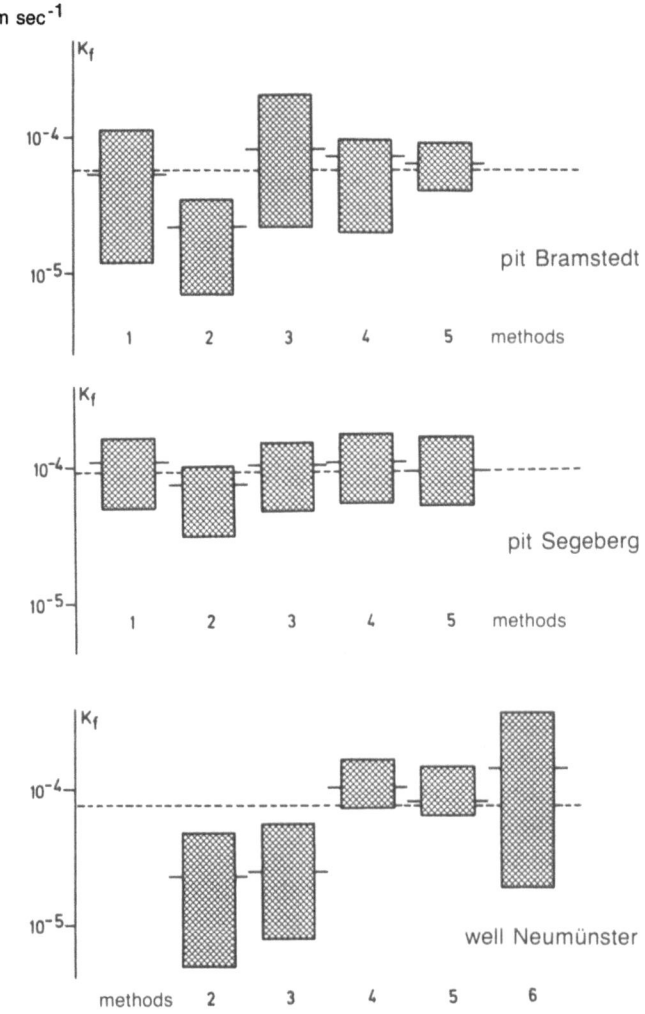

Method 1: Undisturbed sample in STENZEL permeameter, normal seepage
Method 2: Same procedure as in method 1, but sample disturbed
Method 3: Same conditions as in method 2, but waterflow bottom upwards
Method 4: Evaluation of granulometry, after HAZEN
Method 5: Evaluation of granulometry, after BEYER
Method 6: Pumping; different evaluation approaches

Fig. 4.10. Variability of K-estimates as related to different methods. (After Pekdeger and Schulz 1975)

ecosystem. They form a process-response system of high complexity, and inadvertent impact can consequently induce a bewildering number of changes. The remaining questions centre around defining the manifold cause and effect linkages, the time frame of change, and the significance of any changes. The following forest dieback model (Fränzle et al. 1985) is therefore designed to present a comprehensive graph-theoretic model of the relevant interactions (foldout model III). It is based on the comparative evaluation of about 80 hypotheses on relevant tree diseases and forest decline, and underlines the necessity to make studies on complex forest diseases a consistent part of long-term ecosystem research. Since a detailed presentation of the manifold inter-relationships depicted, and the uncertainties that remain, is beyond the scope of the present review, the following sections are limited to an analysis of effects of air-borne pollutants on vegetation which appear to be of particular impor-tance in this context (cf. Linthurst 1983; Treshow 1984; Fränzle et al. 1985; Umweltbundesamt 1986b, d).

4.2.1 Direct Effects of Acid and Basic Deposition and Photo-Chemical Oxidants

With regard to its effects on vegetation, sulphur dioxide is one of the best-investigated atmospheric pollutants. In Europe, it has until the last few years been considered the air pollution component with the highest practical phyto-toxic potential, but now nitrogen oxides and ozone are regarded as exerting an equally damaging influence. SO_2 pollution constitutes one of the most challeng-ing trans-boundary problems because of its widespread distribution and persis-tence in the atmosphere (cf. Chap. 3.4.2, Zölitz 1985, Fränzle 1988). It affects vegetation either directly in gaseous form or via acidification of precipitation or indirectly, i.e. via soil acidification.

Among the anthrophogenic nitrogen oxides the mono- and dioxide are most important vegetation-threatening pollutants. They may affect plants either directly after uptake via the leaf stomata or indirectly after soil nitrogen enrichment or in their capacity as precursors of photochemical oxidants (cf. Chap. 3.4.3). Among these, in Europe, ozone is the most relevant phytotoxic oxidant according to its occurrence and concentration.

It is probably only in California that peroxyacetylnitrate (PAN) as the most important organic peroxide, appears in phytotoxic amounts, while ambient air values in Europe are normally several times lower (Becker et al. 1985).

4.2.1.1 Sulphur Dioxide

Pollutant uptake via the stomata depends on the one hand on external meteorological factors (cf. Chaps. 3.2, 3.3, 3.4), on the other it is controlled by internal morphological, physiological and biochemical factors in leaf or needle. Together they define diffusion barriers to mass transport of SO_2 from ambient air into the plant cells (cf. Fitter and Hay 1987).

In conjunction with light and relative humidity, temperature regulates stomatal opening and thus correspondingly the uptake rate, diffusion in the intercellular space and subsequent chemical reactions, e.g. the formation of sulfite or bisulphite. Since the physiological activity of plants increases with temperature between 4 and 35 °C, it may be assumed that sensitivity to pollutants rises correspondingly (Guderian 1978, 1985). Stomatal opening is directly controlled by relative humidity. Humidities exceeding 75% generally stimulate gas exchange, and enhance the uptake of gaseous pollutants over a wide range of temperature.

Short-term water shortage in a plant normally leads to a reduction of stomatal opening and consequently of gas exchange. If the shortage continues (cf. Sect. 4.1.2), physiological and morphological alterations occur, which further reduce gas exchange. In connection with the influence of soil nutrients on plant growth and vitality, it can furthermore be noted that optimal specific nutrient supply will reduce pollutant effects (Schröder et al. 1986).

SO_2 entering plant cells from the intercellular space and dissolving in the tissue fluid phase in dependence on pH conditions and acidic precipitation may cause injury to foliage. The pollution load is primarily a product of pollutant concentration and exposure duration. Guderian (1985) and Guderian et al. (1985) showed that for equal products (doses) damage to leaves and growth performance increased more with rising SO_2 concentration than with longer duration of exposure. Clearly effects are also influenced by frequency of exposure; i.e. the shorter the interval between successive exposures the stronger the effect (Jäger and Schulze 1988).

SO_2 disturbs the stomatal regulation and causes a general stress situation which affects the whole metabolism of the plant. Consequently, unspecific alterations of enzyme and membrane activities occur which could also be induced by various other environmental stresses like water stress, temperature stress, mineral deficiency etc. Table 4.11 summarizes the reported effects of SO_2 on enzyme activities in plants.

Another direct influence of SO_2 which is of considerable physiological importance concerns the amino acid and polyamine contents of plants. The most remarkable effect on amino acids is the shift of the glutamic acid/glutamine ratio towards the amide (Jäger et al. 1972; Paul and Bassham 1978). In spruce needles there is an additional increase in arginine, ornithine and proline (Jäger and Grill 1975). The latter effect, which is also reported for plants under water stress (Jäger and Meyer 1977), may also be indicative of some pollutant influence on the water balance of the plant.

The unspecific alterations of regulative macromolecules and membrane activities result in a lack of control of certain metabolites. Hoffmann et al. (1976), Harvey and Legge (1979) and Wellburn et al. (1981) showed that the content of purine nucleotides, in particular ATP, is changed considerably in polluted plants. SO_2 and its derivatives inhibit photophosphorylation and oxidative phosphorylation to a variable extent in isolated chloroplasts or mitochondria, respectively (Asada et al. 1968; Ballantyne 1973; Silvius et al. 1975). They further cause a general stress situation within the plant which, in turn, induces ethene production (Bressan et al. 1978; Bucher 1978; Peiser and Yang 1979). There may be a close connection between this ethene production and the accelerated abscission caused by air pollutants (Jäger 1982; Steubing and Fangmeier 1987).

Table 4.11. Effects of SO_2 on enzyme activities in selected plant genera. (After Jäger 1982) ($\uparrow$ = increase, $\downarrow$ = decrease)

Enzyme	Effect	Plant genus	Experimental conditions
GDH (Glutamate dehydrogenase)	$\uparrow$	*Pisum, Vicia, Lolium Hordeum, Lycopericon, Picea* a.o.	0.1–1 ppm, field, plant exposure
GS (glutamine synthetase)	—	*Pisum, Lolium, Poa* a.o	0.1–2 ppm,
GOT, GPT (glutamate-oxaloacetate transaminase, glutamate-pyruvate transaminase)	$\uparrow\downarrow$	*Pisum, Rumex, Hordeum, Tulipa, Medicago* a.o.	0.1–1 ppm, plant exposure
POD (*peroxidase*)	$\uparrow$	*Picea, Pinus, Alnus, Pisum, Lolium* a.o.	0.05–0.2 ppm, field, plant, exposure
RuDPC (ribulose diphosphate carboxylase)	$\downarrow$	*Pisum, Pinus, Spinacia*	0.2–2 ppm, sulfite
G6PDH (glucose-6-phosphate dehydrogenase)	$\uparrow\downarrow$	*Hordeum, Vicia, Medicago* a.o.	0.02–0.3 ppm
NiR (nitrite reductase)	$-\downarrow$	*Lolium, Dactylis, Phleum, Poa*	0.06–1 ppm
SOD (superoxide dismutase)	$\uparrow\downarrow$	*Populus, Pisum Spinacia*	0.1–0.15 ppm, 2 ppm

Comparison

HF/HCL:	GDH $\uparrow$,	POD $\uparrow$,	NiR + SOD ?	
NO_x:	GDH $\uparrow$,	POD-$\uparrow$,	NiR $\uparrow$,	SOD $\uparrow\downarrow$
SO_2:	GDH $\uparrow$,	POD $\uparrow$,	NiR $\downarrow$,	SOD $\uparrow\downarrow$

On the community level, SO_2 impact is highly species-depedent.The most sensitive species may display visible signs of damage; disturbed growth, repro-duction and vitality seem to occur among the next most sensitive (Wittig et al. 1985; Steubing 1987; Steubing and Fangmeier 1987). With longer exposure, the boundary conditions of species competition change so that the less sensitive ones may ultimately spread. Especially in complex forest communities, such changes may gradually change the genetic structure, dominance patterns and relative abundance of species (McClenahen 1978; Scholz and Geburek 1987; Scholz 1981a, b; Gregorius et al. 1985; Taylor et al. 1991). However in dependence on the evolutionary stage attained (cf. Chap. 2.2.2) complex systems can also possess the greatest resilience, i.e. buffering capacity and recovery potential with regard to pollutants, while ecosystems with less species diversity and spatial heterogeneity will tend to respond to pollution with a relatively more marked simplification of structure.

4.2.1.2 Nitrogen Oxides

After predominant uptake via the leaf stomata, the highly soluble nitrogen oxides dissolve on the cell surfaces and form nitrites and nitrates. NO, being less soluble in water than NO_2, is taken up to a lesser degree and is also less phytotoxic (Mansfield and Posthumus 1985). In the cytoplasm, nitrites and nitrates can be reduced by nitrate reductase, and in the chloroplasts to ammonia by nitrite reductase; subsequent incorporation into organic compounds brings about detoxification of the hydrated nitrogen oxides (Guderian et al. 1985, Guderian 1988).

Contrary to the assumption that biomembranes are the first points of NO_2 disturbance, there is conclusive evidence that biochemical effects of nitrogen oxide exposure are mostly due to raised cell nitrate and nitrite. Ultrastructural analyses have revealed frequent invaginations of chloroplasts and mitochondria, fibrillar and crystalloid structures in chloroplast stroma, associated with thylacoid swelling, which suggests that NO_2 damages the chloroplasts and not the plasmalemma (Guderian 1988).

Physiological responses of leaves include reductions in photosynthesis, often at NO_2 concentrations and exposure duration which do not cause visible effects. RuBP-carboxylase exhibits a direct positive relationship with concomitant fluctuations of NO_2 concentration (Malhotra and Kahn 1984). Yoneyama et al. (1980) and Wellburn et al. (1980) found a stimulation influence of both NO and NO_2 on nitrite and nitrate reductase activity. Also the effects of NO_x on plant pigment contents vary with concentration; chlorophyll and carotene are reduced and phaetophytin is formed only in tissues of visibly damaged leaves, while there is no influence on the pigments in the others (Kändler and Ullrich 1964).

Effects of NO_x pollution on plant communities are little known. Scholl (1975) found positive as well as negative effects in the neighbourhood of a fertilizer plant. Legumes proved especially sensitive with leaf damage, significantly reduced growth and fewer and lighter nodules, while mass increase was considerably enhanced as a result of the high nitrogen supply. Soil nitrate was clearly raised and spinach nitrate concentration was augmented to a toxicologically suspect level (Kübler 1965). An increase in nitrogen indicator species in the ground vegetation of several Westphalian millet grass-beach forests is considered the result of longlasting exposure to low NO_x pollutant concentrations (Wittig et al. 1985). Clearly sphagnum communities in ombrotrophic mires are very sensitive to nitrogen input (Ferguson and Lee 1983; Dierßen 1990).

The influence of ambient humidity and temperature on uptake of NO_x is likely to resemble that of SO_2 and O_3, but light affects NO_x responses in a different way. While adverse effects of sulphur dioxide and ozone are relatively low at night, nitrogen dioxide can be more injurious in the dark because of diminished nitrite reduction (van Haut and Stratmann 1967; Zeevaart 1976).

Sensitivity of plants to NO_2 exposure increases with rising soil water content (Benedict and Breen 1955; Kato et al. 1974). The inverse relationship exists between NO_x uptake or effects and substrate nutrient supply (Anderson and Mansfield 1979; Matsumaru et al. 1979; Srivastava and Ormrod 1984) since, with falling nitrogen supply, stimulation due to NO_2 and NO increased or their adverse effects diminished.

Important components in assessing NO_x toxicity are leaf age, developmental stage of plants, and species, varietal and individual resistance. Middle-aged leaves, just fully developed or still growing slightly, are more sensitive than younger or older ones (van Haut and Stratmann 1967). Conifers respond to acutely damaging pollutant concentrations more sensitively in spring and early summer than in the rest of the year. As regards pine and spruce, the previous year's needles remain the most sensitive during the early developmental stages of the new needles, but are replaced by these in early summer when they are completely developed. Fir and larch needles display a contrasting reaction, i.e. they prove particularly liable to damage very early in their development. Low pollutant loads with chronic effects affect fully developed and older needles rather than young ones (Elkiey and Ormrod 1980), with chloroses predominating over necroses, often accompanied by premature needle drop (Spierings 1971; Sinn and Pell 1984).

Also the plant as a whole differs in sensitivity to acute and chronic pollution stress at different development stages. Annual species showed the greatest yield reduction as a consequence of acute exposure at the phase between vegetative and generative growth during fruit formation (van Haut and Stratmann 1967). With regard to chronic pollutant stress, grasses exposed from emergence on proved more sensitive than those first treated at tiller formation (Whitmore and Mansfield 1983).

By analogy with the influence of leaf or needle age and developmental stage also plant species, sorts and varieties, and individual plants in a population vary in NO_x sensitivity. The silver birch and the larches are particularly sensitive while yew, fir, oak and beech are less so (McClenahen 1978; Gregorius et al. 1985; Müller-Starck 1985; Wittig et al. 1985; Scholz and Geburek 1987; Guderian 1988). Among agricultural and horticultural species, legumes are highly sensitive, but the brassicas proved particularly resistant (Scholl 1975).

4.2.1.3 Ammonia

In the course of the last decade, the importance of NH_3 and NH_4^+ effects has grown considerably as a result of increasing ammonia emissions from decomposing animal manure. An ever-expanding amount of such manure has been produced not only by the greater number of cows and other ruminants grazing outdoors, but also by the drastically increased number of cattle, pigs and poultry in highly intensive husbandry. The gaseous NH_3, continuously emitted from fields, pastures, stalls, sheds and storage containers, together with the NH_4^+ ions produced from its reactions with rain, mist, fog, dew or other water surface layers, may affect the vegetation in various ways (Posthumus 1988).

In parts of Lower Saxony (Windhorst 1984; Fabrewitz 1986; Lieth 1987) or in The Netherlands (Roelofs et al. 1985; Buijsman et al. 1987) for instance, the areas with naturally low soil fertility have recently seen a dramatic increase in intensive livestock farming on sandy soils. Here the emission of additional nitrogen compounds is affecting the natural vegetation by causing direct injury to the aerial parts of plants, as has been also observed in the past on several occasions already (Garber 1935; Kühn 1966; Garber and Schürmann 1971; Ewert 1978, 1979; Hunger 1978; Tesche and Schmidtchen 1978; Temple et al. 1979; van der Eerden 1982). The effects of NH_3 and NH_4^+ are similar as regards disturbed nutrient balance and increased frost, disease and pest sensitivity (cf. Chap. 2.3.2), but the uptake routes differ. Gaseous NH_3 is taken up via the stomata (Hutchinson et al. 1972; van Hove 1987) but dissolved NH_4^+ ions through the leaf surface by exchange with K^+, Ca^{2+} and Mg^{2+} (Roelofs et al. 1985). At high concentrations of NH_3, which is dissolved in water and transformed into NH_4^+ within the plants, the process of phosphorylation is inhibited, and consequently also carbohydrate production and plant growth. Both NH_3 and NH_4^+ may destroy cell membranes, leading to necrosis of leaf parts and higher sensitivity to frost damage (van der Eerden 1982). Erosion of the leaf cuticula is another result of NH_3 and NH_4^+ exposure (van der Eerden and Wit 1987). According to Nihlgård (1985) the effects of NH_x may contribute to forest dieback in Europe.

Owing to different genetic predisposition and related morphological or physiological properties, there are marked differences in sensitivity to NH_x among plant species and among varieties or cultivars within species. Broadleaved trees are less sensitive than conifers, and also various cultivars of cauliflower differ in sensitivity (Posthumus 1988).

Among the external factors controlling NH_x uptake, the meteorological situation merits particular attention. Light conditions and relative humidity may stimulate the leaf stomata to open, intensifying gas exchange. Thin water films on leaves or needles may increase the uptake of NH_4^+ by exchange with basic cations like K^+, Ca^{2+} and Mg^{2+}. Both low light intensity and low temperature inhibit carbohydrate synthesis, thus reducing the plant's resilience or capacity to detoxify ammonia by producing amino acids. Therefore winter in temperate climate regions is generally a sensitive period for ammonia damage to plants (cf. also Chap. 3.4.3). As regards eutrophication due to nitrogen compounds, it is clear that it will occur preferentially on poor soils, affecting raised bogs in particular.

4.2.1.4 Photooxidants

Photochemical air pollution was detected for the first time in the Los Angeles area on the basis of vegetation injury (Haagen-Smit et al. 1952). The relative sensitivity of different plant species was defined, and from 1954 programmes were started to introduce pinto beans, annual blue grass and petunia into air pollution effect monitoring (Middleton et al. 1955; Juhrén et al. 1957; Noble 1956; Steubing and Jäger 1982). Somewhat later it became evident that O_3 was also the cause of extensive injury to tobacco in the eastern USA (Heggestad and Middleton 1959). Stephens et al. (1961) discovered that the typical bronzing of the abaxial leaf surfaces was caused by peroxyacetyl nitrate (PAN). In The Netherlands, injury to tobacco, annual blue grass and small stinging nettle was observed for the first time in 1965 and attributed to the combined effect of O_3 and PAN (ten Houten 1966).

PAN is quantitatively the most important member of the peroxyacetyl nitrates. Although other homologues (e.g. PPN, PBN, PBzN) may be much more phytotoxic, only trace amounts of them have been detected (Taylor 1969; Bos et al. 1978); consequently PAN has received the greatest attention as a phytotoxicant. Since it does not occur in natural air (cf. Chap. 3.1), it may be a better indicator of photochemical air pollution than ozone (Posthumus 1977; Posthumus and Tonneijck 1982).

Ozone is now considered, together with sulphur oxide, to be among the preponderant atmospheric pollutants in Europe (Prinz et al. 1982; Guderian et al. 1985; Guderian 1988), while ambient air values for PAN are several to many times lower than in California, where it may appear in "potentially phytotoxic episodes" (Temple 1982; Becker et al. 1985). After stomatal uptake, the biochemical effects of O_3 are primarily due to its oxidative potential and to the formation of free radicals. According to Mudd (1982), the preferred targets in plant organs are the double bonds of unsaturated fatty acids, sulphhydryl groups and disulphide linkages. Cell compartmentation is disturbed since membrane-bound enzymes are released from their sites and cytoplasmic enzyme systems are inhibited, in particular as a result of ion imbalance. In addition to these and related effects on proteins, pigments and nucleic acids, cell function becomes severely disturbed (Tingey and Taylor 1982). This may involve increased production of polyphenols (Kuokol and Dugger 1967) and ethene (Tingey et al. 1976), premature senescence (Thompson and Taylor 1969), reduced vitality, and enhanced sensitivity to biotic and abiotic stresses (cf. Sect. 4.2.3).

The primary effects of O_3 are followed by structural and ultrastructural changes such as development of crystalloids in chloroplast stroma, chloroplast membrane invaginations, swelling or breakdown of thylakoids and an increase in plastoglobuli (Mitchell et al. 1979). Root growth is frequently more retarded than that of the aerial parts of the plant as a result of different "partitioning", i.e. altered distribution of ozone-reduced assimilates to the plant organs (Blum and Tingey 1977; Cooley and Manning 1987; Küppers and Klunepp 1987).

Dicotyledon leaves are most sensitive to O_3 attack during later phases of leaf expansion, i.e. after reaching their maximum growth rate (Evans and Ting 1974). Also conifer needles are most intensely affected by ozone in the extension phase shortly before reaching final length (Davis and Coppolino 1974). The same dependence of sensitivity on the stage of development applies to plants. Short-lived species and tree seedlings proved to be most sensitive early in their development (Richards et al. 1980), but under favourable conditions early leaf damage tends to grow out by harvest more than injury incurred during later stages of development (Townsend and Dochinger 1974).

Investigations on the ozone resistance of native European plants are still sparse, and results are mainly based on short-term exposures to high concentrations with visible leaf damage as effect criterion so that the following allocation to sensitivity categories must be regarded as provisional (Guderian et al. 1985). Among the most sensitive species are perennials such as vines, European larch and fir; the higher resistance of copper beach and Norway spruce appears remarkable. However, there is not such a clear difference in ozone sensitivity between broadleaved and conifer species as exists with regard to acidic pollutants. Wheat, oat, barley and rye, potato and legumes, as well as the fodder plants alfalfa and red clover, are among the agricultural plants most threatened by O_3.

Among the external factors influencing ozone uptake, light intensity plays a particular role. Low intensities during early stages of growth increase sensitivity while, during exposure, adverse effects are intensified by high light intensities (Heck et al. 1967). Consequently, most plants are only little affected by O_3 exposure during darkness.

Also the relative humidity during plant development influences both the anatomy and physiology of leaves and thus, indirectly, pollutant uptake on exposure, during which the amount of ozone incorporated is again directly proportional to humidity because of its effect on stomata. Soil water content exerts a similar influence on stomatal opening, i.e. short-term water shortage reduces hydration and thus ozone uptake, while longer-term adaptations may have continuous consequences for gas exchange and thus for pollutant uptake (Dean and Davis 1967).

4.2.1.5 Combined Effects of Pollutant Gases

Sulphur dioxide and nitrogen oxides are frequently emitted in combination and normally together with other pollutants such as soot and heavy metal particles; in many places photooxidants also occur. In these cases the effects of combinations of pollutants are not simply the sum of the individual component effects, but the relative concentrations of the components or the specific sensitivity of the plants to one of them may determine the response. Thus pollutants in combination may display antagonistic (reduced), additive (sum of individual pollutant effects) or synergistic (super-additive) effects.

Most experiments of the past on exposure to combinations of pollutants were made under laboratory conditions using higher concentrations of gases than are typical in the field. Newer fumigation techniques, for instance, open-top chambers (cf. Guderian et al. 1985) or field fumigation systems (see Steubing and Fangmeier 1987) have now made it possible to expose plants under field conditions with increasingly realistic concentration ranges. From these experiments it ensues that pollutant combination effects depend on pollutant concentrations, with SO_2 and O_3 often antagonistic at levels where the individual gases cause severe leaf injury, while super-additive effects predominate at lower concentrations (i.e. $SO_2 < 140$ $\mu g/m^3$, $O_3 < 100$ $\mu g/m^3$), and in particular near the threshold dosage (Heagle and Johnston 1979; Kress and Skelly 1982). In addition, it is worth mentioning that the injury symptoms for SO_2 and O_3 in combination are not a simple mixture, but frequently the very symptoms of one pollutant occur more clearly or earlier than those of the other (Brennan and Lewis 1978; Jäger and Schulze 1988).

As regards combinations of NO_2, SO_2 and O_3, effects were produced in many plants by lower concentrations of NO_2 and the other gases than if they acted alone, especially at the lower exposure levels (Ashenden 1979; Ashenden and Williams 1980; Guderian 1988a, b). Synergistic effects of NO_2 in combination with SO_2 or O_3 appear to be due to a reduction of the nitrate and nitrite reductase activities. While the latter tend to rise in response to NO_2 acting on its own, the activation of these reducing enzymes is limited or even suppressed by simultaneously present SO_2 or O_3 (Leffler and Cherry 1974; Wellburn et al. 1980; Robinson and Wellburn 1983). This adverse effect is further enhanced by the formation of free radicals. An inhibitory influence on the regulation of transpiration occurred already at less than 20 ppb SO_2 + 20 ppb NO_2 (Wright et al. 1986).

The combination of NO_2 with O_3 is likely to endanger vegetation more than NO_2 + SO_2 and should receive higher attnetion since these pollutants form essential components of photochemical smog (cf. Chap. 3.4). The threat to vegetation may increase still if all three pollutants interact simultaneously (Mooi 1984; Guderian et al. 1985; Guderian 1988a, b).

Combination effects, however, are also dependent on the temporal order in which the pollutants affect the plants. If they are first exposed to NO_2 and thereafter to SO_2, the usual SO_2 effects are reduced, while the opposite sequence causes super-additive effects because SO_2 affects nitrite reductase (Wellburn et al. 1981). Attention must also be drawn to the combination of gaseous SO_2 with acid precipitation; the effects of this combination are discussed in connection with the new types of forest damage in Sect. 4.2.1.7.

Critical Levels

In conclusion, and with some reservations resulting from the above remarks, critical levels for short-term and long-term exposures may be defined. In view of the fact that most horticultural and agricultural plants do not show adverse effects at SO_2 concentrations below 30 $\mu g/m^3$ while particularly sensitive species of trees, mosses, lichens, and of bushy and grassland vegetation are already adversely influenced by concentrations of about 20 $\mu g/m^3$, the critical level for SO_2 acting on its own is to be set at 20 $\mu g/m^3$ (≈ 0.007 ppm) as annual mean value. Experimental fumigation trials indicating first adverse effects from 70 $\mu g/m^3$ SO_2 (≈ 0.025 ppm) for cultivated plants form the basis for setting the short-term value at this concentration (Wentzel 1983; Jäger and Schulze 1988).

Leaf necroses were the predominant effects of short-term, and mainly single exposures of plants to > 1 ppm NO_2, while lower concentrations caused alterations in photosynthesis and respiration as well as in organelle ultra-structures. The particular sensitivity of plants at night has also to be taken into account, and the fact that frequently high NO_2 concentrations are associated with high NO levels. This led Guderian (1988a) to suggest the following critical levels for acutely damaging NO_2 exposures (Table 4.12).

For longer-term exposures, a seasonal differentiation of sensitivity is indicated. Thus for summer, when the physiologically more active plants can rapidly metabolize NO_2, a half-year mean of 0.03 ppm appears tolerable. In winter, however, when plants are particularly sensitive, a half-year mean of 0.02 ppm is considered necessary to protect both short-lived plants and the long-lived arboreal flora.

Experimental investigations into ozone damage (Linzon et al. 1975; Heck and Brandt 1977; Guderian et al. 1985) permit defining the following critical levels for the protection of vegetation against O_3 as a single pollutant (Table 4.13). For longer-lasting exposures, 50 $\mu g/m^3$ (≈ 0.025 ppm) O_3 appear indicated as vegetation period average (composed of 7-h daylight mean values, 9.00–16.00).

For NO_2 in combination with O_3 and SO_2, the World Health Organization has adapted an air quality guideline of 30 $\mu g/m^3$ NO_2 as arithmetic annual mean for areas where SO_2 and O_3 may occur in annual and vegetation period means of up to 30 $\mu g/m^3$ SO_2 and 60 $\mu g/m^3$ O_3. (Guderian 1988b). It should be remembered, however, that nitrogen oxides do not only have direct effects, but may also affect plants indirectly via the soil (cf. Sect. 4.2.2.3), with additional nitrogen input greatly endangering nutrient-poor plant formations such as

Table 4.12. Acutely damaging NO_2 concentrations. (After Guderian 1988)

Plant resistance	1/2-h mean $\mu g/m^3$	ppm
Sensitive species	800	0.4
Average species	1200	0.6
Less sensitive species	1800	0.9

Table 4.13. Acutely damaging O_3 concentrations.
(After Guderian 1988b)

Exposure time (h)	$\mu g/m^3$	ppm
0.5	300	0.150
1	150	0.075
2	110	0.055
4	80	0.040
8	60	0.030

heath and moor communities. Ten μg NO_2/m^3 may already suffice to damage the particularly sensitive ombrotrophic mires (Lee et al. 1985).

On the basis of exposure–effect relationships as determined for a variety of plants of widely differing sensitivity by various authors Posthumus (1988) deduced the following critical levels for NH_3:

100 $\mu g/m^3$ ($\approx$ 0.14 ppm) for a monthly mean concentration,
600 $\mu g/m^3$ ($\approx$ 0.86 ppm) for a 24-h mean concentration,
10 mg/m^3 ($\approx$ 14.3 ppm) for a 1-h mean concentration.

4.2.1.6 Particulate Pollutants

Air-borne suspended particulates comprise a wide variety of pollutants including soot, dusts from cement-kilns, fertilizer and metal processing plants, lead particulates from traffic and industrial sources, dusts loaded with fluoride, sulphuric and other mineral acids, but also polycyclic hydrocarbons (cf. Chaps. 3.1.2, 3.5.5–3.5.7). In the flue dust of a lead smelter, for instance, the specific components include, besides $PbOSO_4$ and $PbSO_4$, ZnO, Fe_3O_3, Fe_3O_4, As_2O_3, sulphate, sulphite, bisulphite, sulphur, chloride, fluoride and phosphate (Eatough et al. 1979).

Particulates normally exert both physical and chemical influences on plants, and not infrequently the latter are more important. A further aspect of particulates is their combination with each other and with gaseous pollutants, and quite frequently the combined effects constitute the very environmental hazard. Therefore high-volume samplers, as established by the US Environmental Protection Agency as the reference method for total suspended particles (Clements 1978), can only indicate the quantity and, after analysis, the quality of particulates in the ambient air, but cannot indicate their biological effects. In plants, the exposed surface reacts not only in dependence on the area affected, but especially in relation to surface structure. Plants with glabrous leaf surfaces (e.g. *Quercus robur*, *Fagus sylvatica*, *Fraxinus excelsior*) collect less particulate matter than trees with glabrous but resinous surfaces (e.g. *Betula alba*, *Alnus glutinosa*), and these, in turn, less than the rough and partly hairy surfaces of leaves of *Tilia cordata*, *Carpinus betulus*, *Ulmus* sp. and *Corylus avellana* (Ernst

1982). The high sensitivity of lichens and bryophytes to particulate pollutants is partly due to their exceedingly high surface area and rough structures (Le Blanc et al. 1974).

Generally, particulates deposited on a leaf or needle surface will reduce quantity and quality of penetrating radiation (Rohde 1962; Auclair 1977). Thus photosynthesis and biomass production are reduced, as is widely known from the effects of cement-kiln dust (Lerman and Darley 1975; Lariland et al. 1978; Oblisami et al. 1978; Borka 1980; Singh and Rao 1981). Besides size and structure of the leaf surface the leaf age, i.e. the duration of exposition, and the growth behaviour, i.e. apical or intercalar growth, control the accumulated amount of particulates.

Another physical effect of particulate deposition is the clogging of the stomata, which reduces gas exchange and stomatal transpiration (Wedding et al. 1975). Dusted leaves transpire less water than clean ones, and this generally causes an overheating of leaf tissues, independent of particulate composition as demonstrated by the effects of road traffic (Eller 1977a, b) and industrial dust (Maier et al. 1979).

In comparison to gaseous pollutants, the chemical effects of particulates may normally be characterized as accumulative, i.e. they exert a physiological influence only after reaching a certain critical level or after reaction with water or aqueous solutions of highly variable pH (Frevert and Klemm 1984; Aniansson 1988a; Klemme 1988). Therefore effects of this type cannot be properly measured by short-term exposure. The following examples must suffice to illustrate, in conjunction with Sect. 4.2.1.7, the role of particulate pollutants in plant injury.

Cement-kiln dust reacts with the water vapour derived from transpiration of plant leaves to form a hydrate which builds up a layer of calcium silicate. All further deposition of cement-kiln dust will then be much less dependent on transpiration activity than on air humidity (Czaja 1961, 1966). Hence short-term exposure of plants will produce results largely differing from those of long-term exposure.

In smelter particles the metals are frequently associated with sulphate and sulphite (Hansen et al. 1974, 1975; Eatough et al. 1975) so that, in the presence of vapour or water, the plant surface may be covered by a slightly to moderately acid metal solution (Jensen 1985). Thus the first reactive cellular compartment is the cell wall. Most of the ion-uptake by lichens and mosses is controlled by ion-exchange processes (Tuominen 1967; Handley and Overstreet 1968; Puckett et al. 1973; Nieboer et al. 1975, 1978; Brown 1976); consequently the cation-exchange capacity (and in some cases also the anion-exchange capacity, for instance for arsenates) of the cell wall will determine the accumulation of metals (Clymo 1963; Spearing 1972; Rejment-Grochowska 1976; Nieboer et al. 1978; Folkeson 1979). With regard to biomonitoring the intraspecific differences in accumulation capacity must be considered. For instance, studies with *Hypogymnia physodes* have shown a marked local variation of this capacity (Laaksovirta et al. 1976).

At the community level, and in particular as regards perennial vegetation, the indirect effects of dust sedimentation may become decisive. Owing to migration of heavy metals through the soil (cf. Sect. 4.4.1.3), there may occur a concentration increase in the rooting zone of plants with a short rooting system such as grasses and herbs.

The resultant stress will select for pollutant-resistant strains in the same way as described for noxious gases (Bell and Clough 1973; Taylor and Murdy 1975; Horsman and Wellburn 1977; Murdy 1979; Ernst 1982; Treshow 1984).

4.2.1.7 Combination Effects of Acid Precipitation

A comparative analysis of wet deposition shows that acid precipitation with increased sulphate and nitrate concentrations occurs in heavily polluted areas as well as in less polluted ones (Perseke 1982; Aniansson 1988a). In the latter, i.e. outside the emission areas, the concentrations of acid substances in rain are reduced by a factor 2–3 only. Furthermore, the spatial pattern of wet deposition is largely determined by the precipitation pattern, which is of particular importance for mountain regions where thus high wet deposition rates occur (cf. Chap. 3.5.4).

In view of the complicated physical and chemical transformation processes preceding and accompanying precipitation (see Chap. 3.3 and 3.4) the chemical composition of acid rain and fog is highly variable in time and space, probably comprising not only the whole set of gases described in the preceding sections, variable proportions of inorganic acids resulting from their partial oxidation and heavy metal dusts, but also many other substances. The knowledge of corresponding effects on vegetation is consequently still fragmentary, and uncertainties remain. This applies in particular to the nature of additive and synergistic effects of the various components of acidic precipitation.

Direct Effects on Leaves

Injury to foliage largely depends on the effective dose of acidic precipitation to which sensitive tissues are exposed; it is controlled by the contact time of the water droplets or films with the foliage surface. Contact time, in turn, is regulated by the wettability of the leaf or by morphological features which retard or prevent rapid runoff from the surface (Martin and Juniper 1970; Evans et al. 1981; Shriner 1981). Evans et al. (1977a, b, 1978) came to the conclusion that about 95% of foliar lesions occurred where water is likely to accumulate, i.e. near bases of specialized epidermal cells, e.g. trichomes, stomatal guard and subsidiary cells, and along veins.

Studies with isolated cuticles have shown a dependence of cuticular penetration on solution pH (McFarlane and Berry 1974; Orgell and Weintrauf 1975; Evans et al. 1981). In continuation of Shriner's work (1974), Hoffman et al. (1980) indicated a mechanism by which precipitation acidity acts as an important factor in *weathering epicuticular waxes*. Strong acid inputs to this system would oxidize and release a wide range of carbon chain acids from the basic wax matrix (cf. Schulten et al. 1986).

Also modifications of leaf structure following exposure to simulated acidic precipitation have been described (Evans and Curry 1979), resulting in formation of galls on adaxial leaf surfaces; but lesions of this type have not been reported after ambient rainfall events in the field (Linthurst 1983).

More important, therefore, are *foliar leaching* and physiological response effects of leaves due to acidic precipitation. Foliar losses of K, Mg and Ca from bean and maple seedlings were found in experiments to increase as the acidity of precipitation increased (Wood and Bormann 1974). Abrahamsen and Dollard (1979) observed that Norway spruce (Picea abies) lost a greater amount of nutrient elements by leaching under most acidic treatments (pH 2.5), and Wood and Bormann (1977) noted similar results for eastern white pine (*Pinus strobus L.*)

Kaupenjohann et al. (1988) determined simulated acid rain effects (sulphuric acid of pH 2.4 and 2.7; 200 ml/tree and day) on nutrient status of *Picea abies* in a 10 days' experiment with 3 years' samplings. The results of base leaching from the canopy support the work of Richter et al. (1983) and Skiba et al. (1986). Contrary to the above observations, K- and Ca-contents of the needles were not affected, but Mg-content decreased with increasing acidity. While this is in agreement with field observations indicating Mg-deficiency as a common symptom of forest decline in SO_2-polluted areas (Schrimpf et al. 1984; Schröder 1989), Mengel et al. (1987) could not induce a decrease in Mg-contents of spruce needles by acid treatment. The difference in reaction may be due to different developmental stages of the plants treated; Kaupenjohann et al. carried out their experiments at the end of the vegetation period, while Mengel et al. treated the specimens during the vegetation period.

These results, and in particular systematic field investigations (Ulrich 1983a; Zech and Popp 1983; Block and Bartels 1985; Matzner 1985; Roelofs and Boxman 1986; Fränzle et al. 1987; Lindberg et al. 1990), suggest that leaching is a very important variable from a nutrient/throughfall chemistry perspective. It exerts a marked influence on precipitation chemistry as it passes through the canopy to the forest floor and ultimately to the aquatic system as Table 4.14 shows. These and similar findings (Seibt et al. 1977; Kramer and Ulrich 1985; Matzner 1985; Reiter et al. 1986; Beyschlag et al. 1987; Wenzel and Ulrich 1988) would suggest that an increased uptake of nutrients can compensate for leaching losses, provided other regulative mechanisms remained intact (cf. Marschner 1986).

In comparison to these results, a conclusive linkage between plant *physiological response* to acidic deposition is more difficult to establish and further comparative studies are still needed. On the one hand, Sheridan and Rosenstreter (1973), Ferenbaugh (1976), Hindawi et al. (1980) and Jaakhola et al. (1980) described reduced chlorophyll content following tissue exposure to acidic solutions, which would be explainable in the light of the above leaching effects. Irving (1979), on the other, reported increased chlorophyll content in leaves exposed to simulated acid precipitation of pH 3.1. While Ferenbaugh (1976)

Table 4.14. Precipitation chemistry of beech and spruce forest stands in the Hochsauerland, Federal Republic of Germany (inputs in $kg\,ha^{-1}$)

Site	Type		H	SO_4	NO_3	Cl	PO_4	NH_4	Ca	Na	K	Mg
Wilzenberg	Throughfall, beech	n	9	9	8	9	9	12	12	12	12	12
		12×	1.56	145.43	57.42	35.09	0.64	6.67	12.94	14.28	24.79	3.66
		s	0.12	11.18	1.49	1.50	0.10	0.36	0.70	1.26	1.20	0.23
		fc	6.8	2.6	0.9	1.4	0.5	0.9	3.1	1.2	4.5	1.7
	Throughfall, spruce	n	9	9	8	9	9	12	12	12	12	12
		12×	0.33	58.64	65.57	28.01	1.65	6.58	7.45	9.15	22.70	3.08
		s	0.02	2.51	1.12	1.81	0.22	0.40	0.47	0.55	1.68	0.13
		fc	1.4	1.1	1.1	1.1	1.3	0.9	1.8	0.8	4.1	1.4
	Open-air precipitation	n	9	8	8	7	7	11	12	11	12	11
		12×	0.23	56.04	61.17	25.92	1.25	7.71	4.23	12.26	5.56	2.22
		s	0.03	2.67	2.37	1.64	0.15	0.34	0.29	1.21	0.40	0.09
Burbecke	Throughfall, beech	n	9	9	8	9	9	12	12	12	12	12
		12×	1.98	170.80	97.52	44.79	1.11	9.10	15.02	19.59	22.99	4.63
		s	0.10	9.59	4.11	1.63	0.22	0.38	0.46	2.10	1.04	0.18
		fc	6.0	3.3	1.2	1.9	0.9	0.8	5.6	2.7	6.1	2.3
	Throughfall, spruce	n	9	9	8	9	9	12	12	12	12	12
		12×	0.43	73.67	90.27	40.92	0.64	10.32	8.17	14.31	30.44	3.76
		s	0.03	3.02	3.61	1.90	0.07	0.60	0.29	1.30	1.96	0.11
		fc	1.3	1.4	1.1	1.7	0.5	0.9	3.1	2.0	8.1	1.8
	Open-air precipitation	n	8	8	8	8	7	11	12	12	12	11
		12×	0.33	51.71	79.73	24.11	1.25	11.72	2.67	7.33	3.77	2.05
		s	0.03	1.74	3.35	0.96	0.18	0.76	0.18	0.50	0.28	0.14

n = number of measurements, 12× = cumulative annual input, s = standard deviation of monthly measurements, fc = concentration factor throughfall/open-air precipitation

found a significant increase of both respiration and photosynthesis at the unrealistically low pH of 2.0, Irving (1979) did not observe adverse effects on plants to rain at pH 3.1 (cf. also Amthor 1984; Matzner and Ulrich 1983, Adriano and Johnson 1989).

A generalized criterion accounting for responses to acidic and acidifying substances is net primary productivity, i.e. measurements of biomass, yield, and incremental size increases etc. (Abrahamsen 1980b; Athari 1980; Kenneweg and Kramer 1984; Kramer and Dong 1985; Blaschke 1986; Dong and Kramer 1987; Koltzenburg and Knigge 1987). It thus appears that further study in this area would be warranted only to better understand the regionally variable combination of factors and mechanisms by which the growth has been altered.

A complete assessment of the influence of acidic precipitation on plant communities must also take into account changes in *life cycle stages* as affected by environment and season (Cowling 1978; Linthurst 1983; Matzner and Thoma 1983; McColl and Firestone 1986). Seed germination, seedling emergence and establishment are early growth phases which proved potentially susceptible to acidic deposition (Matziris and Nakos 1977; Weber and Lee 1979; Linthurst 1983); but also plants which have reached maturity or are in reproductive phases may be quite sensitive (Likens et al. 1976; Wood and Bormann 1977; Jacobsen 1980; Raynal et al. 1980; Evans 1982; Kaupenjohann et al. 1988; Mengel et al. 1987). Thus seedling growth studies have shown that juvenile plants may exhibit reduced or enhanced growth depending on the plant species. Even where growth is stimulated by acidic deposition, as is the case with Aleppo pine (*Pinus halepensis* Mill.), foliar injury may simultaneously occur (Matziris and Nakos 1977).

A better understanding of the various effects of acidic deposition on life cycle stages and reproductive processes is important because of the potential impact on the successional stages and eventual species replacement in forests (Scholz 1981a; Gehrmann 1983). The long-term impact of pollutants induces selection against more sensitive and exposed trees. Whenever individual differences in response to acidic deposition (or air pollution in general) are not only due to environmental effects but also to genetic factors or pre-deposition, such selection will eventually involve losses in genetic variation of the population affected. In the final end this may signify the irreversible loss of genetic information, which, owing to homozygosity, reduces the adaptive capacity of species and populations to natural stress factors and diseases (Scholz 1981b; Scholz and Geburek 1983).

4.2.2 Indirect Effects of Acidic Precipitation

Important indirect effects of acidic precipitation on plant communities are potential alterations of host–insect interactions, host–parasite interrelations, symbiotic associations and soil acidification. The biotic interactions could involve, apart from the above influences of acidic deposition on host plants, direct influences on an insect, microbial pathogen, or microbial symbiont; or an interference with the interactive processes of plant and agent, i.e. infestation or disease of symbiosis.

4.2.2.1 Biotic Interactions

Under appropriate environmental conditions all stages of tree life cycles and all tree tissues and organs may be subject to impact by a great number of microbial pathogens including viroids, viruses, mycoplasmas, bacteria, fungi, and nematodes. Yet the current understanding of the influence of acidic deposition on pathogens and the diseases they cause is still comparatively limited (cf. Linthurst 1983; Treshow 1984; Fränzle et al. 1985; Ulrich and Sumner 1991).

Interactions with Pathogens

Shriner (1974, 1977) has examined the effects of simulated rain acidified to the common range of ambient precipitation pH (i.e. 3.2 and 6.0) on several host–parasite systems under greenhouse and field conditions. In these experiments sulphuric acid increased suspectibility of kidney beans (*Phaseolus vulgaris*) to halo blight, suggesting decreased resistance of the plant to the organisms. The pathogen, however, is also sensitive to acidic conditions, and the same also applies to others. At pH 3.2 the telia production by *Cronartium fusiforme* (fungus) in willow oak, root-knot nematodes (*Meloidogyne* sp.) on kidney bean, and *Uromyces phaseoli* (fungus) were inhibited in dependence on the particular stage of the disease cycle.

Bruck et al. (1981) treated loblolly pines with simulated rain at pH 4.0 and found significantly fewer galls after inoculation with *Cronartium fusiforme*. Lacy et al. (1981) analogously observed that bacterial populations on soybean leaves were reduced due to exposure to acidic deposition.

The complexity of the plant–pathogen system makes it difficult to determine the significance of these and similar laboratory-type studies under field conditions. A potentially increased susceptibility of the plant to the pathogens, and the sensitivity of the pathogens to acidic deposition may ultimately result in little or no net increase or decrease in pathogen activity (Linthurst 1983). This assumption would be compatible with findings of a wide spectrum of viroids, viruses, bacteria and mycoplasmas in injured conifers and broadleaved trees in Germany, Switzerland and eastern France which, isolated, did not prove infectious to conifers under experimental conditions (Elstner and Osswald 1984; Mathé 1985; Frenzel 1986; Nienhaus 1986). Thus the importance of pathogen stress in present forest dieback in Central Europe remains to be established in comparison to other damaging effects while there are some histological indications of virus injuries from earlier centuries (Frenzel 1986).

Interactions with Insects and Microbes

Like pathogens, insects and the diseases they cause may play important roles in succession, species composition, density, and productivity of plant associations. Chemicals leached by acid rain could prove attractive, repulsive, or provide orientation to insects infesting plants. As regards surface microbes, leached compounds may inhibit vegetative growth or spore germination (phenolic compounds, alkaloids) or stimulate vegetative growth (as nutrients) or spore germination (as inducers or nutrients). But to date, comparative studies to examine this issue are still lacking. Also the question to which extent leaching of radiotoxic substances from plant surfaces could also exert a restrictive influence on plant surface biota needs to be resolved.

The fact, however, that neither a direct nor an indirect injury to plants due to natural or anthropogenic radionuclides in ambient air (e.g. radon, thoron, xenon, tritium, ^{14}C) has been established hitherto (König 1986, Metzner 1986; Münnich and Weiss 1986), could be interpreted in this sense.

4.2.2.2 Interactions with Pesticides

The direct effects of herbicides on plants can be considerably enhanced by lowering the pH on leaf surfaces due to acidic precipitation. For compounds of the phenoxyacetic and dinitrophenol type this phenomenon has been observed for many years (Crafts and Reiber 1945; Crafts 1961). Its analysis reveals a dependence on the pK_a value of the compound (cf. Sykes 1979) as related to ambient acidity. If the pH of precipitation is distinctly inferior to the pK_a value of a pesticide, the formation of the unionized species is promoted, which more readily penetrate vegetative membranes than do ionized species. This is most likely to occur in herbicide applications to forests, pastures, minimum-tillage crop production systems, or aquatic systems where the foliage has hard time to accumulate acidifying solutions of aerosols and gases.

Table 4.15 summarizes pK_a values of selected pesticides in the range of current pH values of acid precipitation in Central Europe (cf. Chap. 3.5.4), i.e. between 2.5 and 4.3 (Winkler 1983b; Frevert and Klemm 1984). In the present context, concentration phenomena in fog aerosols merit particular attention (cf.

Table 4.15. pK_a values of selected pesticides

Compound	pK_a value
Amitrole (3-amino-1,2,4-triazole)	4.2
Fuberidazole (2-(furyl-(2))-benzimidazole)	4
2,4-D (2,4-dichlorophenoxyacetic acid)	2.64–2.73
MCPA (2-methyl-4-chlorophenoxy-acetic acid)	2.90

Chap. 3.3 and 3.4). In Maryland, Washington DC and California, for instance, insecticides like parathion, malathion, chloropyifos or herbicides like atrazine, simazine, dinitroaline had concentration factors of 50 to 3000 in fog aerosols as compared to the primary concentration in the ambient air phase (Klemme 1988).

4.2.2.3 Soil Acidification and Related Effects on Soil-Borne Microorganisms

Comprehensive research schemes in Europe and North America have provided ample evidence during the past two decades that soil acidification may be the most significant factor which ultimately affects the growth of vegetation (cf. Ulrich and Sumner 1991). Nutrients losses, sulphate adsorption, accelerated mineral weathering, metal mobilization and availability, and nitrate additions are among the primary issues to be considered when defining long-term effects on forest tree species. As a consequence of acidic deposition, each of these variables can be significant in regulating site development and plant growth (cf. foldout model III).

Soil Acidification

Soil acidification is a natural process in areas with a surplus of precipitation and resultant deep drainage (see Sect. 4.1.2), and many of the compounds deposited from the polluted atmosphere occur naturally in soils. The problem of assessing their effects, in particular with regard to forest dieback, is therefore one of quantification. To this end, a comparison is made of the amount of acid deposited from polluted atmosphere (cf. Chap. 3.5) and the amount produced in forest soil by natural biotic and abiotic processes. By comparing these two amounts with H-ion sinks in the forest, it is possible to determine how the acid/base properties of forest soil are influenced in areas of acidic deposition.

Figure 4.11 provides a schematic synopsis of the origin and fate of H-ions in different parts of, and processes in an ecosystem. In unpolluted air a major source of H-ions is carbonic acid in precipitation. Another important source of carbonic acid is the CO_2 production by soil organisms and plant roots, and H-ions are also produced when N- and S-compounds are mineralized. Nutrient uptake by plants produces H- or bi-carbonate ions in dependence on the ratio of cation to anion uptake. Since forest trees normally take up more cations than anions, H-ion release is usually connected to the nutrient uptake in forest soil. When humus material accumulates, which may involve periods of several thousand and even more than 10,000 years (Scharpenseel 1971), the soil H-ions also accumulate. The main process counteracting acid accumulation in soil is silicate weathering as far as Ca, Mg, K and Na are released and protons used up, but H-ions may also be consumed by volatilization of N- and S-compounds (Abrahamsen and Stuanes 1983; Ulrich 1987a).

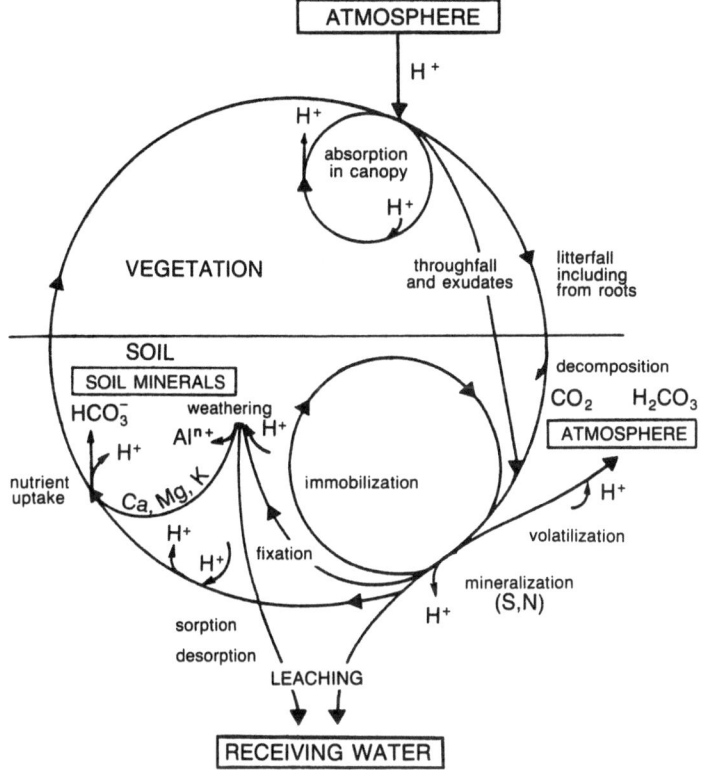

Fig. 4.11. Uptake, production and fate of H-ions in terrestrial ecosystems. (After Abrahamsen and Stuanes 1983)

An accumulation of base cations in the biomass is thus coupled with a corresponding accumulation of H-ions in the soil. Provided, however, that the organic matter produced in an ecosystem is decomposed completely to CO_2, H_2O and inorganic compounds, and that biomass is not removed from the system, there should be no long-term net accumulation of H-ions. Such an idealized state of a terrestrial ecosystem can be described by the following balance equation (Ulrich 1987b):

$$CO_2 + H_2O + xM^+ + yA^- + hv \rightleftarrows (CH_2OM_xA_y) + O_2 . \qquad (4.32)$$

It describes to which extent organisms $(CH_2OM_xA_y)$ are sources or sinks for CO_2, H_2O, cations (M^+), anions (A^-), H-ions and energy (hv), and thus indicates that the acidifying effect of an incomplete decomposition or removal of biomass corresponds to the net uptake of cations over elements taken up as anions (cf. Table 4.16).

Table 4.16. Amounts of H-ions produced in different forest ecosystems. (Abrahamsen and Stuanes 1983)

Location	Site description	H^+ production me/m^2 yr	Calculated as
Sweden	Pine, high productivity	300	Excess cation
	Pine, low productivity	≈ 50	accumulation in trees and humus
Hubbard Brook Experimental Forest USA	Deciduous forest	122	Total H^+ sources in the watershed minus the atmospheric input
Sweden	Pine, low productivity		Cation/anion balance
	Mineralization	223	
	Root uptake	309	
Solling	Beech	216	Cation/anion balace
	Spruce	346	

According to the above relationships, the natural H-ion production in forest ecosystems will vary considerably with soil properties, climate, species composition and age of forest stand, productivity, forest management, etc. As shown by Nilsson et al. (1982), after forest regeneration it increases up to a stand age of about 20 years, whereafter it gradually declines.

Cations exist in three phases of the soil system, i.e. in solution absorbed to the negatively charged soil particles, and in the primary and secondary soil minerals (cf. Sect. 4.4.1.2). Changes in the concentration in one of these phases influence the concentration in the others. If, for instance, the concentration of H- ions and associated anions increase in the soil solution, H-ions will exchange with cations on the surface of soil particles. Other cations may then move into the soil solution, and consequently the H-ion saturation of the soil will increase. Increased soil acidity, in turn, is likely to increase the release of elements from the primary minerals by enhanced weathering rate.

In addition to the occurrence of cations in the soil solution, the existence of mobile anions (Cl^-, NO_3^-, HCO_3^-, and frequently SO_4^{2-}) is a premise for leaching (Johnson and Cole 1980). In connection with acidic deposition, NO_3^- is weakly adsorbed to the soil particles, but it is readily taken up by plant roots and soil organisms; therefore it is normally leached in small quantities in natural forest soils. Abrahamsen (1980) compared the input of N by wet deposition with the runoff from forest catchments and in lysimeters with forest soils in Europe and North America. On the average, the leaching of N amounted to only 30% of the incoming N by wet deposition. Including dry deposition, the relative leaching of N would be even lower, which means, presuming negligible denitrification, that more than 70% of the incoming N will stay in the soil–plant system. NO_3^- thus contributes relatively little to the leaching of cations from the soils referred to, except during snowmelt, when the biological activity is low (Galloway and Dillon 1983; Niestlé 1987).

The situation may be different, however, in areas with high atmospheric deposition or intensive agriculture, such as Germany, The Netherlands, southern Sweden and the United Kingdom. In the UK, the level of nitrate in rivers and in water pumped from deep wells has risen greatly over the last 50 years. The extent of the rise is highly variable and depends on a range of factors, but there is some evidence that this trend is changing. Data from monitoring at 149 stations on rivers in England, Wales and Scotland show no significant change in the average nitrate content of river water from 1977 to 1984 although regional variations exist (Foster et al. 1986; UK-DOE 1986; Aleamo et al. 1987). This apparent plateau has occurred even though fertilizer use has increased considerably especially in the arable farming areas. A similar trend is available from some rivers in the Federal Republic of Germany; i.e. the increasing concentration has halted for the Danube, or even decreased for the Mosel, Weser and Elbe rivers (Umweltbundesamt 1986a). It appears that organic nitrogen is the major source of nitrate; hence reduction of fertilizer application rates will not reduce water nitrate levels substantially in the short term (ECETOC 1988a).

In contrast to NO_3^-, the sulphate ion may be sorbed in soils rich in amorphous oxides and hydroxides of Al and Fe as are the plinthic phases of tropical soils (e.g. Ferralsols). Since such compounds are not as abundant in the soils of the sensitive areas in North America and Europe, and since the biological demand for S is much less than for N (cf. Chap. 3.1.1), SO_4^{2-} is only weakly retained in soil–plant systems of the temperate zone. This again is evident from the relationship between wet deposition and leaching of sulphate from forest catchments or lysimeters. Although Abrahamsen's (1980) conclusion that the leaching of S is slightly less than the total deposition is not in agreement with measurements in the Schönbuch area (Germany), it is clear that SO_4^{2-} leaching is a crucial factor for the removal of cations from soils.

The efficiency of the exchange processes described is dependent on the base saturation, soil pH and the relative proportions of variable and permanent charges of the soil colloids. A low base saturation, combined with a low soil pH and a relatively high proportion of permanent charges will result in a lower replacing efficiency of H-ions (Wiklander and Andersson 1972; Wiklander 1980; Schachtschabel et al. 1989). Table 4.17 illustrates the serious base depletion of selected German soils as a result of over-utilization of the ecosystems for centuries, and in some cases even for millenia. In this table, median values are given, which means that half of the representative soils studied have stocks below the values indicated. The Hamburg soils are classified according to buffer ranges (Ulrich 1981a), and the percentages indicate the relative acreage of the soils in the respective range.

According to the data available the acidification front of sandy soils in Central Europe, i.e. the transitional zone from base depletion to the release of

Table 4.17. Stocks of exchangeable K, Ca and Mg in selected mineral soils of Germany. (After Ulrich 1987a)

	Median values		
	K	Ca (kg/ha)	Mg
Hamburg forestry area, 0–60 cm			
Buffer ranges			
Carbonates, 1%	580	> 10 000	> 500
Silicates, 2%	160	8 000	300
Exchangeable substances, 12%	160	1 200	64
Aluminium, 14%	160	270	33
Al/Fe, 71%	120	170	28
North Rhine-Westphalia, 0–80 cm			
Soils on limestone	1 300	20 000	550
Soils on shales	360	500	100
Loess soils	480	2 200	440
Sandy soils	200	380	40

toxic cation acids, lies below the rooted zone in the percolating layer at a depth of 1–6 m. This indicates that the effect of acids on soil largely depends on the pattern of infiltration and flow through the soil, which is mainly governed by soil structure, the distribution of large and small pores, and by the quantity of percolating water (cf. foldout model II and Sect. 4.1.2).

Interactions with Soil-Borne Organisms

Soil acidification can cause serious damage to soil organisms and plant roots. Populations of most root pathogens are greatest within the uppermost 10–15 cm of soil, where they have maximum impact on feeder roots of established plants and young seedlings in particular (Garrett 1981; Florenzano 1983; Mathé 1985). Also soil microorganisms are under the greatest influence of precipitation chemistry in this zone since frequently distinctly greater soil depths may be required to substantially alter severely acidic rain through exchange phenomena (cf. Sect. 4.4.1.2). In analogy with the sensitivity of higher plants to pollution, also the reaction of soil-borne pathogens under different pH conditions is related to life cycle phases. A drop in pH from 5.0. to 4.5 caused only an 8% reduction in *Phytophthora cinnamoni* chlamydospore germination (Mircetich et al. 1968) but caused a 94% depression of mycelial growth after 28 days (Linthurst 1983).

Soil-borne microorganisms can form beneficial symbioses with plant roots, the best known of which is probably the infection and modulation of legume roots by various strains of the bacterium genus *Rhizobium* (Campbell 1980).

This mutualistic symbiosis allows fixation of atmospheric nitrogen. Shriner (1974) found that kidney beans and soybeans exposed to simulated rains of pH 3.2 developed an average of 75% fewer nodules than plants exposed to pH 6.0 treatments, and exhibited lower nitrogenase activity. Waldron (1978) found similar inhibitory effects on nodulation, but nitrogenase activity was similar to that of plants exposed to deionized water, which may suggest a kind of compensatory effect by the plant–bacterium system.

Mycorrhizae, as the most common plant root–soil microorganism symbioses in nature, benefit their host plants by effectively increasing the absorptive surface area of the root system (Harley and Smith 1983), thus providing for increased uptake of nutrients and in some cases water. Presumably mycorrhizal fungi might respond in ways similar to the above fungal pathogens to acidified soil solutions. Feicht (1980) infested soybeans with the endomycorrhizal fungus *Glamus macrocarpus* and exposed the plants to simulated rains over the pH range 5.5–2.8 twice a week for 15 weeks. Although no changes in the fungal colonization of roots could be observed, sporulation of the fungus exposed to pH 2.8 was inhibited by approximately 40% compared to plants exposed to pH 5–6 treatments.

Pollutant gases such as SO_2 or O_3 in large concentrations which may displace soil air because of their higher specific gravity (cf. Chap. 2.1.2.2) have deleterious effects on mycorrhizal fungi. Some of them, however, are able to sustain moderate concentrations. Air-borne or mobilized heavy metals such as copper, lead, nickel and cadmium, may damage the mycorrhizal fungi, but the susceptibility to such pollutants varies from one fungus species to another (Høiland 1986). Also aluminium mobilized in soil through acidification processes is suspected of damaging the mycorrhizae. Again, the various fungal species may react differently to Al, probably reflecting their ability to form harmless metal–organic complexes (cf. Jansen et al. 1988).

Certain forest tree species seem to be protected from root diseases by ectotrophic mycorrhizae which form a physical barrier to parasite infection (Marx 1969, 1972). Damage to the fungal mantle by acid rain or high NH_4^+ deposition (van Aalst 1984; Arnolds 1985; Høiland 1986) might consequently allow pathogens to enter the root. In a comparative study on physiological effects of NH_4^+ and Al^{3+} on pine forest ecosystems in The Netherlands, Boxman and Roelofs (1986) found ectomycorrhizal extension growth severely inhibited by increasing NH_4^+/K^+ ratios which were due to enhanced deposition of zoogenic ammonium sulphate (cf. Chap. 3.4.2). The NH_4^+-ion exerts an immediate negative influence on the nutrients status of soil fungi like *Amanita*, *Canococcum*, *Hebeloma*, *Laccaria*, *Pisolithus*, *Rhizopogon*, *Suillus* and *Telephora* by reducing the uptake of K^+, Ca^{2+}, Mg^{2+} and PO_4^{3-}.

Impacts on Forest and Crop Productivity

A major area of both scientific and economic interest in the last two decades has been the potential of forest trees and crops for indirect effects of acidic precipitation via impacts on soil nutrient availability or soil leaching rates. The comparative analysis of these problems requires both large-scale field experiments and long-term field observations.

Cogbill (1976) used dendrochronology to analyse historical growth patterns of beech, birch, and maple in the White Mountains of New Hampshire and red spruce in the Smoky Mountains of Tennessee and concluded that no synchronized regional decrease in radial growth had occurred. Research in Norway (Abrahamsen et al. 1976, Stand 1980), however, indicated a regional damage pattern with decrease in growth of white spruce and Scots pine occurring principally in the eastern third of the country. Later studies in New York (Raynal et al. 1980) with red spruce and white pine and in the New Jersey pine barrens (Johnsson et al. 1981) with pitch, short-leaf, and loblolly pine have shown comparable patterns of decline for these species over the past 30 years.

A decline of red spruce, but not for fir or white birch, has been quantitatively documented in the Green Mountains of Vermont (Siccama et al. 1982) and also observed in New York and New Hampshire (Linthurst 1983). In the Green Mountains, trees in all size classes proved to be affected between 1965 and 1979, with an overall reduction of approximately 50% in basal area and density. In Roman and Raynal's (1980) opinion, the primary cause of decline is not likely to be successional dynamics, climatic changes, insect damage or primary pathogens. Following Manion (1981), the spruce damage has the characteristics of a complex biotic–abiotic disease related to environmental stress to which acidic deposition may be contributing.

Similar reductions in growth beginning around 1950 were found in Sweden by Johnsson (1976), using Scots pine and Norway spruce. The effect was most pronounced in areas receiving the heaviest loading of acidic deposition. Subsequent research has shown that soils in forest areas have changed dramatically. While nutrients have been halved in the last 30 years, acidity has increased between five and tenfold in southern Sweden, and these effects have penetrated to depths of 1 m. The related extent of forest damage is a matter of grave concern. One spruce in four and one pine in seven have lost over 20% of their needles. In spite of favourable (i.e. cool and wet) weather for trees in the last few years, forests have recovered very slowly or not at all, and damage has even been recorded in young trees. From forest damage surveys (beginning in 1984) it can now be concluded that the vitality of a substantial proportion of Sweden's coniferous forests is impaired after previous unique growth rates. These seem to be due to enhanced nitrogen deposition which contributed to acidification on the one hand, but on the other increased production in ecosystems where nitrogen was previously the growth-limiting factor (Aniansson 1988a).

In Germany, damage of ecosystems attributed to flue gases, SO_2 in particular, was already described in the neighbourhood of industrial plants more than a century ago. The situation did not change essentially until the late 1960s, when a new type of forest injury became evident which was no longer related to local emitters but occurred to an increasing extent in presumably clean air areas far away from conurbation and industrial centres (Bauch et al. 1979; Athari 1981, 1983; Nogler 1981; Athari and Kramer 1983a; Kenk 1983, 1984). In the middle 1970s, severe dieback of silver fir (*Albies alba*) was observed in southern Germany. In the early 1980s, other conifers, especially Scots pine (*Pinus sylvestris*) and Norway spruce, (*Picea abies*) showed similar symptoms on rapidly increasing areas, and recently also deciduous species like beech (*Fagus sylvatica*) and oak (*Quercus robur*) proved affected. According to the 1985 damage inventory, 52% (i.e. 3.8 million ha) of the forest area in the 58 growth districts of the Federal Republic of Germany were injured. Level 1 of damage (reduction of vitality) characterizes about two-thirds of this portion while medium to high damage (levels 2–4) occurs on 19% of the forested area (Bundesminister für Ernährung, Landwirtschaft und Forsten 1985).

As a result of comprehensive studies (cf. Bundesminister für Forschung und Technologie 1985; Fränzle et al. 1985; Umweltbundesamt 1986b, d), it is clear that lack of nutrients, climatic factors (Cramer 1986), mistakes in forest management, infections and pests may play an important role in the present complex forest disease; but there is a rapidly growing number of findings which indicate that the above anthropogenic air pollutants, and possibly also others like lead triethyl or nitrophenols (Faulstich 1986; Rippen et al. 1987b) and related soil acidification are the essential direct and indirect causes of damage (Ulrich 1984; Prinz 1984; Elstner et al. 1985; Hüttermann 1986; Guderian 1988).

At the cellular level the pollutants may lead to a disturbance of photosynthesis and respiration and the following primary and secondary metabolisms. In some cases, impairments of photosynthesis occur already before visible damage can be observed (Arndt and Kaufmann 1985). At the same time, photooxidants and acid rain will cause weathering of cuticular structures, destruction of membrane systems, and leaching of nutrients out of the leaves. Quantitative and qualitative changes in photosynthesis (Benner and Wild 1987) lead to reduced net primary production, increased respiration, and a changed assimilate allocation to the root–mycorrhiza system, stem, branches and leaves and also to the reproduction and phytogenous defense systems. Such stressed trees emit unsaturated hydrocarbons like ethene, isoprene and terpenes (cf. Chap. 3.1.2) which react with secondary pollutants, e.g. ozone (Chap. 3.4.4) to highly phytotoxic tertiary pollutants in the immediate neighbourhood of the plants. Weakening by reduced assimilate allocation to the roots will furthermore lead to reduced uptake of nutrients from soil which is already disturbed by acidification, i.e. impoverished in K and Mg.

Murach (1984) determined a Ca deficiency in the roots of declining spruce and fir, but the same levels of Al as in healthy trees; Rehfuess et al. (1982) noted Mg and possible Ca deficiencies by foliar analysis even in base-rich soils. Contrary to Ulrich's (1981b, 1983b) contention that a change in fine root biomass is paralleled by a toxically high Al/Ca ratio, Rehfuess points out that the two phenomena are not neccessarily synchronized, so that marked decreases in fine root biomass can precede the increase in soil solution Al (cf. also Lindberg and Johnson 1989).

Mineral deficiency due to leaching and soil acidification is frequently coupled with a relative surplus of nitrogen with respect to assimilated carbon as a consequence of enhanced deposition of nitrate and ammonia and the intensity of subsequent nitrification processes. Kriebitzsch (1978), who studied nitrification in many types of acidic forest soils, divided them into the groups A, B, C and D. In group A there was no nitrification and ammonium was the only nitrogen source. In soils of groups B and C, there was partial nitrification while group D soils had complete nitrification. Former heathland soils in Germany now cultivated with various *Pinus* species and *Pseudotsuga* mainly belong to group A, where the nitrate level is low whereas the ammonium concentration is high.

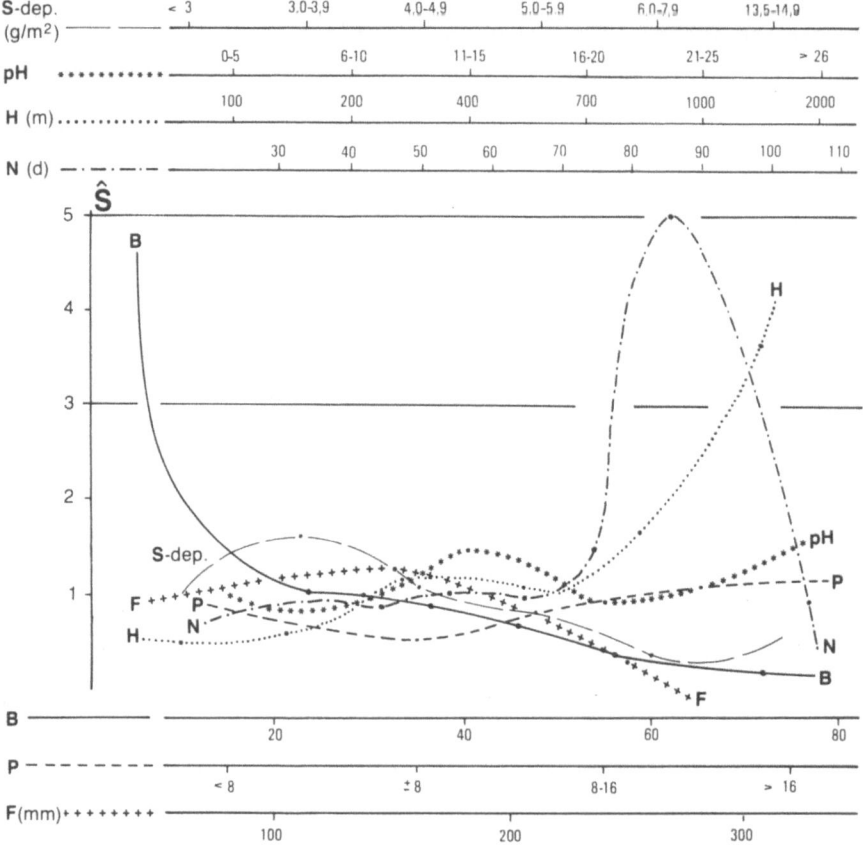

$\hat{S}$ = ratio of damaged/undamaged forest area; H = altitude (m); N = number of foggy days; B = soil quality (nutrient supply); P = porosity (percentage of pores >50 μm $\varnothing$); F = plant available field capacity (mm); pH = percentage frequency of acid rain (i.e. pH <4.5); S-dep = sulphur deposition

Fig. 4.12. Forest decline as related to selected site factors

Quantitative and qualitative changes in the living biomass of forest trees will bring about changes in the food-web of the forest ecosystem, some consumer (e.g. insects) or decomposer organisms (e.g. fungi) may exhibit an explosive growth pattern while others are completely suppressed. The disturbance of soil biocenoses may then induce a reduced litter decomposition which leads to a further shortage in mineral nutrients. Trees subject to these stress factors are very likely extremely sensitive to secondary damage by frost, drought or insect infestation.

An analysis of the spatial correlation pattern of site qualities and damage levels (Fränzle et al. 1985; Schröder et al. 1986) corroborates the importance of pedogenic nutrient supply and acidification processes for forest decline. Figure 4.12 illustrates on the basis of comprehensive cross tabulations of several hundred German stands, the extent of forest damage in terms of relevant site factors.

The standardized damage level D (i.e. proportion of injured/unaffected forest area) correlates best with pedogenic nutrient supply, frequency of fog situations and elavation, while the correlation with the other site factors is also significant on the 99.9% level although of little statistical relevance. It should be noted that $D = f(S, E)$ is a monotonous function, while $D = f(F)$ has a marked maximum.

Thus stands on acidified members of the ranker, cambisol and podzol groups with a pH < 4, marked nutrient deficiency and concomitant reduction of buffering capacity down to the Al- and Fe-ranges are most liable to dieback. Among the air-borne pollutants, SO_2 is particularly important. In combination with other pollutants (cf. Sects. 4.2.1.5 and 4.2.1.6) its effect on both vegetation and soil is enhanced by fog (contingency coefficient after Pearson $C = 0.53$). The amount of sulphur deposition is mainly controlled by precipitation ($C = 0.80$). The comparison of the pH- and sulphur deposition curves ($C = 0.89$) summarizes the major influence of gaseous and particulate pollutants on forests on acid soils as a result of the above mechanisms. Their inherent complexity implies the recognition that critical levels for tolerable pollutant concentrations can be adequately defined only with regard to the whole set of concomitant stress factors affecting a plant community (cf. Chap. 2.2).

Crop Productivity

Visible injury to crops as a result of ambient acidic precipitation has not been reported from modern high productivity agrarian ecosystems. Therefore attempts to evaluate impacts of acidic deposition have to be based on simulated precipitation in field experiments or controlled environments.

Yet most of the crop cultivars studied in the field and the crop varieties studied in controlled environments exihibited no effect on growth or yield as a result of exposure to simulated rain usually up to ten times more acidic than ambient (Linthurst 1983). Likewise, out of five cultivars of soybeans studied, only one exhibited reductions in yield (Evans et al. 1981). Genetic variations, possibly in combination with characteristics of the soil or simulated rain applied in these experiments, may account for the observed variability.

4.2.3 Dynamic Models of Forest Dieback

Simulation models of forest decline which can, for the time being, hardly cover more than essential parts of the introductory graph-theoretical model have to deal with two types of change – change in complexity through increasing or decreasing differentiation, and adjustment to extrasystemic changes. Negative and positive feedback processes can be correlated with systemic continuity or discontinuity, respectively. Continuity results when deviation-correcting mech-anisms are and remain operative so as to ensure that structure and function are kept viable within the given parameters of the system. Quantization occurs when deviation is amplified to the very point where no deviation-correcting mechanism can prevent the rupturing of the basic systemic framework – in other terms, when the latter are no longer able to contain and canalize the energies and thrust which have been generated. The process results in altering the relationships of the system with its environment, creating new spatiotem-poral, structural and functional boundaries – in short, the system is transformed to a new level of internal organization and environmental integration. There-fore, and specifically with regard to forest decline, models must employ systems concepts to explain the circumstances in which quantification can occur from one developmental stage to another. Hence, they have to account for both systematic levels of biocenotic organization and cybernetic processes which demonstrate (1) systemic self-stabilization within a given organizational level, and (2) systemic transformation resulting in a biocenotic quantum across an environmental frontier.

4.2.3.1 Models with Non-Linear Dose-Effect Relationships

Following models of reduced photosynthesis and leaf formation, and increased leaf ageing and shedding under air pollution stress (Kohlmaier et al. 1983/84; Bossel et al. 1985), Kohlmaier and Plöchl (1986) developed a more comprehen-sive model concept for Norwegian spruce damage. It includes the root system with reduced nutrient uptake and higher feeder root turnover, thus coupling the canopy and the soil root system. Reduced photosynthesis and changed assimil-ate allocation are simulated during the entire exposition to air-borne pollutants which may be part of or the whole life span of a tree. Assimilate demand and supply have a hierarchic structure, as described in the previous sections, i.e. respiration, growth of foliage, roots and living wood. The effect of an air pollutant on leaves and roots depends on both the specificity of instantaneous effects and the accumulative effects during the whole exposure time (cf.

Chap. 4.2.1.7) involving elimination and repair mechanisms of the plant. The complicated interaction pattern of these response mechanisms normally results in nonlinear dose–effect relationships with respect to concentration and accumulated pollutant and damage level of the plant.

During one time-step, the model first calculates the photosynthesis rate as dependent on external factors such as light intensity and air temperature, and internal parameters such as photosynthesis leaf efficiency; water uptake by the roots is the limiting factor. Pollutants can enter the system by two different paths, i.e. via leaves and roots; in the first case, the photosynthetic efficiency of a particular needle age class is reduced, in the second, fine root turnover and consequently assimilate demand are increased. The momentary pollutant flux into the leaf is described in terms of gas exchange between the intercellular air and the ambient air by Fick's first diffusion law:

$$I_{in}^{leaf} = (c_a - c_i)/r_{ai} , \tag{4.33}$$

where the flux I_{in}^{leaf} is proportional to the concentration difference $(c_a - c_i)$ between ambient and intercellular air, and r_{ai} is a specific resistance factor for a particular pollutant (cf. Chap. 3.5.3). The corresponding flux per unit canopy or unit ground area is then obtained by multiplication with the leaf area index (LAI). For the boundary condition $c_a \gg c_i$ Eq. (4.33) reduces to

$$I_{in} = (r_{ai}/LAI)c_a = v_{dep}c_a , \tag{4.34}$$

which is approximately equivalent to the dry deposition rate with v_{dep} being the deposition velocity (cf. Chap. 3.5.4).

It may be assumed that the pollutant load in the leaf or at its surface is eliminated either by physical processes (i.e. leading or litter fall) or by transport to other parts of the plant, or by metabolic processes. It is more appropriate, however, to model the elimination process following a Michaelis–Menten kinetics of the type:

$$d\bar{c}_1/dt = v_{dep}c_a - k_{out}\bar{c}_i/[1 + (\bar{c}_i/K_c)] , \tag{4.35}$$

where $\bar{c}_i$ = concentration per unit ground area, k_{out} = characteristic elimination constant, while the other symbols have the above meaning. The mode of interaction and effect of a pollutant inside the plant is likely to be specific for the particular pollutant "j" (cf. Sect. 4.3.1). In the simplest case, the effect, above the specific no-effect level, will be proportional to the interior concentration $\bar{c}_i^j$:

$$eff^j (t) = a_j \bar{c}_i^j (t) , \tag{4.36}$$

where a_j is a specific effect constant relating to a particular process, e.g. photosynthesis.

The effect variable eff(t) may be composed of linear and higher-order terms of n interacting species:

$$eff(t) = \sum_{j=1}^{n} a_j \bar{c}_i^j (t) + \sum_{j=1}^{n} \sum_{k=1}^{n} a_{jk} \bar{c}_i^j \bar{c}_i^k , \tag{4.37}$$

where synergetic effects are included in the second term with $j \neq k$.

According to Bossel et al. (1985), the effect variable affects the leaf senescence function

$$ALT(t) = 0.1[1 + eff(t)] , \tag{4.38}$$

with $0 \leq eff \leq 1.5$ and $0.1 \leq ALT \leq 0.25$, implying the value 0.1 as the natural senescence rate without pollutant damage. The senescence rate influences the number of needle age classes:

$$NJ = 0.8/ALT(t) \tag{4.39}$$

and both determine the leaf's photosynthetic efficiency. Hence the photosynthetic efficiency of a particular needle age class is:

$$E(N) = 1 - ALT(t)[(N - 1) + TJ] , \tag{4.40}$$

where $0 < TJ < 1$ characterizes the season and $1 < N < J$ the age of a needle class.

The average photosynthetic efficiency is then expressed by:

$$EFF = \sum E(N) L(N)/LEAF , \tag{4.41}$$

where $L(N)$ is the leaf mass of a particular needle age class and LEAF the total leaf mass. In unaffected systems EFF ranges between 0.55 and 0.65 (Kohlmaier and Plöchl 1986).

Acidification of soils and related leaching of mono- and bivalent cationic nutrients (Sect. 4.2.2.3) leads to higher feeder root turnover rates while a gradual loss of feeder root mass occurs. Following a titration curve of a buffered chemical system, it is suggested (Bossel et al. 1985) that this effect may be described by a logistic curve of the type:

$$W_{to}(t) = f[W(t)] = W_{max}/[1 + F_k(e^{-KIt})] , \tag{4.42}$$

where $W(t)$ = accumulated pollutant at time t, W_{max} = maximum accumulation, F_k = correction factor $(W_{max} - 1)$, K = buffer constant, I = constant input, $W_{to}(t)$ = fine root turnover factor at time t.

This supply–demand model of photosynthetic efficiency and assimilate allocation appropriately describes the decline of a forest ecosystem under the influence of chronic air pollution stress. It clearly indicates that non-linear effects may drastically accelerate the observed dieback phenomena; but more experimental base data on both the biochemical and physiological levels are required to better account for the combined effect of two or more pollutants (cf. Bossel 1986; Godbold and Hüttermann 1986; Schultz 1987; Ulrich 1990).

4.2.3.2 Time Series Analysis and Mapping

It is a common feature of the various methods of time series analysis of dynamic systems such as ecosystems that they involve the observation of a complete set of state variables. This may constitute a major drawback in studies of empirical systems whose inherent dynamics is not known a priori. Under these circumstances, however, use can be made of the fact that in dynamical systems with a finite number of state variables the information on the momentary value of all state variables can be substituted by information on the recent history of a part of the variables. Grossmann et al. (1984) therefore used time series analysis in combination with geographic information systems (GIS) to develop a scenario method which describes forest damage in sequential form by means of maps.

A geographic information system is a computer hardware and software system designed to collect, manage, analyse and display spatially referenced data. GIS have emerged as the major spatial data handling tools for solving complex natural resource, planning problems, and the use of GIS technology has revolutionary implications for conducting research and presenting research results (cf. ACSM-ASPRS 1986).

An advanced GIS like ESRI's ARC/INFO consists of two major subsystems. ARC maintains the topologic structure of the data base. Points and lines, initially defined in terms of digitizer coordinates, are used to represent mapped features by transforming these primary values to a user-defined map projection, e.g. the Universal Transverse Mercator projection. INFO, the data base management sub-system, is a relational data base management system used to store and process attribute information associated with the geographic features maintained by ARC. These two sub-systems (instead of INFO other relational data bank systems like ORACLE may be conveniently used) are integrated so the user can access the data base either through spatial (ARC) characteristics or descriptive (INFO, ORACLE) attributes of the data.

Grossmann's et al. (1984) approach POLLAPSE is a combination of dynamic models writing information into the data bank of GIS, which thus produces a time series of maps of forest growth or damage. These maps can be compared with the actual development so that deviations become readily discernible. The hypotheses underlying the dynamic models are:

1. photochemical oxidants cause foliar damage (a), and subsequent attack of air-borne pollutants (b) leads to leaching eventually resulting in nutrient deficiency (cf. Prinz et al. 1982)
2. Ulrich's soil acidification hypothesis (Ulrich 1981b, 1983b) or
3. a combination of the two preceding assumptions.

The predictive quality of these hypotheses was tested in the Pfaffenhofen area (Bavaria) on the basis of 462 forest stands, and the POLLAPSE methodology showed that the combined hypothesis (3) yielded the relatively best interpretation of the forest damage phenomena observed in terms of both spatial pattern and degree of damage. In conclusion it should be mentioned that the above methodology of time series analysis and mapping also offers excellent possibilities for the management of forests or other biotic systems as a base for sustained viability.

4.3 Pollutant Impact on Materials

The economic consequences of anthropogenic pollution in industrialized countries reach far beyond the realms of forest and crop productivity. Unclean air contributes to the decay of structures through the corrosion of metals and the disfigurement of buildings, it soils clothing and household furnishings, and it damages irreplaceable historic and cultural monuments and artefacts. Tax revenue suffers in cities where deteriorated buildings and degraded neighbourhoods have reduced property values. Another expense of air and water pollution is that of monitoring, surveying and policing suspected sources of emissions. In addition to control agencies, many other government agencies, such as departments of health, planning and public works, also spend money on air pollution problems. Furthermore the enforcement of pollution ordinances and the increased burden placed on the courts in hearing cases of alleged pollution nuisance contribute to swell the costs of pollution control to municipalities and states.

The effects of pollution are often inextricably interwoven with natural environmental influences like humidity and solar radiation, and are thus difficult to assess. In the Federal Republic of Germany the Federal Environmental Agency has published conservative estimates of air pollution effects for 1980. Damage to utility buildings was over 1500 million DM per annum, corrosion damage amounted to more than 2000 million DM p.a., while the additional expenses for washing and cleaning were approximately 1000 million DM p.a. By far higher, and probably more realistic, are estimates by Heinz (VDI 1985) who quoted figures in the order of 50,000 million DM p.a. This corresponds well to OECD estimates, according to which the total air pollution costs in the member states of the Organization, including medical expenses, amount to 3–5% of the gross national product (Fränzle 1988). In contrast to this material damage, in 1982 total forest damage, including lost profits in the timber industry, was estimated to be about 700 million DM.

To rescue our cultural heritage from air pollution consequently implies comparable costs. They are likely to be highest in Italy and Greece, but cautious estimates in the Federal Republic of Germany have also come up with figures of at least 50–60 million p.a. It will cost at least 500 million Danish kroner just to make good the damage in Copenhagen. Preserving a single ornamental sandstone or limestone doorway in central Stockholm could cost Skr 150,000–200,000, and conservation and restoration work on the Karolinska Chapel of Riddarholmen Church (Stockholm) is likely to amount to more than 6 million Skr (Aniansson 1988b).

4.3.1 Pollutant Effects

Air pollution seems to exert damaging influences on nearly all materials which are of technical or artistic relevance. Of particular importance are iron and steel, non-ferrous metals and alloys, natural stones, concrete and glass (especially mediaeval windows), and synthetic polymers including coatings and paints. Airborne pollutants affecting these materials are: SO_2, NO_x, O_3, photochemical oxidants, H_2S, NH_3, CO_x, solid particles, acid precipitation, chlorine, fluorine, bromine and cyano compounds, and some other substances.

They promote or speed up natural weathering or ageing of materials by acting synergetically with the relevant elements of boundary layer climate, but they can also induce corrosion processes. Antagonistic effects may also occur, e.g. corrosion layers on the surfaces of exposed metals may increase resistance to further atmospheric attack. In the fields of deterioration, conservation and maintenance research it is therefore essential to take account of the influence of macro- and microclimate (Haberland and Sperlich 1983; Kemp 1986) and biological deterioration processes caused by bacteria, fungi, algae, lichens and mosses, and synergic and catalytic effects.

4.3.1.1 Corrosion of Metals and Metal Alloys

Corrosive deterioration of metals is an electrochemical, and mainly an oxidation process whose equilibrium depends largely on temperature and pollutant concentration. Figure 4.13 illustrates the commonest cases of oxygen and acid

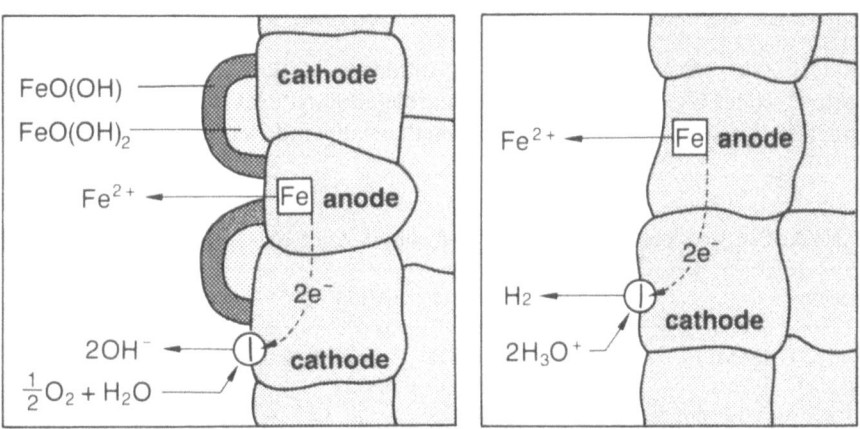

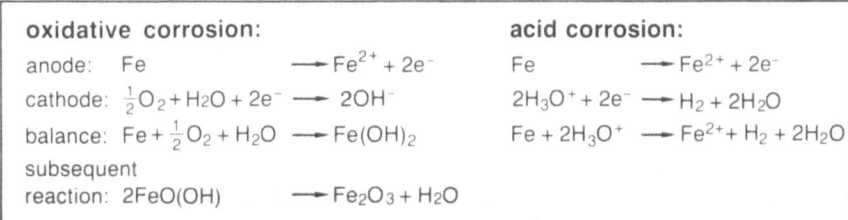

oxidative corrosion:

anode: $Fe \longrightarrow Fe^{2+} + 2e^-$

cathode: $\frac{1}{2}O_2 + H_2O + 2e^- \longrightarrow 2OH^-$

balance: $Fe + \frac{1}{2}O_2 + H_2O \longrightarrow Fe(OH)_2$

subsequent
reaction: $2FeO(OH) \longrightarrow Fe_2O_3 + H_2O$

acid corrosion:

$Fe \longrightarrow Fe^{2+} + 2e^-$

$2H_3O^+ + 2e^- \longrightarrow H_2 + 2H_2O$

$Fe + 2H_3O^+ \longrightarrow Fe^{2+} + H_2 + 2H_2O$

Fig. 4.13. Corrosion of an iron surface. (After Folienserie des Fonds der Chemischen Industrie 1987)

corrosion. High relative humidity and wetting agents such as rain, fog or dew are the main corrosion-determining factors, and consequently many damage functions contain terms like "duration of wetness" or "duration of humidity > 80%". This figure points to the fact that below 70% relative humidity condensation of water vapour on fat-free metal surfaces normally leads to the formation of monomolecular water layers only. This practically inhibits the formation of electrolytes, because their ionic constituents require several water molecules for hydration. Thus corrosion becomes exceedingly slow below the above threshold value. Other parameters influencing the corrosion process are: intensity of air circulation, orientation of material surface, and solar radiation.

Tables 4.18 and 4.19 summarize the results of empirical determinations of dose–response relationships for SO_2 and iron alloys or various steels, respectively.

Attempts to differentiate corrosive effects of NO_x and SO_x were made in The Netherlands (Ministerie van Volkshuisvesting 1984); 80% are attributed to SO_2, 20% to NO_x. In particular Ni-alloys used as electric contact metals showed corrosion effects already at NO_x concentrations of 2–3µg NO_x/m^2.

Long-term effects of corrosive atmospheres on various technical metals are summarized in Table 4.20; the protective role of zinc coatings is illustrated by Table 4.21.

4.3.1.2 Degradation of Inorganic Non-Metallic Materials

Materials at relevant risk under the influence of polluted atmospheres are natural stones like limestone, calcite, marble and artificial stones like reinforced concrete, and glasses, in particular mediaeval windows and their paintings

Table 4.18. Corrosion rates of various iron alloys. (After Nriagu 1978)

Locality	Empirical equation
Tokyo (1 month)	$r = (0.083S_1 + 0.066R + 0.028H - 1.63)t$
Sheffield (1 year)	$r = (0.70S + 0.035)t$
Moscow (2–12 months)	$r = (0.025S_2 + 0.031TS_2 + 0.015T + 0.062)kWtP$
Ottawa (1–18 months)	$r = -0.647 (S - 0.679) (t_w)^{0.677}$
NASN network (1–2 years)	$r = 183.5t^{1/2} \exp (0.0642S_3 - 163.2/H)$
NASN network (1–2 years)	$r = 325t^{1/2} \exp (0.00272S - 163.2/H)$
Czechoslovakia (9–15 months)	$r = k_1(H_2)^{k_2} \exp (k_3T + k_4S)$

r = corrosion rate; H = relative humidity; H_2 = frequency of occurence of relative humidity exceeding 80%; R = rainfall; t = duration of exposure; t^w = time of wetness; T = temperature; S = SO_2 content of air; S_1 = SO_3 collected by the lead candle method; S_2 = SO_3 in moisture film; S_3 = sulfate content of suspended particulates P = rust; W = (time of wetness)/(elapsed time); $k, k_1, k_2, k_3, k_4,$ = empirical constants.

Table 4.19. Corrosion depth (y) and loss (z) of different steel types (UN Economic Commission For Europe 1984)

Survey details	Significant parameters	Equations
T SO₂ t_w 2 sites 8 months PbO₂ method Monthly exposures Rate mg/dm²/day of wetness SO₂ mg SO₂/dm²/day	T SO₂ assumed meteor. conditions to be constant	$y = 0.131X + 0.180Z + 0.787$ $y = $ log corr. rate mg/dm²/d $X = SO_2$ mg/dm²/d $Z = $ temperature °F during wetness (monthly av.)
7 sites 1, 12, 18 months t_w panel T, ambient T, SO₂, Cl	t_w SO₂	$z = 0.16\, t_w^{0.7}\,(SO_2 + 1.78)$
6 sites SO₂ NO NO₂ CO H/C oxidant weather particulate	SO₂ t Oxidant	$y = 9.013\,(e^{0.0016 SO_2}\,(4.768t)^{0.7012} - 0.00582\,ox)$ different equations for each steel
57 sites 1, 2 years RH T SO₂, parts SO₄²⁻, NO₃⁻	t $x = $ SO₄ or SO₂ RH	$y = a_0\, t\,(e^{(a_1 x - a_2/RH)})$

Table 4.19. (Contd.)

Survey details	Significant parameters	Equations
9 sites 30 months Windspeed, direction T, O_3, H/C, NO_x, total S, SO_2, RH, SO_4^{2-}, NO_3^-, particulates	Galvanized steel Weathering steel Al 7079–7651 House paint Marble Silver Nylon time Total Sulphur (TS) SO_4^{2-} Total Suspended Particulates (TSP)	Weathering Steel log rate = $0.702 - 0.588 \log t - 0.004$ TS $+ 0.006\ SO_2 + 0.011\ H_2S - 0.010\ NO_x$ $+ 0.006\ TS - 0.005\ SO_4 - 0.001\ NO_3$
3 sites 1974–79 25 monthly measurements	SO_2 µg/m³ No. of days with precipitation H^+	Monthly rate $y = 1.54\ SO_2 + 2.34\ \text{NPREC} + 0.05\ H^+ - 15.2$ y = monthly corrosion rate of steel
Western Ruhr, 22 sites 14 days exposure SO_2 max 155 mg SO_2/m²/d	Carbon steel	$y = 0.0106\ SO_2 + 2.0$
1975 32 sites: 5 Finland 6 Sweden, 19 Norway Met. data Air pollution levels Zinc, steel	SO_2 Annual rates SO_2 g/m³ Cl g/m²/yr	y Steel = $5.28\ SO_2 + 176.6$

1980 5 sites Steel, zinc Al 4 yr exposure Cu 2 yr averages SS	t_w SO_2	$y = 1.17\, t_w^{0.66}\,(SO_2 + 0.048)$
CMEA 8 sites $-15\,°C < T < 30\,°C$ $80\% < RH < 100\%$ SO_2, Cl	Zn Temperature(T) Steel SO_2 Cu t_w Al $y = g/m^2/yr$ Mg Alloy	$y = [(2.0 \times 10^{-3} + 7.3 \times 10^{-3}\,T)\,t_w$ $+ (1.43 + 6.0 \times 10^{-2}\,T)\,(SO_2)]$
As above	SO_2 mg/m²/d t_w hrs RH > 80% and $t > 0\,°C$	$y = 71.99\, t_w^{0.386}\, SO_2^{0.556}$ Corrosion loss over 3.650 days $y = 0.0152\, t_w\ 0.428\ SO_2\ 0.570$ Steady corrosion rate
5 sites 5-year test CMEA countries	H_2O = no. hrs RH > 80% = t_w SO_2 mg/m²/d	$y = 1.445 \times 10^{-2}\,(H_2O)\ 0.824\ SO_2^{0.458}$ mg/m²/d
y = corrosion depth (μm) SO_2 μg/m³	T = mean panel temperature when wet (K)	$R = 1.9872$ cal/g mol t = exposure time RH = Average relative t_w = wetness time (years) humidity
z = corrosion loss (mg/area) ox = oxidant (mg/m³)		

Table **4.20.** Weight loss of metal panels [a] after 20 years' exposure in various atmospheres (ca. 1930–1954)

City	Exposure classification	Average loss in weight, %					
		Commercial copper (99.9% + Cu)	Commercial aluminium (99% + Al)	Brass (85% Cu, 15% Zn)	Nickel (99% + Ni)	Commercial lead (99.92% Pb, 0.06% Cu)	Commercial zinc (99% Zn, 0.85% Pb)
Altoona, Pennsylvania	Industrial	6.1	—	8.5	25.2	1.8	30.7
New York, New York	Industrial	6.4	3.4	8.7	16.6	—	25.1
La Jolla, California	Seacoast	5.4	2.6	1.3	0.6	2.1	6.9
Key West, Florida	Seacoast	2.4	—	2.5	0.5	—	2.9
State College Pennsylvania	Rural	1.9	0.4	2.0	1.0	1.4	5.0
Phoenix, Arizona	Rural	0.6	0.3	0.5	0.2	0.4	0.8

[a] Panels 22.86 × 30.48 × 0.089 cm

Table 4.21. Corrosion rates of zinc and iron ingot and service lives of zinc coatings in various atmospheres. (After Nriagu 1978)

Service lives of zinc coatings in various atmospheres[a]

Site	Type of Atmosphere	Corrosion Rate (μm year)		Estimated life of 610 g/m² zinc coating (years)
		Zinc	Iron ingot	
Motherwell, Scotland	Industrial	5	61	17
Woolwich, Kent	Industrial	4	69	19
Sheffield (University)	Industrial	5	86	15
Sheffield (Attercliffe)	Industrial	15	119	5
Llanwrtyd Wells	Rural	2	56	34
Calshot, Hants	Marine	3	114	23
Apapa,Nigeria	Marine tropical	0.8	20	100
Basrah, Iran	Dry, subtropical	0.3	8	300
Congella, Durban	Marine industrial	5	76	17

[a] These data are based on 5-year-exposure tests.

(Riederer 1973, 1977; Luckat 1981; Jörg et al. 1985). Natural and artificial stones containing variable amounts of calcareous cement or consisting of limestone are liable to degradation by a series of processes:

- decomposition by carbonic acid,
- acidic precipitation,
- frost weathering,
- dry deposition,
- microbial deterioration,
- capillary ascending groundwater,
- rust formation within metal-reinforced concrete,
- corrosion by air-borne suspended particles.

In principle, the same climatic factors controlling the impact of air pollutants on metals (Sect. 4.3.1.1) also constitute essential boundary conditions for the attack of inorganic nonmetallic materials. Humidity may accelerate pollutant effects dramatically.

Skoulikidis (1983) defined the following logarithmic dose–effect relationship for the SO_2-induced transformation of $CaCO_3$ into $CaSO_4 * 2H_2O$ (gypsum):

$$y = \frac{(2Eu\mu' + \mu'e\ V_m)t}{n_e \cdot F'} \tag{4.43.1}$$

where y is the rate of gypsum formation, which is considered as an analogue of a

Table 4.22. Empirical stone damage functions. (After Jörg et al. 1985, from various sources)

Sulphate formed (g) = 0.05 (porosity) + 0.07 (permeability) + 0.04 (water adsorption)
$\qquad$ + 0.07 (calcite/dolomite) + 0.32

y (weathering in μm a^{-1}) = 3.31 + 0.78 rel. humidity (%) + 2.95 × 10^{-3} SO$_2$ (μg m^{-3})
$\qquad$ + 3.33 × 10^{-3} O$_3$ (μm^{-3})

L_1 (weight loss in %/a) = 0.03D + 0.5 (r_2 = 0.36)
(Bauberg sandstone)

L_2 (Krensheim Muschelkalk) = 0.018D + 0.6 (r^2 = 0.8)

D = average annual dry deposition of SO$_2$(mg SO$_2$/m^2 d),
$\qquad$ determined by IRMA absorption method

---

galvanic cell whose potential E depends on pollutant concentration as follows:

$$E = E_0 - \frac{2.3RT}{n_e F'} \log \frac{P_{CO_2}}{P_{SO_2} \cdot \frac{1}{2}P_{O_2} \cdot 2P_{H_2O}} \qquad (4.43.2)$$

In either equation the symbols have the following meaning: u = conductivity of gypsum; μ', μ'_e = transition numbers of Ca ions or electrons, respectively; V_m = molar volume of gypsum; n_e = number of electrons exchanged with a Ca ion (i.e. 2); F' = Faraday constant; E_0 = standard potential; P_{CO_2}, P_{SO_2}, P_{O_2}, P_{H_2O} = partial pressures of CO$_2$, etc. Other stone damage functions are summarized in Table 4.22.

4.3.1.3 Degradation of Organic Materials

Organic materials such as wood, paper, leather, wool, cotton and synthetic polymers are also affected by air pollutants (Baer and Banks 1985; Riederer 1985; Verein Deutscher Ingenieure 1985). The principal degradation process is chain scission with concomitant oxidation, followed by fading of colours and loss of gloss and mechanical strength. However, in view of the large number of different polymer types there exist a comparably high number of degradation mechanisms, and generalizations about environmental effects are not easy. In addition to humidity, solar radiation is a particularly important factor in the degradation of polymers (cf. Sect. 4.3.4.4.). Jellinek (1977) summarizes laboratory experiments whereby degradation effects resulting from SO$_2$ and NO$_2$ were defined in terms of $\bar{s}$ values, where $\bar{s}$ is the ratio polymer chain-length before and after fumigation (Table 4.23).

The results of controlled environmental laboratory exposures of different paint types are contained in Table 4.24.

Table 4.23. Survey of the effect of SO_2 and NO_2, respectively, on polymers in terms of $\bar{s}$ values for exposure time of 1 h at 25 and 35 °C, respectively, in presence of 1 atm of air and near ultraviolet radiation ($\lambda > 2800$ Å); $\bar{s}$ values obtained in presence of 1 atm of air and near ultraviolet radiation only are shown

	PE	PPRO	Atactic PST	Isotactic PST	Atactic PMMA	Isotactic PMMA	PαMS	PVP
$\bar{s}$, UV, 1 atm air	Some CL	0.0	8×10^{-3}	0	7×10^{-3}	0.0	Some CL	CL
$\bar{s}$, UV, 1 atm air and 1 cm Hg of LSO	Appreciable CL	CL	2.1×10^{-2}	1.27×10^{-1}	4×10^{-2}	1.26×10^{-1}	6.7×10^{-2}	Some CL
$\bar{s}$, UV, 1 atm air and 1 cm Hg of NO_2	Some CL	Some CL			5×10^{-3}			

	PVF	PAN	PC	Nylon	BUR	PIP	PBD
$\bar{s}$, UV, 1 atm air	Some CL	Some CL	1×10^{-2}	0	3.6×10^{-1}	1.32	CL
$\bar{s}$, UV, 1 atm air and 1 cm Hg of SO_2	3.6×10^{-2}		0	0	1.4	1×10^{-1}	CL
$\bar{s}$, UV, 1 atm air and 1 cm Hg of NO_2		0		1.3	1.4×10^{-1}	CL	Some CL

CL = cross-linking; PE = polyethylene; PPRO = polypropylene; PST = polystyrene; PMMA = polymethyl methacrylate; PαMS = poly-α-methylstyrene; PVP = polyvinyl pyrrolidone; PVF = polyvinyl fluoride; PAN = polyacrylonitrile; PC = polycarbonate; BUR = butyl rubber; PIP = polyisoprene; PBD = polybutadiene

Table 4.24a. Paint erosion rates and probability data (T-test) for controlled environmental laboratory exposures

Type of paint	Mean erosion rate (nm/hr with 95% confidence limits) for unshaded panels		
	Clean air control	SO_2 (1.0 ppm)	O_3 (1.0 ppm)
House paint			
Oil	5.11 ± 1.8	35.81 ± 4.83[b]	11.35 ± 2.67[b]
Latex	0.89 ± 0.38	2.82 ± 0.25[b]	2.16 ± 1.50[a]
Coil Coating	3.02 ± 0.58	8.66 ± 1.19[b]	3.78 ± 0.64[a]
Automotive refinish	0.46 ± 0.02	0.79 ± 0.66	1.30 ± 0.33[b]
Industrial maintenance	4.72 ± 1.30	5.69 ± 1.78	7.14 ± 3.56

[a] Significantly different from control at an α of 0.05.
[b] Significantly different from control at an α of 0.01.

Table 4.24b. Paint erosion rates and probability data (T-test) for field exposures

Type of paint	Mean erosion rate (nm/month with 95% confidence limits) for panels facing south			
	Rural Clean air	Suburban	Urban (SO_2 dominant, $\sim 60\ \mu/m_3$)	Urban (oxidant dominant $\sim 40\ \mu/m_3$)
House paint				
Oil	109 ± 19.1	376 ± 124[b]	361 ± 124[a]	533 ± 157[b]
Latex	46 ± 13	76 ± 18[b]	97 ± 8[a]	165 ± 142
Coil coating	53 ± 20	254 ± 48[b]	241 ± 20[b]	233 ± 43[b]
Automotive refinish	23 ± 28	58 ± 18[a]	41 ± 10	43 ± 10[b]
Industrial maintenance	91 ± 41	208 ± 361[a]	168 ± 99	198 ± 61[b]

[a] Significantly different from control at an α of 0.05.
[b] Significantly different from control at an α of 0.01.

4.3.2 Critical Levels of Pollution

In contrast to biotic systems, which dispose of efficient homoeostatic resilience and stability mechanisms (cf. Chap. 2.2 and 2.3), materials have no inherent self-repair mechanisms which allow compensation of a certain level of chemical stress. Even a minimum quantity of a pollutant can induce reactions, provided they are kinetically possible (Chap. 2.1.3). It is a secondary consideration to

Table 4.25. Indoor air quality criteria for museum and library archives. (After Baer and Banks 1985)

	Concentration (μg m^{-3})		
	SO$_x$	NO$_x$	O$_3$
British Museum Libraries	0.0	0.0	0.0
National Bureau of Standards	1.0	5.0	25.0
Newberry Library	10.0	10.0	2.0
G.Thompson	10.0	10.0	2.0
Canadian Conservation Institute	25.0	25.0	25.0

decide if the reaction is detrimental to the material under discussion or to objects made of this material.

Material damage is an irreversible cumulative process, and a deterioration of a cultural monument like a statue or a mediaeval window can at best be stopped or restored. The necessary consequence would be to demand a reduction of pollutant concentrations to no-effect levels, i.e. a hypothetical zero pollution. In practice, however, critical levels may be defined in terms of concentrations which appear to be low enough to make any residual damage acceptable (Baer and Banks 1985). Thus, museum authorities have set critical levels as indoor air quality criteria for archives (Table 4.25).

When considering the values in Table 4.25 as guidelines for SO$_x$, NO$_x$ and ozone concentrations it is worth mentioning that they refer to indoor situations where relative humidity is rarely higher than 60%.

4.3.3 European Research Programmes on Pollutants Effects and Restoration

One essential offspring of the European research co-operation programme EUREKA is a specialized though comprehensive project in the sphere of cultural heritage conservation. Its official title is EUROCARE, which stands for European Project on Conservation and Restoration. Officially speaking "EUROCARE's aim is to develop presently not available industrial products and technologies as well as craft skills for conservation and restoration work based and supported by scientific knowledge. This would lead to the establishment of advanced technical standards and European guidelines for the examination and treatment of objects and monuments". Thus EUROCARE's aim is to set up a strong European task force of chemists, geoscientists, biologists, engineers, architects, conservators and cultural historians who co-operate with industry to produce new conservation techniques and products employing the most modern aspects of high technology. An important element in the EUROCARE project will be the organization of a European data base on objects, materials, environmental conditions, analytical and conservation methods.

Another major international research project, the International Co-operative Programme on Effects on Materials, Including Historic and Cultural Monuments, has been launched in the ECE framework, and is related to the Convention on Low-Range Transboundary Air Pollution. At 39 test sites in 13 countries, identical measurements will be performed on structural metals (steel, weathering steel, zinc, aluminium, copper and cast bronze), sandstones, and limestones, painting coats (i.e. coil-coated steel with alkyd melamine, steel and wood with silicon alkyd paint, and wood with primer and acrylate), and electric contact materials (nickel, copper, silver and tin). The following test parameters are measured (Kucera 1988):

– temperature and relative humidity, combining to give the time during which relative humidity $> 80\%$ and temperature $> 0\,°C$,
– solar radiation,
– SO_2, NO_2 and O_3 concentrations,
– amount, pH and conductivity of precipitation,
– concentrations of sulphate, nitrate, chloride, ammonium and calcium ions in precipitation.

It is the major aim of this programme to systematically assess how acidifying air pollutants affect atmospheric corrosion of a wide range of materials, used for both purely functional structures and in buildings and artefacts of cultural value. Distinguishing the effects of dry and wet deposition, more reliable dose-response relationships should be able to be established. This, in turn, would make it possible to work out reliably the costs of corrosion and provide a sound basis for future political decisions on emission cuts.

4.4 Interactions of Environmental Chemicals with Soil and Subsoil

Chemical compounds introduced into soils can be adsorbed by soil constituents or transformed by soil organisms. Furthermore they may be taken up by plants, washed out by rain or irrigation water or evaporated in gaseous form. The extent of adsorption is controlled by various soil properties, including composition and content of organic matter, type and content of clay, exchange capacity, and acidity; it is also closely related to the physical and chemical properties of the substance in solution (cf. Chaps. 2.1.2 and 2.1.3). In view of the complex situation, our empirical knowledge of the behaviour of environmental chemicals is still fairly limited and the possible danger of excessive fertilizer or manure application, pesticides, and heavy metal contamination of soils and crops has been recognized only in the last decades, mostly in relation to the agricultural use of sewage sludge or in connection with atmospheric deposition.

As a result of the need to assess potential risks, inventories have been established (e.g. BUA 1989), and various models have been developed to address a number of specific situations and to describe the complex interactions involved. These are equilibrium or speciation models, because they deal with the distribution of various chemical species (Bonazountas 1987; Richter 1986; Fränzle et al. 1989). Two basic approaches to the solution of the species distribution problem exist which are based on equilibrium constants and the Gibbs free energy (cf. Chap. 2.3). In both cases, the most stable condition is sought by means of solutions to sets of non-linear equations. Aquatic equilibrium models are at a developmental stage whose current versions are steady-state models formulated for one or multiple compartments (cf. Sposito and Mattigod 1980; Mattheß 1990).

4.4.1 Environmental Factors and Chemistry

The chemical, physical and biological properties of a chemical, in conjunction with the environmental characteristics of a site result in a series of processes associated with the transport and transformation of the substance in soil, aeration zone and groundwater (Fränzle et al. 1989). For analytical and modelling purposes, it is indicated to distinguish the following sets of physical, chemical and biotic processes affecting dissolved contaminant migration and fate in soil systems (see foldout model II):

1. advection, dispersion, volatilization,
2. sorption, ion exchange,
3. ionization,
4. hydrolysis,
5. oxidation–reduction,
6. complexation,
7. bioaccumulation,
8. biotransformation.

In sand and gravel aquifers, the dominant factor in the migration of a dissolved chemical is advection, the process transporting solutes by the bulk motion of flowing groundwater. Groundwater velocities normally range between 1 and 1000 m/a, while matrix throughflow attains velocities of 50–2500 m/a, pipeflow and overland flow 50–100 m/h, and channel flow 300–10,000 m/h (Sect. 4.1.2.2).

In the soil and aeration zones the processes of advection, dispersion, volatilization and sorption are of importance to both trace-level analyses and large-scale release analyses. Bulk properties such as viscosity and solubility are usually only important in simulations involving large amounts of contaminants (Darimont 1985; Fränzle et al. 1987).

Information on the above chemical and biotic processes is largely presented in Chap. 2.1.3; therefore emphasis here is on physical processes among which dispersion, sorption, ion exchange and biotransformation are particularly important.

4.4.1.1 Dispersion

Dispersion of dissolved contaminants in water results from two basic processes, molecular diffusion in solution and turbulent mixing, which brings about an overall net flux of solutes from a zone of high concentration to a zone of lower concentration.

Diffusion

Diffusion is the process whereby ionic or molecular substances move under the influence of their kinetic activity in the direction of their concentration gradient. The mass s of diffusing substance passing through a given cross-section dA per unit of time is proportional to the concentration gradient (dc/dx) (Fick's first law):

$$s \cdot dA = D \frac{dc}{dx}. \tag{4.44}$$

where c = number of molecules in unit volume, and D = diffusion coefficient which depends on the medium and external boundary conditions.

Fick's second law is derived from Eq. (4.44) in the linearized form by defining the temporal variation of concentration in a volume $d\tau$ at the point x:

$$\frac{\partial c}{\partial t} = D \frac{\partial^2 c}{\partial x^2} \tag{4.45}$$

Considering the three-dimensional case yields

$$\frac{\partial c}{\partial t} = D \left(\frac{\partial^2 c}{\partial x^2} + \frac{\partial^2 c}{\partial y^2} + \frac{\partial^2 c}{\partial z^2} \right). \tag{4.46}$$

Linear solutions of Fick's law are provided by Anderson et al. (1965).

In general terms, the diffusion coefficient $D = \mu kT$, where k = Boltzmann constant and T = absolute temperature while μ = mobility. μ, a scalar in isotropic media and a tensor in general, is inversely proportional to molar mass and intensity of Brownian motion. Numerical values of D at room temperature are $0.1–1$ cm^2 s^{-1} for gases, around 10^{-5} cm^2 s^{-1} for solutes of low molecular mass, $10^{-7}–10^{-8}$ cm^2 s^{-1} for high molecular solutes, and $10^{-9}–10^{-12}$ cm^2 s^{-1} for plasticizers in PVC (e.g. phthalates). Modern experimental devices for determining diffusion coefficients by means of radiotracers in tubes are described by van der Sloot et al. (1988).

Hydrodynamic Dispersion

Porous flow and diffusion produce a hydrodynamic dispersion of a chemical in solution. Owing to tortuous or smaller passages in the porous medium some of the liquid accomplishes less direct line movement than a solution taking a more direct path. In addition, the chemical may also diffuse into stagnant pores to be released slowly when the main body of chemical has passed.

Thus the hydrodynamic dispersion is a function of the mean pore water velocity v_p. In addition, diffusion contributes to dispersion. Therefore, the apparent dispersion coefficient D_a comprises a velocity-dependent and a velocity-independent term which may be combined in a linear form:

$$D_a = \theta\left(\frac{1}{\tau}D + v_p D_v\right),$$ (4.47)

where θ = volumetric water content, $(1/\tau)$ = dimensionless continuity, D_a = diffusion coefficient (as above), v_p = mean pore water velocity, D_v = dispersion factor (cm). In case that θ is independent of depth, the effective dispersion coefficient D_e becomes

$$D_e = \frac{D_a}{\theta} = \frac{1}{\tau}D + v_p D_v .$$ (4.48)

The dispersion factor D_v, i.e. the geometric dispersivity after Scheidegger (1957a), is of approximately the same magnitude as the mean diameter of soil aggregates (Bear 1972). For loess soils in southern Lower Saxony with a field capacity of ≈ 0.35 relative volume, a mean winter rainfall of ≈ 2 mm/d and a corresponding $v_p \approx 0.6$ cm/d, a mean $D_v = 2$ cm, $(1/\tau) = 0.2$, and D 2 cm²/d, the above Eq. (4.48) yields a $D_e = 1.6$ cm²/d. Close to field capacity, where continuity is not likely to change much, the effective dispersion coefficient is largely controlled by porous flow, while diffusion is most important under conditions of reduced soil moisture.

Considering the vertical translocation I_v of dissolved chemicals which do not interact with the soil matrix, the above formulae yield for both diffusion or dispersion (and also convection) the following generalized transport formulation (Richter 1987):

$$I_v = D_a\frac{\partial c}{\partial z} + cq_w .$$ (4.49)

The first term defines diffusion or, with non-vanishing water flow v_w or q_w, dispersion, the second describes convection. Together with the local balance

$$\frac{\partial(\theta c)}{\partial t} = -\frac{\partial I_v}{\partial z} ,$$ (4.50)

the transport equation for the dissolved compound can be formulated in two equivalent versions:

$$\frac{\partial(\theta c)}{\partial t} = \frac{\partial(D_a \partial c)}{\partial z^2} - \frac{\partial(cq_w)}{\partial z} \text{ or} \tag{4.51.1}$$

$$\frac{\theta \partial c}{\partial t} = \frac{\partial(D_a \partial c)}{\partial z^2} - \frac{q_w \partial c}{\partial z}. \tag{4.51.2}$$

For homogeneous soil, or when considering homogeneous sections of a soil column, Eq. (4.51.1) is simplified to:

$$\frac{\partial c}{\partial t} = \frac{D_a}{\theta(\partial^2 c/\partial z^2)} - \frac{q_w}{\theta(\partial c/\partial z)}. \tag{4.52}$$

In many cases, dispersion may be of minor importance. If it is important, however, and in the absence of detailed studies to determine the dispersion characteristics of a given field situation, longitudinal and transverse dispersivities must be estimated based on prior field work in comparable pedological and hydrogeological systems.

4.4.1.2 Sorption, and Ion Exchange

Adsorption is the adhesion of ions or molecules to surfaces or solid–liquid, liquid–gas, and liquid–liquid interfaces, causing an increase in the concentrations of chemicals on the surface or interface over the concentration in solution. Adsorption occurs as a result of a variety of processes with a multitude of mechanisms when such a concentration lowers the free interfacial energy. The higher the free surface energy at an interface, the greater the number of substances capable of lowering it. Those solids held together by valence forces have surfaces of such high energy, for instance, that the most drastic experimental conditions are required to maintain them pure. Clean surfaces of supposedly "inert" metals such as platinum or iridium require about 1 s for complete contamination by adsorption of atmospheric gases in common high vacua at room temperature.

Physical and Chemical Adsorption

Molecules of a given material may be adsorbed from the gas or liquid phase to a solid by van der Waals forces. This process is called *physical adsorption* and resembles a condensation reaching equilibrium in a time limited chiefly by molecular diffusion rates. Van der Waals forces is a collective term for the relatively weak intermolecular forces holding liquids or crystals of hydrocarbons, water, alcohols etc. together or bonding gases or dissolved substances to solid surfaces; they can be subdivided into dispersion forces (i.e. forces resulting from momentary dissymmetries in electron clouds about the atoms), dipole forces and hydrogen bonds (Hauffe and Morrison 1974). They cause two molecules to attract each other with a force inversely proportional to the seventh power of the distance between them. The weakness of these forces is reflected in the low adsorption enthalpy of physiosorptive (or unspecific) bonding; it amounts to hardly more than 40 kJ/mol (Brdicka 1971). Reversibility is a characteristic of physical adsorption which results from rapid equilibration, i.e. at a given temperature and pressure the same amount of material is found to be adsorbed regardless of the direction from which equilibrium pressure and temperature have been approached. From the environmental point of view it is very important, however, that physical adsorption of gases by highly porous media such as silica gel, charcoal, and soil often exhibits "hysteresis", a form of irreversibility due to condensation of liquid in pores. Analogous phenomena occur with chemicals bonded at solid–liquid interfaces.

Molecules of a given substance may also be adsorbed from the liquid or gas phase by a solid through the activation of valence forces. Consequently, they may be considered as reacting with atoms or molecules in the surface of the solid in a manner analogous to any other chemical reaction, and it is usually convenient to think of the products of this reaction as species quite different from the original reacting molecules or atoms. This kind of adsorption is called *chemisorption* or specific bonding in terms of pedochemistry. Temperatures at which it occurs are not related to the boiling point of the adsorbate (material being adsorbed), as is the case with physical adsorption, but to the stability of the surface compound and the rate of its formation. Therefore rates of chemisorption vary enormously with material and temperature, just as do those of ordinary chemical reactions, and may be very much slower than rates corresponding to diffusion of gases or solutes to the surface. Furthermore, the rates are likely to vary strongly with the amount of material adsorbed. Because of these rate characteristics, the amount of adsorbate at a given pressure or concentration and temperature – when observed over practical time periods – may depend strongly on the manner in which these boundary conditions have been reached. Therefore irreversibility is extremely common in chemisorption which is characterized by enthalpies > 100 kJ/mol (Barrow 1974; Brdicka 1971).

It ensues from the adsorption enthalpies that for both physical and chemical adsorption the amount of material adsorbed at equilibrium at fixed pressure or concentration must decrease as temperature increases. This is easy to observe in the case of physical adsorption. Rates of chemisorption, like those of all chemical reactions, increase as temperature increases; but they are often so low that equilibrium is not closely approximated over a normal period of observation. Therefore the amount of a substance chemisorbed in a fixed period of time by an initially clean surface at a fixed pressure or concentration often increases with increasing temperature at low temperatures, reaches a maximum at some intermediate temperature, and finally decreases at high temperatures. Owing to its specificity (covalent bonds, complexation) chemisorption is an essentially unimolecular process, i.e. the layer of adsorbed material is one molecule thick (Barrow 1974), while physical adsorption (following to or competing with chemisorption) normally leads to the formation of a plurimolecular adsorbate layer.

Fundamentals of Ion Exchange

Ionization is the process of dissociation of a molecule into particles of opposite electrical charge. The presence and extent of ionization has a large influence on the chemical behaviour of a substance, in particular on solubility, sorption, toxicity and other biological characteristics. Inorganic and organic acids, bases and salts may be ionized under a wide range of environmental conditions. A weak acid, e.g. fulvic acid or 2,4-D, will ionize to some extent in water according to the reaction scheme:

$$HA + H_2O \rightleftharpoons H_3O^+ + A^-. \tag{4.53}$$

The acid dissociation constant K_a is defined as the equilibrium constant of Eq. (4.53)

$$K_a = \frac{[H_3O^+][A^-]}{[HA][H_2O]}. \tag{4.54}$$

Consequently, a compound is 50% dissociated when the pH of the solution equals the pK_a ($pK_a = -\log K_a$).

Ion exchange as an important sorption mechanism for substances liable to ionization is viewed as an exchange with some other ion that initially occupies an adsorption site on the solid (cf. Helfferich 1959; Greenland and Hayes 1981; Schachtschabel et al. 1989). For example,

$$H^+ \boxed{\begin{array}{c} Ca^{2+}Mg^{2+} \\ \\ K^+Na^+ \end{array}} Al^{3+} + 10NH_4^+ \rightleftharpoons NH_4^+ \boxed{\begin{array}{c} 2NH_4^+ \\ \\ 4NH_4^+ \end{array}} 3NH_4^+ + H^+$$
$$+ K^+ + Na^+ + Ca^{2+} + Mg^{2+} + Al^{3+}. \tag{4.55}$$

The ion exchange property of a soil is mostly due to the clay and silt fractions, the organic matter and hydrous oxides of iron. The soil particles have an amphoteric character but normally carry a net negative charge. It results from two different processes: (1) isomorphic ion substitutions, and (2) ionization of hydroxyl groups attached to silicon of broken tetrahedron planes, in the same way as for ordinary silicic acid. The negative charges due to the former process are more uniformly distributed in the plate- or lath-shaped clay particles, while the latter are at corners and along edges (Wiklander 1964; Schachtschabel et al. 1989). Negative charges may furthermore originate from humic (–COOH, –OH), phosphoric and silicic acids which form a more or less integral part of the clay particle surface.

Positive charges, as indicated by the amphoteric nature of the clay fraction, may originate from hydrous oxides of iron (ferrihydrites), aluminium and manganese, and from exposed octahedral groups which react as bases by attracting protons from the surrounding soil solution. The basic groups of humus are due to nitrogen (cf. Ziechmann 1980). Both the electric charge and the surface charge density vary more or less with soil pH. The negative charge grows, and the positive charge decreases, with rising pH as a result of increasing ionization of the acid groups and decreasing protonation of the basic groups. For decreasing pH the corresponding change goes in the opposite direction. The carboxyl groups of humus ionize under acid conditions, the phenolic hydroxyls mainly above pH 6. Because of the highly irregular shape of many clay minerals, and clay–humus complexes in particular, and the non-uniform distribution of charges in the particles, the surface charge density is quite variable. Thus the charge density and the potential are higher on edges, corners and in furrows or cavities than on plane surfaces.

The electric charge of the soil particles is neutralized by an equivalent amount of oppositely charged ions (counter- or gegenions), attracted mainly by Coulomb forces. These gegenions are exchangeable and comprise, under natural conditions, Ca^{2+}, Mg^{2+}, H^+, K^+, Na^+ and NH_4^+, the relative abundance of which may vary greatly. In very acid soils $Al(OH)_x^{3-x}$ and $Al(H_2O)_6^{3+}$ may constitute a considerable part of the counterions, the proportion increasing with falling pH (Kramer et al. 1983; Ulrich 1983b; Dietze 1985), while in alkali soils the content of Na^+ is exceptionally high. Common anions are SO_4^{2-}, Cl^-, NO_3^-, $H_2PO_4^-$, HPO_4^{2-}, HCO_3^-, and anions of the great group of humic acids.

Because of thermal motion, the gegenions are distributed in the form of a diffuse layer on the adsorbing soil colloids, the structure of which is dependent on the surface charge density, kind of gegenions, temperature, and concentration of electrolytes in the solution. Since the exchangeable ions are hydrated they may be considered as forming a specific kind of solution, often called inner (or micellar) solution in distinction to the outer (or intermicellar) solution of free electrolytes. A particle surface with gegenions can be presumed to behave like a flat molecular condenser, the charge of the particle forming the inner layer and the gegenions the outer layer close to the former (Verwey and Overbeek 1948). The concentration of gegenions is highest in the immediate vicinity of the charged particle surface and decreases at first rapidly and then asymptotically to the intermicellar solution of uniform composition, following Boltzmann's distribution law

$$n_j = n_{jo} \exp\left(-\frac{z_j e \psi}{kT} \right). \tag{4.56}$$

where n_j is the number of gegenious of kind j per cubic centimeter at an arbitrary point in the diffuse double layer, where the potential is ψ; n_{jo} is the corresponding concentration in the outer solution, z_j is the valence of ion j, e the charge of the proton, k the Boltzmann constant, and T the temperature.

It ensues from Eq. (4.56) and the potential distribution given by the Poisson equation that the thickness of the diffuse double layer decreases with increasing valence and concentration of the electrolyte in solution. Experimentally, the volume of the micellar solution may be approximately determined from the anion distribution and the increase in the osmotic pressure when passing from the intermicellar to the micellar solution (Wiklander 1964). Primarily, the above discussion of the structure and properties of the double layer applies only to the relatively flat surfaces of clay minerals, but the ion arrangement around edges and corners may be assumed to show about the same pattern. For the pure soil organic matter, however, the theories of the diffuse double layer are probably not valid, partly because of the complicated structure of humus compounds and partly because of the operation of non-electrostatic forces, leading to complex bondings.

Cation Exchange

It ensues from structure and properties of the diffuse double layer that the intensity of sorptive bonding depends on both size and charge of the ions as shown in Table 4.26.

The influence of valence results from the fact that the electrostatic attraction of cations is proportional to their charge. Charges being equal, a cation will experience a stronger attraction the more it can approach the negatively charged surface of the particle, i.e. attraction increases with decreasing size of the hydrated cation. Hydration, in turn, is a function of charge density, i.e. small and highly charged ions are strongly hydrated, while large ones of equal valence are weakly hydrated (cf. hydration energy in Table 4.26).

As a result of the structure of soil colloids some cations, and in particular anions, experience a *specific adsorption* due to additional van der Waals forces. They may attain the intensity of Coulombic forces which are at the base of *non-specific adsorption*. Specific cation adsorption is reflected in the characteristic selectivity of several clay minerals for K, NH_4^+, Rb and Cs (Schachtschabel et al. 1989).

Specific adsorption may become unimportant for the environmental behaviour of heavy metals which hydrolize partially in the normal pH range of soils. Thus they occur in the soil solution not only in form of M^{2+} ions but also, and increasingly with rising soil pH, as MOH^+ ions. In the course of adsorption frequently protons are liberated:

$$> Fe\text{–}OH + MOH^+ = Fe\text{–}O\text{–}MOH + H^+. \tag{4.57}$$

The adsorption of heavy metals increases with their tendency to forming hydroxo-complexes (MOH^+ and others), which can be defined in terms of the hydrolysis constant K_h:

$$K_h = \frac{[MOH^+][H^+]}{[M^{2+}]} \tag{4.58}$$

Table 4.26. Size, polarizability and hydration energy of cations and their relative exchange to NH_4-smectite. (After Schachtschabel et al. 1989)

	Li^+	Na^+	K^+	Rb^+	Cs^+	Mg^{2+}	Ca^{2+}	Sr^{2+}	Ba^{2+}	Al^{3+}	La^{3+}	Th^{4+}
Ionic radius (Å)	0.76	1.02	1.38	1.52	1.67	0.72	1.00	1.18	1.35	0.54	1.03	0.94
Polarizability (cm^3)	0.025	0.17	0.80	1.42	2.35	0.10	0.54	0.87	1.68	—	—	—
Hydratation energy (kJ/mol)	503	419	356	335	314	1802	1571	1425	1341	4647	3268	—
Relative exchange (%)[a]												
Sch.[b]	24	26	41	74	78	72	74	74	75	—	—	—
M.& L.	—	17	41	76	92	—	—	—	—	—	—	—
J.	32	33	51	63	69	69	71	74	73	85	86	98

[a] To NH_4-smectite for a 1:1 equilibrium solution of M/NH_4.
[b] Sch. = data from Schachtschabel.
M.&L. = data from Martin and Laudelot.
J. = data from Jenny.

derived from reactions of the type

$$M^{2+} + H_2O = MOH^+ + H^+ .$$ (4.59)

Normally, pK_h values are given. A low one indicates a marked tendency to forming hydroxo-complexes at low pH values. This means that the specific adsorption of heavy metals normally increases considerably with falling pK_h values (in parentheses): Cd (10.1) < Ni (9.9) < Co (9.7) < Zn (9.0) ≪ Cu (7.7) < Pb (7.7) ≪ Hg (3.4).

In other words, Cd, Co, Ni and Zn are specifically adsorbed only at relatively high pH values, Cu, Pb and, in particular, Hg already at distinctly lower pH values. pK_h values being equal, as is the case with Cu and Pb, normally the larger ion (Pb) experiences stronger adsorption (cf. Table 4.26).

Two other important factors merit attention. The first is the geometric fit of ions into mineral lattice structures. If an ion fits well, for reasons of size, into an exchange spot, it will be more firmly adsorbed than ions of different size, and the higher the number of such specific lattice structures, the stronger will be the dominance of this particular ion over other ions at equal outer concentration. For example, the diameters of the non-hydrated K^+, NH_4^+ and H_3O^+ are all about 2.7 Å, and therefore they fit well into the hexagonal cavities of the oxygen sheets of layer clay minerals, resulting in a particularly firm bonding (Herms and Brümmer 1984). The second mechanism to be considered is the formation of complex chelates and related types of bondings between certain cations and organic components of soil. Copper ions, for instance, are known to be bound mainly in this way (Wiklander 1964).

Anion Exchange

In comparison to cation exchange, the environmentally important anion exchange on clay minerals, soils and biotic surfaces has been far less studied, and existing knowledge is therefore in some respects still rather limited. It seems, however, that the effect of concentration, mole fraction, and complementary ions on the distribution of exchangeable anions is similar to that for cations. At low pH values the mineral lattice is normally unstable, resulting in dissolution of Al, Fe, Mg and Mn, which are held partly in exchangeable form. Some anions, particularly phosphate, easily form difficultly soluble or complex compounds with these minerogenic cations, which greatly complicates the study of anion exchange.

If a dilute neutral solution of a salt (e.g. KCl) is added to a clay or soil having no adsorbing capacity of the anion at the prevailing pH, the concentration of this same anion in the equilibrium solution will show an increase. This so-called negative adsorption is due to an unequal ion distribution in the diffuse double layer according to the Donnan equilibrium (cf. following sub-section). Contributing factors are primarily the hydration of the exchangeable ions and the nature of the adsorber. Thus the negative adsorption increases with the amount of colloid and the exchange capacity, while it may decrease with increasing volume of the micellar solution, all other things being equal. Furthermore, negative adsorption decreases with increasing salt concentration since addition of salts suppresses the unequal ion distribution. Working with a carboxyl resin, Wiklander (1964) found, for instance, that raising the pH, i.e. increasing the exchange capacity, resulted in a considerable increase in negative adsorption of phosphate. From these and similar experiments, it may be concluded that negative adsorption by pushing the anions out of the micellar solution increases losses by leaching.

Anion adsorption is intimately connected with the nature of soil colloids, i.e. organic matter, clay minerals, contents of hydrous oxides and with the pH of the system (Fränzle 1984b; Herrmann 1985; Herrmann and Lagaly 1985; Schachtschabel et al. 1989). Lowering of pH induces an activation of basic groups due to increased protonation; e.g. in water, $R-OH + H_2O = ROH_2OH$, and in presence of HCl, $R-OH + HCl = ROH_2Cl$. The OH originates in clay minerals from broken bonds and in soils mainly from hydrous oxides of Al and Fe (Muljadi et al. 1966a, b; Flaig-Baumann et al. 1970; Graf and Lagaly 1980; Lagaly 1984). Among the anions bound in this way Cl^-, NO_3^-, SO_4^{2-} and $H_2PO_4^-$ may play a particular role since they function as gegenions and are exchangeable with other anions in the same manner as cations are exchanged, e.g. $R-OH_2^+ \, Cl^- + NO_3^- \rightleftharpoons R-OH_2^+ NO_3^- + Cl^-$. This mechanism explains the distinctly higher anion-exchange capacity of soils than of clay minerals by the content of hydrous oxides, and the great reduction of this capacity if soils undergo a dissolution of these oxides, for example under the influence of acid stemflow (Fränzle et al. 1989).

It must be noted, however, that several anions, such as phosphate, become bound by mechanisms quantitatively more important in natural soils. Thus exchange or displacement of lattice ions may occur, e.g. if an ion has about the same size as OH. There is also evidence that the phosphate ion, though not having the same size as OH, is also bound this way as a structural unit, and this similarity in bonding of hydroxyl and phosphate may also explain the strong displacement of PO_4^{3-} by alkali. This type of specific adsorption can be exemplified by the following set of formulas (Schachtschabel et al. 1989):

$$(\text{Al, Fe})-OH_2^+ + H_2PO_4^- \rightleftharpoons (\text{Al, Fe})-O-PO(OH)_2 + H_2O \qquad (4.60)$$

$$O\!\!\begin{array}{c}-(\text{Al, Fe})-OH\\-(\text{Al, Fe})-OH\end{array} + H_2PO_4^- \rightleftharpoons O\!\!\begin{array}{c}-(\text{Al, Fe})-O\\-(\text{Al, Fe})-O\end{array}\!\!P\!\!\begin{array}{c}\diagup O\\\diagdown OH\end{array} + OH^- + H_2O \, . \qquad (4.61)$$

Since this type of bonding takes place also above the isoelectric point, it can contribute to the cation-exchange capacity according to the valence, i.e. whether $H_2PO_4^-$, HPO_4^{2-}, PO_4^{3-} or arsenate are bound. By this process, which is particularly important in oxisols, the isoelectric point is lowered.

Phosphate is also, and partly in connection with the above processes, retained in soils by precipitation since it forms difficultly soluble compounds

with Fe and Al at low pH, more soluble ones with Ca and Mg at slightly acid
and alkaline reactions, and distinctly less soluble compounds with these two
ions in the alkaline region. When ions such as oxalate, citrate or tartrate release
phosphate in appreciable quantities from soils, this is mainly due to the
formation of soluble complexes, especially with Fe and Al (Wiklander 1964),
rather than to true anion exchange. Also humic acids take part in this exchange,
as evidenced by their capacity of releasing adsorbed phosphate, but the exact
mechanism requires further elucidation. The above and related experience
indicates the importance to make a distinction, for the study of anion exchange
and the application of exchange formulae, between true exchange processes and
other sorption processes.

Ion-Exchange and Adsorption Formulae

Owing to the thermal motion of the adsorbed ions or neutral molecules and the
diffusible free electrolytes or solutes, there must be a continuous exchange of
ions between the exchanger surface (solid phase) with its surrounding micellar
and the intermicellar solutions. At the time- and temperature-dependent equilibrium, the same number of gegenions migrate, per unit time, into and out of the
diffuse double layer; i.e. adsorption and desorption balance each other. In
addition to this permanently occurring process, the term "ion exchange" implies
reactions involving displacement of the equilibrium, caused either by altering
the concentration of the outer solution (cf. Sect. 4.4.3) or by addition of other
ions.

To describe ion-exchange processes, several formulae have been proposed in
the course of exchange studies (Morrill et al. 1982). The first group is essentially
empirical and mainly intended to provide mathematical formulations best
fitting the experimental data. The second group of formulae is related to the
principle of the "law of mass action", according to which the rate of a chemical
reaction is directly proportional to the concentrations of the reactants.

Among the first group the Freundlich (1930), Langmuir and BET (i.e.
Brunauer, Emmett, Teller) approaches merit particular attention (cf. Sandstede
1961, Mayer 1978).

Empirical Adsorption Formulae

Freundlich relates the concentration of a substance in solution to that adsorbed
as follows:

$$x/m = kc^{1/n} , \tag{4.62}$$

where x = amount of substance adsorbed, m = weight of the adsorbent, c
= concentration of substance left in solution, k, n = specific constants.

In logarithmic form Eq. (4.62) gives

$$\log x/m = \log k + 1/n \log C \,,$$

where $1/n$ represents the slope of the straight-line adsorption isotherm while the sorption constant k is given by the intercept with the ordinate at unit concentration. Isotherms of the Freundlich type have the inherent disadvantage of being monotonous functions but describe adsorption and desorption phenomena in very dilute solutions appropriately. Isotherm tests also afford convenient means of studying the effects of different adsorbents and the effects of pH, temperature adsorbent/solution ratios, and salts in solution.

Figures 4.14 and 4.15 illustrate the adsorption behaviour of a pesticide in different soils as related to adsorbent/solution ratio and electrolyte concentration. The temperature dependence of adsorption and desorption is exemplarily summarized for the same soil samples in Table 4.27.

Since granulometry (as an indicator variable of specific surface), organic matter, and the oxalate- and dithionite-soluble Fe, Al, Mn and the Si fractions (which are indicative of pedogenic clay minerals and oxides), and potential acidity may account for over 90% of the observed variance of sorption rates, the Freundlich isotherms of a substance vary considerably in relation to both soil type and soil horizon (Fränzle et al. 1989). Table 4.28 summarizes the sorption

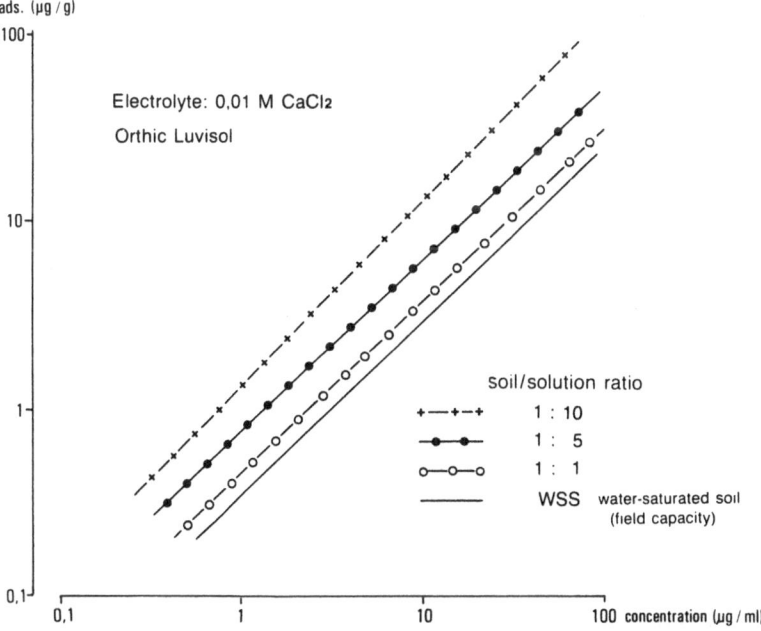

Fig. 4.14. Adsorption of 2,4-D on Ap-material of an Orthic Luvisol as a function of different soil/solution ratios. (After Brümmer et al. 1987)

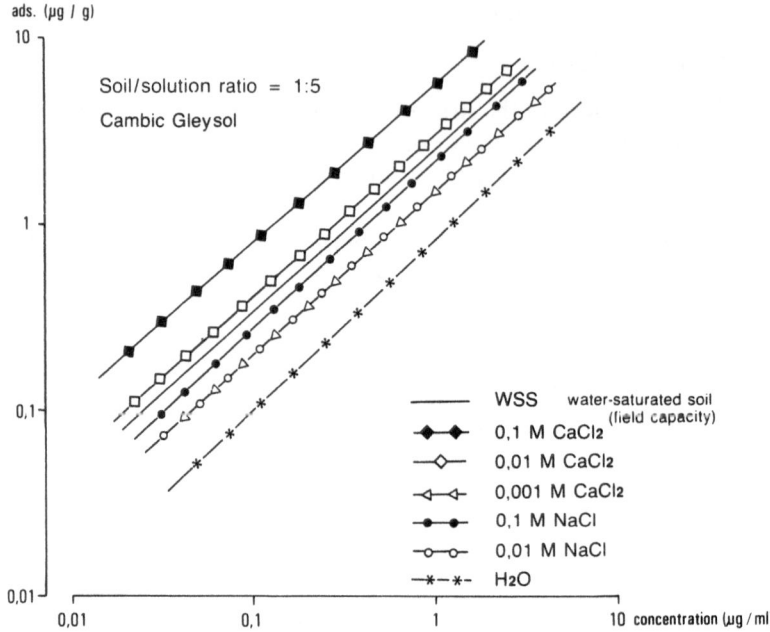

Fig. 4.15. Adsorption of 2,4-D on Ah-material of a Cambic gleysol under the influence of different salt concentrations. (After Brümmer et al. 1987)

characteristics of a series of Holsatian soils with regard to the chemicals 2,4-D and 2,4,5-T.

A simplified adsorption model related to the Freundlich theory is the above-mentioned distribution coefficient, frequently defined in terms of adsorbability of a chemical to activated carbon (P_{AC}). Table 4.29 gives an exemplary synopsis of the adsorbability of organic compounds to activated carbon. It illustrates that as solubility decreases, there is a corresponding increase in amenability to adsorption. It should be noted, however, that the inherent adsorbability of a chemical in pure component tests does not necessarily predict its degree of removal from a dynamic, multicomponent mixture. Yet pure component data may serve as a useful background for an understanding of multicomponent interactions. Tests with individual solutions showed that aromatics, even those containing polar substituents, were quite adsorbable, while substituted aliphatics (alcohols, amines, aldehydes, ketones, carboxylic acids) decreased sharply in amenability to adsorption with decreasing molecular weight (Verschueren 1983). Relatively poor is the adsorption of glycols and glycol ethers.

Furthermore, it should be mentioned that the adsorption of organic compounds, e.g. monosubstituted phenols by soil may well contrast with their previously reported adsorption by activated carbon (Al-Bahrani and Martin

Table 4.27. K and $1/n$ values of Freundlich isotherms for 2,4-D and atrazine as related to temperature. (After Brümmer et al. 1987)

Chemical	Soil type	Temperature °C	Adsorption		Sequential desorption					
			K	$1/n$	K_{Des1}	$1/n$	K_{Des2}	$1/n$	K_{Des3}	$1/n$
2,4-D	Orthic Luvisol	5	0.59	0.98	0.93	1.12	2.59	1.19	9.62	1.29
		15	1.51	0.97	3.25	0.92	7.73	0.95	19.47	0.96
		20	0.78	0.92	—	—	—	—	—	—
		25	0.56	0.95	1.04	0.96	2.89	0.94	2.07	1.12
		35	0.24	0.83	—	—	—	—	—	—
	Cambic Gleysol	5	3.21	0.87	4.89	0.77	6.23	0.87	9.32	0.90
		15	4.62	0.89	5.99	0.91	8.58	0.96	12.72	1.02
		20	3.10	0.85	—	—	—	—	—	—
		25	2.98	0.86	3.99	0.86	5.35	0.90	6.56	0.80
		35	2.32	0.86	2.89	0.78	3.38	0.69	3.37	0.50
Atrazine	Orthic Luvisol	5	1.39	0.93	1.93	0.99	3.44	1.06	6.62	1.11
		15	1.31	0.89	1.81	0.90	3.17	0.95	5.43	1.01
		20	1.61	0.91	—	—	—	—	—	—
		25	1.55	1.06	2.48	1.05	4.47	1.12	8.65	1.09
		35	1.01	0.94	1.30	1.06	2.32	1.21	4.33	1.07
	Cambic Gleysol	5	6.72	0.91	8.78	0.94	10.51	0.92	12.38	0.93
		15	7.23	0.91	8.52	0.90	9.63	0.90	10.56	0.92
		20	8.32	0.92	—	—	—	—	—	—
		25	9.90	1.09	12.67	1.06	14.85	1.05	17.38	1.03
		35	6.70	0.95	9.23	0.94	11.67	0.94	15.57	0.95

— = not calculated; $r = 0.93-0.99$

Table 4.28. Adsorption constants (K, K') and slope $(1/n)$ values of Freundlich isotherms for 2,4,5-T and atrazine. (After Brümmer et al. 1987)

Number	Soil type (locality, land use)	Horizon	% C_{org}	pH_{CaCl_2}	Texture
1	Orthic Luvisol (Krummbek, A)	Ap	1.62	6.81	sL
2	Orthic Luvisol (Siggen, W)	Ah	2.60	3.61	lS
3	Orthic Luvisol (Siggen, W)	Bt	0.37	4.73	cL
4	Gleyic Luvisol (Köhn, W)	Ah	3.66	3.21	lS
5	Dystric Gleysol (Köhn, W)	Ah	2.85	4.56	lS
6	Dystric Cambisol (Hohenwestedt, A)	Ap	1.20	4.92	uS
7	Cambic Gleysol (Hohenwestedt, G)	Ah	3.82	5.19	lS
8	Eutric Gleysol, colluvial (Havighorst, G)	Ah	5.64	5.65	sL
9	Orthic Podzol (Segeberg, W)	Ahe	8.11	2.79	S
10	Orthic Podzol (Segeberg, W)	Bhs	0.98	4.36	S
11	Orthic Podzol (Wasbek, A)	Ap	2.36	4.77	S
12	Eutric Cambisol (Finkhhkg., A)	Ap	2.35	7.01	sL
13	Gleyic Cambisol (Südermarsch, G)	Ah	8.42	3.99	sC
14	Gleyic Cambisol (Südermarsch, G)	SqGo	0.97	5.93	lC
15	Eutric Histosol (Havighorst, G)	HA	16.64	6.43	

A, Field; G, Grassland; W, Forest

1976; Boyd 1982; Kuhnt 1987). For instance, the effect of introducing ortho-CH_3, $-OCH_3$, $-Cl$, and $-NO_2$ groups to phenol on adsorption by activated carbon was

$$OCH_3 > CH_3 > Cl > NO_2 ,$$

when compared on a molar basis. In contrast, adsorption of the same ortho-substituted phenols by soil decreased in the order

$$NO_2 > Cl > OCH_3 > CH_3$$

Table 4.28. (Contd.)

Adsorption 2,4,5-T			Adsorption atrazine		
K	$1/n$	K'	K	$1/n$	K'
1.84	0.96	1.56–2.20	1.63	0.93	1.39–2.12
14.48	0.90	11.50–24.01	5.33	0.94	4.42–6.65
1.51	0.91	1.28–2.26	2.29	0.88	1.62–3.24
26.24	0.91	21.88–43.00	8.63	0.94	7.29–10.90
8.55	0.87	6.15–15.55	6.91	0.87	4.66–11.40
2.48	0.93	2.05–3.38	1.88	0.91	1.37–2.37
8.12	0.91	6.31–12.41	8.72	0.92	7.04–11.70
7.91	0.90	6.31–13.16	9.59	0.91	7.49–13.50
108.14	0.89	97.43–231.54	34.70	0.86	28.00–76.80
5.25	0.89	3.91–8.69	1.07	0.91	0.72–1.40
5.68	0.89	4.23–9.16	4.94	0.89	3.39–7.21
2.37	0.91	1.95–3.45	2.08	0.89	1.52–2.96
47.78	0.93	41.57–71.45	21.32	0.91	17.80–31.80
2.67	0.87	1.89–4.60	3.41	0.89	2.47–4.81
13.65	0.91	11.47–21.31	25.20	0.91	21.20–37.10

K' (concentration-controlled distribution coefficient) = ratio of adsorbed to dissolved chemical

(Boyd 1982). Substituent position effects were also not observed when comparing the adsorption of *o*- and *p*-chlorophenol and *o*-, *m*- and *p*-cresol by activated carbon (Al-Bahrani and Martin 1976).

The Freundlich formula is of parabolic type and cannot therefore give a maximum adsorption value as does the *Langmuir theory*. According to it adsorption is considered proportional to the pressure of gas (or concentration of a solute) and to the amount of surface not yet covered with adsorbed molecules. The rate of desorption (or evaporation) of adsorbed molecules is considered

Table 4.29. Adsorption values (log P_{AC}) and water solubilities (log S_{ppb}) of organic compounds in an activated carbon/water system. (After Dobbs and Cohen 1980; Guisti et al. 1974)

Compound	log P_{AC}	pH	log S_{ppb}
Acetates and acrylates			
Methylacetate	1.85	—	8.5
Ethylacetate	2.31	—	7.9
Propylacetate	2.78	—	7.3
Butylacetate	3.04	—	6.8
Isopropylacetate	2.63	—	7.5
Isobutylacetate	2.96	—	6.8
Vinylacetate	2.56	—	7.4
Ethylacrylate	2.84	—	7.3
Butylacrylate	3.65	—	6.3
Alcoholamines			
Monoethanolamine	1.20	—	9
Diethanolamine	1.90	—	8.9
Triethanolamine	2.00	—	9
Monoisopropanolamine	1.70	—	9
Diisopropanolamine	2.22	—	8.9
Alcohols			
Methanol	0.86	—	9
Ethanol	1.35	—	9
Propanol	1.67	—	9
Butanol	2.36	—	7.9
n-Amylalcohol	2.74	—	7.2
n-Hexanol	3.63	—	6.8
Isopropanol	1.46	—	9
Aldehydes			
Formaldehyde	1.30	—	9
Acetaldehyde	1.40	—	9
Propionaldehyde	1.90	—	8.3
Butyraldehyde	2.35	—	7.9
Benzaldehyde	3.50	—	6.5
Amines			
Butylamine	2.33	—	9
Allylamine	1.96	—	9
Ethylenediamine	1.37	—	9
Diethylenediamine	1.94	—	9
Aniline	2.78	—	7.5
o-Anisidine	4.0–4.8	—	—
p-Nitroaniline	4.5–6.7	7.0	—
Diphenylamine	4.4–6.1	7.0	—
3,3-Dichlorobenzidine	5.3–7.2	7.2	—

Table 4.29. (Contd.)

Compound	$\log P_{AC}$	pH	$\log S_{ppb}$
Aromatics			
Benzene	3.3–3.7	5.3	6.2
	3.58	—	5.8
Toluene	3.7–4.3	5.6	5.7
	2.88	—	5.7
Ethylbenzene	4.3–4.7	7.3	5.2
	3.02	—	5.3
p-Xylene	4.0–4.7	7.3	5.3
Styrene	4.9–5.4	7.0	5.5
	3.19		
Naphthalene	4.7–5.5	5.6	4.5
Anthracene	6.2–6.8	5.3	3.0
Phenanthrene	5.8–6.8	5.3	3.2
Fluoranthene	6.5–7.0	5.3	2.4
Benzo(a)pyrene	6.5–7.0	—	0.5
4-Aminodiphenyl	4.6–6.7	—	—
4-Nitrodiphenyl	5.5–6.7	7.0	—
Polychlorinated biphenyl 1221	5.5–6.3	5.3	4.0
Chlorinated benzenes			
Chlorobenzene	4.8–5.0	7.0	5.7
1,2-Dichlorobenzene	4.6–5.2	5.5	5.2
1,3-Dichlorobenzene	4.2–5.0	5.1	5.0
1,4-Dichlorobenzene	4.9–5.4	5.1	4.9
1,2,4-Trichlorobenzene	4.4–5.3	5.3	4.5
Hexachlorobenzene	4.0–4.3	5.3	0.8
Chlorinated pesticides			
x-Endosulfan	5.7–8.9	5.3	2.7
Endrin	5.9–6.3	5.3	2.4
Aldrin	5.9–6.1	5.3	1.0
Dieldrin	6.1–7.2	5.3	2.3
Chlordane	5.8–7.3	5.3	1.7–3.3
DDT	7.4–8.2	5.3	0.3
Ethers			
Isopropylether	2.90	—	7.1
Butylether	4.59	—	5.5
2-Chloroethylvinylether	3.4–3.5	5.3	7.2
4-Chlorophenylether	5.2–5.5	5.3	4.8
Glycols and glycolethers			
Ethyleneglycol	1.16	—	9
Diethyleneglycol	1.86	—	9
Triethyleneglycol	2.34	—	9
Tetraethyleneglycol	2.44	—	9
Propyleneglycol	1.43	—	9

Table 4.29. (Contd.)

Compound	$\log P_{AC}$	pH	$\log S_{ppb}$
Dipropyleneglycol	1.60	—	9
Ethoxytriglycol	2.63	—	9
Halogenated paraffins and olefins			
Chloroform	3.4–3.8	5.3	6.9
Carbon tetrachloride	4.2–4.4	5.3	5.9
Dichlorobromomethane	3.9–4.7	5.3	—
Dibromochloromethane	3.7–5.7	5.3	—
Bromoform	4.4–5.1	5.3	6.1
Chloroethane	2.8	5.3	6.8
1,1-Dichloroethane	3.4–4.0	5.3	6.7
1,2-Dichloroethane	3.7–3.9	5.3	6.9
	2.94	—	6.9
1,1,1-Trichloroethane	3.5–5.0	5.3	6.6
1,1,2-Trichloroethane	4.1–4.6	5.3	6.7
1,1,2,2,-Tetrachloroethane	4.2–5.5	5.3	6.5
Hexachloroethane	5.0–6.7	5.3	4.7
Trichloroethene	4.7–5.2	5.3	—
Tetrachloroethene	5.0–5.7	5.3	5.2
Dichloropropane	3.7–4.4	5.3	6.4
Ketones			
Acetone	1.74	—	9
Methylethylketone	2.25	—	8.4
Methylpropylketone	2.66	—	7.6
Cyclohexanone	2.61	—	7.4
Organic acids			
Formic acid	1.79	—	9
Acetic acid	1.80	—	9
Propionic acid	1.98	—	9
Bytric acid	2.47	—	9
Benzoic acid	3.6–3.8	—	6.4
Phenols			
Phenol	3.8–4.3	7.0	7.8
2-Chlorophenol	4.1–4.6	5.8	7.5
2,4-Dichlorophenol	4.1–6.5	5.3	6.7
2,4,6-Trichlorophenol	4.5–5.2	6.0	5.9
Pentachlorophenol	4.6–5.5	7.0	4.1
2,4-Dimethylphenol	4.1–5.2	5.8	7.2
2-Nitrophenol	4.2–5.5	5.5	6.3
2,4-Dinitrophenol	4.1–4.6	7.0	6.7
Phthalates			
Dimethylphthalate	4.5–5.7	7.0	6.6
Diethylphthalate	4.1–5.0	5.4	6.0

proportional to the number of such molecules. At equilibrium these two rates must be equal whence it ensues that

$$n/n_m = bp/(1 + bp) \,.$$ (4.63)

This equation, in which n is the number of molecules adsorbed, n_m is the number of molecules required for a complete unimolecular surface film, p is the gas pressure (or concentration of a chemical in solution), and b is a constant at a given temperature, is the "Langmuir adsorption equation" or "Langmuir isotherm".

Many examples of unimolecular adsorption have been found, including most cases of chemisorption which are qualitatively represented by the Langmuir formula; but in many systems with unimolecular adsorption deviations from Langmuir's equation are very pronounced. They seem to result chiefly from two factors, the first of which is the heterogeneity of adsorbent surfaces, and the second, adsorbed molecules or atoms may interact with each other either directly or by a cumulative effect on the energy-binding adsorbed species to the surface (Yariv and Heller-Kallai 1975; Samii and Lagaly 1987).

A third theory of adsorption widely used for representing the adsorption of gases or vapours on solids was advanced by Brunauer, Emmet and Teller (cf. Sandstede 1961). In contrast to Langmuir's theory, the *BET theory* makes allowance for multimolecular adsorption as is commonly observed when nonporous solids adsorb vapours at pressures approaching the saturated vapour pressure. The BET equation therefore is

$$\frac{n}{n_m} = \frac{b\left(\dfrac{p}{p_0}\right)}{\left(1 - \dfrac{p}{p_0}\right)\left[1 + (b - 1)\dfrac{p}{p_0}\right]} \,,$$ (4.64)

in which p_0 is the saturated vapour pressure, while the other quantities have the same meaning as in Eq. (4.63). The equation is designed for physical adsorption and predicts a very sharp increase in the amount of material adsorbed as the saturated vapour pressure is approached, and this is found to be the case (cf. Israelachvili 1985).

Law of Mass Action and Ion Exchange

The principle of the law of mass action has been important for deriving ion-exchange formulas. Former objections that ion exchange is not a real chemical reaction leading to new compounds are now considered irrelevant in the light of the structure of the diffuse double layer and its relationships with the surroundings as dependent on the activity of the diffusible ions and connected with changes in free energy (Bolt and Bruggenwert 1976/82). Therefore in homovalent exchange such as

$$RK + Na^+ \rightarrow RNa + K^+ \,,$$

the equilibrium equation may be written as

$$\frac{[Na^+]_i \, (K^+)_o}{[K^+]_i \, (Na^+)_o} = k_{K,Na} \, , \tag{4.65.1}$$

and for monovalent and divalent ions

$$RCa + 2K^+ \;\rightarrow\; 2RK + Ca^{2+}$$

$$\frac{[K^+]_i^2 \, (Ca^{2+})_o}{[Ca^{2+}]_i \, (K^+)_o^2} = k_{Ca,K} \, . \tag{4.65.2}$$

Brackets denote concentration, parentheses activity; i is the exchanger phase or inner solution, o the outer solution. k, the equilibrium constant, is no real constant but varies more or less with the mole fraction of the two ions.

Donnan Equilibrium

In a system comprising a non-diffusible electrolyte RM and the diffusible electrolyte MA, a constant ion product exists which is usually recognized as the Donnan membrane equilibrium

$$(M^+)_i (A^-)_i = (M^+)_o (A^-)_o \, . \tag{4.66.1}$$

It characterizes any system in which at least one ion species is restrained from freely diffusing throughout the whole system. Thus a clay particle, with its surrounding diffuse double layer may be considered as a micro-Donnan system, where the attractive Coulomb forces between particle surface and gegenions act as a restraint, causing a non-uniform distribution of the gegenions in the micellar solution.

Thermodynamically speaking, the Donnan equilibrium is characterized by the fact that the chemical potential μ of a diffusible electrolyte, and the electrochemical potential of every diffusible ion species, are constant throughout the system:

$$\mu_{MAi} = \mu_{MAo} \, . \tag{4.66.2}$$

For a strong electrolyte, and temperature and pressure being constant, Eq. (4.66.2) can be written

$$\mu_{MA}^0 + RT \ln (MA)_i = \mu_{MA}^0 + RT \ln (MA)_o \, , \tag{4.67.1}$$

whence

$$(MA)_i = (MA)_o \tag{4.67.2}$$

$$(M)_i \, (A)_i = (M)_o \, (A)_o \, . \tag{4.67.3}$$

The above equations infer that the activity of cations and anions varies in opposite directions, so that the activity of the electrolyte is constant at any point of the liquid phase (cf. Nordstrom and Munoz 1986).

For electrolytes of the general type M_nA_m, the ion product formula is

$$(M)_i^n (A)_i^m = (M)_o^n (A)_o^m \tag{4.68.1}$$

or

$$\left[\frac{(M)_i}{(M)_o}\right]^{1/m} = \left[\frac{(A)_o}{(A)_i}\right]^{1/n}. \tag{4.68.2}$$

For systems with more than one diffusible electrolyte, the activity of each must be constant in the aqueous phase. An example of such a multispecies system in equilibrium is given in the following set of equations, describing four cations and three anions (Wiklander 1964).

$$\frac{(K^+)_i}{(K^+)_o} = \frac{(H^+)_i}{(H^+)_o} = \sqrt{\frac{(Ca^{2+})_i}{(Ca^{2+})_o}} = \sqrt{\frac{(La^{3+})_i}{(La^{3+})_o}}$$

$$= \frac{(Cl^-)_o}{(Cl^-)_i} = \frac{(OH^-)_o}{(OH^-)_i} = \sqrt{\frac{(SO_4^{2-})_o}{(SO_4^{2-})_i}}. \tag{4.69}$$

If the equilibrium constant in Eq. (4.69) is equal to 1, the equation becomes equivalent to the mass-action formula (4.65). For instance, for Na and K

$$\frac{(Na)_i\,(K)_o}{(Na)_o\,(K)_i} = 1\,, \tag{4.70.1}$$

or by introducing concentration, f signifying the activity coefficient

$$\frac{[Na]_i\,[K]_o}{[Na]_o\,[K]_i} = \frac{f_{K_i} \cdot f_{Na_o}}{f_{Na_i} \cdot f_{K_o}} = \alpha_{K,Na}\,. \tag{4.70.2}$$

In this equation the equilibrium quotient α is a measure of the relative adsorption energy of the K and Na ions. Thus for the ion species A and B $\alpha_{A,B} > 1$ if A is more weakly adsorbed than B; $\alpha_{A,B} = 1$ if A and B are adsorbed with equal intensity; and finally $\alpha_{A,B} < 1$ if A is more firmly adsorbed than B.

With regard to the validity of the k and α models, numerous experiments have shown that the equilibrium quotients vary more or less with the mole fraction, the variation being the greater the more unequal the chemical properties on the ions, and the greater the selectivity of the exchanger. For natural exchangers, in particular soils, it is only within limited concentration ranges that the quotients have proved to be fairly constant. This is clearly reflected in empirical sorption isotherms (cf. Sect. 4.4.1.3) covering concentration ranges far beyond the realm of validity of the Freundlich theory.

4.4.1.3 Heavy Metals and Organics in Soil

When chemicals are added to soils, either inadvertently or deliberately, their fates are normally partitioned among four possible pathways. Including degradation products, they may leach rapidly through the soil into ground and surface waters, they may be taken up by vegetation, they may volatilize into the atmosphere, or they may be retained and stored in the soil (cf. foldout models I and II). For the first three pathways, if the chemicals are deleterious to the environment, observable effects may be noted within a short time after application.

Nitrogen fertilizers are examples of easily leachable compounds that have caused very serious water quality problems (Fleischer et al. 1987; Wehrmann and Scharpf 1988). Considering the second pathway, the uptake by crops of toxic materials has long been a public health concern, especially in farm areas fertilized by sewage sludge and animal wastes contaminated with such chemicals (MacLean 1976; Williams and David 1976; Wallace et al. 1977; Olsthoorn and Thomas 1986; Sauerbeck 1987).

Volatile pesticides used in soil fumigation are a widely used class of chemicals following the third pathway. Bromomethane, for instance, is mainly used as an insecticide and fumigant in the protection of stores and plants. After fumigation of the ground, it easily diffuses into the air; if introduced deeply enough, it might enter the groundwater (Huygen and van Ijssel 1981; Beratergremium für umweltrelevante Altstoffe 1987e). Retention and storage in soil as the fourth pathway, however, is particularly interesting in terms of the potential for non-linear and time-delayed environmental effects (Blume et al. 1983; Stigliani 1988).

Heavy Metals in Agricultural Soils

Agricultural soils have received major inputs of toxic materials from the application of pesticides and contaminated fertilizers. While some of this material has a rather short life-time in soil, a substantial amount remains stored. This applies in particular to certain heavy metals and organic pesticides present in the soils either as insoluble products or as adsorbates on the surfaces of soil substrates (IPS 1982; de Haan 1987; Sauerbeck 1987).

Heavy metals commonly found in European or North American agricultural soils include cadmium, chromium, copper, mercury, nickel, lead and zinc (cf. Chap. 3.5.5). Table 4.30 provides data on normal background and maximum concentrations. As regards heavy metal limit values, the CEC Directive departs from national regulations in that it does not set a limit value for chromium, since Cr is bound particularly firmly in soil and, therefore, all the less available to plants. The same applies to lead and, basically, also to mercury, but in view of the exceedingly high zootoxic potential of Hg there is little prospect of any further relaxation of the current very cautious limit values. On the other hand, the bioavailability of cadmium, copper, nickel and zinc depends far more on soil acidity. Particularly with regard to zinc and cadmium, the vegetative plant parts not infrequently contain almost as much, if not more, of these heavy metals than the soil on which they grow (cf. McGrath 1984; Jones et al. 1987).

Table 4.30. Geogenic background (Gbgd) and maximum permissible (perm) heavy metal concentrations of selected EC Member States. (Webber et al. 1984; CEC 1986).

Element	Germany		France perm	United Kingdon			CEC Directive perm
	gbgd	perm		gbgd	perm nc	c	
Cd	0.2	3	2	1	3.5	3.5	1–3
Cr	30	100	150	100	600	600	–
Cu	30	100	100	5	140	280	50–140
Hg	0.1	2	1	< 0.1	1	1	1–1.5
Ni	30	50	50	1	35	70	30–75
Pb	30	100	100	50	550	550	50–300
Zn	50	300	300	2.5	280	560	150–300
As				5	10	10	
B				1	3.25	3.25	
F				200	500	500	
Mo				2	4	4	
Se			10	0.5	3	3	

Values are total concentrations in soil except for the United Kingdom. Zn, Cu and Ni extracted by EDTA and B extracted by hot water. nc = non-calcareous, c = calcareous.

Batch experiments under standardized boundary conditions allow to define the relative affinities of heavy metals for different soil clay fractions and metal hydroxides such as α-FeOOH, $Al(OH)_3$ and $Fe(OH)_3$ (Gerth 1985). Sorption reactions are characterized by pH_{50} values (the pH at which 50% of the original ion concentration is adsorbed), by shapes of adsorption curves, and by measuring separation factors or distribution coefficients. Three reaction types can be identified:

1. those associated with soil adsorbing surfaces under the domination of iron hydroxides, in particular goethite; these appear to be controlled by mechanisms involving metal-ion hydrolysis and resulting in relative sorption affinities of Zn > Ni > Cd;
2. those associated with organic surfaces for which metal-ion hydrolysis is of little significance and where little difference in metal-ion affinity is evident;
3. those associated with 2 : 1 layer clay minerals which exhibit greater preference for Zn than for Ni and Cd and higher affinities for each metal at lower pH values (< 5) than is shown by clays dominated by iron hydroxides (Tiller et al. 1984).

The concentrations of Zn in equilibrium solutions with soil clay fractions and whole soil samples at pH-values below 7 are determined exclusively by adsorption–desorption reactions. At neutral to alkaline pH values, precipitation–dissolution processes may take place, and formation of zinc silicates may control the Zn concentration in solution provided natural complexing agents are absent (Brümmer et al. 1983). Model experiments in $CaCO_3$-buffered systems showed that the adsorption capacity for specifically adsorbed Zn (in $\mu mol\ g^{-1}$) by the following components increased in the order: $CaCO_3$ (0.44), bentonite (44), humic acid (842), amorphous Fe- and Al-oxides (1190, 1310) and δ-MnO_2 (1540), which indicates the special role of these components in limiting precipitation reactions.

The adsorption of Cd, Ni and Zn by goethite at concentrations of 0.5 to 5 $\mu mol\ g^{-1}$ α-FeOOH strongly increases with pH in the range from 4 to 8, reaction time and temperature (Gerth and Brümmer 1983). The specificity of adsorption is characterized by the pH_{50} values of 4.9, 5.6 and 5.8 found for Zn, Ni and Cd, respectively, which is in relatively good agreement with comparative analyses by McKenzie (1980) which include the pH_{50} values for Pb (3.0) and Cu (4.3) as adsorbed to Fe $(OH)_3$ gel. The differences between the pH_{50} values closely agree with the differences between the pK values of the first hydrolysis constants of the heavy metals, thus indicating that adsorption on the hydroxylated surfaces of sesquioxides is strongly affected by the hydroxocomplex formation of trace metals (Gerth 1985).

Besides adsorption to the surface, solid state diffusion is also observed, which may be related to defective structures in the lattice of goethite. Linear plots of the amount of metal adsorbed versus the square root of reaction time indicate that the effect of diffusion on the binding of metals by α-FeOOH is increased in the order Cd < Zn < Ni. Considerable metal amounts appear to be occluded and can therefore only be remobilized by complete dissolution of goethite. In relatively long-term experiments (Gerth 1985) this strongly bound fraction of zinc, cadmium and nickel comprised 39.0, 29.0 and 10.5%, respectively, of the total amount of heavy metals added. Neglect of occlusion in the framework of short-term experiments seems to be the reason for deducing affinity sequences different from those above (cf. Gadde and Laitinen 1974; Forbes et al. 1976; Kinniburgh et al. 1976; McKenzie 1980; Benjamin and Leckie 1981).

Sensitivity of German Soils to Heavy Metal Pollution

A regionally differentiating interpretation of experimental data from batch and leaching experiments of complementary complexity allows to define the relative affinity of heavy metals to soil in relating to organic matter, clay mineralogy, metal oxides and pH (cf. Grove and Ellis 1980; Fränzle 1982; Brümmer et al. 1986; König et al. 1986; Blume and Brümmer 1987a; Fränzle et al. 1989). Table 4.31 summarizes the results of such a comprehensive interpretation of sorption-relevant soil properties in score form. In the sandy soils with a low humus content that often occur in north Germany, the pH value is decisive, as indicated in Table 4.32.

In order to allow for higher humus content and finer texture, the adjustments shown in Table 4.33 are required. Table functions of the above type provide the information for a sorption-oriented interpretation of the taxonomic units of soil maps. In the present context the 1:1,000,000 Soil Map of Germany (Hollstein 1963) proved better suited to this end than Roeschmann's (1986) soil association map. Thus the elements of Table 4.34 resulting from such a comprehensive

Table 4.31. Average intensity of physio- and chemosorption of heavy metals to soil constituents. (After Blume and Brümmer 1987)

Metal	Adsorption[a] by			Enhanced sorption above pH
	Humus	Clay	Sesquioxides	
Cd	4[b]	2	3	6
Mn	2	3	3	5.5
Ni	3–4	2	3	5.5
Co	3	2	3	5.5
Zn	2	3	3	5.5
Cu	5	3	4	4.5
Cr(III)	5	4	5	4.5
Pb	5	4	5	4
Hg	5	4	5	4
Fe(III)	5	5	—	3.5
(Al)	5	4	4	4.5

[a] Relevant for pH values below the threshold value given in column 5; above this value strong bonding due to oxide formation (Al, Fe and Mn) or hydroxo-complexes (others).
[b] Scores: 1 very low, 2 low, 3 medium, 4 high, 5 very high.

Table 4.32. Influence of soil acidity on metal sorption in sandy soils of low (0–2%) humus content. (After Blume and Brümmer 1987a)

pH CaCl$_2$ Metal	2.5	3	3.5	4	4.5	5	5.5	6	6.5	7
Cd	0[a]	0–1	1	1–2	2	3	3–4	4	4–5	5
Mn	0	1	1–2	2	3	3–4	4	4–5	5	5
Ni	0	1	1–2	2	3	3–4	4	4–5	5	5
Co	0	1	1–2	2	3	3–4	4	4–5	5	5
Zn	0	1	1–2	2	3	3–4	4	4–5	5	5
Cu	1	1–2	2	3	4	4–5	5	5	5	5
Cr(III)	1	1–2	2	3	4	4–5	5	5	5	5
Pb	1	2	3	4	5	5	5	5	5	5
Hg	1	2	3	4	5	5	5	5	5	5
Fe(III)	1–2	2–3	3–4	5	5	5	5	5	5	5
Al	1	1–2	2	3	4	4–5	5	5	5	5

[a] Scores: 0 none, 1 very low, 2 low, 3 medium, 4 high, 5 very high.

Table 4.33. Adjustments to sorption values of Table 4.31 in consideration of humus content and texture. (After Blume and Brümmer 1987a)

Sorptivity according to Table 4.31	2–3	3	3–4	4	5
Humus content (%)					
0–2	0	0	0	0	0
2–8	0	0–1	0–1	0–1	1
8–15	0–1	0–1	1	1	1–2
> 15	0–1	1	1	1–2	2
Texture					
S[a], u'S	0	0	0	0	0
t'S, l'S, uS, sU, U	0	0	0–1	0–1	0–1
lU, slU, uL, sL, sC	0–1	0–1	0–1	0–1	1
lC, sC, uC, scL, cL	0–1	0–1	1	1	1–2
C	0–1	1	1	1–2	2

[a] S = sand, U = silt, L = loam, C = clay s = sandy, u = silty, l = loamy, c = clayey u' = slightly silty, l' = slightly loamy, c' = slightly clayey.

interpretation procedure, contain the basic information for the construction of small-scale maps of soil sensitivity to heavy metals by means of geographical information systems such as ARC/INFO (Schaller 1988).

The following maps (Fig. 4.16, 4.17) show the spatial differentiation of Cd and Pb sorptivity of German soils. The maps clearly reflect the dominant influence of both pH and organic matter content on solubility and adsorptive bonding of heavy metals. A comparative analysis of the situation shows that the pH influence on solubility decreases in the order Cd > Zn > Ni > Cu > Pb (Herms and Brümmer 1980; Herms 1982). In the pH range 3–7 covering the majority of forest soils (pH mostly 3–6) and agricultural soils (pH mostly 5–7), the Cd and Zn concentrations in soil solution can be approximated by the following equations:

$$\log \text{Cd (mg/l)} = -0.46 \text{ pH} + 2.11$$

$$r = 0.96; n = 49$$

$$\log \text{Zn (mg/l)} = -0.47 \text{ pH} + 2.97$$

$$r = 0.97; n = 49$$

In contrast, the mobilizing effect of organic matter increases in the order Cd < Zn < Pb < Cu for neutral to weakly alkaline soil conditions. In the pH range 3–6, however, organic matter contributes to reduce heavy metal mobility (Lichtfuß 1977; Herms and Brümmer 1980).

In comparison to Cd and Zn the other heavy metals exhibit a distinctly lesser degree of spatial variability owing to minor pH and organic-matter control (cf.

Table 4.34. Sorption capacity of German soils to heavy metal ions

No.[a]	Soil subtype[a]	pH	Evaluation[b] Cd	Zn	Cu	Pb	Humus (%)	Evaluation[c] Cd	Zn	Cu	Pb	Granulometry	Evaluation[c] Cd	Zn	Cu	Pb	Sum[d] Cd	Zn	Cu	Pb
1	Rendzinas	6.5–8	5	5	5	5											5	5	5	5
2	Chromic Luvisols	7.5	5	5	5	5											5	5	5	5
4	Calcaric Regosols	6–7	5	5	5	5											5	5	5	5
5	Calcaric Rankers	7.7	5	5	5	5											5	5	5	5
6	Chernozems	7.8	5	5	5	5											5	5	5	5
7	Phaeozems	7.5	5	5	5	5											5	5	5	5
8	Cambisol I	6–7.5	5	5	5	5							0	0	0	0	5	5	5	5
9	Cambisol II	7–7.5	5	5	5	5							0	0	0	0	5	5	5	5
10	Cambisol III	4.2	1	2	3	4	1.0	0	0	1	1	sL	0	0	0	0	1	2	4	5
11	Cambisol IV	7.0	5	5	5	5							0	0	0	0	5	5	5	5
12	Cambisol V	4.3–5	2	3	4	5	6–10	1	0	1	1	lS-lC	0	0	0	0	3	3	5	5
13	Cambisol VI	4–5	2	2	3	4	4–8	0	0	1	1	lS	0	0	0	0	2	2	4	5
14	Cambisol VII	5.1	3	3	4	5	0.3	0	0	0	0	lS	0	0	0	0	3	3	4	5
15	Cambisol VIII	4.4	2	3	4	5	0.7	0	0	0	0	lS-sL	0	0	0	0	2	3	4	5
16	Cambisol IX	4–5	2	2	3	4	4–8	0	0	1	1	lS-sL	0	0	0	0	2	2	4	5
17	Cambisol X	4.2	1	2	3	4	0.7	0	0	0	0	S	0	0	0	0	1	2	3	4
18	Orthic Luvisol I	5–6	3	3	4	5	1.5–2.5	0	0	1	1	lS-sL	0	0	0	0	3	3	5	5
19	Orthic Luvisol II	5.3	3	4	5	5	0.65	0	0	0	0	lS-sL	0	0	0	0	3	4	5	5
20	Orthic Luvisol III	5–8	3	3	4	5	2–4	1	0	1	1	uL	0	0	0	0	4	3	5	5
21	Orthic Luvisol IV	5.6	3	4	5	5	0.28	0	0	0	0	uL	0	0	0	0	3	4	5	5
22	Orthic Luvisol V	7.0	5	5	5	5							0	0	0	0	5	5	5	5
23	Orthic Luvisol VI	4.8	2	3	4	5	2.1	0	0	1	1	uS-uL	0	0	0	0	2	3	5	5
24	Orthic Luvisol VII	4.7	2	3	4	5	0.32	0	0	0	0	S-uL	0	0	0	0	2	3	4	5
25	Orthic Luvisol VIII	4.1	1	2	3	4	2.54	0	0	1	1	S-lS	0	0	0	0	1	2	4	5

Table 4.34. (Contd.)

No.[a]	Soil subtype[a]	pH	Evaluation[b]				Humus (%)	Evaluation[c]				Granulometry	Evaluation[c]				Sum[d]			
			Cd	Zn	Cu	Pb		Cd	Zn	Cu	Pb		Cd	Zn	Cu	Pb	Cd	Zn	Cu	Pb
26	Dystric Gleysol I	5.5–6	3	3	4	5	2.5–5	0	0	1	1	IS-L	0	0	0	0	3	3	5	5
27	Dystric Gelysol II	4.6	2	3	4	5	1.36	0	0	1	1	IS-uL	0	0	0	0	2	3	5	5
28	Dystric Gleysol III	5.7	3	4	5	5	0.66	0	0	0	0	L-C	0	1	1	1	3	5	5	5
29	Podzol I	5.0	3	3	4	5	0.71	0	0	0	0	S	0	0	0	0	3	3	4	5
30	Podzol II	4.4	2	3	4	5	1.1	0	0	0	0	S-IS	0	0	0	0	2	3	4	5
31	Podzol III	4.5	2	3	4	5	1.2	0	0	1	1	S	0	0	0	0	2	3	5	5
32	Podzol IV	4.2	1	2	3	4	0.7	0	0	0	0	S-IS	0	0	0	0	1	2	3	4
33	Podzol V	3.9	1	2	3	4	5.37	1	0	1	1	S-IS	0	0	0	0	2	2	4	5
34	Podzol VI	4.8	3	3	4	5	0.28	0	0	0	0	S-IS	0	0	0	0	3	3	4	5
35	Various soil types I	6–7.5	4	4	5	5	2–15	1	0	1	1	L-uL	0	0	0	0	5	4	5	5
36	Various soil types II	4–7.5	1	2	3	4	2–15	1	0	1	1	variable	0	0	0	0	2	2	4	5
37	Various soil types III	5.5–7.5	3	3	4	5	2–5	1	0	1	1	IS-IC	0	0	0	0	4	3	5	5
38	Rock outcrops	7.5–8	5	5	5	5											5	5	5	5
39	Fluvisol I	5–7	3	3	4	5	8–10	1	0	1	1	L-C	0	1	1	1	4	4	5	5
40	Fluvisol II	3.6–5	1	1	2	3	4–30	0	0	1	1	S	0	0	0	0	1	1	3	4
41	Fluvisol III	3.6–7	1	1	2	3	4–30	0	0	1	1	—	0	0	0	0	1	1	3	4
42	Eutric Cambisols	7.7	5	5	5	5	0.88	0	0	0	0	sU-C	0	0	0	0	5	5	5	5
43	Eutric Fluvisols	6.5	4	5	5	5	0.81	0	0	0	0	tU-C	0	0	0	0	4	5	5	5
44	Eutric Histosols	5.4	3	4	5	5	67.48	1	0	2	2	—	0	0	0	0	4	4	5	5
45	Calacaric Fluvisols	7.5–8	5	5	5	5	2.5–4.5	0	0	1	1	S	0	0	0	0	5	5	5	5
46	Dystric Histosols	3.8	1	1	2	3	80	1	0	2	2	—	0	0	0	0	2	1	4	5
47	Ferralsols/Acrisols	4.5–5.5	2	3	4	5	3–5	0	0	1	1	C	0	1	1	1	2	4	5	5

[a] After Hollstein (1963), translated. [b] Cf. Tables 4.31, 4.32. [c] Cf. Table 4.33. [d] Cumulative evaluation, 5 = maximum value.

Fig. 4.17). An analysis of more than 3000 stream sediments in Schleswig-Holstein yielded additional data for a correlation analysis of the behaviour of geochemically relevant heavy metals. The following matrix indicates that Pb and Cu on the one hand, and Cd and Zn on the other, co-variate most in regard of their distribution patterns, while the other elements listed display an intermediate behaviour.

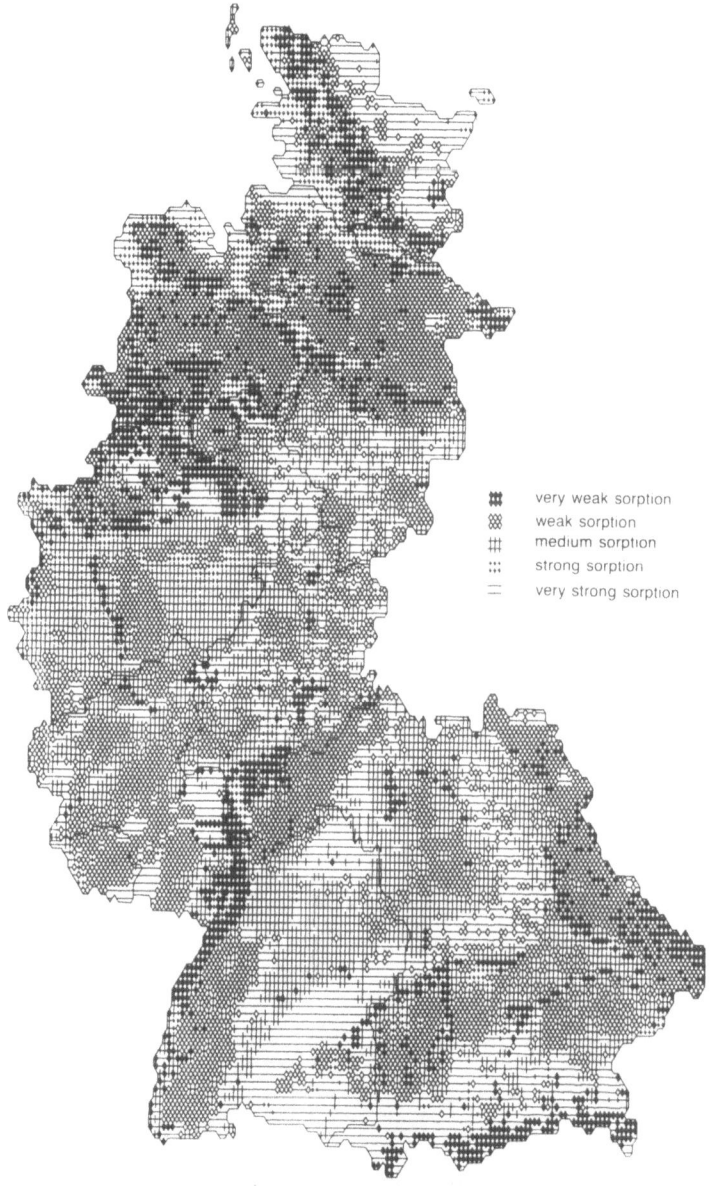

▓	very weak sorption
▓	weak sorption
⧻	medium sorption
⋮⋮	strong sorption
=	very strong sorption

Fig. 4.16. Spatial differentiation of Cd sorptivity of German soils

In conclusion, this means that the degree of spatial differentiation of soil sensitivity to heavy metal input attains a minimum for Pb and Cu while it is distinctly higher for Ni and Cr, and maximum for Cd and Zn. Furthermore both maps and Table 4.35 indicate the importance of many loess soils as potential heavy metals sinks.

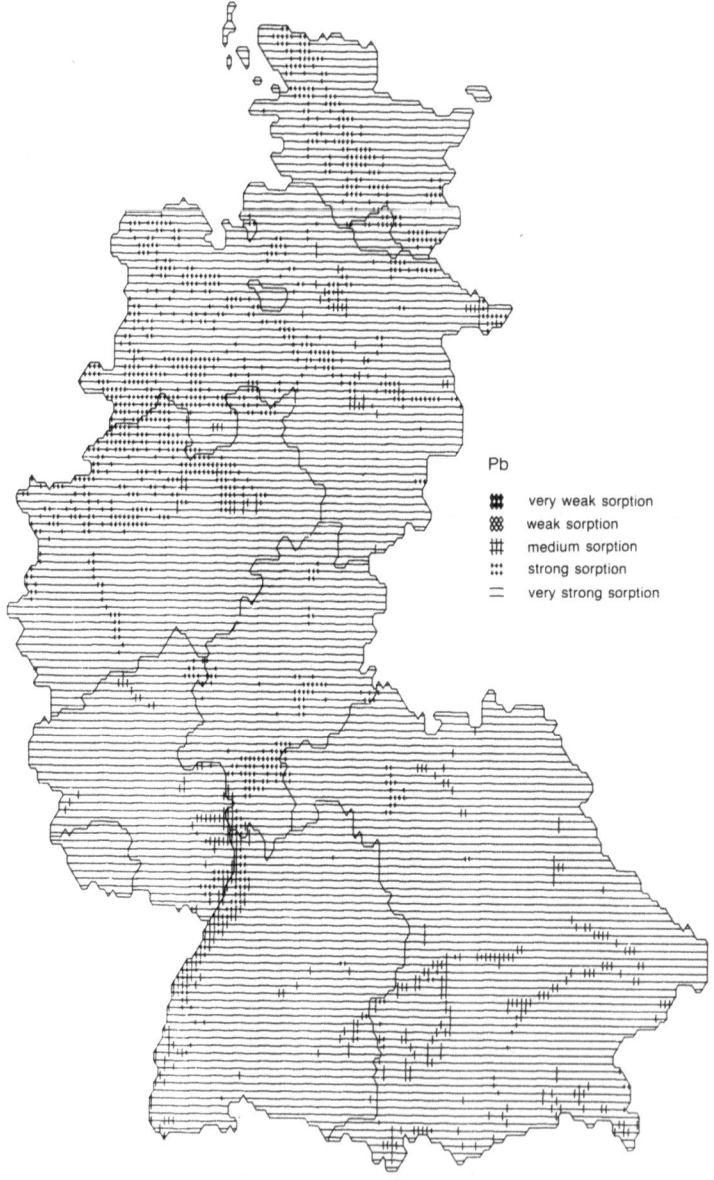

Pb

▦	very weak sorption
▨	weak sorption
♯	medium sorption
⠿	strong sorption
=	very strong sorption

Fig. 4.17. Spatial differentiation of Pb sorptivity of German soils

With regard to the risk assessment of groundwater bodies, the relative affinity of soils and sediments to different heavy metal species have to be matched to additional data on soil-water balance and microbial activity as described in the following sections. The importance of compound or element speciation is derived from the fact that different species will exhibit different mobility and reactivity in soil and sediment, different availability for plant uptake and toxicity for organisms (cf. Gerth 1985; Taylor 1989). Important reactions governing the chemical speciation of heavy metals in soil solution are hydrolysis (Chap. 2.1.3.1), formation of chloride complexes or complexation with organic compounds. Thus synthetic organics like EDTA, DTPA or NTA have frequently been used as model compounds for the analysis of the latter process which is also of high practical importance. It may, for example, serve to improve heavy metal availability in soil with respect to plant uptake and growth

Table 4.35. Correlation matrix of six selected heavy metals in stream sediments of Schleswig-Holstein

Pb	Cu	Zn	Cd	Ni	Cr
Pb	0.723[a]	0.675	0.512	0.338	0.148
Cu		0.534	0.527	0.296	0.124
Zn			0.729	0.395	0.123
Cd				0.406	0.115
Ni					0.117
Cr					
$\alpha = 0.001$					

[a] Underlining = maximum value.

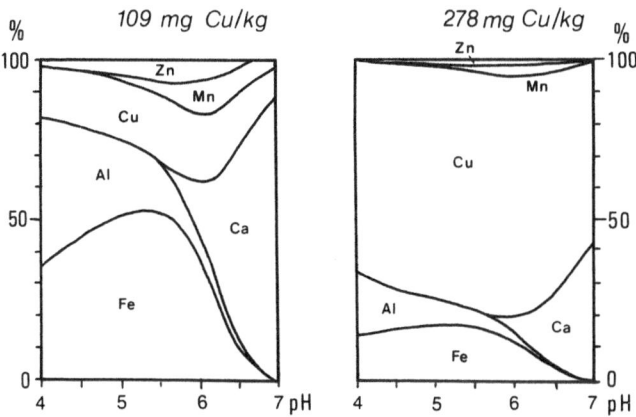

Fig. 4.18. Calculated distribution of EDTA over six metals as a function of pH, for heavy metal extraction of soil at two different Cu contents. (After De Haan 1987)

or to facilitate the removal of heavy metal from polluted soil (Bernhardt 1984). Figure 4.18 summarizes the results of heavy metal speciation calculations for NTA.

Organics in Soil

In the same way as modern agricultural production systems rely on fertilizer use with respect to plant nutrition (Fig. 1.3), they are dependent on the application of pesticides to plants and soil for crop protection and disease control purposes. In addition, many industrial processes release compounds into the atmosphere in particulate or gaseous forms which may eventually be deposited on soil.

Because of the large number of organic chemicals involved in these processes (e.g. roughly 300 are applied as pesticide components on a large scale in EC Member States) a limitation to example treatment is indicated. Therefore reference to specific chemicals like phenolic compounds, non-polar organics, pesticides and polynuclear aromatic hydrocarbons is given merely by way of paradigmatic illustration [cf. Sect. 4.4.3.4(3)].

Adsorption of Phenol and Substituted Phenols

Phenol is the basic structural unit for a wide variety of synthetic organics including many agricultural chemicals, and is therefore one of the most widely used organic compounds in existence, annual world production being approximately 3×10^6 t (Rippen et al. 1987a). Phenol and substituted phenols are degradation products of several pesticides, and may be produced during the hydrolytic cleavage of various organophosphorus compounds or may be the result of the degradation of chlorinated phenoxyacetic acid (Hill and Wright 1978; Müller and Lingens 1988). Adsorption of phenol by soil yielded essentially linear equilibrium adsorption isotherms, and Freundlich K values (ranging from 0.57 to 1.19) were largely dependent on soil type. Adsorption of substituted phenols on activated carbon and the adverse effect of other similar solutes have also been studied (Al-Bahrani and Martin 1976). Batch equilibrations under anaerobic conditions and using a 1:10 soil/solution ratio were set up by Boyd (1982) for phenol and a series of substituted phenols yielding simple isotherms for the individual substances and competitive adsorption isotherms for two- and three-component admixtures. The soil was the top of a typic Argiaquoll (Brookston clay loam) with 5.1% organic matter, pH = 5.7, and CEC = 22.2 mEq/100 g.

The equilibrium adsorption data could be described by the Freundlich equation. Table 4.36 summarizes the results which permit to evaluate the effect of substituent type and position on the adsorption of monosubstituted phenols. The Freundlich isotherms corroborate that introducing a $-CH_3$, $-OCH_3$, $-Cl$, or $-NO_2$ group into the ortho, meta or para position results (for the soil studied)

Table 4.36. Freundlich isotherm constants for adsorption of phenol and substituted phenols by soil. (After Boyd 1982)

Compound	K	$1/n$	r^2
Phenol	0.48	0.79	0.963
o-Cresol	0.59	0.66	0.986
m-Cresol	0.92	0.87	0.982
p-Cresol	1.31	0.68	0.992
o-Mehoxy-	1.07	0.56	0.994
m-Methoxy-	0.94	0.89	0.991
p-Methoxy-	1.50	0.50	0.988
o-Chloro-	1.37	0.80	0.998
m-Chloro-	1.78	0.83	0.994
p-Chloro-	1.88	0.70	0.982
2,4-Dichloro-	3.38	0.67	0.984
2,4,5-Trichloro-	9.75	0.71	0.998
o-Nitro-	3.05	0.89	0.998
m-Nitro-	1.42	0.73	0.991
p-Nitro-	1.48	0.72	0.998
Catechol	3.18	0.36	0.998
Resorcinol	0.28	0.40	0.941

in increased adsorption compared with phenol. The same applies to introducing an –OH group into the ortho position (catechol), whereas placing an –OH group in the meta position (resorcinol) resulted in decreased adsorption compared with phenol. Consequently, of the 14 monosubstituted phenols, only resorcinol gave a Freundlich constant K less than phenol. Thus Table 4.37 suggests, by way of comparison of Freundlich K values and water solubilities, that, with the exception of catechol, the amenability of mono-substituted phenols to adsorption co-variates with decreased water solubilities. Similarly for resorcinol, increased solubility results in decreased adsorption (cf. Sect. 4.4.1.2).

The mechanism of phenol adsorption on soil presumably involves H-bond formation between the phenolic OH-group and H-bonding sites on organic matter and clay surfaces. The addition of substituents to phenol can affect adsorption as related to H-bonding by their effects on (1) acidity or basicity of the phenolic hydroxyl, and (2) steric hindrance of H-bond formation. Adsorption of the meta- and para-substituted phenols which are not sterically hindered seems to be related to substituent effects on the basicity of the phenolic hydroxyl. These effects can be described in terms of Hammet σ constants which relate the electron withdrawing or releasing effect of a substituent (cf. Chap. 2.1.1). While σ values are generally not used for the ortho position, α comparison of Hammet constants and Freundlich K values for meta- and para-substituted phenols is given in Table 4.38.

With the exception of the nitrophenols, the smaller σ values for para substitution correspond to larger K values, i.e. greater adsorption. When comparing the meta and para isomers of cresol, methoxyphenol, and chlorophenol, greater substituent electron-donating ability results in greater adsorption; only for the nitrophenols are differences in both the values and the

Table 4.37. Measured and predicted K_{oc} values for phenol and substituted phenols. (After Boyd 1982)

Compound	S, g/100 g	K_{oc} [a]	K_{oc} [b]
Phenol	9.3	7.322	16.10
o-Cresol	2.5	18.031	21.88
m-Cresol	2.6	17.552	34.58
p-Cresol	2.3	19.092	48.66
o-Methoxy-	1.84	22.250	40.01
m-Methoxy-	—	—	35.10
p-Methoxy-	4.0	13.061	55.70
o-Chloro-	2.85	16.481	51.15
m-Chloro-	2.6	17.552	69.17
p-Chloro-	2.71	17.060	70.04
o-Nitro	0.21	98.61	120.02
m-Nitro	1.35	27.51	52.83
p-Nitro	1.69	23.58	55.25
2,4-Dichloro-	0.46	57.59	125.97
2,4,5-Trichloro-Catechol	< 0.2	101.97	363.26
(o-Dihydroxybenzene)	45.1	2.479	118.44
Resorcinol			
(m-Dihydroxybenzene)	123	1.245	10.36

[a] Predicted $\log K_{oc} = -0.686 \log S \ (\mu g \ ml^{-1}) + 4.273$.
[b] Measured $K_{oc} = (K/2.684) \times 100$; $2.684 = \%OC = \%OM/1.9$.

Table 4.38. Hammet σ constants and Freundlich K values of monosubstituted phenols. (After Boyd 1982)

Substituent	σ		K	
	meta	para	meta	para
–CH$_3$	− 0.069	− 0.17	0.93	1.31
–OCH$_3$	+ 0.115	− 0.268	0.94	1.50
–Cl	+ 0.373	+ 0.227	1.78	1.88
–NO$_2$	+ 0.710	+ 0.778	1.42	1.48
–OH	+ 0.121		0.28	

corresponding K values negligible. Greater substituent electron-donating ability increases the basicity of the phenolic –OH and thereby enhances its ability to H-bond by acting as a proton acceptor.

Ortho substitution of $-CH_3-OCH_3$, and –Cl results in less adsorption than para susbstitution (cf. Table 4.38), which is probably due to steric hindrance of H-bonding involving phenolic –OH. Such a hindrance was not observed for the –OH and $-NO_2$ groups. Ortho-substitution resulted in greater adsorption for both these groups, which is particularly marked for catechol, whose water solubility is greater than that of phenol. Presumably, chelating effects on mineral surfaces increasing the sorption-controlling specific surface are of importance for this phenomenon (Fränzle 1971b).

In the case of o-nitrophenol, the formation of multiple H-bonds is unlikely, because this compound forms a strong intramolecular H-bond between the –OH and $-NO_2$ groups (Morrison and Boyd 1973). Its enhanced adsorption may be attributed to decreased water solubility; it appears to occur via a hydrophobic mechanism rather than H-bonding. The latter mechanism, however, seems to be the reason that adsorption of the di- and trichlorophenols was greater than predicted for hydrophobic sorption.

Comparing adsorption from multicomponent systems with adsorption from single-component systems shows that adsorption is reduced by the presence of other solutes. Table 4.39 summarizes the results of competitive adsorption experiments. The data suggest that adsorption would be affected more adversely as the number of solutes increases. Cumulative K values in Boyd's experiments were: 1.62 (p-chlorophenol + p-nitrophenol), 1.68 (p-cresol + p-chlorophenol + p-nitrophenol) and 1.80 (o-, m- and p-nitrophenol). In the light of frequent demands for activated carbon to replace natural soils in sorption testing systems, it should be noted that the adsorption of monosubstituted phenols by soil may contrast considerably with their previously reported adsorption by activated carbon. The effect of introducing ortho

Table 4.39. Competitive adsorption of monosubstituted phenols from binary and ternary systems. (After Boyd 1982)

Admixtures	K	K_i[a]	$1/n$	r^2
p-Chloro-	1.22	1.37	0.59	0.996
p-Nitro-	1.33	1.48	0.71	0.995
p-Cresol-	1.42	1.31	0.71	0.996
p-Chloro-	1.06	1.37	0.67	0.984
p-Nitro-	1.01	1.48	0.66	0.984
o-Nitro-	2.05	3.05	0.82	0.999
m-Nitro-	0.95	1.42	0.65	0.993
p-Nitro-	1.07	1.48	0.63	0.995

[a] K_i refers to Freundlich K values for adsorption from single-component system (Table 4.36).

–CH$_3$, –OCH$_3$, –Cl, and –NO$_2$ groups to phenol on adsorption by activated carbon was

$$OCH_3 > CH_3 > Cl > NO_2,$$

when compared on a molar basis. In contrast, adsorption of the same o-substituted phenols by soil decreased in the following order

$$NO_2 > Cl > OCH_3 > CH_3.$$

Substituents position effects were also not observed when comparing the adsorption of o- and p-chlorophenol and o-, m- and p-cresol by activated carbon (Al-Bahrani and Martin 1976).

Sorption Behaviour of Halogenated Alkenes and Benzenes

Laboratory batch and column experiments were conducted by Schwarzenbach and Westall (1981) to elucidate the sorption behaviour of non-polar organic compounds in river water–groundwater infiltration systems. Table 4.40 summarizes the test substances used and their sorption behaviour in natural aquifer

Table 4.40. Sorption of non-polar organic compounds on natural aquifer material. (After Schwarzenbach and Westall 1981)

comp Z	C_0^a µg l^{-1}	log K_{ow}^Z	$\bar{K}_p^Z \pm s$ ($n = 6$)
Toluene	20	2.69	0.37 ± 0.12
1,4-Dimethylbenzene	20	3.15	0.50 ± 0.10
1,3,5-Trimethylbenzene	20	3.60	1.00 ± 0.16
1,2,3-Trimethylbenzene	20	3.60	0.95 ± 0.11
1,2,4,5-Tetramethylbenzene	20	4.05	1.96 ± 0.45
n-Butylbenzene	20	4.13	3.69 ± 0.98
Tetrachloroethylene	100	2.60	0.56 ± 0.09
Chlorobenzene	20	2.71	0.39 ± 0.12
1,4-Dichlorobenzene	20	3.38	1.10 ± 0.16
1,2,4-Trichlorobenzene	40	4.05	3.52 ± 0.39
1,2,3-Trichlorobenzene	40	4.05	3.97 ± 0.64
1,2,4,5-Tetrachlorobenzene	40	4.72	12.74 ± 2.52
1,2,3,4-Tetrachlorobenzene	40	4.72	10.48 ± 1.66

a = initial concentrations in most batch and column experiments; $K_p^Z = K_f/[K_b \rho \, (1 - \varepsilon)/\varepsilon]$, where K_f is the forward reaction rate constant (s^{-1}), K_b is the reverse reaction rate constant (s^{-1}), ρ is the density of the sorbent (g$_S$/cm$_S^3$), and ε is the total porosity (cm$_l^3$/cm$_T^3$). (The subscripts l, s and T refer to liquid phase, solid phase, and total porous medium, respectively.) $\bar{K}_p^Z$ = average value of K_p from six measurements with different initial concentrations and different amounts of sorbent: $(2C_0, 5\,g)$, $(2C_0, 2.5\,g)'$ $(C_0, 5\,g)$, $(C_0, 2.5\,g)$, $(C_0/2, 5\,g)$, $(C_0/2, 2.5\,g)$; sorbent is no. 6 in Table 4.41.

material. It ensues from these data that, for the low concentrations typical of the environment, sorption equilibria can be described by the equation $S = K_p C$, where S = concentration in the solid phase, K_p = partition coefficient, and C = concentration in the liquid phase. For a variety of sorbents, Table 4.41 shows that the partition coefficient K_p^Z of a particular compound Z can be estimated from its octanol/water partition coefficient K_{ow}^Z and from the organic carbon (OC) content f_{oc} (fraction organic carbon) of the sorbents if f_{oc} is greater than 0.001.

For the wide range of natural sorbents defined in Table 4.41 a highly significant linear correlation was found between the average K_{oc} and K_{ow} values:

$$\log K_{oc}^Z = 0.72 \log K_{ow}^Z + 0.49 \quad (r^2 = 0.95). \tag{4.71}$$

Table 4.41. Correlation of K_p^Z with K_{ow}^Z for different sorbents. (After Schwarzenbach and Westall 1981)

No.	Description	BET spec. surf. (m²/g)	f_{oc}	Correlation[d]: $\log K_p = a \log K_{ow} + b$		
				a	b	r^2 ($n = 10$)
1	River sediment[a]	1.2	0.0056	0.69	− 1.66	0.95
2	KS1[a]	4.4	0.0073	0.70	− 1.76	0.97
3	KB1H[a]	3.2	0.0008	0.57	− 1.96	0.97
4	KB2[a]	2.1	0.0006	0.55	− 1.73	0.96
5	KB1T[a]	2.6	0.0004	0.50	− 2.05	0.97
6	KB1H[b]	4.9[c]	0.0015	0.71	− 2.31	0.97
7	Sediment from a eutrophic lake	18	0.019	0.67	− 0.86	0.98
8	Sediment from a highly eutrophic lake	nd[f]	0.058	0.87	− 1.40	0.98
9	Kaolin	12	0.0006	0.43	− 1.37	0.96
10	γ-Al$_2$O$_3$	120	< 0.001	0.25	− 0.83	0.95
11	SiO$_2$	500	< 0.0001	0.12	+ 0.40	0.29
12	Activated sewage sludge	nd[f]	0.33	0.67	+ 0.40	0.99[e]

[a] Sample from field site in the lower Glatt Valley, Switzerland, prepared by dry sieving, $\Phi < 125$ μm.
[b] Sample from same field site, prepared by wet sieving, $63 < \Phi < 125$ μm degassed at 220 °C.
[c] Value probably too high.
[d] Correlation made for compounds with $2.6 < \log K_{ow} < 4.7$.
[e] $n = 6$.
[f] nd = not determined.

A comparison of this linear free-energy relationship with other correlations found in the literature and with other experimental results published for other natural organic sorbents and nonpolar solutes shows that both Eq. (4.69) and the corresponding equation derived by Karickhoff et al. (1979) are consistent with experimental data for a wide range of non-polar compounds. Deviations from either equation are noted for substances of high lipophilicity. Introducing the organic-carbon content of the sorbents into the above Eq. (4.71) allows to define the partition coefficient K_p for many nonpolar organics Z:

$$\log K_p^Z = 0.72 \log K_{ow}^Z + \log f_{oc}(s) + 0.49. \tag{4.72}$$

This equation is valid for sorbents containing more than 0.1% organic carbon ($f_{oc} > 0.001$); their K_p values can be predicted within a factor of 2.

In contrast to the above findings on substituted phenols, the sorption behaviour of halogenated alkenes and benzenes does not appear to be influenced when studied in a mixture at low concentrations. Further experiments are required to define more precisely the validity of these statements with regard to higher concentrations.

Sorption of Low-Polarity Organic Compounds in Low-Carbon Environments

In comparison with the numerous studies showing the dependency of organic contaminant sorption on the natural organic carbon content of soils and sediments (e.g., Briggs 1973; Karickhoff et al. 1979; Kenaga and Goring 1980; Schwarzenbach and Westall 1981), attempts to assess the relative importance of mineral contributions to sorption are less in number. Hassett et al. (1980) noted that the relative contribution of the clay mineral fractions becomes significant as the ratio of swelling minerals to organic carbon increases. Karickhoff (1984) found a ratio threshold value of > 60 (clay minerals/TOC) for mineral surfaces to make a discernible contribution to sorption of non-polar organics while the ratio decreased to 25–60 for compounds with polar functional groups.

From these and other studies (McCall et al. 1980; Means and Wijayaratne 1982; Mingelgrin and Gerstl 1983; Gschwend and Wu 1985), it appears that for very hydrophobic molecules, total organic carbon content of the sorbent controls sorption down to carbon levels that defy quantification. Conversely, sorption of less hydrophobic organics will be distinctly more affected by clay minerals and amorphous metal oxides in many unsaturated and most saturated subsurface environments where organic carbon is quite low. For instance, Stauffer and MacIntyre (1986) determined batch sorption isotherms using C-labelled compounds; the sorbents used were Al_2O_3, $Al(OH)_3$, α-FeOOH and aquifer material. A surface soil (Ap) was introduced for comparison with the sorption of the low-carbon materials; the sorbates used were 1-methylnaphtalene, naphtalene, tri-chlorothene and *o*-dichlorobenzene. The acidity and ionic strength were adjusted to different groundwater conditions. The order of sorption coefficients for all sorbents was 1-methylnaphtalene > naphtalene > *o*-dichlorobenzene > trichloroethene.

In general, sorption of each compound was strongest for the Ap horizon material and decreased in the order: Ap ≫ FeOOH > aquifer material > aluminium oxides. Under basic conditions, the sorption coefficients were significantly reduced in all cases. At high ionic strengths sorption increased slightly.

A part of the above sorption isotherm investigations run with low-carbon sorbents and sorbates of low to moderate hydrophobicity at low soil/solution ratios showed that extraction of the sorbate from the sorbent provides a much more precise approach to determining Freundlich K values than does solution phase difference (Banerjee et al. 1985; Fränzle et al. 1989).

Co-Adsorption Phenomena

The adsorption of nuclein bases from aqueous solutions exhibits unusual features, specially if smectites are the sorbents. Lailach et al. (1986a, b), Lailach and Brindley (1969) and Thompson and Brindley (1969) measured the adsorption of various pyrimidines, purines and nucleosides on bentonite and illite in the presence of gegenions [Li, Na, Mg, Ca, Co, Ni, Cu, Fe(III)] and at different pH values. Lailach and Brindley (1969) also found pronounced co-adsorption from mixtures of two bases on bentonite.

The H-bonding of one base to an organic cation already attached to an exchange site was documented by Doner and Mortland (1969). One nuclein molecule bridges between a hydrated interlayer cation and the protonated base on an exchange site; consequently, nuclein bases with several polar groups are particularly suited for such interactions. Compatibility between the equivalent area and the area of base-pairs allows optimal aggregation of protonated and unprotonated species, probably in cluster form. Clustering appears to be similar to the phenomenon of surface condensation (Mingelgrin and Tsetkov 1985), but adsorbate clusters are strongly controlled by the surface-charge density. Lower densities (equivalent area > base-pair area) favour the adsorption of non-associated bases, and the promoting effect of clustering is lost. At higher charge densities, however, the formation of base pairs becomes less probable because of denser cation packing.

Samii and Lagaly (1987) found that not only the extent of co-adsorption but also the amounts of bases adsorbed from one-component solutions were highly dependent on the layer-charge density and very sensitive to both type and concentration of salts added. Sodium salts increased the adenine adsorption on a Wyoming montmorillonite considerably beyond a threshold concentration of 0.5×10^{-3} mol/l. The efficacy of salts increased in the order: NaCl < $NaNO_3$ < NaI. Conversely, for a Texas montmorillonite, only NaI enhanced adsorption. This is likely to be due to different surface-charge densities of the two smectites, which results in notable differences of the gegenion concentrations involved.

The threshold concentrations for the adsorption of cytosine to the Wyoming montmorillonite were higher than for adenine, and plateaus were not reached. The reversed order of salt influence is interesting: $NaNO_3 < NaI < NaCl$. The promoting effects of the salts is apparently related to a "breaking" action of the anions I^-, Cl^-, and NO_3^- on the water structure (Luck 1984). They increase the proportion of free OH groups (i.e. OH groups not involved in H bonds), which promotes hydration of the bases; the stronger hydration of the bases then reduces their adsorption. A comparable behaviour was found for polyethylene glycols. Agents breaking the water structure promote hydration of the polyether chain, and the turbidity points are consequently shifted to higher temperatures. Thus the admixture of salts seems to exert two opposite effects: (1) it decreases adsorption if hydration effects are dominant, and (2) it enhances adsorption if base–base associations are made easier or intensified.

The importance of base–base associations is also reflected by co-adsorption phenomena (Samii and Lagaly 1987). Only small amounts of uracil and thymine were adsorbed on montmorillonite, but adsorption increased by an addition of adenine or cytosine. The behaviour was complex, however, so that the results for a particular smectite cannot be generalized. The reduced co-adsorption of cytosine–thymine is comparable to the base-paring in nucleic acids (complementary bases: adenine–thymine, non-complementary: cytosine–thymine). The association of cytosine molecules was very strong and probably even stronger than for cytosine–uracil and cytosine–thymine pairs. Thus the adsorption of thymine and uracil is less promoted by cytosine. Formation of pairs in which two organic bases share the same proton seems to be of a more general importance (Farmer and Mortland 1966; Mortland 1966); it explains also the relationship between maximum adsorption and pH value (cf., e.g. Yariv and Heller-Kallai 1975).

4.4.2 Bioavailability and Biotransformation

Usually no simple relationship exists between the total concentration of a compound in soil and its biological action which may reveal itself in the development of plants and reactions of soil organisms. This is due to the fact that only part of the total amount of the compound is bioavailable, i.e. apt to induce a reaction.

The concept of bio-availability has played an important part in the field of plant nutrition in relation to crop production, but is no less applicable in soil pollution studies. This points to the evident advantage of relying in soil protection considerations on experience derived from conventional soil fertility (de Haan 1987). Bache (1977) has developed a more theoretical approach based on the concepts of quantity and intensity, combined with compound mobility as a determining factor. Quantity (Q) is an extensive variable referring to the mass of the compound under consideration, while intensity (I) is an intensive variable, related to the (electro-)chemical potential of the compound. Expressed in joules per mole, it reflects the amount of free energy per unit amount of compound. Correspondingly, for compounds in solution, the activity or concentration provides an appropriate means to express the value of I. The relationship between Q and I, as indicated in the form of the $Q-I$ curve, yields important information via its slope, which is referred to as the buffering intensity for the compound under the boundary conditions of the system considered. Thus the buffering intensity may vary considerably with soil properties and system parameters, as well as with the position on the $Q-I$ curve.

Knowledge of the $Q-I$ relationship may greatly contribute to the evaluation of contaminant availability for organisms, even where in addition to quantity and intensity mobility may play a role. Mobility then refers to the transport rate of a contaminant through soil from the point of its prevalence to the point of biological action, e.g. plant roots (cf. Sect. 4.2.2). While the compound may exert a marked influence on organisms via uptake or other forms of exposure, there is also a reciprocal influence of the organisms affected on compound behaviour because many soil or water properties controlling compound behaviour are at least in part biologically controlled, e.g. redox potential and pH.

In fact, the chief and possibly the sole biological agents for the total conversion of organic compounds to inorganic products in soil and water are microorganisms. Incomplete transformation is often of environmental concern because the products of these partial reactions may be (1) more toxic than the original substance (Scheerer 1988), (2) toxic, while the parent compound was nontoxic at usual environmental concentrations, (3) more persistent than the original substance, or (4) subject to biomagnification (cf. Chap. 2.1.3.4 and 2.1.3.5).

In their action on natural or anthropogenic compounds, microorganisms may bring about mineralization, detoxication, cometabolism, activation, or defusing (Alexander 1980). Mineralization means the complete conversion of an organic compound to inorganic products. Detoxication is the transformation of a toxic substance into innocuous metabolites; it characteristically includes mineralization, but many other kinds of detoxication are known. Cometabolism is the microbial metabolism of a substance which the responsible organism cannot use as a nutrient; it does not result in mineralization. In activation, a primarily non-toxic compound is transformed into a toxic one, or the toxic potential is enhanced. Defusing, i.e. the conversion of a potentially hazardous compound into an innocuous metabolite before its potential for harm is expressed, is so far known only in laboratory cultures.

4.4.2.1 Effects of Chemicals on Microorganisms and Soil Fauna

Microorganisms participate in the formation of soil aggregates both directly and indirectly. Fungal or actinomycetal hyphae link dispersed mineral particles

Table 4.42. Population densities and diversity of edaphon in relation to land use. (After Ottow 1985)

	Tillage	Grassland
Total biomass (m^2/30 cm)		
Bacteria	20–80 g DW	50–120 g DW
Fungi	50–100 g DW	50–150 g DW
Individuals (per g dry soil)		
Bacteria	10^5–10^8	10^6–10^9
Actinomycetes	10^4–10^6	10^5–10^7
Fungi (Hyphae)	10^3–10^4	10^5–10^6
Algae	10^3–10^6	10^3–10^5
Protozoa	10^3–10^6	10^4–10^7

in their mucous layer; the formation of extracellular products such as polysaccharides or polyuronic acids yields "soft" organic cements while aromatic polymers of the organic matter produce "harder" cements. Thus the organic matter composition of soil aggregates is normally quite different from that of a non-aggregated soil, and this difference is normally further increased by the subsequent metabolism of microorganisms in the aggregates (Balloni and Favilli 1987).

Within the aggregates there are many microsites, each having a distinct niche of microbial colonization. These niches constitute "elementary microecosystems" (Nikitin 1973) which are populated by micro-aggregates of microbial cells in the form of rosettes, spirals and stars. These elementary microcolonies have a diameter of 5–10 µm and multiply through repeated aggregation until they form "population granules" whose dimensions vary from 300 to 500 µm. The further association of these granules with the mineral particles then leads to the formation of microplots whose diameter may attain 2.5 mm. The composition, density and activity, or conversely, the sensitiveness of the microflora largely depend upon the micromorphological and structural aspects of these microhabitats (Ottow 1985). Their diversity is thus reflected in both the abundance and complexity of microflora and fauna, as indicated in Table 4.42.

Inhibitory Effects of Heavy Metals

The impact of exhaust gases and the treatment of soils with sewage sludge may lead to heavy metal concentrations far beyond established tolerance levels (cf. Sect. 4.4.1.3) and induce a considerable reduction in microbial biomass and its functions (Brookes et al. 1984; Herms and Brümmer 1980; Doelman 1986; McGrath and Brookes 1986). Extending the scope of the relatively numerous short-term experiments on heavy metal effects on microbial populations and their adaptive potentials (e.g. Babich and Stotzky 1980; Tyler 1981; Wilke 1986), Doelman and Haanstra (1979, 1984) investigated the influence of Cd, Cr, Cu, Ni, Pb and Zn in long-term experiments. These heavy metals brought about a decreasing inhibition of microbial CO_2 production with length of experiments while the inhibition of urease activity increased with Cd, Cu and Zn concentration.

Wilke (1988) studied long-term effects of 12 inorganic pollutants on soil microbial activity in a sandy cambisol. As, Be, Br, Cd, Cr, F, Hg, Pb, Se, Sn and V were added in form of inorganic salts to cambisol field plots in 1975 and 1976. Samples were taken in 1984 and 1985 to determine biomass, ATP content, hydrogenase, alkaline phosphatase, saccharase and protease activities.

Table 4.43. Long-term effects of inorganic pollutants on biomass (BM) and ATP contents, dehydrogenase (DHA), alkaline phosphatase (APA), saccharase (SA) and protease activities (PA) of a sandy Cambisol in terms of percentage of control. (After Wilke 1988)

Elements	Addition (mg kg^{-1})	PE (mg l^{-1})	DHA	ATP	BM	APA	SA	PA
			Activities in % of control					
Control	0	—	100a	100a	100a	100a	100a	100a
As I	50	0.68	91a	91a	119a	119a	96a	92a
As II	300	1.92	95a	96a	105a	80b	94a	91a
Be I	30	0.009	80a	117a	68b	117a	95a	92a
Be II	80	0.032	74b	99a	46c	99a	72b	79b
Br I	60	0.024	95a	110a	n.d.	110a	78b	70b
Br II	240	0.050	88a	105a	71b	105a	89a	75b
Cd I	50	0.08	76b	89a	74b	78b	100a	82b
Cd II	200	0.23	59c	75b	53c	45c	89a	50c
Cr I	300	0.051	67b	91a	99a	91a	86b	81b
Cr II	800	0.096	63b	82b	95a	82b	80b	73b
F I	500	19	65b	68b	64b	77b	78b	108a
F II	2000	37	35c	37c	n.d.	63b	61c	91a
Hg I	50	0.009	98a	113a	75b	113a	69b	107b
Hg II	200	0.046	46b	70b	68b	70b	52c	83c
Ni I	100	0.23	73b	115a	n.d.	73b	95a	91a
Ni II	400	1.60	26b	76b	n.d.	52c	50b	51b
Pb I	1000	0.07	96a	89a	n.d.	83a	97a	129b
Pb II	4000	0.23	66b	62b	75b	66b	87a	92a
Se I	20	0.06	84a	111a	105a	91a	96a	126b
Se II	40	0.16	103a	116a	64b	105a	96a	120a
Sn I	117	< 0.01	88a	105a	114a	105a	80b	88a
Sn II	467	< 0.01	90a	95a	92a	95a	80b	91a
V I	100	4.3	78a	98a	97a	98a	100a	69b
V II	400	6.4	44b	93a	99a	93a	60b	57b

PE = Percolation extract. Differences are significant if percentages have indices other than a.

Table 4.43 shows that all pollutants except Se and Sn inhibited the microbial activity in the soil. Be, F, Ni and V reduced microbial biomass, protease and dehydrogenase activities, although their total concentrations were still below the maximum values recommended in the Federal Republic of Germany; Br proved to be even effective below the officially "tolerable" concentration of 10 mg/kg.

A comparison of Table 4.43 with the following correlation matrix (Table 4.44) shows that dehydrogenase activity and microbial biomass were the most sensitive parameters to define inhibitory effects of inorganic pollutants on soil microflora while there is only a medium, although highly significant correlation between these two properties.

Figure 4.19 summarizes the characteristic damage threshold values for the key biological processes of respiration, nitrification and nitrogen mineralization (Doelman 1986; Sauerbeck 1987) which have to be taken into account when heavy metal-containing formulations or sewage sludges are applied. These figures are comparable in magnitude to the values defined in CEC Directive 1986 (cf. Führ et al. 1983; Delschen and Werner 1989; Ruck 1989).

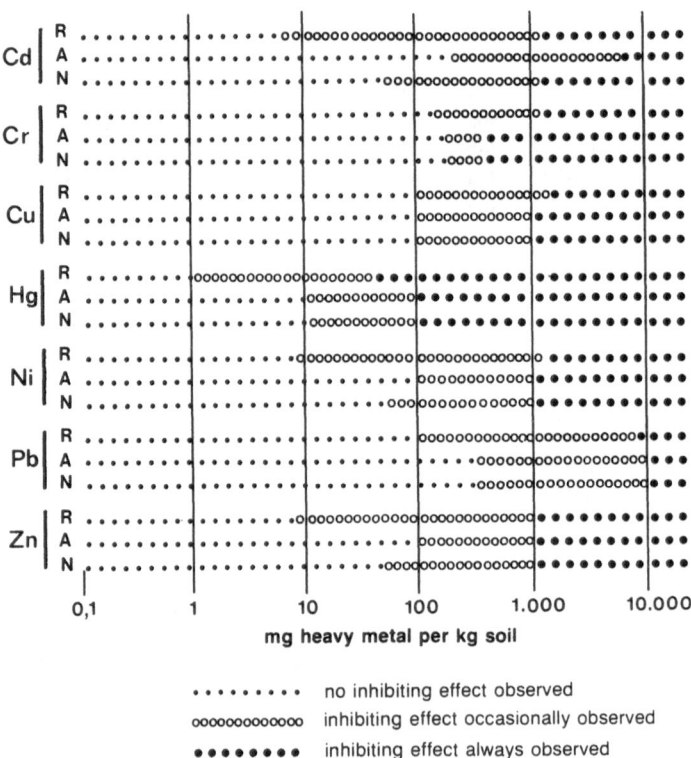

Fig. 4.19. Impact of heavy metals on microbial respiration (*R*), nitrogen mineralization (*A*) and nitrification (*N*) in soil. (After Doelman 1986)

Table 4.44. Correlation coefficients of microbial properties. (After Wilke 1988)

	DHA	ATP	BM	APA	SA	PA
DHA	—	0.73	0.49	0.74	0.68	0.65
ATP	—	—	0.56	0.71	0.48	0.45
BM	—	—	—	0.31*	0.33*	0.25*
APA	—	—	—	—	0.50	0.55
SA	—	—	—	—	—	0.56
PA	—	—	—	—	—	—

Significance: $p < 0.001$, $r > 0.58$; $p < 0.01$, $r > 0.47$; $p < 0.05$, $r > 0.37$.
* = not significant.

Effects of Pesticides

Numerous soil biologists have examined the possibilities of *suppressions of the total population* of bacteria, actinomycetes and fungi or of *specific groups* such as mycorrhizal fungi, rhizosphere inhabitants, algae, protozoa, *Nitrobacter*, *Nitrosomonas*, *Pseudomonas*, *Bacillus*, *Azotobacter*, and *Thiobacillus* (cf. Alexander 1969; Hance 1980; Simon-Sylvestre and Beaumont 1982; Costa et al. 1987). To draw a representative picture rather than attempting an exhaustive survey, only a few examples can be cited, therefore.

Herbicides used for weed control purposes in cropped land are normally applied to soil at low rates, the final concentration rarely being more than several $mg\,kg^{-1}$. This is equivalent to $1–5\,kg\,ha^{-1}$ of the active component, which amounts to about $22,000\,t/a^{-1}$ for the (pre-1989) Federal Republic of Germany (Ottow 1985). In view of the fact that these concentrations are far (i.e. at least ten times) lower than those required for inhibition, it would appear that these compounds would not materially affect the dominant microbial groups. Slight biological disturbances are occasionally observed with certain of the compounds applied at low concentrations, but the effect is typically small and frequently transitory (Wildförster 1985) (Table 4.45). Possible exceptions may be found among photosynthetic microorganisms; e.g., Fitzgerald (1962) has reported a marked sensitivity of algae to monuron.

It is pertinent to point out that pH may alter pesticide toxicity considerably. For instance, 2,4-D, 2,4,5-T and MCPA at far greater than recommended concentrations reduced the radial growth of fungi at acid pH but not at neutral reaction (cf. Sect. 4.2.2.2). IPC (Isopropyl *N*-phenylcarbonate) toxicity, by contrast, was not influenced by pH (Newman 1947; Zsoldos and Haunold 1979). pH influence may, in part at least, be indirect, i.e. related to the pH control of pesticide sorption, which is apt to drastically reduce pesticide availability (Ottow 1985; Fränzle et al. 1989).

Insecticides represent a class of compounds whose mode of action is entirely different from that of herbicides (IPS 1982) and they are often applied to soil at

Table 4.45. Pesticide concentrations failing to inhibit soil microorganisms.[a] (After Alexander 1969)

Bacteria	Actinomycetes	Fungi	*Rhizobium*
Aldrin (100)	Arsenite (500)	Aldrin (100)	2,4-D (25)
AMS (500)	Atrazine (70)	Arsenite (500)	4-(2,4-DB) (25)
Arsenite (500)	Dalapon (34)	Atrazine (75)	MH (250)
Atrazine (70)	Demeton (1500)	Chlordane (100)	2,4,5-T (25)
Chlordane (100)	Diazinon (40)	2,4-D (25)	TMTD (1000)
2,4-D (25)	DNOC (200)	Dalapon (34)	
Dalapon (34)	HCH (1000)	DDT (100)	
DDT (100)	Malathion (1500)	Demeton (1500)	
Demeton (1500)	Phorate (1500)	Diazinon (40)	
Diazinon (40)	Simazine (70)	Dieldrin (100)	
Dieldrin (100)	TCA (10)	DNOC (200)	
DNOC (200)		HCH (1000)	
HCH (1000)		Heptachlor (100)	
Heptachlor (100)		MCPA (25)	
MCPA (25)		Methoxychlor (100)	
MH (100)		Malathion (1500)	
Methoxychlor (100)		Simazine (70)	
Phorate (1500)		2,4,5-T (25)	
Simazine (70)		TCA (10)	
2,4,5-T (25)		Toxaphene (100)	
TCA (10)		Zineb (45)	
TMTD (50)			
Toxaphene (100)			
Zineb (45)			

[a] Values indicate ppm, unless otherwise indicated.

appreciably higher concentrations. Nevertheless, Table 4.45 indicates that relatively large quantities are required to induce great consequences. Aldrin, chlordane, DDT, dieldrin, HCH, heptachlor, malathion, methoxychlor and phorate, for instance, would appear to be relatively non-hazardous. Yet partial inhibition of one or another microbial group has been reported in certain soils with many insecticides although these changes are rarely dramatic (Eno 1958). Because of the prolonged persistence of many of the chlorinated hydrocarbons, however, alterations of the soil microflora could be maintained for long periods (cf. Sect. 4.4.1.3).

Fungicides behave entirely different from insecticides and herbicides with regard to their impact on soil microflora, since they are specifically screened to destroy plant-pathogenic microorganisms, but their action is rarely confined to one or a limited group of microbial strains or genera. Commonly fungicides kill large segments of the saprophytic population and eliminate the many reactions

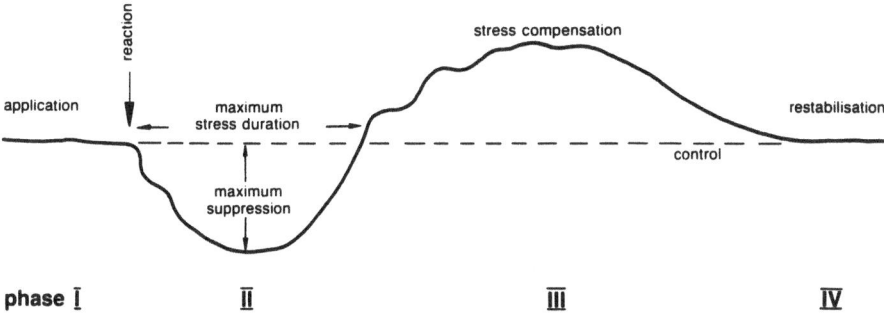

Fig. 4.20. Schematic representation of pesticide impact on soil microflora. (After Ottow 1985)

these microorganisms catalyze. The extent of the suppression of actinomycetes, fungi, protozoa and heterotrophic and autotrophic bacteria depends upon the physical and toxicological properties and the concentration of the pesticide applied, the soil qualities and many environmental variables (cf. Sect. 4.2).

Figure 4.20 shows that the microfloral response to pesticides, and in particular to fungicides and fumigants, can be divided into four phases: (I) initial phase comprising the period during and immediately following pesticide application, (II) stress phase in which the population begins to adjust itself to its altered environment, (III) the subsequent period in which a temporary new biological equilibrium is frequently established as a compensation for pesticide application, and (IV) the final phase which is essentially the climax community characteristic of the original soil. It should be noted that maximum duration of stress is of higher importance than maximum stress intensity for assessing the physiological relevance of a pesticide. Normally depressions of less than 60 days are considered as tolerable or negligible because they are readily compensated for and may also be due to natural stress situations (Domsch et al. 1983; Ottow 1985). Furthermore, the length of the initial or lag phase is of importance as a general observation on microbial response to pesticide (or other environmental chemicals) concentration. In many cases, the only response observed at higher concentration is an extension of the lag period, the degree of suppression and subsequent compensation remaining unchanged. In other cases, both the stress rate and the lag time vary. In terms of biodegradation of a pesticide this means that, in general, the rate of biodegradation and the shape of the degradation curve may be more significant than the percentage degradation at the end of an arbitrarily defined period (ECETOC 1986b).

In the case of fungicide applications, the population of fungi is drastically reduced as a result of the treatment, but also the numbers of bacteria and actinomycetes commonly fall. This stage (II) is followed by a period (III) which may extend for eight months or more in which a variety of microbial types multiplies in a way characteristic of r-strategists, and these strains and species may attain cell densities in excess of the unamended soil. The extent of this increase in microbial biomass depends upon soil properties (Filip 1975), concentration and toxicokinetic character of the fungicide employed (Hance 1980; Huber 1982; Morrill et al. 1982). Among the explanations given for the large bacterial numbers appearing after fumigation, the following are particularly important (Martin and Ervin 1952; Martin et al. 1957; Martin 1963; Nowak 1984; Pestemer 1985): (1) the microorganisms killed are used as food sources, (2) the fungicide remaining in the soil is an additional source of carbon for the recolonizers, (3) microorganisms are eliminated which are effective in nutrient competition or antagonism, (4) changes in the availability of organic matter to decomposition as a result of the treatment.

The dominant organisms may retain their superiority for appreciable times ranging from weeks to a year or more. *Aspergillus, Fusarium, Mucor, Penicillium, Pyrenochaeta* and *Trichoderma* are genera which contain strains active in the recolonization of fumigated soils. Their dominance may result from lesser sensitivity to agents of partial sterilization or higher growth or recolonization rates (Alexander 1969).

In comparison with the suppressive effects of pesticides *inhibition of microbial transformations* normally occurs at distinctly lower concentrations already. Table 4.46 illustrates, however, that most *herbicides* fail to retard soil respiration, organic matter turnover and nitrification if employed in the common weak control concentrations.

Among microbial transformations, nitrification seems to be one of the most sensitive conversions in soil, and the rate of ammonium oxidation is frequently diminished at pesticide concentrations which do not influence many of the other important biochemical reactions. The doses required to reduce the activity of autotrophic bacteria, in particular *Nitrosomonas*- and *Nitrobacter*-related genera dominating the oxidation, may therefore serve as indicators of the lowest

Table 4.46. Inhibition of microbial processes by pesticides.[a] (After Alexander 1969)

Respiration	Nitrification	Nitrification (Contd.)	Nodulation
DD (3500)	Allyl alc (70)	EDB (23 gal/acre)	Aldrin (50)
EDB (7600)	CDAA (12)	Ferbam (187)	HCH (12.5)
Mylone (150)	CDEC (12)	Heptachlor (50)	Heptachlor (100)
Nabam (100)	Chlorate (90)	Methyl bromide (435)	
	Chloropicrin (435)	Monuron (25)	
	CIPC (12)	Mylone (150)	
	CuSO$_4$ (50)	Nabam (100)	
	Cyanamide (200)	PCP (5)	
	DD (23 gal/acre)	Telone (133)	
	DNBP (10)	Vapam (75)	

[a] Values indicate ppm, unless otherwise indicated.

pesticide level needed to induce an unfavourable microbial response (cf. Stangen and Kerkhoff 1984; Jenkinson and Smith 1988). Nevertheless, in view of the appreciable quantities of *herbicides* required for a significant decline of nitrification, one may conclude that proper herbicidal practices will not affect substantially those microbial processes which the maintenance of soil fertility depends upon (Ottow 1985; Blume 1990).

Occasionally, the situation may be different for *insecticides*, since higher concentrations of these chemicals are often required for an effective control of subterranean insect pests, which may reduce nitrification and the nodulation of leguminous plants. By contrast, certain *fungicides* often markedly alter the biological activity, and again, nitrification is apparently among the more sensitive of the biological processes (Chandra and Bollen 1961). Thus, as a result of fumigant application, the loss of nitrogen through leaching or denitrification of nitrate may be diminished; i.e. the slowly leachable, non-denitrifiable ammonium ion is not converted to the rapidly leachable and denitrifiable nitrate ion, while ammonium levels in soil continue to rise, demonstrating that ammonification is not abolished. A resumption of nitrate production is noted after inhibitory periods ranging from 3 to 8 weeks (Alexander 1969), depending largely on the specific toxicity of the chemical and soil properties. Furthermore, toxicity varies with the individual microbial species, the morphological and physiological stage of the organism affected (endospore, conidia, sclerotia, hyphae, vegetative cells etc.) and the time of exposure of the organism to the toxicant. Frequently, fungal spores and sclerotia (Aypar 1990) are more resistant than the naked hyphae, active protozoan forms are more readily killed than cysts, bacterial endospores are less influenced than the vegetative cells whose resistance varies appreciably with the age of the organisms (cf. Chap. 2.2.2).

In comparison with heavy metals and pesticides, relatively little is known about the fate and biological effects of *surfactants* in soil (Cordon 1966; Hartmann 1966; Horowitz and Givelberg 1979; Figge and Schöberl 1989; Kimerle 1989; Litz et al. 1987). Dependent on the cationic, anionic or non-ionic character and concentration of the tenside, negative influences on the activity and vitality of soil microorganisms are described by Litz et al. (1987). In particular, nitrifying bacteria, which are essential for soil fertility, are affected (Bundesgesundheitsamt 1973). The adsorption of detergents causes a depolarization of the cell membranes, which impairs exchange processes. Linear alkylbenzenesulfonates (LAS), and even more their polar secondary products, are strongly adsorbed to soil organic matter (Figge and Schöberl 1989; Kimerle 1989; DVWK 1989). Together with rapid LAS degradation, this largely reduces the transport in percolating water. The increased amenability of the secondary products to adsorption, however, then impairs the further mineralization of these substances because of their appreciably reduced bioavailability. Due to adsorption of LAS to sediments sufficiently rich in organic carbon, only a limited fraction of the total LAS is available to sediment-dwelling organisms (Kimerle 1989). Similar results for marine benthic invertebrates are reported by Bressan et al. (1989).

Table 4.47. Components of soil fauna (per m^2 topsoil). (After Ottow 1985)

	Tillage	Grassland
Collembola	10^3–10^4	10^4–10^5
Mites (Oribatei)	10^2–10^4	10^4–10^6
Worms (Chaetopoda)	10^3–10^5	10^4–10^5
Insect larvae	10^2–10^3	10^3–10^4
Earth-worms	5–100	50–300

In comparison to soil microorganisms, the *faunal assemblages* (micro-, meso-
and macroorganisms) of soils are distinctly more sensitive to pesticides,
in particular to insecticides, some fungicides and herbicides (Edwards and
Thomson 1973). Comprehensive, though not exhaustive, studies have shown
that the taxa listed in Table 4.47 are liable to pesticide impact both directly and
indirectly, the resultant biological effects being positive or negative and of
different duration. In consideration of the ecological or foodweb position of
edaphic metazoan species it is easily intelligible that the reduction of competing
or predating species may cause a rapid increase of one or several other taxa
while an elimination of food resources (protozoa, mites, insect larvae) leads to an
indirect weakening of selected predators.

Changes of the subterranean animal populations are justfiably of considerable concern with the long-lived hydrocarbons (Atri 1985), not only
because of the immediate biological response to treatment but also because of
the possible duration of action and bioaccumulation effects (cf. Chap. 2.1.3.5).
For instance, the chlorinated hydrocarbons heptachlor, aldrin, dieldrin or
lindane proved highly toxic to wire worms at the normal concentrations
< 5 kg ha^{-1}, worms in the treated areas being killed rapidly, while worms
nearby tended to avoid the affected sites (Fredericksen and Lilly 1955). Similarly, DDT reduced the abundance of mites, and HCH has had an adverse effect
on the number of *Collembola* (Sheals 1955). As a consequence, their application
is either prohibited or strictly regulated. Earth worms are extremely sensitive to
some compounds like benomyl and carbaryl (Table 2.7) but are hardly influenced by a great many others even at concentrations of 1000 mg kg^{-1} soil
(Ottow 1983; Friesel et al. 1984).

Under modern agricultural practices, therefore, lack of organic matter and
regularly repeated disturbances of the habitats by ploughing, harrowing etc., are
definitely more important for the soil fauna than the application of non-
persistent pesticides, provided the intervals between sequential use are long
enough (cf. Blume 1990). The frequently repeated application of the same
pesticides, however, e.g. for spraying purposes in orchards or vineyards up to ten
times a year, may eliminate certain faunal components.

The overall effect of an accumulation of toxic wastes is to reduce the
structural diversity of ecosystems. This is equivalent to shortening foodweb
interrelationships and favours (1) populations of small hardy plants, (2) small-

bodied herbivores that reproduce rapidly, and the foodweb of decay. The loss of structure also implies a loss of regulation, i.e. the simplified communities are subject to rapid changes in the density of these smaller, more rapidly reproducing r-strategic organisms that have been released from their normal control.

In the light of the problems related to persistent, less specific pesticides, *current development* aims at providing selectively acting crop protection agents against harmful insects, plant diseases and weeds. They must be readily biodegradable (cf. OECD 1981b, 1984) and they must have only minimum influence on the environment. This requires extensive ecotoxicological tests as well as residue analyses in addition to tests for biological efficacy, which cover a period of 10 years.

One important step in this direction are products and processes that involve the use of pesticides only once an attack from a harmful organism has been observed or once the threshold level has already been exceeded. Another step is the modification of natural plant defences. These are partly special plant ingredients formed during the course of evolution, or phytoalexines formed by many plants in response to infection from fungi, bacteria or viruses. An illustrative example of the modification technique are the pyrethroids derived from phyrethrines, insecticidally acting substances from the petals of chrysanthemums. Natural pyrethrum is liable to rapid photodegradation (cf. Chap. 2.1.3.2) and cannot therefore be used in agricultural practice. Attempts have consequently been made to change the structure of the active ingredients of pyrethrum in order to make them sufficiently light-stable. The most effective product to date is Decis, which remains active even in low doses and is thus not harmful to the environment. Apart from its good plant tolerance, it has an excellent insecticidal action (Hoechst 1987).

4.4.2.2 Chemical Structure and Biodegradation

Biodegradation is the most important degradative mechanism for organic compounds in nature. It can occur under conditions in which oxygen is present (aerobic) or absent (anaerobic). In the former process, organic carbon is oxidized to CO_2, while in the latter it may be ultimately reduced to CH_4. In either case, the more important environmental variables affecting the rate and the extent of biodegradation are: temperature, pH, salinity, oxygen concentration, concentration of chemical, composition of microbial associations and concentration of viable microorganisms, quantity and quality of nutrients (other than xenobiotic substances), trace metals and vitamins (cf. foldout Model II: transformation system I).

In general, microbial degradation is impeded by ether bonds, triple substituted and condensed phenyl groups; in particular, if the latter have methoxyl-, sulfon-, nitro- or chlorine substituents in meta position. Among the nitrophenols and chloroanilines the ortho-isomers are relatively recalcitrant. Multiple substitution increases the recalcitrance of aromatic molecules considerably, since their aerobic degradation is effected by relatively specific mono- and dioxygenases (Hamaker 1972; Alexander 1973, 1980; Knackmuss 1979; Ottow 1982). Halogen substituents delay the electrophilic attack of these oxygenases considerably by reducing the enzyme affinity. Therefore multiple chlorine-substituted aromatics form relatively persistent compounds, the more so as the delayed co-metabolic attack yields "dead-end" metabolites, which can neither be used for cell-building purposes nor as an energy resource (Helling et al. 1974).

As a consequence reliable conclusions about biodegradation are not generally possible on the basis of chemical structure alone (Verschueren 1983), therefore many laboratory methods have been developed for studying the aerobic biodegradative potentials of chemicals, (cf. OECD Test Guidelines 301 A–E, 302 A, 303 A, 304 A, 1981). In comparison, tests for assessing anaerobic biodegradion have received less attention (ECETOC 1988b). Nevertheless, it must be considered to be an important process for substances which are not aerobically biodegradable or whose physico-chemical properties are such that their occurrence in an aerobic environment is restricted, for instance chemicals which are strongly adsorbed or insoluble. For these substances it is possible that anaerobic biodegradation is the major process responsible for breakdown in the environment.

The process is complex and is usually considered to occur in at least three concurrent main stages, the first of which involves hydrolysis of complex organic molecules like carbohydrates, proteins and lipids by the action of extracellular enzymes. In the subsequent stage (acidogenic step), the hydrolysis products are fermented yielding mainly short chain fatty acids, alcohols, hydrogen and carbon dioxide. Alcohols and acids are then converted to acetate hydrogen in the acetogenic step. The last degradation step is effected by methanogenic bacteria (methanogens) which utilize acetate and hydrogen to form methane. A schematic diagram illustrating the various steps and microorganisms involved collectively referred to by the element MI in foldout Model II, is represented in Fig. 4.21.

Anaerobic processes tend to be self-inhibitory, since the hydrolysis of complex substrates such as fats, proteins and carbohydrates yields volatile organic acids. By lowering the pH of the system, these may inhibit the growth of methanogens, unless the system has sufficient buffer capacity (ECETOC 1988b). Inhibition of gas production as a result of the toxicity of the chemical is also possible. Finally, there are also some indications that certain compounds, e.g. nitrilo-triacetic acid, are only degraded anaerobically when aerobic organisms acclimatized to the compound are incorporated in the system (Moore and Barth 1976; Bernhardt 1984).

When the organic compound liable to degradation or enzymatically induced metabolic transformations is utilized as a carbon source, the growth rate of the responsible microorganism is dependent upon the concentration of the former. The rate of substrate utilization then becomes (Moore and Ramamoorthy 1984):

$$- \, dc/dt = \mu x/Y = (\mu_m/Y) \cdot CX/(K_s + C) = k_b \cdot CX/(K_s + C) \,, \qquad (4.73)$$

where μ = specific growth rate, X = biomass per unit volume, μ_m = maximum specific growth rate, K_s = concentration of the substrate to support half-maximum specific growth rate $(0.5\mu_m)$, $k_b = (\mu_m/Y)$ = biodegradation constant, and Y = biomass produced from a unit amount of substrate consumed. These constants μ_m, K_s and Y are dependent on the above environmental variables.

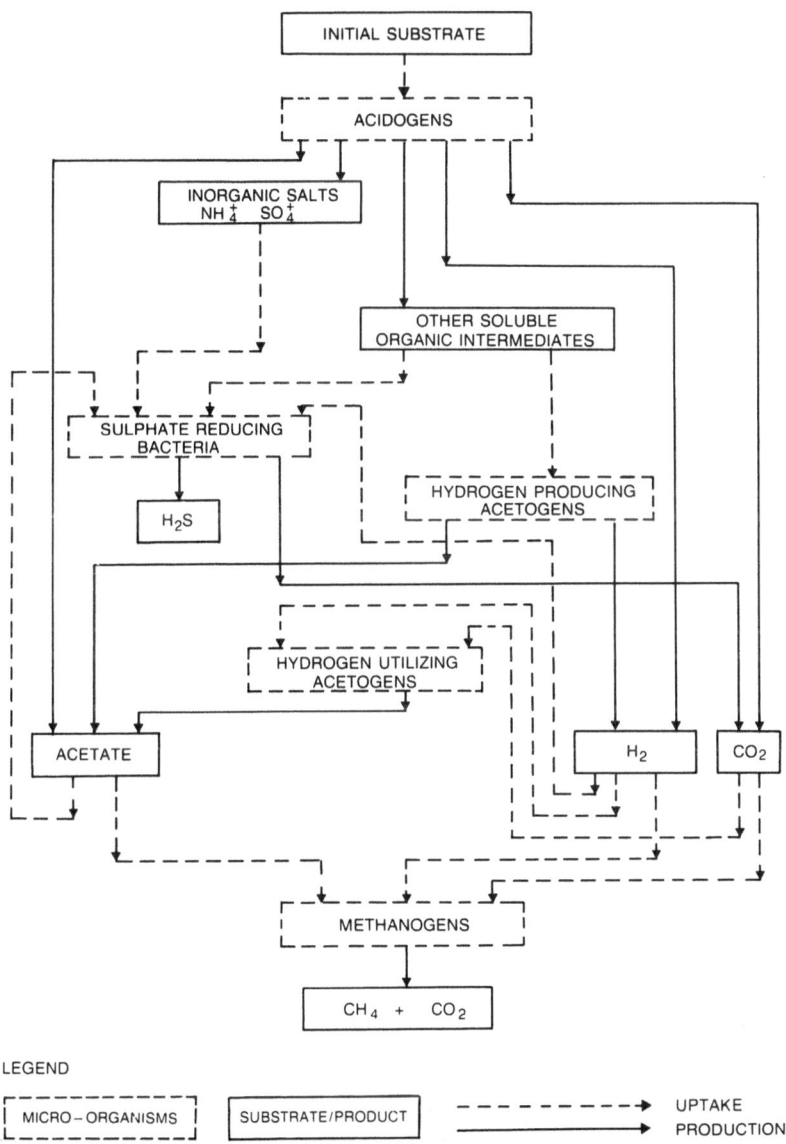

Fig. 4.21. Substrate dissimilation in anaerobic biotransformation processes. (After Donnelly 1984)

Aliphatic Hydrocarbons

Aliphatic hydrocarbons are characterized by an open-chain structure and a variable number of single, double and triple bonds (cf. Chap. 3.4.4). Alkanes with four or more carbon atoms can exist in both straight-chain and branched-chain isomers. The first two members of the alkene series, characterized by one double bond, exist in only one form, whereas the next higher homologue, C_4H_8, has two straight-chain isomers and one branched-chain isomer. Aliphatic hydrocarbons may undergo halogenation, which results in derivatives containing a variable number of chloride, bromide, fluoride, and iodide ions. Halogenation is both a natural and an industrial process, the former leading to toxic compounds in the environment like halomethanes, while the latter yields polyhalogenated derivatives that find wide application as solvents, degreasers, dry-cleaning agents, refrigerants and organic syntheses agents. In general, linear aliphatic hydrocarbons are most susceptible to biodegradation while alkyl branching and halogenation usually reduce the amenability to biodegradation (Tou et al. 1974; McConnell et al. 1975; Pearson and McConnell 1975; Thom and Agg 1975).

Aliphatic alcohols and fatty acids are most liable to biodegradation unless they have tertiary or quaternary carbon atoms in their molecules which are resistant to oxidation. In comparison to alcohols, esters and acids *aliphatic ethers* are normally less biodegradable; in particular if they have branched carbon atoms or halogen atoms. *Aliphatic ketones* normally are readily biodegradable, but branching and halogen substitution may render ketones more resistant to degradation or biotransformation (Kitano 1983). The biodegradability of esters is related to the stability of the ester bond; by way of hydrolysis (Chap. 2.1.3.1) the corresponding alcohols and acids are formed. More transformation-resistant esters have a tertiary carbon atom in the acid or alcohol moiety. *Aldehydes* and *carboxilic acids* are comparable to ethers and ketones, i.e. halogenation may make the derivative less susceptible to degradation.

The biotransformation of *acid amides* seems to depend on the stability of the amide bond and the substitution of amino hydrogen, while the degradability of *amines* is largely affected by the primary, secondary or tertiary nature of the aminogroup. *Imides* undergo hydrolysis, which yields amide-acids. Usually the subsequent step of hydrolysis transforming the amide-acid into amine and acid, is rate-determining (Bonse and Metzler 1978).

Mono- and Bicyclic Hydrocarbons

Monocyclic aromatics consist of a benzene ring with three double bonds. Substitution of the hydrogen atoms yields chlorobenzenes, nitrobenzenes, toluene, ethyl benzene and a great number of other derivatives. Although the carbon atom is bonded to three other atoms rather than four, the ring is not considered unsaturated and its stability results from the specific electron structure (Sykes 1976). The larger the number of alternative arrangements of the electrons, the greater the stability of the molecule, which is exemplified by the increase in stability of chlorobenzenes with increasing halogenation (Moore and Ramamoorthy 1984).

Monosubstituted benzenes (α-form) and *N*-alkyl *substituted anilines* are thus distinctly less biodegradable than the parent compounds. With regard to disubstituted benzenes the effect of "negative substituents" is stronger than that of "positive substituents". Among nitrobenzoic acid, aminobenzoic acid and dichlorobenzenes, biodegradability is greatly affected by the substituent position. For the former homologues the decomposition periods by a soil microflora are 4 d (*p*-nitrobenzoic acid), 8 d (*o*-nitrobenzoic acid), and > 64 d (*m*-nitrobenzoic acid). For *o*-aminobenzoic acid biodegradation in soil takes place in 2 d, while *p*- and *m*-aminobenzoic acids have decomposition periods of 8 d or > 64 d, respectively (Verschueren 1983). Methyl and amino groups of *mono or disubstituted anilines* and *disubstituted nitrobenzenes* are generally considered as positive substituents. All toluidines are biodegradable but phenylene diamines are degradation-resistant (Kitano 1983). Compounds with two substituents other than the amino group are degradation-resistant.

Chlorobenzenes and *hexasubstituted benzenes* are degradation resistant with very few exceptions, whereas the microbial metabolism of *chlorophenols* is controlled by several factors. An aromatic ring with the halogen atom in the m-position is resistant to breakdown. Fungi may produce the enzymes tyrosinase and peroxidases which can degrade PCP. Inactivation of the hydroxyl group is the primary detoxifying process.

The degradation-resistant *biphenyls* (PCBs) have either "negative substituents" or more than two "positive substituents" on the same benzene ring (DFG 1988). *Biphenyl ethers*, *biphenyl amines* and *biphenyl alkanes* are all degradation-resistant, and the same applies to their derivatives (Koch 1989).

In aerobic activated sludge in a semi-continuous operation mono-, di- and trichlorobiphenyls are significantly biodegraded and volatilized, whereas PCBs with >5 Cl atoms/molecule tend to sorb to suspended particulates and sediments and resist biodegradation (Kimbrough and Jensen 1989). A faster degradation rate is reported for commercial PCB mixtures than for single components; furthermore, addition of biphenyl to the substrate seems to enhance the biotransformation of certain PCBs, indicating possible cometabolic mechanisms. From these and similar findings (ECETOC 1988, 1991), it follows that the prediction of the biodegradability of a mixture from pure components may be quite imprecise since the physico-chemical properties may be quite different (ECETOC 1983, 1985, 1986).

Polychlorinated dibenzo-p-dioxines (PCDDs) are now suspected to be ubiquitous contaminants in both terrestrial and aquatic ecosystems. It is not possible, however, to estimate accurately the amounts released into the environment or to measure residues except in cases of extreme ambient contamination, and thus the overall threat posed by PCDDs has not yet been determined at the present time (Kearney et al. 1972; Mercier 1977; Rappe and Buser 1989). PCDDs tend to be rapidly and strongly sorbed by most soils, in particular the organic phase, which suggests relative immobility in most matrices. They are highly resistant to microbial degradation, and consequently the half-life of the most toxic isomer 2, 3, 7, 8 TCDD is likely to exceed 10 years (Mercier 1977; Koch 1989). Recently, Rast found a *Brevibacterium* strain in Rhine water which can be cultivated on a phenolic nutrient solution and proved capable of degrading simpler dioxins and dibenzofurans in soil (Naturwissenschaftliche Rundschau 1990).

The biodegradability of *aromatic ketones* is comparable to that of aliphatic ketones, i.e. largely affected by the structure adjacent to the carbonyl group (Kitano 1983; Koch 1989). *Naphtalene* and most of its derivatives are degradation-resistant (Kimbrough and Jensen 1989).

Polycyclic Hydrocarbons

Polynuclear or polycyclic aromatic hydrocarbons (PAHs) are fused compounds built on benzene rings. Fusion imparts chemical properties in between those of the highly aromatic benzene and those of alkenes. The environmentally significant PAHs range between naphtalene ($C_{10}H_8$), consisting of two fused rings, and the highly symmetrical coronene ($C_{24}H_{12}$), built of seven rings. In this range, there is a large number of PAHs differing in both the number and positions of aromatic rings with varying number, position, and chemistry of substituents on the basic ring system. Physical and chemical properties of PAHs vary in a comparatively regular trend with molecular mass. Susceptibility to redox reactions increases with increasing molecular mass, while aqueous solubility and vapour pressure decrease almost logarithmically with increasing molecular mass (Neff 1979; Verschueren 1983; cf.Chap. 2.1.2.4 and 2.1.2.5).

PAHs originate from natural or anthropogenic sources and are widely distributed in plant and animal tissues, surface water, soils, sediments and air (Johnston 1989). Several compounds are known or suspected carcinogens, while PAHs are not acutely toxic to most forms of life (Kraybill et al. 1978; Kraybill 1980; Chu et al. 1981). The addition of alkyl substituents generally enhances the carcinogenic potency, while hydrogenation and methylation bring about a decrease in potency (Moore and Ramamoorthy 1984). Although many PAHs proved teratogenic in experimental animals, there is no evidence of similar effects in humans (Heidelberger 1976).

Microbial transformation of PAHs in soil or sewage involves the introduction of two (ortho or para) hydroxyl groups through a *cis*-dihydrodiol intermediate into the aromatic nucleus (Gibson 1976, 1977). Lower molecular mass PAHs can be degraded completely to H_2O and CO_2 (Jerina et al. 1971), whereas higher molecular mass PAHs form various phenolic and acidic metabolites (Barnsley 1975; Gibson et al. 1975). In comparison with these condensed ring aromatics, the corresponding *cycloalkanes* are usually less susceptible to bio-degradation. The biodegradability of heterocyclic and bicyclocompounds is quite variable. Whereas part of the pyridines is readily biodegradable, benzo-thiazole and related compounds are degradation-resistant, and the same applies to biocyclo compounds (Kitano 1983; BUA 1991).

Petroleum Hydrocarbons

The highest levels of petroleum hydrocarbons in water have been reported following oil spills. Prior to the unique oil catastrophe in the Persian Gulf in 1991, probably the largest event was the Ixtoc I blowout in the Gulf of Mexico, which led to an average concentration of high molecular mass ($> C_{10}$) hydro-carbons of 10,600 $\mu g l^{-1}$ within several hundred metres of the emission point (Boehm and Fiest 1982). After the wreck of the Amoco Cadiz in 1978 (Nounou 1979) residual concentrations up to 240 $\mu g l^{-1}$ were found in sea water from Northwest Brittany, which can be compared to an average concentration of $\approx 3 \mu g l^{-1}$ as reported for coastal waters off Guernsey, i.e. some 200 km from the spill site (Law 1981; Fränzle and Zilling 1984; Kremer 1989). The natural background level of hydrocarbons in sediments normally ranges from ≈ 1 to $> 500 \ mg l^{-1}$, whereas in the aftermath of spills residues have exceeded 5000 $mg \ kg^{-1}$.

Virtually all crude oils and hydrocarbons may undergo microbial trans-formation under favourable conditions. *Bacillus, Candida, Arthrobacter, Pseu-domonas* and *Vibrio* species exhibited the widest range of hydrocarbon-utilizing potentials (cf. Sect. 4.4.2.3). Degradation studies of crude oil in Arctic tundra ponds showed that oil alone did not increase the number of heterotrophic or oil-degrading microflora over a period of 28 days, whereas weekly addition of oleophilic phosphate at 0.1 mM significantly stimulated microbial growth in the presence or absence of oil (Bergstein and Vestal 1978). Addition of inorganic phosphate failed to induce this effect, and the microorganisms mineralized the hydrocarbon part of the phosphate before the polyaromatic fraction.

Bacteria isolated from Atlantic Ocean sediments, mainly various species of *Pseudomonas* and *Actinobacter*, utilized an artificial mixture of *n*-alkanes, cyclohexanes and aromatic hydrocarbons. All of these, including naphtalene, phenomthrene, benz[a]anthrocene, perylene and pyrene showed some biotrans-formation (Walker et al. 1976a, b). The rivers of Athabasca oil sands areas can, under aerobic conditions and suitable nutritional regime, support an active hydrocarbon-oxidizing microbial community (Wyndham and Costerton 1981a). In situ colonization of bitumen surfaces composed of complex cyclic and branched aliphatic, aromatic hydrocarbons and heteroatomic high molecular mass compounds showed that bituminous hydrocarbons were readily colonized in both summer and winter by microbial populations native to Athabasca oil sands. In addition, isolates from the sediments of the Athabasca river and its tributaries can utilize all fractions of Athabasca bitumen except the asphaltene fraction, whereas saturated, aromatic, and first polar fractions of the bitumen were preferentially degraded (Wyndham and Costerton 1981b).

It ensues from these and other investigations (Kremer 1989) that mainly abiotic environmental conditions control the biodegradation rates of petroleum hydrocarbons. In particular, the degree of dispersion, availability of oxygen, concentration of nutritional, nitrogen and phosphorus compounds, and temperature are relevant. Therefore production of surface-active compounds by many microorganisms is considered to be concomitant with their growth on oil and hydrocarbons (Cooper and Zajic 1980). But while some chemical dispersants used in treating oil spills enhance bacterial transformation of crude oils (Traxler and Bhattacharya 1978; Marty et al. 1979), others interfere with biodegradation causing population shifts (Griffiths et al. 1981).

Since hydrocarbons are subjected to sorption, volatilisation, weathering and microbial metabolism as Fig. 4.22 illustrates, most low molecular compounds disappear from spilled oil, leaving mainly tar-like substances. This material is to a certain degree washed up on beaches and thus provides an opportunity to monitor the amount of tar in the oceans (Fränzle and Zilling 1984). Knap et al. (1980) concluded, for instance, that during the 1970s there was a 15% increase in the amount of tar balls on Bermuda beaches despite reported improvements in tanker design and operation. This possibly indicates that the amount of oil discharged to the Atlantic Ocean has increased in recent years. Zilling (1987) came to similar conclusions for the German Bight.

4.4.2.3 Active Microorganisms

Microorganisms are versatile in their catabolic activities, and therefore a wide array of compounds is subject to microbial transformation or cometabolism in water and soil. However, microorganisms are selective, and their enzymes are unable to catalyze reactions involving every one of the synthetic chemicals of economic importance. If microorganisms have no enzymes active in a needed metabolic sequence or if the substrate is protected from microbial attack because of special properties of the environment, the compound tends to persist (cf. Chap. 2.3.1).

Degradation of Hydrocarbons and Pesticides

A list of microbial genera reported to degrade *aliphatic hydroaromatic* and *aromatic hydrocarbons* is presented in Table 4.48. It shows that the active species include representatives of almost every important genus of bacteria, the chief actinomycete genera, and a host of fungi typical of soil environments. This variety of microbial types is to be expected, however, in view of the structural diversity of natural or man-made hydrocarbons. Yet it is remarkable that some species apt to degrade aliphatic and naphthenic hydrocarbons may also transform aromatic ones.

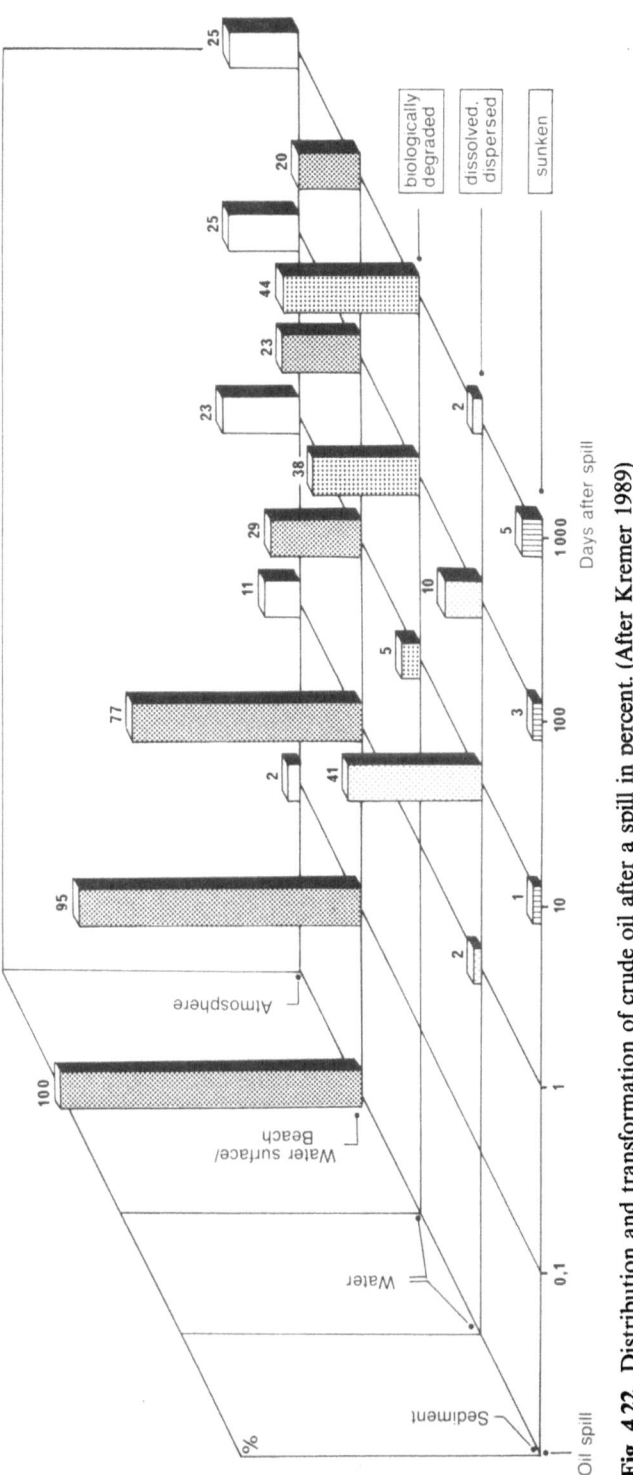

Fig. 4.22. Distribution and transformation of crude oil after a spill in percent. (After Kremer 1989)

Table 4.48. Hydrocarbon-degrading microorganism genera. (Fränzle 1991)

Achromobacter	Endomyces
Acinetobacter	Flavobacterium
Acremonium	Hyphomicrobium
Actinomyces	Methanomonas
Aeromonas	Mycrococcus
Alcaligenes	Micromonospora
Aspergillus	Monilia
Bacillus	Mycococcus
Bacterium	Penicillium
Candida	Pseudomonas
Caulobacter	Torula
Corynebacterium	Torulopsis
Debaryomyces	

Among the transformation processes of hydrocarbons, oxidation, hydroxylation, conjugation and hydrolysis are of particular importance, while for pesticides and other environmental chemicals also reduction and dehalogenation must be mentioned. Figure 4.23 illustrates the sequential transformation of naphthalene as an example.

The anaerobic microbial attack of hydrocarbons has been much less studied than the oxidative transformation with its related processes (cf. Fuhs 1961; Yanagita 1990). The following formula indicates the fundamental mechanism.

$$4C_nH_{2n+2} + (2n - 2)H_2O \rightarrow (3n + 1)CH_4 + (n - 1)CO_2 \tag{4.74}$$

Under standard conditions the corresponding changes in free energy are: $-44\,kJ/mol$ ethane, $-59.4\,kJ/mol$ propane, $-141.8\,kJ/mol$ hexane, and $-254.5\,kJ/mol$ undecane. Table 4.49 summarizes microbial taxa degrading pesticides.

Regarding this exemplary list, it is interesting to speculate on the physiological mechanisms by which a microorganism metabolizes synthetic compounds which it and its progenitors have probably never encountered. It appears possible that many of the detoxifying enzymes degrading xenobiotics have, as their usual substrates, molecules involved in common cell reactions. Yet, the structures of many pesticides resemble no known microbial metabolites, and the obvious scarcity of detoxifiers indicates unique metabolic systems for the inactivation reactions.

Often the mere removal of a single substituent such as a halogen atom will eliminate toxicity, but frequently inactivation requires a major structural change in the parent molecule. Moreover, a single compound may be transformed by different species in diverse ways. For instance, phenoxyalkanoic herbicides are inactivated by certain species by beta-oxidation of the side chain (Gutenmann et al. 1964) while a *Fluviobacterium* brings about detoxication by cleaving the ether linkage of the molecule (MacRae and Alexander 1963).

Enzymatic dehalogenation is responsible for the inactivation of several herbicides; but microbial dehalogenases liberate the bound halogen from only a few closely related organic compounds, and they do not appear to act on many

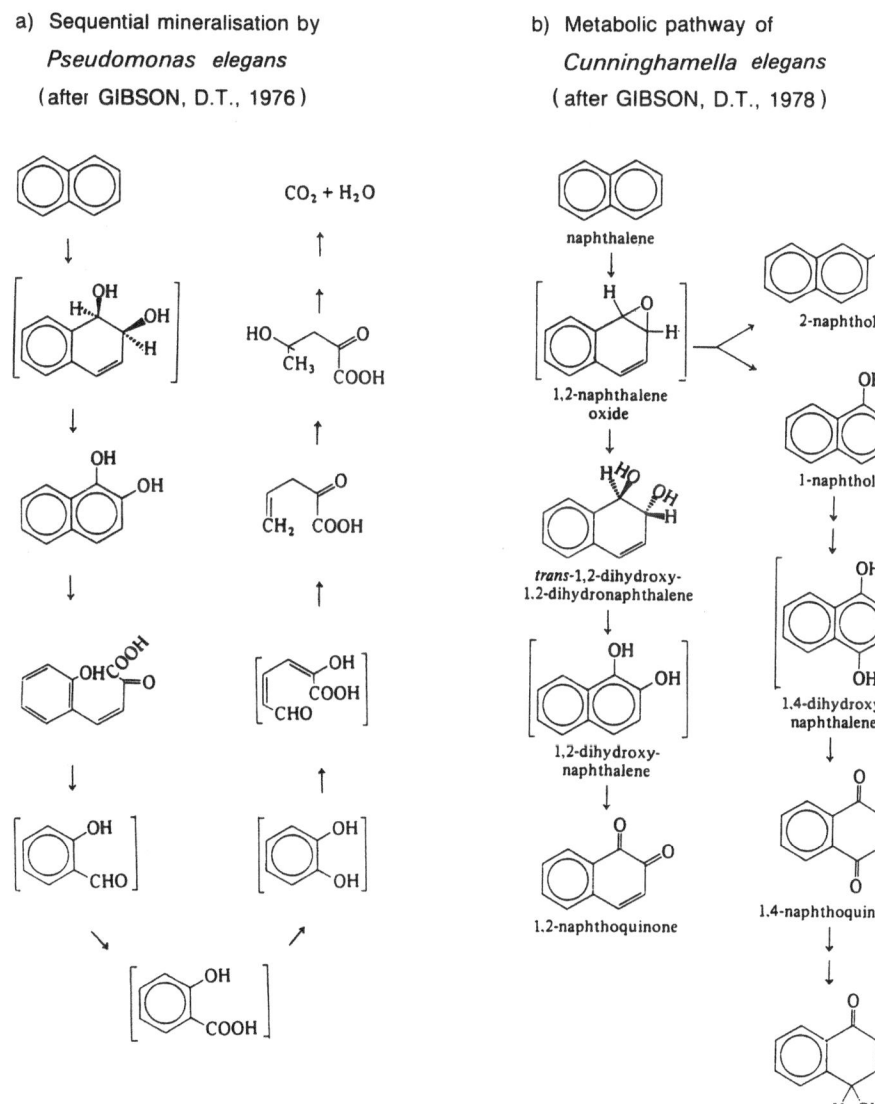

a) Sequential mineralisation by

   Pseudomonas elegans

   (after GIBSON, D.T., 1976)

b) Metabolic pathway of

   Cunninghamella elegans

   (after GIBSON, D.T., 1978)

Fig. 4.23. Models of organism-specific transformation of naphthalene. (After Gibson 1976, 1978)

Table 4.49. Microorganisms degrading pesticides. (After Alexander 1969)

Compound	Active microorganism
Allyl alcohol	*Pseudomonas fluorescens*
	Nocardia corallina
	Trichoderma viride
Alpha-amino-2,6-dichlorobenzaldoxime	*Pseudomonas putrefaciens*
CIPC	*Pseudomonas*
4-CPA	*Pseudomonas*
	Aspergillus niger
	Mycoplana
	Achromobacter
2,4-D	*Arthrobacter globiforme*
	Achromobacter, Flavobacterium
	Pseudomonas
	Flavobacterium, Corynebacterium
	Mycoplana
Dalapon	*Alternaria* sp.
	Nocardia, Pseudomonas
	Streptomyces aureofaciens
	Pseudomonas dehalogenans
	Arthrobacter, Agrobacterium
	Bacillus, Alcaligenes, Penicillium
	Corynebacterium spp.
4-(2,4-DB)	*Flavobacterium*
	Nocardia opaca
DD	*Arthrobacter, Bacillus*
Dichloroacetate	*Trichoderma, Penicillium, Fusarium*
	Agrobacterium
	Pseudomonas, Nocardia
DNBP	*Pseudomonas* spp.
DNOC	*Pseudomonas*
	Corynebacterium simplex
Endothal	*Arthrobacter globiformis*
Maleic hydrazide	*Acaligenes faecalis*
	Flavobacterium diffusum
MCPA	*Achromobacter*
	Pseudomonas
	Corynebacterium
	Mycoplana
MCPB	*Nocardia opaca*
Monochloroacetate	*Pseudomonas, Nocardia*
	Trichoderma, Penicillium, Fusarium
Monuron	*Pseudomonas* spp.
Panodrench	*Bacillus*

Table 4.49. (Contd.)

Compound	Active microorganism
PCP	Basidiomycetes
	Basidiomycetes
	Pseudomonas crucivae
Semesan	*Aspergillus, Penicillium*
TCA	*Micromonospora*
	Pseudomonas
	Agrobacterium, Arthrobacter, Pseudomonas
Thiourea	*Aspergillus, Penicillium, Trichoderma*
TMTD	*Glomerella cingulata*
Zineb	*Saccharomyces*

of the same substrates modified to contain different halogen substituents. Hence dehalogenases may be inoperative upon compounds with the same organic moiety but different halogen substituents or compounds with identical halogen substituents but different organic moieties (Alexander 1969).

Microbial Activation and Production of Stimulatory Compounds

In many cases the above description of the sequential microbial conversion of pesticides to a series of innocuous products adequately depicts their fate in terrestrial ecosystems. Nevertheless, sufficient instances of other biochemical reactions are now available to suggest that three other types of pesticide transformations may occur in soil: (1) the chemical added to the soil has no pesticidal activity, but is transformed microbially to yield substances which act upon pest or non-pest species; (2) the toxic primary substance is converted to derivatives which are also capable of affecting organisms in some toxic manner; and (3) the parent compound is metabolized with the formation of stimulatory products.

1. An example of microbial activation is the hydrolysis and subsequent oxidation of the inactive sesone to 2,4-D, a reaction effected by *Bacillus cereus* var. *mycoides* and probably also other microorganisms (Audus 1953). Similarly, 2,4-D is formed during the breakdown in soil of a variety of 2,4-dichlori-phenoxyalkanoic acids, apparently by β-oxidation of the fatty acid portion of the molecules (Gutenmann et al. 1964).

2. The widespread epoxidation of the chlorinated hydrocarbon insecticides aldrin and heptachlor to yield dieldrin and heptachlor epoxide results in the formation of products more toxic to insects. The epoxides resist degradation and remain active for many years (Alexander 1969). Basically comparable is the fate

of certain phenylurea derivatives, acylanilines and phenylcarbamates that are microbially transformed into mono- and polysubstituted chloroanilines which either condense in soil to form azocompounds or undergo covalent bonding in the organic matter forming bound residues (Hsu and Bartha 1974).

3. Transformation products of several pesticides exert a stimulatory influence on the growth of higher plants or other organisms. Thus, stimulation has been observed during the decomposition of dinitro herbicides (Crafts 1949), and 2,4-D transformation products appear to favour breakdown of organic matter and plant growth (Jensen and Petersen 1952; Newman et al. 1952).

The foregoing has only highlighted some of the major issues concerned with biotically effected changes in organic chemicals, and no attempt has been made to go into experimental details or to review the comprehensive literature (cf., for instance, Erikson et al. 1990; Yanagita 1990). With a view to the central role of biotransformation as depicted in foldout model II (MI), the purpose has been to cite the major problems and concepts that have come to the fore as environmental scientists, microbiologists and toxicologists view the vast array of chemicals that have been, or are likely to be introduced into terrestrial ecosystems.

4.4.3 Chemical Fate Modelling

The fate and behaviour of chemicals in soil is affected by both abiotic and biotic degradation or transformation processes inside the soil compartments on the one hand, and transport phenomena, both within the soil and to other environmental compartments such as air, water and subsoil on the other (cf. foldout model II). The transformation reactions determine the exposure pattern and duration while the transfer modes indicate the area and targets likely to be contaminated (cf. Sects. 4.4.1 and 4.4.2). Mathematical modelling of these processes is a field of current intensive work, and correspondingly the variety of models has increased very much during the last decade. But although the variety of models seems to be large, if not to say bewildering, only very few "really different" modelling concepts exist, and the number of elementary physical or chemical processes modelled remains rather limited.

In general soil/groundwater modelling concepts refer to point source pollution and can be categorized into (1) unsaturated soil zone or soil, (2) saturated soil and subsoil (groundwater), (3) geochemical and (4) ranking. The first two categories are designed to describe the behaviour of organic and inorganic dissolved chemicals, or immiscible fluids for non-aqueous phase compounds; they follow comparable patterns of physical–chemical approach and mathematics. The third enters into chemistry and speciation modelling, while the fourth follows a screening approach (Bonazountas 1987).

A comparison of the foldout models I and II indicates that soil modelling is a very complex issue and a major characteristic of a soil compartment – as compared to an air or water compartments – is that its temporal physical and chemical behaviour is much more governed by "out-compartmental" forces such as solar radiation, air temperature and precipitation. In the light of the resultant mathematical complexity, most of the soil models account for vertical flows, groundwater models for horizontal flows. They can be further classified into deterministic models which describe the system as a cause/effect relationship, and stochastic models which incorporate some concept of probability. Either type may be developed from observation, semi-empirical approaches and theoretical concepts. Deterministic formulations can be further classified into simulation models applying an empirical equation which is forced, by means of calibration coefficients, to adequately describe the system, and analytical models in which the derived equations describe the relevant physical and chemical reactions of a system.

Normally a deterministic or stochastic soil quality model consists of two modules:

- the moisture module aiming to predict soil water behaviour; and
- the solute module aiming to describe chemical transport, transformation and soil or subsoil quality.

If the flow module driving the solute module is absent from the model the user has to provide for relevant input information to the solute module from observed data at a representative site.

Four solution procedures are mainly applied to formulate mathematical systems (models): the analytical, the numerical, the statistical, and the iterative. Analytical techniques are usually employed for simplified and idealized situations; numerical approaches have become standard practice in soil quality modelling (cf. Richter 1986). Also statistical techniques have academic respect, and iterative solutions are developed for special cases.

4.4.3.1 Transport Modelling of Dissolved Chemicals

The prevailing modelling concepts of the behaviour of dissolved or miscible chemicals in the unsaturated and saturated soil and subsoil zones follow three different mathematical formulation patterns (Freeze and Cherry 1979; Bachmat et al. 1980; Duynisveld 1983; Faust 1984; Abriola and Pinder 1985; Reiche 1990). In conventional terms these may be designated as: (i) differential equation modelling, (ii) compartmental modelling, and (iii) stochastic modelling.

Unsaturated Soil Zone Modelling

Differential Equation Models for the unsaturated soil zone comprise a moisture and a solute module and have been presented by various researchers (e.g. Huff 1977; Mackay 1979; Enfield et al. 1980; Schwartz and Growe 1980). The moisture module is derived from the following equations: (1) the water mass balance equation, (2) the water momentum and (3) the Darcy equation. In some cases also other equations such as the surface tension or potential energy equation (Bonazountas 1987) are employed. Moisture movement in soil is defined by a one-dimensional, vertical, unsteady and isotropic formulation as

$$\delta[K(\psi)\cdot(\delta\psi/\delta z + 1)]/\delta z = C(\psi)\delta\psi/\delta t + S \qquad (4.75)$$

$$v_z = -K(z, \psi)\delta\Phi/\delta z , \qquad (4.76)$$

where z = elevation (cm), ψ = pressure head or soil moisture tension head in the unsaturated zone (cm), $K(\psi)$ = hydraulic conductivity (cm/min), $C(\psi) = d\theta/d\psi$ = slope of the moisture (θ) versus pressure head (ψ) (cm^{-1}), t = time (min), S = water source or sink term (min^{-1}), $\Phi = z + \psi$, and v_z = velocity of vertical moisture flow (cm/s). The output of the moisture module are the parameters θ and v_z as input to the solute module.

The latter describes the mass balance of a chemical species in a representative soil volume $dV = dx\, dy\, dz$. The one-dimensional formulation of the solute module, frequently called convective dispersive-differential mass transport equation, is

$$\partial(\theta c)/\partial t = [\partial(\theta K_0 c)/\partial z] - [\partial(vz)/\partial z] - [\rho \partial s/\partial t] \pm \Sigma P, \qquad (4.77)$$

where c = dissolved chemical concentration in soil moisture, K_0 = apparent diffusion coefficient in soil-air, v = Darcy velocity of soil moisture, ρ = soil density, s = adsorbed concentration of chemical on soil particles, z = depth, ΣP = sum of sources or sinks of the chemical within the soil volume.

Following Bonazountas (1987) some principal deficiencies must be mentioned when modelling solute transport on the basis of the above approach:

– Only diffusion, convection, adsorption and (possibly) degradation are described, whereas important processes like fixation or cation exchange (cf. Sects. 4.4.1.1 and 4.4.1.2) are either neglected or simply incorporated into the sources and sinks term of the equation because of mathematical complexity.
– The model is preferably applicable to the transport of organics, while transport of metals which may be subject to other processes cannot be directly modelled (cf. Sect. 4.4.1.2).
– Volatilization (cf. Chap. 2.1.2.4) is only implicitly predicted by means of boundary diffusion constraints, which frequently leads to an over-estimation or under-estimation of the theoretical volatilization rate.
– No empirical equation for a process can be incorporated since the model family has its own predictive mechanism.
– Chemical concentrations are estimated only in soil moisture and on soil particles, but concentrations in soil air are omitted (cf. Chiou and Shoup 1985).
– The discretized model version with numerical solutions has a pre-set temporal and spatial discretization grid which involves high operational cost, since input data have to be entered for each node of the grid.

In addition, modelling evaluations have shown (Bonazountas 1987) that the currently available coupled models of hydrologic flow and geochemical interactions are adversely influenced by: (1) insufficient consistency between the basic theoretical frameworks, laboratory experiments and field research, (2) limited knowledge about non-equilibrium conditions, (3) inadequacy of geochemical submodels to couple with hydrologic transport submodels, (4) lack of reliable input data, in particular for dispersion and chemical reaction rate coefficients, and (5) numerical difficulties with model solutions.

Compartmental Modelling

Compartmental models may avoid the deficiencies of the above modelling approach since they may handle geochemical issues in a more sophisticated way. The solute fate module frequently applies the law of pollutant mass conservation to representative user-specified soil elements. The principle of mass conservation is employed over a specific time step, either to the entire soil matrix or to its different phases such as solids, moisture and air. Labouring under the equilibrium assumption, the chemicals concentration in the other phases can be calculated once the concentration in one phase is known.

Taking the SESOIL model (Bonazountas and Wagner 1984) as an example the basic mathematical equations governing compartmental soil quality modelling can be summarized as follows.

The law of pollutant mass concentration for a representative element is formulated over a small time step as

$$M = M_{in} - M_{out} - M_{trans}. \tag{4.78}$$

By means of Henry's law (cf. Chap. 2.1.2.4) the solute concentration of a compound can be related to its soil air concentration

$$c_{sa} = c \cdot H/R \cdot (T + 273), \tag{4.79}$$

where c_{sa} = pollutant concentration in soil air, c = dissolved pollutant concentration, H = Henry's law constant, R = universal gas constant, and T = temperature in °C. Adsorption isotherms or coefficients (cf. Sect. 4.4.1.2) are used to determine the pollutant concentration of the soil from the sum of the concentrations of the pollutant adsorbed, cation exchanged or otherwise associated with the soil particles. The total concentration of a chemical in the soil matrix is then calculated from the concentration of pollutants in each phase and the related volume of each phase by

$$c_t = (n - \theta)c_{sa} + \theta c + \rho_b s, \tag{4.80}$$

where c_t = total concentration of chemical in soil matrix, n = soil porosity, θ = soil moisture content, c_{sa} = chemical concentration in soil-air, c = chemical concentration in soil moisture, ρ_b = soil bulk density, and s = chemical concentration on soil particles.

The above formulations define input terms to the fundamental Eq. (4.78), which is then sequentially applied for each time step and each sub-compartment of the user-specified matrix. Thus the term M_{in} may relate to chemical input from rain (upper layer), from soil moisture, from an upper layer or from a lower one. The term M_{out} designates chemical exports from the individual sub-compartment, whereas the term M_{trans} describes all transformations and chemical reactions taking place in the sub-compartments considered (cf. Sects. 4.4.1 and 4.4.2). All terms can be normalized to the soil moisture concentration by means of interconnecting equations which specifically describe processes such as volatilization, ion exchange, fixation, photolysis, hydrolysis, biological activity, plant uptake, biotransformation, etc. (cf. Chaps. 2.1.2 and 2.1.3). The solution of the resulting system of equations can be a complicated issue and may require development of new numerical solution techniques or algorithms (e.g. WASMOD/STOMOD modules in: Reiche 1990; Fränzle et al. 1989).

Stochastic, Probabilistic and Other Modelling Concepts

Stochastic or probabilistic techniques (P.H. Müller 1975) can be employed to the solution of Eq. (4.78) and to the definition of the moisture module (cf. Schwartz and Growe 1980) or can lead to new modelling concepts (Jurry 1982). Stochastic approaches are mainly aimed at simulating "break through" times of overall concentration threshold levels, rather than individual processes or concentrations in individual soil compartments. Since major processes are described by means of black-box or response function approaches and not individually, the coefficients or response functions have to be calibrated to field data. The Schultz (1982) approach may finally be referred to as an illustrative example of other modelling concepts relating to soil models for solid waste sites and specialized pollutant leachate issues.

Saturated Soil and Subsoil Modelling

Saturated soil or subsoil modelling normally involves differential equations defining the flow and the solute modules. Following Bachmat et al. (1980) the two modules can be written as

$$\bar{v}(\rho k/\mu)(\bar{v}p - \rho g \bar{v}z) - q = \partial(\Phi\rho)/\partial t \qquad (4.81)$$

$$\bar{v}[\rho C(k/\mu)(\bar{v}p - \rho g \bar{v}z)] + \bar{v}(\rho E)\bar{v}C = \partial(\rho\Phi C)/\partial t, \qquad (4.82)$$

where $\bar{v}$ = mean pore water velocity, C = concentration, mass fraction, E = dispersion coefficient, g = gravity acceleration, k = permeability, p = pressure, q = mass rate of production or injection of liquid per unit volume, t = time, z = elevation above a reference plane, Φ = porosity, ρ = density and μ = viscosity.

The proliferation of literature models related to the above basic formulation is mainly due to different model dimensionalities (1, 2, 3), model features (e.g. with adsorption, ion exchange, single ionic or multi-ion transport), sources and sinks described, the variability of the physical–chemical boundary conditions imposed, and the solution procedures employed (e.g. analytic, finite difference, finite element, stochastic). Interested readers are referred to the comprehensive reviews by Custodio et al. (1987), Kinzelbach (1987) and van Genuchten and Alves (1982).

Model diversity as related to ever-increasing conceptual features may be illustrated by an exemplary account of modelling ion transport in saturated soil. An early model of solute movement in salt-affected soils (Tanji et al. 1972) considers cation exchange and precipitation–dissolution, but describes the processes in a rather simple way. Melamed et al. (1977) developed a salt flow model representing the precipitation–dissolution sequence of salts by a first-order reaction, but neglecting cation exchange processes. In addition, the authors dealt with salts in terms of total salinity rather than with concentrations of individual ionic species. A certain improvement is due to Glas et al. (1979), who considered individual species in the process of precipitation–dissolution during solute flow through soil but analyzed gypsum transport only. An extension of the model of Tanji et al. (1972) was given by Robbins et al. (1980a, b), who considered more ionic species in solution and extended the application range of the model but based it on principles of chemical equilibrium to define the solute interactions with the soil matrix. The same holds for the hydrosalinity models developed by the US Bureau of Reclamation (cf. Shaffer and Gupta 1981).

Persaud and Wierenga (1982) presented an entirely different approach by considering the soil solution as a Donnan system (Sect. 4.4.1.2), and thus developed a new type of expression for the retardation factor in the convective–dispersive solute transport model. Their conclusion that this new ion exchange model was an improvement as compared with the previously employed Freundlich or Langmuir equations, was to a certain extent corroborated by Robin and Elrick (1985). A further modification was introduced by Utermann and Richter (1988), who suggested the use of variable Grapon coefficients for modelling cation exchange. Neither Persaud and Wierenga nor Utermann and Richter, however, took the process of precipitation–dissolution into account.

Combining the main features of the above models of Glas et al. (1979) and Persaud and Wierenga (1982), Chen et al. (1990a) developed a coupled convective–dispersive model to describe the solute movement of four ionic species (Ca^{2+}, Na^+, SO_4^{2-}, Cl^-) through soil. The model considers precipitation and dissolution of $CaSO_4$ as well as cation exchange or adsorption during steady-state water flow in saturated soil columns. The governing partial differential equations were solved numerically with an implicit Crank–Nicolson scheme. Hypothetical miscible displacement experiments were carried out with the model and compared with the results of transport experiments in soil columns. The experimental data were used to evaluate the model, and it was found that the observed breakthrough curves were described satisfactorily when the soil matrix/solute interaction was defined in terms of an exchange process instead of an adsorption process (Chen et al. 1990b).

Selected Soil and Groundwater Models

A selection of available models of solute transport in the unsaturated and saturated soil and subsoil zones is listed in Table 4.50. They are well-documented, operational and representative of the various structures, features and capabilities developed during the last two decades.

– At − 123D (Yeh and Ward 1981) is a series of pollutant transport sub-models in one, two or three dimensions for chemical and radioactive waste pollutants and for different types of releases. The model can provide up to 450 sub-model combinations for different boundary conditions.
– Chen et al. (1990a, b) employed a coupled convective–dispersive model to represent multi-ion transport in saturated soil as described above.
– Duynisveld (1983) has developed a flow and solute model which permits to plot time–depth curves of water movement and to determine the residence time of water in the unsaturated soil, taking into account the uptake of water by roots. The transport of solute is simulated by means of finite-difference models, considering the influence of adsorption processes and nutrient uptake by roots.
– EXSOL (Schernewski et al. 1990) is a compartmental model describing chemical transport in soil assuming rapid equilibration processes between the liquid, gaseous and solid phases which are defined in terms of sorption coefficients. The elimination of the chemical is modelled specifically as plant uptake, volatilization, degradation and leaching.
– MMT/VVT (Foote 1982), originally developed for radionuclide transport, is a one or two-dimensional solute transport model (MMT) which is driven off-line by a flow transport such as VVT (i.e. variable thickness transport). It employs the numerical random-walk technique and accounts for advection, sorption and decay.
– PATHS (Nelson and Schur 1980) is an analytical groundwater model which provides only a rough estimate of the temporal and spatial status of a pollutant fate.
– PESTAN (Enfield et al. 1980) is based on the analytic solution of Eq. (4.77). Its applicability is relatively limited, unless model coefficients (e.g. adsorption rate) can be estimated from monitoring studies. The steady-state moisture behaviour components are user input and can be obtained by any model of the literature.
– SCRAM (Adams and Kurisu 1976) is a dynamic finite-difference soil model with a DE flow module and a DE solute module which can handle moisture

Table 4.50. Selected soil and groundwater models and features

Models	Model type — Unsaturated zone	Saturated zone	Aquatic equilibrium	Model formulation — Ranking	Flow module	Solute module	DE approach	Compartmental	Mathematics — Statistical, other	Analytical	Numerical	Chemistry issues — Organics	Inorganics	Metals	Gaseous phase	Increased chemistry	Source/Information
ADL/LeGrand	X	X		X													LeGrand (1980)
AT-123D	X	X				X	X			X		X	X	X	X		Yeh and Ward (1981)
Chen		X				X					X		X	X			Chen et al. (1990a, b)
Duynisveld	X				X	X		X			X	X	X	X			Duynisveld (1983)
EXSOL	X	X			X	X					X	X	X	X	X		Schernewski et al. (1990)
FEMWASTE	X	X			X	X		X			X	X					Yeh and Ward (1981)
GEOCHEM			X			X	X	X			X		X	X		X	Sposito and Mattigod (1980)
MITRE/JRB	X	X		X													JRB Associates (1980)
MMT/VVT	X	X			X	X	X				X	X					Foote (1982)
PATHS		X				X	X		X		X	X	X	X			Nelson and Schur (1980)
PESTAN	X					X	X			X		X					Enfield et al. (1980)
PLUME		X				X	X			X		X		X			Bonazountas (1987)
R. WALK		X				X	X				X	X					Prickett et al. (1981)
SCRAM	X				X	X					X	X					Adams and Kurisu (1976)
SESOIL	X				X	X		X	X	X	X	X	X	X	X	X	Bonazountas and Wagner (1984)
USGS Models	X	X			X	X	X	X			X	X	X	X	X		Appel and Bredehoeft (1978)
WASMOD/ STOMOD	X	X			X	X	X	X	X		X	X	X	X		X	Fränzle et al. (1989), Reiche (1990)

DE = differential equation

behaviour, surface runoff, organic pollutant advection, dispersion, adsorption; but no computer code has been developed to simulate volatilization and degradation. Owing to the large number of input data required, this model does not seem to have found wide application.

– SESOIL (Bonazountas and Wagner 1984) is a flexible compartmental model with a hydrological cycle and a pollutant cycle module which can be easily adjusted to the temporal and spatial resolution of the study objectives. The model estimates the water budget components from available data, and simulates the pollutant cycle by accounting for a considerable number of chemical processes for both inorganic (metal) and organic pollutants.

– WASMOD/STOMOD (Fränzle et al. 1989; Reiche 1990) is a dynamic soil compartmental model, composed of a soil moisture/groundwater module and a solute transport module. Calibrated and validated on the basis of batch, lysimeter and field experiments the variable proportions of adsorption and capillary soil water of the organic matter, clay and silt phases are computed. Sorption kinetics, water-borne vertical translocation, irreversible fixation and microbial degradation of organic and inorganic chemicals are simulated for each successive horizon or layer of the unsaturated zone including input rates into groundwater.

The remaining models of Table 4.50 follow the basic patterns described above, and with the exception of Chen et al. (1990), they all handle one species at a time (cf. Sect. 4.4.3.3), while two soil models (SESOIL, AT-123D) can also handle pollutants.

Ranking Modelling

Ranking models have the purpose of assessing environmental impacts of waste disposal sites. Their primary focus was on groundwater contamination; later models included health considerations. These models rate or rank contaminant migration at different sites in the light of hydrogeologic, soil, waste type, density and site design features. Since ranking approaches are based on questions and answers, and on weighting factors that the user has to specify, they are liable to subjectivity in use. Yet they have received wide dissemination, because they are easy to employ and do not require use of computers. Well-known examples of ranking models are LeGrand (1980) and MITRE/-JRB (1980).

4.4.3.2 Immiscible Contaminant Modelling

The simultaneous movement of two immiscible liquids in a porous medium is traditionally called two-phase flow or three-phase flow, respectively, if air is explicitly considered as the third phase. Two-phase flow has been of particular interest in the field of petroleum geology since the 1930s (cf. Wyckhoff et al.

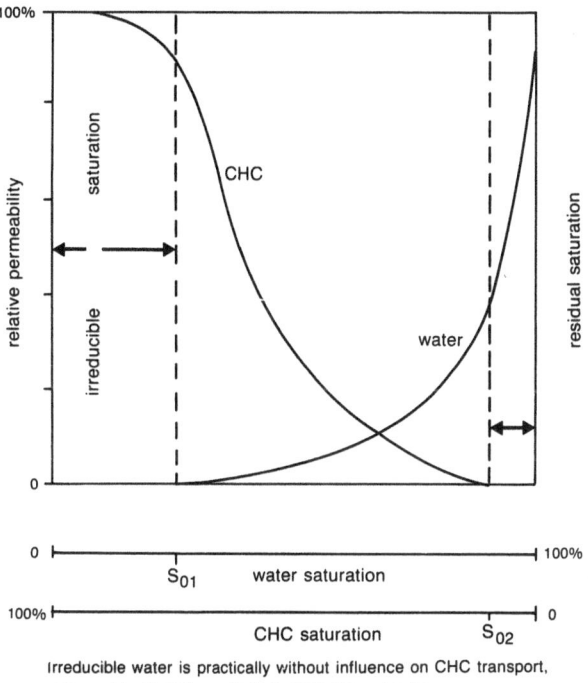

Irreducible water is practically without influence on CHC transport,
while residual CHC substantially impedes water flow.

Fig. 4.24. Relationship of pore water content and relative permeability of porous medium. (After Schwille 1982)

1932), but in the more recent past the class of halogenated hydrocarbons has assumed a comparable importance (Schwille 1981, 1982; Leo 1983).

Figure 4.24 illustrates the typical relationship between the water saturation of pores and the relative permeability. A simultaneous flow of either fluid is limited to the field between irreducible saturation and residual saturation, which has two important implications. On the one hand, a certain quantity of an immiscible liquid is limited in migration through a porous medium by the effect of residual saturation. On the other, the immiscible fluid in the state of residual saturation cannot be replaced by water as would be the case with a miscible one; a transport in water can take place only in relation to (the normally reduced) solubility in this medium.

The following soil (or sediment) and fluid parameters are necessary to model flow rate and pattern of immiscible fluids in porous media:

– physical and chemical properties of the fluids such as molecular mass, density, viscosity, solubility, polarity, surface wetting, and contaminants in fluid (e.g. metal species); and
– chemical and physical properties of soil or sediment such as soil structure and texture, horizontation, organic matter and clay mineralogy, field capacity, effective porosity and pore size distribution (cf. Sect. 4.4.1).

Basically two concepts exist in modelling flow rates of immiscible fluids in soil:

1. by employing the three-phase equations in porous media jointly with a set of relationships; and
2. by applying the pertinent two-phase flow equations relating to water/ immiscible contaminant or air/immiscible fluid or air/water sub-systems, and by appropriately adjusting the coefficients (e.g. relative permeability of fluid) of the governing equations to describe the specific problem.

Abriola and Pinder (1985) follow the first concept to describe the simultaneous transport of a chemical contaminant in three physical forms: as non-aqueous phase, as a soluble component of an aqueous phase and as mobile fraction of the gas phase. The contaminant may be composed of at most two components, one of which may be volatile and slightly water-soluble while the other is non-volatile and insoluble in water. The model equations are derived from basic conservation of mass principles and the incorporation of various constitutive relations and approximations. Effects of matrix and fluid compressibilities, gravity, phase composition, interphase mass exchange, capillarity, dispersion and diffusion are taken into account. The resulting mathematical formulation consists of a system of three non-linear partial differential equations subject to two equilibrium constraints. The solution is obtained by means of a Newton–Raphson iteration scheme.

The model presented by Faust (1984) is a realization of the second concept. It thus eliminates the need for an equation governing the third phase but requires knowledge about the "relative" permeability of the "second" phase (e.g. water/ immiscible liquid) to the "first" phase (e.g. air) instead. Example applications of the model are given to demonstrate model function in relation to solving governing equations and model sensitivity to fluid properties. No field model application or validation is presented, however, since data such as relative permeabilities and capillary pressures are not available for the type of immiscible fluid and the sites considered.

In general, two-phase relative permeabilities are determined experimentally, then these results are often fitted to polynomial functions of saturation. Two-phase permeabilities can further be estimated from analytic functions relating to water or dissolved chemical characteristics, surface tension, porosity, intrinsic permeability and a number of other parameters (Mercer 1984).

4.4.3.3 Aquatic Equilibrium Modelling

Fast reactions of inorganic dissolved species in aqueous media can be classified under the following general categories (cf. Chap. 2.1.3):

– reactions with solvent molecules, in particular dissociation,
– substitution reactions with solvent or dissolved species, and
– redox reactions with dissolved gases or other ionic species.

These reactions can lead to the formation of new complex ions with different ligands in the coordination sphere or ions with different oxidation state of the metal centres or still other ions. They are essential, since the medium the particular inorganic substance is dissolved in, will largely determine the speciation of the particular metal ion. In addition, upon mixing of solutions of these metals with the environment chemical modification of the species will occur initially by thermodynamically favoured rapid reactions and subsequently by slower reactions (Dahmke 1988; Dahmke et al. 1988; Peterson 1988; Boening 1989). Among the latter, the following reactions are particularly important in aqueous media (Bonazountas 1987):

– ligand substitution reactions of kinetically relatively inert ions,
– electron transfer reactions involving inner sphere mechanisms for relatively inert ions and some outer sphere reactions,
– reactions with dissolved gaseous species or bacterially induced reduction,
– precipitation of solids by formation of insoluble species through substitution reactions,
– formation of metal hydroxides and sulphides by oxidation reactions.

These slower reactions impact the speciation and concentration in the aqueous phase and determine both type and extent of further interactions in a manner comparable to the fast reactions. Therefore the time frame for some of the above reactions may indicate that characterizing the speciation in the original aqueous phase will also define the chemical behaviour of the compound in the environment, while in other cases understanding speciation implies an exact knowledge of environmental interactions. To this end, a larger number of speciation models have been developed to address a series of specific situations (Lyman et al. 1982). They are based on chemical thermodynamic reactions and principles, and do also account for additional processes, such as redox reactions (Lindberg and Runnels 1984), adsorption (Langmuir 1979), complexation (Kirkner et al. 1984; Daniele et al. 1985; Katlein 1986), ligand exchange (Stumm et al. 1980) and others (James and Rubin 1979).

Modelling Concepts

Excellent state-of-the-art reviews on chemical equilibria models of inorganic pollutants in soils are presented by Cederberg et al. (1985), Jenne (1979), Kincaid et al. (1984), Miller and Benson (1983), Nordstrom et al. (1979), Sposito (1985), Theis et al. (1982). Readers interested in details should refer to these publications.

Following Kincaid et al. (1984), four families of speciation models can be distinguished which may be further grouped into the following categories:

1. "speciation" codes defining speciation equilibria of inorganics for a terrestrial water compartment, and

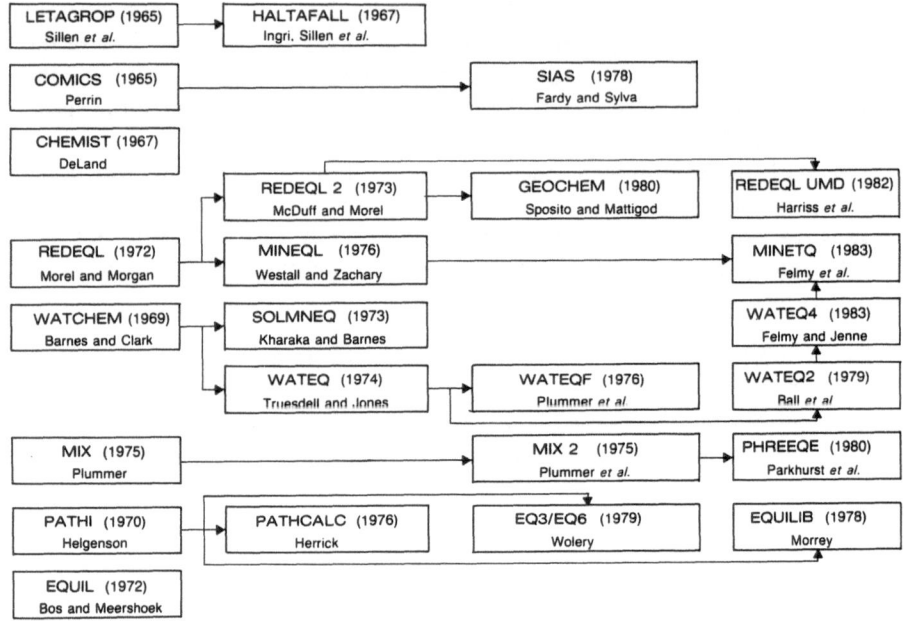

Fig. 4.25. Evolution of geochemical models. (After Kincaid et al. 1984)

2. "coupled speciation" codes which can simulate both speciation equilibria and transport of individual dissolved species or multi-component solutions (Jennings et al. 1982) in the terrestrial environment, both in space and time.

Figure 4.25 summarizes the evolution of geochemical models from 1965 on.

Coupled speciation models such as CHEMTRN (Miller and Benson 1983), FIESTA (Kirkner et al. 1984) or TRANQL (Cederberg et al. 1985) are a major subject of current research. CHEMTRN is a one-dimensional transport/speciation model for solutes in saturated porous media. It includes dispersion, diffusion, advection, ion exchange, formation of complexes, speciation in the aqueous phase, and the dissociation of water. The mass action, transport and site constraint equations are introduced in differential/algebraic form and are solved simultaneously. TRANQL is a groundwater mass transport and chemical equilibrium model for multicomponent systems. The equilibrium interaction chemistry involves complexation, ion exchange, competitive adsorption and dissociation of water; it is posed independently of the mass transport equations, which leads to a set of algebraic equations for the chemistry coupled to a set of differential equations for the mass transport. According to model developers, results proved the model to be versatile with potential for extension to a wide range of equilibrium reactions.

Judging from these examples, coupled speciation models appear to be powerful geochemical tools. Yet they have the inherent disadvantages of being very large and requiring extensive input data; thus they not infrequently prove inefficient for practical solutions to problems (Bonazountas 1987). Therefore speciation equilibrium models have been more extensively applied and validated, with laboratory and field activity coefficient as input data. They reflect the significant advances made in aquatic chemistry (Stumm and Morgan 1981; Sposito 1985; Mattheß 1990). In this respect many similarities exist between models conceived for soil or groundwater and surface waters.

Two basic approaches can be distinguished (Sposito 1985):

– in the first case the composition of an electrolyte solution is described with the total number of molalities of the stoichiometric components of neutral solutes, while the thermodynamic properties of the solution are formulated with "mean ionic" activity coefficients for neutral solutes;
– in the second case, the composition of an electrolyte solution is expressed with the molalities of molecular "species" presumed to exist in solution, whereas the thermodynamic properties of the solution are described with "single species" activity coefficients for the presumed molecular constituents.

Geochemical data bases associated with these two model categories and the underlying equilibria processes such as adsorption/desorption, precipitation/dissolution, reduction/oxidation etc., are the most important features of a code. Since reasonably good data bases are not available for all of the models listed in the following Table 4.51, decision-makers have to first review a number of codes that may meet project requirements and eventually select the model which suits the study objectives best.

Selected Speciation Models

A selective summary of computer codes suitable for geochemical modelling is provided by Table 4.51. Two of the best known geochemical and speciation models are GEOCHEM and MINEQL. GEOCHEM is based on the code REDEQL 2 for calculating equilibrium speciation of elements in a soil solution. After identifying the component species as uncomplexed metal cations, the free proton, uncomplexed ligands and the free electron, single-species activity coefficients are calculated. The model contains critical thermodynamic data for soils, a procedure for calculating cation exchange, and a correcting method for non-zero ionic strength up to 3 moles.

MINEQL is similar in overall structure to GEOCHEM, since it also originates from REDQL. MINTEQ, in turn, is formed from MINEQL and the data base of WATEQ; it includes ion speciation, redox equilibria, calculation of activity coefficients, adsorption, solubility and mass trnasfer. A principal difference between MINEQL or MINTEQ and GEOCHEM is that MINEQL can accept the concentration of any free ionic species, soluble complex or dissolved gas as input data to be held fixed during a calculation, while GEOCHEM can do this only for the activities of H^+, e^-, $CO_2(g)$ and $N_2(g)$.

Table 4.51. Summary of geochemical code capabilities, adaptability and availability

Models	EQ3/EQ6	GEOCHEM	MINEQL	MIX2	PHREEQE	REDEQL UMD
Source/Information	Wolery (1979)	Sposito and Mattigod (1980)	Westall et al. (1976)	Plummer et al. (1975)	Parkhurst et al. (1980)	Ingle et al. (1978)
Language	FORTRAN	FORTRAN	FORTRAN	FORTRAN	FORTRAN	FORTRAN
System	CDC 6600/7600 VAX	IBM 370, 4314 VAX	IBM 370	IBM 370	Amdahl 470 V/7 DEC VAX	IBM 370 CDC
Number of elements	18	44	38	7	19	42
Number of dissol. species	140	2000	Variable	18	123	150
Gases	81	2	Variable	1	3	2
Org. species	0	889	Variable	0	0	34
Redox reactions	4	20	Variable	0	Variable	24
Minerals	130	185	Variable	8	21	154
Range of temperature (°C)	0–300	25	25	—	0–100	25
Range of pressure (100 kPa)	1 or 500	1	1	—	1	1

Table 4.51. (Contd.)

Models	SOLMNEQ	WATEQF	WATEQ2	WATEQ3	WATEQ4F	WATSPEC
Source/information	Kharaka and Barnes (1973)	Plummer et al. (1976)	Ball et al. (1980)	Ball et al. (1981)	Ball et al. (1987)	Wigley (1977)
Language	PL/1	FORTRAN IV	PL/1	PL/1	FORTRAN 77	FORTRAN
System	IBM 370	IBM 370 UNIVAC 1110	IBM 370 Honeywell	IBM 370 Honeywell	IBM PC	ICL 190 3T IBM 370
Number of elements	24	19	29	30		16
Number of dissol. species	181	105	220	227	220	69
Gases	3	3	3	3	3	2
Org. species	10	0	12	12	12	0
Redox reactions	12	8	12	12	12	6
Minerals	158	101	309	309	309	40
Range of temperature (°C)	0–350	0–100	0–100	0–100	0–100	0–100
Range of pressure (100 kPa)	1–1000	1	1	1	1	1

Essential to any soil modelling effort is model output validation, which can be defined for the purpose of this section as the process analysing the validity of the predicted chemical concentrations or mass in the soil column or in groundwater, as compared to measured concentrations from monitoring data. A disagreement in absolute levels of concentration (i.e. predicted versus measured) does not necessarily indicate that either the modelling approach or field sampling are incorrect or that either data set needs revision. It simply points to the important fact that field sampling and modelling approaches rely on two different perspectives of the same situation.

4.4.3.4 Regionalized Fate Modelling

After the focus of environmental interest and concern evolved from water and air pollution in the 1950s to the 1970s, soil pollution has come to prominence in the 1980s, partly as the result of the discovery of many locations with contaminated soil. Causes of soil contamination can be burial of hazardous wastes, spills and leaks of chemicals and fuels, excessive application of liquid manure or pesticides, short-range air pollution, etc. In all these cases, realistic hazard predictions involve matching the basic chemical and toxicological data of the pollutant to additional data on the properties of the different soils where the substance may ultimately occur.

Based on models like those described in the preceding sections, this is accomplished by defining the specific migration of the chemical in, and the resultant impact on the soil–vegetation complex in the light of spatio-temporally variable boundary conditions. Models, however, are generally calibrated and validated on the basis of a more or less punctiform input data set only, and therefore the expedient areal extrapolation of a model poses particular problems. An adequate solution involves the combination of the model with a geographic information system which maintains the topologic structure of the environment-specific climatologic, hydrologic and soil data used as input to the model and to validate model output. The following examples illustrate this assessment approach at both the large-scale and small-scale levels.

Large-Scale Assessment of Acceptable Nitrate Rates in Soil

The approach is based on the combination of the chemical fate model WASMOD/STOMOD (Fränzle et al. 1989; Reiche 1990) with the geographic information system ARC/INFO (ESRI 1987). The basic structure of WASMOD/STOMOD is summarized in the following two figures.

With the aid of the minimum set of input data listed in Figure 4.27, the submodules (1) and (2) define effective precipitation as stand-specific hydrologic input variable. Sub-modules (3) to (8) then describe deep drainage through the soil and aeration zones, and the last two groundwater discharge into a receiving

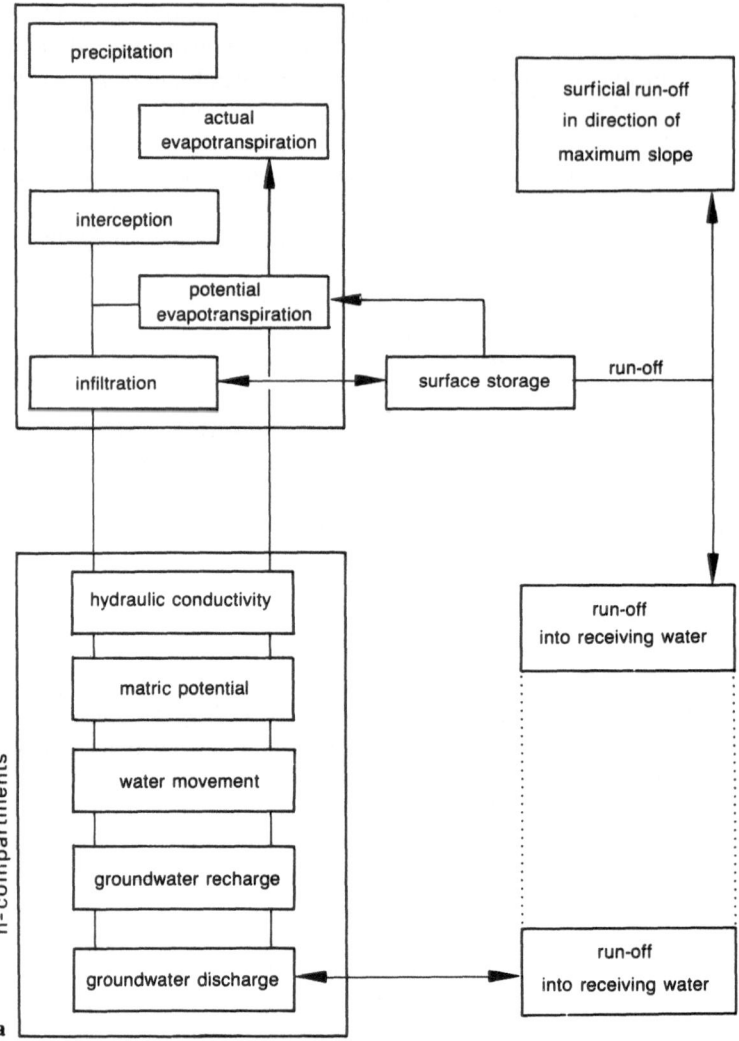

Fig. 4.26a. Soil moisture module of the integral WASMOD/STOMOD model. (After Reiche 1991)

water. Based on these input data, the solute module defines the variable proportions of adsorptive and capillary soil moisture as related to the organic matter, clay and silt phases. Thereafter adsorption/desorption equilibria, irreversible fixation, microbial degradation and water-borne vertical translocation are calculated for each successive horizon or layer of the unsaturated soil and aeration zone. The equilibrium concentrations in the mobile and immobile

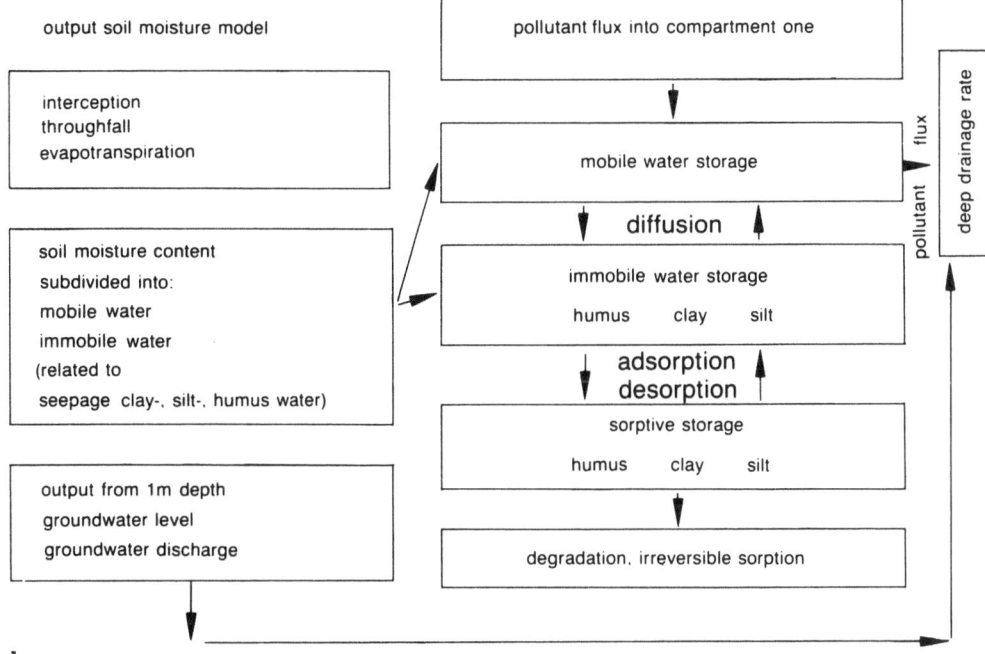

b

Fig. 4.26b. Solute model of the integral WASMOD/STOMOD model. (After Reiche 1991)

water fractions and soil matrix are determined on the basis of varying soil-moisture content on the one hand and time-dependent diffusion and sorption kinetics on the other. The model is free from limiting meso- and macro-scale assumptions on homogeneity (i.e. larger homogeneous compartments and uniform distribution of chemicals) and stationarity (i.e. large-scale equilibria in extensive compartments).

For specific purposes of nitrate modelling an extended STOMOD version was developed (Reiche 1990) which includes a nitrogen sub-module as depicted in Fig. 4.28. The calibration of the extended model was based on comprehensive data sets from batch, lysimeter and field experiments on 28 plots in different parts of Schleswig-Holstein.

The transfer of the model to two different areas of the Bornhöved Lake District (30 km S of Kiel) was accomplished by means of ARC/INFO, whose software consists of two major sub-systems. ARC maintains the topologic structure of the data base used to represent mapped features. The data base management sub-system INFO is a relational data bank used to store and process attribute information associated with the geographic features maintained by ARC. In this case, the necessary soil data were derived from large-scale maps of the German Soil Taxation Survey (Reichsbodenschätzung) by means of

1. **Partial model potential evapotranspiration (ETpot) after HAUDE**
   ETpot = f (temperature, relative humidity)

2. **Partial model interception (ETi) after v. HOYNINGEN-HUENE**
   ETi = f (leaf area index, precipitation)

3. **Partial model suction (Ψ)**
   Ψ = f (organic matter, silt, clay)
   reference data: pF curves

4. **Partial model hydraulic conductivity (Ku)**
   Ku = f (saturated conductivity, suction)
   reference data: field data of saturated conductivity after DARCY

5. **Partial model evapotranspiration (ETa), modified after BRAUN**
   ETa = f (suction, root density, ETp (HAUDE), ETi)

6. **Partial model soil water movement**
   δθ/δt = f (suction gradient, Ku, groundwater level)
   reference: DARCY's equation

7. **Partial model change in water storage**
   δBF(vol)/δt = f (δθ/δt, precipitation, ETa)

8. **Partial model lowermost compartment**
   δBF(vol)u/δt = f (δBf(vol)/δt, groundwater level

9. **Partial model groundwater discharge**
   δGW/δt = f (Kf,α[receiving water])

10. **Partial model groundwater level**
   δGH/δt = f (δBF(vol)u/δt, δGW/δt)

Initial data (daily):	Temperature, precipitation, relative humidity
Initial data (constant):	o % org. matter, % silt, % clay o leaf area index o saturated hydraulic conductivity o root density o slope of groundwater surface
Initial data (primary):	o soil moisture o groundwater level

Fig. 4.27. Basic structure of WASMOD module

an automatic translation procedure (Reiche 1990). Figures 4.29 and 4.30 illustrate the application of the extended WASMOD/STOMOD version to modelling different aspects of the nitrogen balance of arable land.

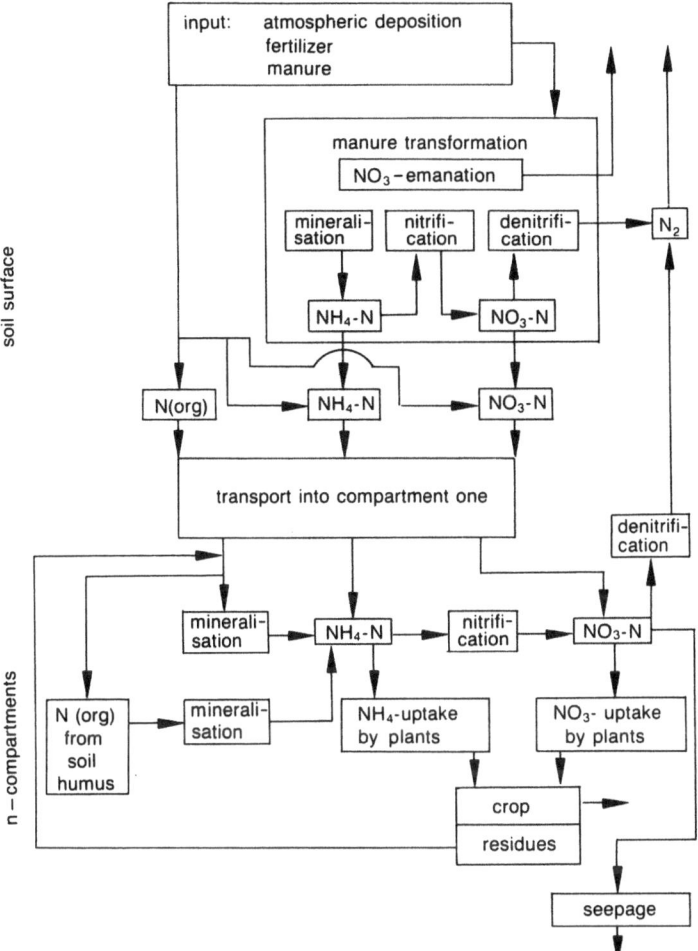

Fig. 4.28. Nitrogen sub-module of STOMOD. (After Reiche 1990)

Small-Scale Assessment of Heavy Metal Sorptivity

Table functions in score form (DVWK 1989) (cf. Sect. 4.4.1.3) provide the information for a sorption-oriented interpretation of the taxonomic units of soil maps which have to be appropriately transformed into binary data sets by means of digitizing procedures. In the present context the 1 : 1,000,000 Soil Map of Germany (Hollstein 1963) proved better suited to this end than Roeschmann's (1986) soil association map. Thus the elements of Table 4.34 resulting from such a comprehensive interpretation procedure allow the construction of small-scale maps of soil sensitivity to heavy metals by means of geographical information systems such as ARC/INFO (ESRI 1987).

OTHER AREAS
▥   low - fen
⊟   forest
☐   lakes

0   500 m

Fig. 4.29. Nitrate leaching potential of different soils in percent of N_{min} content. (After Reiche 1990)

Figure 4.31 illustrates in a comparative way the spatial differentiation of cadmium, copper and zinc sorptivity of Schleswig-Holstein soils.

Small-Scale Assessment of Pesticide Impact on Soil

On the same basis as above also fate and behaviour of organics can be approximately described by means of table functions (cf. Blume and Brümmer 1987b; Litz et al. 1987; DVWK 1989). In combination with geographic information systems, they permit the rapid production of generalized smaller-scale single species sensitivity maps (e.g. Rüstmann 1990). A more detailed simulation of chemical behaviour should be based on more sophisticated models which take into account synergetic effects, metabolites, specific site structures and their temporal changes, etc. Six parameters, defined in score form on a 0 (practically not) to 5 (very much, very high) scale, are employed to describe chemical reactions in soil:

– solubility in mg l^{-1} (20° C)

     0: < 0.1, 1: 0.1–1, 2: 1–10, 3: 10–100,
     4: 100–1000, 5: > 1000;

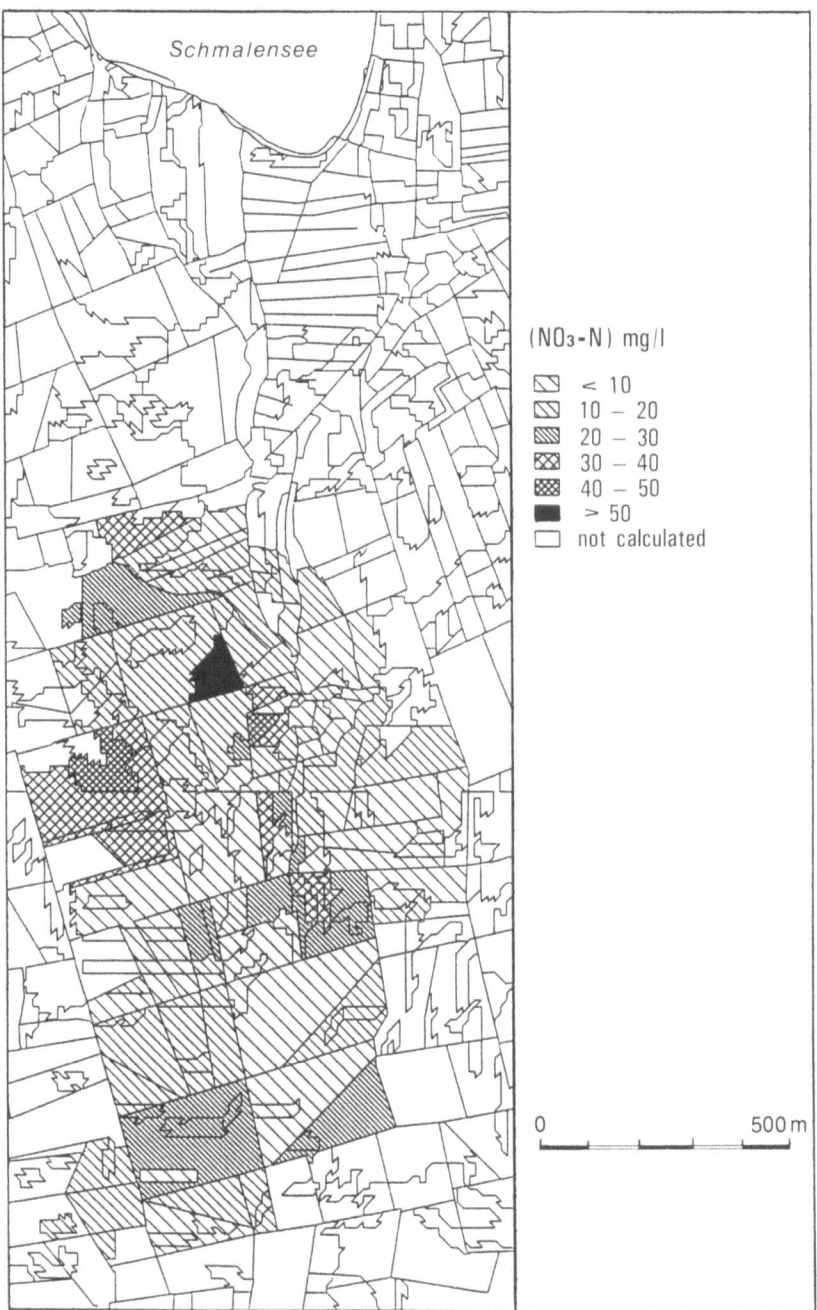

Fig. 4.30. Calculated nitrate concentrations in seepage water as related to soil and land use patterns in the Schmalenseefelder Au catchment. (After Reiche 1990)

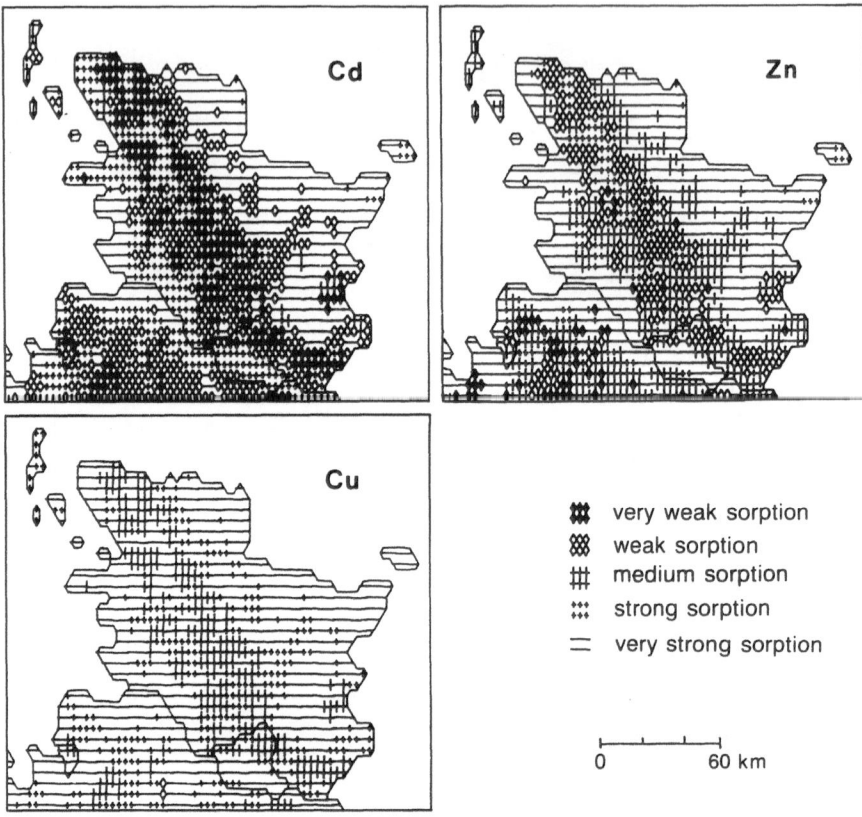

Fig. 4.31. Sorption capacities of Schleswig-Holstein soils for cadmium, copper and zinc

– adsorption, as defined by the Freundlich k-values for organic matter (k_{oc}) and clay (k_c) at a soil/solution ratio of $1:2$ (5)

    0: < 0.5, 1: 0.5–50, 2: 50–150, 3: 150–500, 4: 500–5000, 5: > 5000.

pH influence $+$ or $-$, i.e. adsorption increases with increasing or decreasing pH value;

– degradability, defined in terms of time required to produce an approximately 90% transformation at a soil temperature ranging from 11 to 16° C:

    1: > 3 years, 2: 1–3 years, 3: 18 weeks to 1 year, 4: < 18–6 weeks,
    5: < 6 weeks;

– volatility, defined by the vapour pressure of the pure substance in hPa at 20–25° C:

    1: $< 10^{-3}$, 2: 10^{-3}–100, 3: 100–500, 4: > 500

or Henry constants (cf. Chap. 2.1.2.4)

    1: $< 4 \times 10^{-6}$, 2: $4 \times 10^{-6} - 4 \times 10^{-4}$, 3: $4 \times 10^{-4} - 0.04$, 4: > 0.04.

Table 4.52. Behaviour of organic chemicals in soil. (After DVWK 1990)

Chemicals	Aqueous solubility	Adsorption to org. matter	Adsorption to clay	pH influence	Degradation aerobic	Degradation anaerobic	Volatility	Concentrations[a] A (mg kg⁻¹)	B	C
a) Environmental chemicals										
Chlorinated alkanes and alkenes										
1. Dichloromethane	4	1	1	0	3–4	2	4	0.1	5	50
2. Trichloromethane	4	2	2	0	2–3	1–2	4	0.1	5	50
3. Tetrachloromethane	4	1–2	2	0	2–3	1–2	4	0.1	5	50
4. Dichloroethane	4	2–3	2	0	3	2	4	0.1	5	50
5. Monochloroethene	4	1	1	0	4	4	3			
6. Trichloroethene	4	2	1	0	3	2	4	0.1	5	50
7. Tetrachloroethene	3	1–2	1	0	3	2	3	0.1	5	50
Carbocyclic compounds										
8. Benzene	4	1	1	0	3	2	4	0.01	0.5	5
9. 1,2-Dichlorobenzene	3	3	2	0	2	2	3	0.05	1	10
10. 1,2,4-Trichlorobenzene	2	3–4	1–2	0	2–3	2–4	3	0.05	1	10
11. Phenol	4	1–2	2	–	4	3	2	0.05	1	10
12. 2,5-Dichlorophenol	4	3	1	–	3	4		0.01	0.5	5
13. 2,4,5-Trichlorophenol	4	3	2	–	4	4		0.01	0.5	5
14. Pentachlorophenol	2	4	1	–	3–4	4	2	0.01	0.5	5
15. Aniline	4	1–2	2–3	+	4–5	4	2			
16. 4-Chloroaniline	4	3	1	+	2	2	2			
17. Toluene	3	1	2–3	0	4	2–3	3	0.05	3	30
18. Toluidine	4	2–3	1	+	2	1				
19. Xylene	3	2–3	1–2	0	3		2	0.05	1	10

Table 4.52. (Contd.)

Chemicals	Aqueous solubility	Adsorption to org. matter	clay	pH influence	Degradation aerobic	anaerobic	Volatility	Concentrations[a] A	B	C mg kg⁻¹
Polynuclear aromatics										
20. Naphthalene	2	4	2	0	4	(1)	3	0.1	5	50
21. Naphthylamine		4	3	+	2	1	1			
22. Anthracene	1	4–5	3	0	2	3	1	0.1	10	100
23. Phenanthrene	2	4–5	3	0	3	3	1	0.1	10	100
24. Pyrene	1	4–5	3	0	1–2		1	0.1	10	100
25. Benzo-(a)pyrene	1	5	3	0	1		1	0.05	1	10
26. Dibenzo-(a,h)pyrene	1	5	3	0	1		1			
27. Fluoranthene	1	4–5	2–3	0	2–3		1–2	0.1	10	100
28. 2,2'-Dichlorobiphenyl	2	5	3	–	3	2	2			
29. 2,4,5-Trichlorobiphenyl	1	5	3	–	2–3	2	1			
30. 2,2',4,5,5'-Pentachloro-biphenyl		5	3	–	1	1				
31. Hexabromobiphenyl	1	5	3	–	1	1	1			
32. 4,4'-Diaminobiphenyl	4	3	4	+	2	1	1			
Heterocyclic compounds										
33. Pyridine	4	1	2	+	4–5	3	3	0.1	2	20
34. 2,3,7,8-Tetrachloro-dibenzo-p-dioxin	1	5	3	0	1	1	1			

No.	Compound										
35.	Octachlorodibenzo-p-dioxin	1	5	3	0	2	2	1			
36.	1,2,3,7,8-Pentachloro-dibenzofuran	1	5	3	0	2	2	2			
	Mineral oil derivatives										
37.	Petroleum	2	4	3	0	2	1				
38.	Kerosine	2	4	3	0	4	3		20	100	800
39.	Diesel oil and light fuel oil	2	4	3	0	4	3		100	1000	5000
40.	Heavy fuel oil	2	4	3	0	1	2		100	1000	5000
	Further organics										
41.	Phthalic acid	4	1–2	1	—	4	4	2			
42.	LAS	3	3–4	1	—	4	3				
43.	NTA, Trilon A	4	2–3	1	—	4	3–4				
44.	Nonylphenol	4	2–3	2	(—)	4	2–3	3			
45.	Diethylhexylphthalate	1	4–5	3–4	0	3–4	2	1			
46.	Dibutylphthalate	2	3–4	2	0	4	3	2			
47.	Acrylamide	4	1	1	(—)	4–5	4	2			

Table 4.52. (Contd.)

Chemicals	Aqueous solubility	Adsorption to org. matter	clay	pH influence	Degradation aerobic	anaerobic	Volatility	Licence[b]
b) Pesticides[c]								
Bipyridyl derivatives								
1. Diquat (H)	4	5	4-5	+	3		1	+
2. Paraquat (H)	4	5	4-5	+	1		1	+
Diazines								
3. Chloridazon (H)	3	3	2	(−)	4-5		1	+
S-triazines								
4. Prometryn (H)	4	3	2	(−)	4		1	(+)
5. Atrazine (H)	2-3	2	2	−	3-4	3-4	1	(+)
6. Simazine (H)	2	2	2	(−)	3-4		1	(+)
7. Propazine (H)	2	2		(−)	3-4		1	+
8. Anilazine (F)	1				5		(1)	+
9. Terbuthylazine (H)	2	2-3	2	(−)	3-4		1	+
S-triazoles								
10. Amitrole (H)	4	2	2	(−)	4		1	(+)
11. Propiconazole (F)	3			(−)			1	+
12. Triadimenole (F)	3			(−)	(1)		1	+
Triazinones								
13. Metribuzin (H)	4	2	(2)	(−)	4		1	(+)
14. Metamitron (H)	4	1		(−)	5		1	+

Phenoxyfatty acids								
15. 2,4-D (H)	4	1–2	1	−	4		1	+
16. MCPA (H)	4	2	1	−	4		1	+
17. Mecoprop (H)	4	1		−	4–5		1	+
18. 2,4,5-T (H)	3	2	1	−	3	3	1	+
19. Fluazifop-butyl (H)		2		(−)	5			(+)
20. Dichlorprop (H)	3–4	2		(−)	4–5		1	(+)
Benzoic acids								
21. Dicamba (H)	4	2	1	−	5		1	(+)
22. Chloramben (H)	4			(−)	2		2	(+)
23. 2,3,6-TBA (H)	4	2	1	(−)	3		2	
Picolinic acid derivatives								
24. Picloram (H)	4	2	1	−	2–3		1	(+)
Halogenated aromatic hydrocarbons								
25. Dinoseb acetate (H)	4	2		(−)	5		2	(+)
26. DNOC (H)	3	2–3		(−)	4		2	(+)
27. Pentachlorophenol (I)	2	4	2	−	3–4	4	2	
28. Quintozene (F)	1			(−)	2–3	4	2	+
29. Hexachlorobenzene (F)	1	4	2	(−)	2	1	2	
Halogenated aliphatics and derivatives								
30. Endosulfan (I)	1	4		0	3	(2)	2	+
31. Aldrin (I)	1	(5)	(3)	0	3	(2)	1	
32. Dieldrin (I)	1	5	3	0	1–2	1–2	1	
33. Heptachlor (I)	1	5	3	0	2–3	2–3	2	
34. Chlordane (I)	1	5		0	3		2	

Table 4.52. (Contd.)

Chemicals	Aqueous solubility	Adsorption to org. matter	clay	pH influence	Degradation aerobic	anaerobic	Volatility	Licence[b]
Halogenated aliphatics and derivatives								
35. Lindane (I)	2	3–4	3	0	2–3	5	2	(+)
36. DDT	1	5	(4)	0	1–2	3	1	
37. 1,3-Dichloropropene (N)	2	2	2	0	5	4	3	(+)
38. Methylbromide (J, N)	4						4	
Organic phosphorus esters								
39. Melathion (I)	3	2–3		0	3	3–4	2	+
40. Parathion (A, I)	2	4–5	2–3	0	4–5	4	2	+
41. Phorate (A, I)	2–3	4	(2)	0	4		2	
42. Diazinon (I)	2	3	(2)	0	4	4	2	
Methylcarbamates								
43. Carbaryl (I)	2–3	2–3		0	3–4	4	2	
44. Carbofuran (A, I, N)	4	3		0	4	3–4	1	(+)
45. Propoxur (I)	4	2		0	(4)		2	(+)
46. Phenmedipham (H)	1–2	3–4		0	4		1	
Phenylcarbamates								
47. Propham (H)	3	2	1–2	0	4–5		1	+
48. Chloropropham (H)	3	3	1–2	0	4		1	+
Thiocarbomates								
49. Diallate (H)	2	4		0	4–5		1	
50. Triallate (H)	2	4	2	0	3–4		1	+

Compound								
51. Butylate (H)	2	3		0	4		2	+
52. EPTC (H)	3	2		0	4		2	+
Dithiocarbamates								
53. Maneb (F)	1	1–2		0	5		1	+
54. Metam, Na (F, H, N)	4	3–4		0	4		1	(+)
55. Zineb (F)	2	3		0	4–5		1	+
56. Thiram (F)	2	2		0	4–5		1	+
57. Propanil (H)	3		1–2	0	4		1–2	
58. Mancozeb (F)	1	2		0	4–5		1	+
Oxicarbamates								
59. Aldicarb (I, N)	4	2	1	0	3–4	4	2	(+)
Acetanilides								
60. Propachlor (H)	4	2	1–2	0	4		1	(+)
61. Carboxin (F)	3	5	3	0	4–5		2	+
62. Metolachlor (H)	4	2–3		0	4		1	+
63. Metazachlor (H)	2	2–3		0	4–5		1	(+)
Urea derivatives								
64. Fenuron (H)	4	2–3	2	0	3		1	+
65. Monuron (H)	3	3	2	0	4		1	+
66. Diuron (H)	2	3	2	0	3		1	+
67. Monolinuron (H)	3–4	3	2	0	4		1	+
68. Methabenzo-thiazuron (H)	3	3	2	0	3		1–2	+
69. Metoxuron (H)	4	2	(2)	0	4–5		1	+
70. Linuron (H)	3	2–3	2	0	4		1	+
71. Chlortoluron (H)	3	2–3	2	0	4–5		1	+ −

Table 4.52. (Contd.)

Chemicals	Aqueous solubility	Adsorption to org. matter	clay	pH influence	Degradation aerobic	anaerobic	Volatility	Licence[b]
Dinitroanilines								
73. Nitralin (H)	1	3–4	(3)	0	3–4		1	+
74. Trifluralin (H)	1	4–5	(3)	0	3–4		1	+
75. Pendimethalin (H)	1			0	4		1	+
Benzimidazoles								
76. Benomyl (F)	2	5	3	0	4		1	+
77. Thiophanate methyl (F)	2			0	4–5		1	+
78. Carbendazim (F)	2	4		–	3–4		1	+
79. Prochloraz (F)	4						1	+
Phthalimide derivatives								
80. Captafol (F)	1–2	1–2		0	4–5		1	+
81. Captan (F)	1	1–2		0	4–5		1	+
Aromatic cyanides								
82. Dichlobenil (H)	2	2		0	3		2	+
83. Ioxynil (H)	2	2		0	5		1	+
84. Promoxynil (H)	3	2		0	4–5		1	+

[a]Critical concentrations after Leidraad Bodemsanering (1993); A reference level—can be accepted as unpolluted; B indicative value for further investigation; C clean up necessary.

[b] – = prohibited, + = licensed, (+) = prohibited in water-supply areas.

[c] A = acaricide, F = fungicide, H = herbicide, I = insecticide, N = nematocide, M = molluscocide, W = growth regulator.

Considering the influence of organic matter content on the adsorption process the scores of Table 4.52 have to be more precisely defined according to Table 4.53. Further refinements relate to texture, pH value and temperature, etc. (Tables 4.54, 4.55, 4.56).

Table 4.53. Sorption capacity of soil related to organic matter content. (After DVWK 1989)

Organic matter content of soil[a] (%)	Relative sorption capacity (Table 4.52)				
	1	2	3	4	5
0.5–1	0	0	1	1.5	2
1–2	0	0.5	1.5	2	3
2–8	0.5	1	2	3	4
8–15	0.5	1.5	2.5	3.5	4.5
> 15	1	2	3	4	5

[a] Average value of the uppermost 3 dm of soil, i.e. the Ah or Ap horizons.

Table 4.54. Sorption capacity of soil related to texture. (After DVWK 1989)

Texture classes[a]	Relative sorption capacity (Table 4.53)			
	1	2	3	4
1. S Ss'2[a]	0	0.5	1	1.5
2. Sc2 Sl2, 3 Ss' 3,4 S's S'	0.5	1	1.5	2
3. Sls' Sl4 S'l2, 3 S'ls Ls Ls' Sc3 Cs4 S'C2, 3	0.5	1	2	3
4. S'l4 Cl Cs2, 3 C's Lcs Lc2, 3 Lcs' S'c4	1	1.5	2.5	3.5
5. C	1	2	3	4

[a] Medium texture of the uppermost 3 dm of soil, i.e. the Ah of Ap horizons.
S = sand, S' = silt, L = loam, C = clay, s = sandy, s' = silty, l = loamy, c = clayey.

Table 4.55. pH control of relative sorption capacity of soil. (DVWK 1990)

pH influence	(Table 4.53)	pH ($CaCl_2$)		
	> 6.5	6.5–5.5	5.5–4	< 4
+	+ 0.5	0	− 0.5	− 1
−	− 0.5	0	+ 0.5	+ 1

A further refinement relates to temperature.

Table 4.56. Degradability of organic chemicals related to mean air temperature. (DVWK 1989)

Degradability (Table 4.52)	Mean temperature of vegetation period (°C)			Mean annual temperature (°C)		
	21–16	16–11	11–6	12–9	9–6	6–3
1	1.5	1	0.5	0.5	0.5	0
2	2.5	2	1.5	1.5	1	0.5
3	3.5	3	2.5	2.5	2	1.5
4	4.5	4	3.5	3.5	3	2.5
4–5	5	4.5	4	4	3.5	3

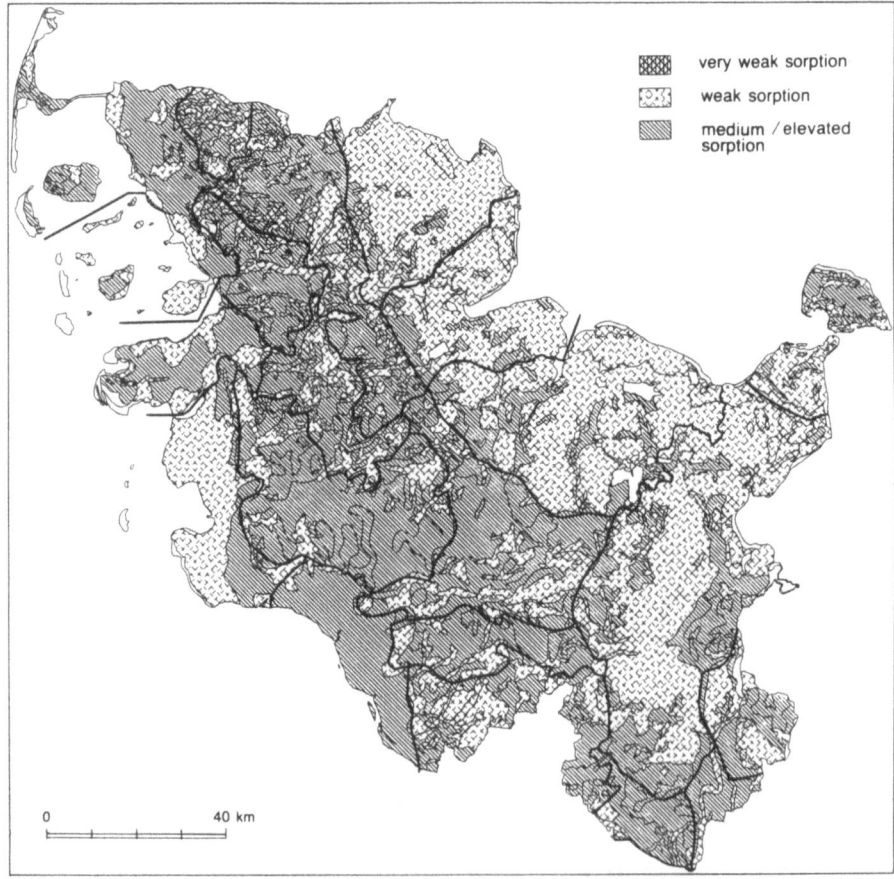

Fig. 4.32. Potential adsorption of atrazine/terbuthylazine to Schleswig-Holstein soils. (After Rüstmann 1990)

A digital evaluation of the 1:500,000 soil map of Schleswig-Holstein (Stremme 1981), which differentiates 12 soil taxa in combination with 8 texture classes, provides the basic data for an aereal application of the above table functions. It is based on the hypothetic assumption that the maximum content of organic matter of each soil taxon occurred in all of its representatives. Thus the resultant small-scale ARC-INFO map of sorption capacities for the herbicide terbuthylazine (Rüstmann 1990) is a reflection of both the generalisations of the primary soil map and this pedochemical assumption.

A deliberately conservative evaluation would be based on the opposite assumption of overall minimum values of organic matter content in the different soils. In either case, maps of this type are idealizations from the real-world situation but nevertheless suited for small-scale planning purposes. More precise larger-scale data sets on organic matter and clay contents, pH values etc. would permit the production of correspondingly more differentiated maps for multiple purposes. In the final end only the lesser degree of inherent model sophistication would distinguish them from maps based on the combination of more complex modelling approaches with high-resolution geographic information systems.

References

Abrahamsen G (1980a) Impact of atmospheric sulfur deposition on forest ecosystems. In: Shriner et al. (eds) Atmospheric sulfur: environmental impact and health effects. Ann Arbor Sci Publ MI: 397–416

Abrahamsen G (1980b) Acid precipitation, plant nutrients and forest growth. In: Drabløs D, Tollan A (eds) Proc Inst Conf Ecol Impact Acid Precipitation. SNSF, Project Norway, pp 58–63

Abrahamsen G, Bjor K, Horntvedt R, Tveite B (1976) Effects of acid precipitation on coniferous forests. In: Brakke FH (ed) Impact of acid precipitation on forest and fresh water ecosystems in Norway. Research Report 6, SNSF-Project, pp 37–63

Abrahamsen G, Dollard GJ (1979) Effects of acid precipitation on forest vegetation and soil. In: Ecological effects of acid precipitation. EP Workshop EA-79-6LD

Abrahamsen G, Stuanes AO (1983) Effects of acid deposition on soil: an overview. VDI-Berichte 500, Saure Niederschläge: pp 279–287. VDI-Verlag, Düsseldorf

Abriola LM, Pinder GF (1985) A multiphase approach to modeling of porous media contamination by organic compounds: 1. equation development, 2. numerical solution. Water Resour Res 21: 11–32

ACSM-ASPRS (American Congress on Surveying and Mapping – American Society for Photogrammetry and Remote Sensing) (1986) Geographic information systems. ACSM-ASPRS, Falls Church, Va

Adams RT (1973) Factors influencing soil adsorption and bioactivity of pesticides. Residue Rev 47: 1–54

Adams RT, Kurisu FM (1976) Simulation of pesticide movement on small agricultural watersheds. Final Report. Environmental Research Laboratory EPA, Athens, GA

Adema EH, van Ham J (eds) (1984) Zure regen, oorzaken, effecten en beleid. Pudoc, Wageningen

Adriano DC, Johnson AH (1989) Biological and ecological effects. In: Adriano DC, Salomons W (eds) Acidic precipitation, vol 2. Springer, Berlin Heidelberg New York London Paris Tokyo Hong Kong Barcelona

AFNOR T-90-302 Test (1977) Méthode d'évaluation en milieu aqueux de biodégradabilité des produits organiques

Agster G (1986) Ein- und Austrag sowie Umsatz gelöster Stoffe in den Einzugsgebieten des Schönbuchs. In: Einsele G (ed) Das landschaftsökologische Forschungsprojekt Naturpark Schönbuch. VCH Verlag, Weinheim, pp 343–356

Alabaster JS, Lloyd R (1980) Water quality criteria for freshwater fish. Butterworth, Washington DC

Al-Bahrani KS, Martin RJ (1976) Adsorption studies using gas-liquid chromatography – 1. Effect of molecular structure. Water Res 10: 731–736

Albrecht F (1941) Report from the scientific expedition to the UW Province of China under the leadership of Dr. Sven Hedin. The Sino-Swedish expedition 14, IX Meteorology 2, Stockholm

Alcamo J, Amann M, Hettelingh J-P, Holmberg M et al. (1987) Acidification in Europe: a simulation model for evaluating control strategies. Ambio 16: 232–245

Alexander M (1967) Pollutants that resist the microbes. New Sci 35: 439–440

Alexander M (1969) Microbial degradation and biological effects of pesticides in soil. In: Unesco (ed) Soil biology. Paris, pp 209–240

Alexander M (1971) Microbial ecology. Wiley, London

Alexander M (1973) Nonbiodegradable and other recalcitrant molecules. Biotechnol Bioeng 15: 611–647

Alexander M (1980) Biodegradation of toxic chemicals in water and soil. In: Haque R (ed) Dynamics, exposure and hazard assessment of toxic chemicals. Ann Arbor Science, Ann Arbor, pp 179–190

Althaus W, Sörensen O (1971) Untersuchungen über den Gehalt der Vorfluter an krebserregenden Stoffen und die Beeinflussung dieses Gehaltes durch Abschwemmungen von bestimmten Straßenbelägen. Schriftenr Dokumentationszentr Wasser 19: 1–72

Altshuller AP, Cohen IR (1963) Structural effects on the rate of nitrogen dioxide formation in the photooxidation of organic compound–nitric oxide mixtures in air. Int J Air Water Pollut 7:787–797

Alvim P de T (1978) Perspectivas de produção agricola na regiâo amazônica. Interciencia 3: 243–251

Amthor JS (1984) The role of maintenance respiration in plant growth. Plant Cell Environ 7: 561–569

Anderson KB, Bayer J, Hofer LJE (1965) Linear solutions of Fick's law. Ind Eng Chem Process Design Dev 4: 167–171

Anderson LS, Mansfield TA (1979) The effects of nitric oxide pollution on the growth of tomato. Environmental Pollution 20: 113–121

Andersson F, Fagerstrøm T, Nilsson I (1980) Forest ecosystem responses to acid deposition – hydrogen and nitrogen/tree growth model approaches. In: Hutchinson TC, Havas M (eds) Effects of acid precipitation on terrestrial ecosystems. NATO Conference Series, Series I, Ecology, vol 4, pp 319–334. Plenum, New York

Aniansson B (1988a) "Critical loads" basis for action. Acid Mag 6: 10–13

Aniansson B (1988b) Stopping our cultural heritage from crumbling. Acid Mag 6: 14–16

Appel BR, Wesolowski JJ, Hidy GM (1974) Analysis of the sulfate and nitrate data from the aerosol characterization study. AIHL Rep No 166 Air and Industrial Hygiene Laboratory, California State Department of Health Berkeley, California, pp 1–16

Appel CA, Bredehoeft JD (1978) Status of groundwater modeling in the US geological Survey. US Department of the Interior, Washington, DC

Arbeitskreis Grundwasserneubildung (1977) Methoden zur Bestimmung der Grundwasserneubildungsrate. Geol Jahrb C19: 3–98

Arber W (1986) Biotechnologie im Aufwind der Molekularbiologie. Naturwiss Rundsch 9: 371–379

Arcos JC (1983) Comparative requirements for premarketing/premanufacture notifications in the EC countries and the USA. J Am Coll Toxicol 2: 131

Arkley KJ, Ulrich R (1962) The use of calculated actual and potential evapotranspiration for estimating potential plant growth. Hilgardia 32: 443–461

Arndt U, Kaufmann M (1985) Wirkungen von Ozon auf die apparente Photosynthese von Tanne und Buche. Allg Forstz 156: 19–20

Arnolds E (1985) Verandringen in de paddestoelen flora (Mycoflora). Wet Meded KNNV 167

Arthur D Little, Inc (1981) Prepared revisions to MITRE Model. Arthur D Little, Cambridge, MA

Asada K, Deura R, Kasai Z (1968) Effect of sulfate ions on photophosphorylation by spinach chloroplasts. Plant Cell Physiol 9: 143–146

Ashenden TW (1979) The effects of long-term exposure to SO_2 and NO_2 pollution on the growth of Dactylis glomerata L. and Poa pratensis L. Environmental Pollution 18: 249–258

Ashenden TW, Williams IAD (1980) Growth reduction in Lolium multiflorum Lam. and Phleum pratense L. as a result of SO_2 and NO_2 pollution. Environmental Pollution (Series A) 21: 131–139

Asman WAH, Jonker PJ, Slanina J, Baard JH (1982) Neutralization of acid in precipitation and some results of sequential rain sampling. In: Georgii HW, Pankrath J (eds) Deposition of atmospheric pollutants. Reidel, Dordrecht, pp 115–123

Athari S (1980) Untersuchungen über die Zuwachsentwicklung rauchgeschädigter Fichtenbestände. Dissertation, Forstl Fak Univ Göttingen

Athari S (1981) Jahrringausfall, ein meist unbeachtetes Problem bei Zuwachsuntersuchungen in rauchgeschädigten und gesunden Fichtenbeständen. Mitt Forstl Bundesversuchsanst Wien 139: 7–27

Athari S (1983) Zuwachsvergleich von Fichten mit unterschiedlich starken Schadsymptomen. Allg Forstztg 38: 653–655

Athari S, Kramer H (1983a) Zur Problematik der Erfassung von umweltbedingten Zuwachsverlusten in Fichtenbeständen. Forst Holzwirt 38: 204–206

Athari S, Kramer H (1983b) Erfassung des Holzzuwachses als Bioindikator beim Fichtensterben. Allg Forstztg 38: 767–769

Athari S, Kramer H (1985) Ergebnisse von Wachstums- und Strukturanalysen in erkrankten Fichtenbeständen. Schr Forstl Fak Univ Göttingen Niedersächs Forstl Versuchsanst 82: 6–96

Atkinson R, Pitts JN (1980) Kinetics and mechanisms of gas phase ozone and hydroxyl radical reactions in polluted atmospheres. In: Funcke W, König J, Stöber W, Klein AW, Schmidt-Bleek F (eds) Proc Int Worksh on test methods and assessment procedures for the determination of the photochemical degradation behaviour of chemical substances. Dec 2–4, 1980, Reichstag, Berlin. FhG, UBA, Münster, pp 125–140

Atkinson R, Darnall KR, Lloyd AC, Winer AM, Pitts JN (1979) Kinetics and mechanisms of the reactions of the hydroxyl radical with organic compounds in the gas phase. Adv Photochem 11: 375–488

Atlas E, Giam CS (1981) Global transport of organic pollutants: Ambient concentrations in the remote marine atmosphere. Science 211: 163–165

Atri FR (1985) Chlorierte Kohlenwasserstoffe in der Umwelt. Fischer, Stuttgart

Aubréville A (1961) Etude écologique des principales formations végétales du Brésil. Centre Technique Forestier Tropical, Nogent-sur-Marne

Auclair D (1977) Effects of dusts on photosynthesis: II. Effects of particulate matter on photosynthesis of Scots pine and poplar. Am Sci For 34: 47–58

Audus LJ (1953) Fate of sodium 2,4-dichlorophenoxyethyl sulphate in soil. Nature 171: 523–524

Aypar C (1990) Untersuchung der Inhaltsstoffe der Sclerotien von Sclerotinia sclerotiorum (Lib) de Bary. Dissertation, Univ Kiel

Baader HR (1978) Eine Klimatologie des ersten Hauptsatzes. Dissertation, Meteorol Inst Univ Kiel

Bache BW (1977) Practical implications of quantity-intensity relationships. Proc SEFMIA, Tokyo

Bachmat Y, Bredehoeft J, Andrews B, Holz D, Sebastian S (1980) Groundwater management: the use of numerical models. Am Geophys Union, Washington, DC

Baer NS, Banks PN (1985) Indoor air pollution – effects on cultural and historic materials. Int J Mus Manage Curatorship 4: 9–20

Bailey GW, White JL (1970) Factors influencing the adsorption, desorption, and movement of pesticides in soil. Residue Rev 32: 29–92

Baker FG (1978) Variability of hydraulic conductivity within and between nine Wisconsin soil series. Water Resour Res 14: 103–108

Baldwin AC, Barker JR, Golden DM, Hendry DG (1977) Photochemical smog. Rate parameter estimates and computer simulations. J Phys Chem 81: 2483–2492

Ball JW, Jenne EA, Cantrell MW (1981) WATEQ3 – geochemical model with uranium added. US Geol Surv 81-1183, Washington, DC

Ball JW, Nordstrom DK, Jenne EA (1980) Additional and revised thermochemical data and computer code for WATEQ2 – a computerized chemical model for trace and major element speciation and mineral equilibria of natural waters. US Geol Surv Water Res Invest 78-116

Ball JW, Nordstrom DK, Zachmann DW (1987) A personal computer FORTRAN translation of the geochemical model WATEQ2 with revised data base. US Geol Surv Open-File Rep 87-50, Menlo Park, Calif

Ballantyne DJ (1973) Sulphite inhibition of ATP formation in plant mitochondria. Phytochemistry 12: 1207–1209

Balloni W, Favilli F (1987) Effects of agricultural practices on the physical, chemical and biological properties of soils: Part I – Effect of some agricultural practices on the biological soil fertility. In: Barth H, L'Hermite P (eds) Scientific basis for soil protection in the European Community. Elsevier, London, pp 161–179

Ballschmiter K (1981) Ausbreitungswege der Xenobiotika in der Umwelt. Ökol Vögel 3 Sonderh: 149–160

Banerjee P, Piwoni MD, Ebeid K (1985) Sorption of organic contaminants to a low carbon subsurface core. Chemosphere 14: 1057–1067

Barnsley EA (1975) The bacterial degradation of fluoranthene and benzo(a)pyrene. Can J Microbiol 21: 1004–1008

Barrow GM (1974) Physikalische Chemie Teil III. Bohmann, Wien

Barsch D, Flügel WA (1978) Das hydrologisch–geomorphologische Versuchsgebiet Hollmuth des Geographischen Instituts der Universität Heidelberg. Erdkunde 32: 61–70

Bartelli LJ, Odell RT (1960a) Field studies of a clay-enriched horizon in the lowest part of the solum of some brunizem and gray-brown podzolic soils in Illinois. Soil Sci Soc Am Proc 24: 388–391

Bartelli LJ, Odell RT (1960b) Laboratory studies and genesis of a clay-enriched horizon in the lowest part of the solum of some brunizem and gray-brown podzolic soils in Illinois. Soil Sci Soc Am Proc 24: 391–395

Battelle-Institut Frankfurt (1976, 1977) Anforderungen an Emissionskataster und meteorologische Daten im Hinblick auf ihre Verwendung als Eingabedaten für Rechenmodelle zur Ausbreitungsrechnung. Forschungsbericht 104 02 408 des Umweltforschungsplanes des Bundesministers des Innern im Auftrag des Umweltbundesamtes. Teil 1 (1976), Teil 2 (1977), Berlin

Battelle-Institut Frankfurt (1982) Austausch von Luftverunreinigungen an der Grenz-fläche Atmosphäre/Erdoberfläche (trockene Deposition). Forschungsbericht 104 02 609 des Umweltforschungsplanes des Bundesministers des Innern im Auftrag des Umweltbundesamtes, Berlin

Bauch J (1983) Biological alterations in the stem and root of fir and spruce due to pollution influences. In: Workshop on the effects of accumulation of air pollutants in forest ecosystems. D Reidel Co, Hingham, MA

Bauch J, Klein P, Frühwald A, Brill H (1979) Alterations of wood characteristics in *Abies alba* Will. due to "fir dying" and considerations concerning its origin. Eur J For Pathol 6: 321–331

Baumgartner A (1966) Energetic bases for differential vaporization from forest and agricultural stands. Repr Int Symp Forest Hydrol, pp 381–389

Baumgartner A (1967) Entwicklungslinien der forstlichen Meteorologie. Forstwiss Centralbl 86: 156–175

Baur JR, Bovey RW (1974) Ultraviolet and volatility loss of herbicides. Arch Environ Cont Toxicol 2: 275–288

Baur JR, Bovey RW, McCall MG (1973) Thermal and ultraviolet loss of herbicides. Arch Environ Cont Toxicol 1: 289–302

Bear J (1972) Dynamics of fluid in porous media. Elsevier, New York

Beck T (1970) Der mikrobielle Abbau von Herbiziden und der Einfluß auf die Mikroflora des Bodens. Zentralbl Bakteriol Abt II 124: 304–313

Becker KH, Fricke W, Löbel J, Schurath U (1985) Formation, transport and control of photochemical oxidants. In: Guderian R (ed) Air pollution by photochemical oxidants. Ecological Studies. Springer Berlin Heidelberg New York 52: 3–125

Beilke S (1969) Neue Ergebnisse über das Auswaschen atmosphärischer Spurengase und Aerosole. Ann d Met NF 4: 122–125

Beilke S (1983) Saure Niederschläge – die heutige Situation in Europa. Elektrizitäts-wirtschaft 17/18: 638–642

Bell JNB, Clough WS (1973) Depression of yield in ryegrass exposed to sulphur dioxide. Nature 241: 47–49

Benarie MM (ed) (1976) Atmospheric pollution. Elsevier, Amsterdam

Benarie MM, Detrie P (1978) Assessment of an OECD study on long-range transport of air pollutants (LRTAB) involving some aspects of air chemistry. In: Benarie MM (ed) Studies in environmental science. Atmospheric pollution 1978, proc of the 13th Int Colloq, Paris, April 25–28, vol 1. Elsevier, Amsterdam, pp 207–215

Benecke P (1984) Der Wasserumsatz eines Buchen- und eines Fichtenwaldökosystems im Hochsolling. Schr Forstl Fak Univ Göttingen Niedersächs Forstl Versuchsanst 77. Sauerländer, Frankfurt/M

Benecke P, van der Ploeg RR (1978) Wald und Wasser. I. Komponenten des Wasserhaus-haltes von Waldökosystemen. II. Quantifizierung des Wasserumsatzes am Beispiel eines Buchen- und Fichtenaltbestandes im Solling. Forstarchiv 49: 1–7, 26–32

Benedict HM, Breen WH (1955) The use of weeds as a means of evaluating vegetation damage caused by air pollution. In: Proc 3rd Int Air Poll Symp SRI: 177–190

Benet JC, Jouanna P (1983) Non-équilibre thermodynamique dans les milieux poreux non saturés avec changement de phase. J Heat Mass Transfer 26: 1585–1595

Benjamin MM, Leckie JO (1981) Multiple site adsorption of Cd, Cu, Zn and Pb on amorphous iron oxyhydroxide. J Colloid Interface Sci 79: 209–221

Benner P, Wild A (1987) Measurement of photosynthesis and transpiration in spruce trees with various degrees of damage. J Plant Physiol 129: 59–72

Beratergremium für umweltrelevante Altstoffe (BUA) der Gesellschaft Deutscher Chemiker (1985a) Chloroform. BUA-Stoffbericht 1. VCH, Weinheim
- (1985b) o-Chlornitrobenzol. BUA-Stoffbericht 2. VCH, Weinheim
- (1985c) Pentachlorphenol. BUA-Stoffbericht 3. VCH, Weinheim
- (1986a) Umweltrelevante Alte Stoffe: Auswahlkriterien und Stoffliste. VCH, Weinheim
- (1986b) Di-(2-ethylhexyl)-phthalat. BUA-Stoffbericht 4. VCH, Weinheim
- (1986c) Nitrilotriessigsäure. BUA-Stoffbericht 5. VCH, Weinheim
- (1986d) Dichlormethan. BUA-Stoffbericht 6. VCH, Weinheim
- (1986e) Chlormethan. BUA-Stoffbericht 7. VCH, Weinheim
- (1987a) m-Dichlorbenzol. BUA-Stoffbericht 8. VCH, Weinheim
- (1987b) o-Nitroanisol. BUA-Stoffbericht 9. VCH, Weinheim
- (1987c) p-Nitroanisol. BUA-Stoffbericht 10. VCH, Weinheim
- (1987d) Dinitrotoluole. BUA-Stoffbericht 12. VCH, Weinheim
- (1987e) Brommethan. BUA-Stoffbericht 14. VCH, Weinheim
- (1987f) 1,2,4-Trichlorbenzol. BUA-Stoffbericht 17. VCH, Weinheim
- (1987g) p-Nitroanilin. BUA-Stoffbericht 19. VCH, Weinheim
- (1987h) Tris(2-chlorethyl)-phosphat. BUA-Stoffbericht 20. VCH, Weinheim
- (1987i) Bis(2-chlorethyl)ether. BUA-Stoffbericht 21. VCH, Weinheim
- (1987j) Dibutylphthalat. BUA-Stoffbericht 22. VCH, Weinheim
- (1988a) p-Chlorinitrobenzol, m-Chlornitrobenzol. BUA-Stoffbericht 11. VCH, Weinheim
- (1988b) Nonylphenol. BUA-Stoffbericht 13. VCH, Weinheim
- (1988c) Diphenylamin. BUA-Stoffbericht 15. VCH, Weinheim
- (1988d) 1,3,5-Trichlorbenzol. BUA-Stoffbericht 16. VCH, Weinheim
- (1988e) Ditolylether. BUA-Stoffbericht 18. VCH, Weinheim
- (1988f) Tributylamin. BUA-Stoffbericht 23. VCH, Weinheim
- (1988g) Benzol. BUA-Stoffbericht 24. VCH, Weinheim
- (1988h) Hexachlorcyclopentadien. BUA-Stoffbericht 25. VCH, Weinheim
- (1991) 2-Mercaptobenzothiazol and Salze. BUA-Stoffbericht 74
Berg H (1948) Allgemeine Meteorologie. Dümmler, Bonn
Bergstein PE, Vestal JR (1978) Crude oil biodegradation in arctic tundra ponds. Arctic 31: 158–169
Bernhardt H (ed) (1984) NTA: Studie über die aquatische Umweltverträglichkeit von Nitrilotriacetat (NTA). Richarz, Sankt Augustin
Berry BJL, Marble DF (eds) (1968) Spatial analysis. Prentice-Hall, Englewood Cliffs, NJ
Bertalanffy L von (1968) General system theory. Penguin Press, New York
Bertsch W, Schwille F, Ubell K (1979) Versickerungsversuche mit Import-Rohöl und niedrig-viskosem schwerem Heizöl. Ber Dtsch Ges Mineralölwiss Kohlechem 144
Besch WK (1982) Kritische Bewertung der Rolle von toxikologischen Tests bei der Gewässerüberwachung. Decheniana Beih 26: 67–77
Bettleheim J, Littler A (1979) Historic trends in sulfur oxide emissions in Europe since 1865. CEGB Report PL-GSIE/1/79
Beven K, Germann P (1982) Macropores and water flow in soils. Water Resour Res 18: 1311–1325
Beyer W (1964) Zur Bestimmung der Wasserdurchlässigkeit von Kiesen und Sanden aus der Kornverteilungskurve. Wasserwirtsch-Wassertech 14 (6): 165–169
Beyschlag W, Wedler M, Lange OL, Heber U (1987) Einfluß einer Magnesiumdüngung auf Photosynthese und Transpiration von Fichten an einem Magnesium-Mangelstandort im Fichtelgebirge. Allg Forst-Jagdztg 42: 738–741

Bhosle NB, Mavinkurve S (1980) Hydrocarbon-utilising microorganisms from Dona Paula Bay, Goa. Mar Environ Res 4: 53–58

Bick H, Neumann D (eds) (1982) Bioindikatoren. Decheniana Beih 26: 1–198

Bidleman TF, Olney CE (1975) Long-range transport of toxaphene insecticide in the atmosphere of western North Atlantic. Nature 257: 475–477

Billings CE, Gussman RA (1976) Dynamic behaviour of aerosols. In: Dennis R (ed) Handbook on aerosols. Bedford, Mass, pp 40–65

Bjor K, Teigen O (1980) Effects of acid precipitation on soil and forest. 6. Lysimeter experiment in greenhouse. In: Drabløs D, Tollan A (eds) Proc Int Conf Ecol Impact Acid Precipitation. SNSF – Project Norway, Oslo pp 200–201

Blackburn PS (ed) (1983) Acid precipitation. A bibliography. US Dept of Energy, Technical Information Center, Tennessee

Bläsig H, Merkel B, Udluft P (1984) Untersuchungen zum Entwässerungsverhalten von quartären Kies- und Sandgemischen. Z Wasser Abwasser Forsch 17: 1–6

Blair EH (1981) A framework of consideration for setting priorities for the testing of chemical substances. In: Proc of the Worksh on the Control of Existing Chemicals under the Patronage of the Organisation for Economic Co-operation and Development. Berlin, pp 252–260

Blaschke H (1986) Einfluß von saurer Beregnung und Kalkung auf die Biomasse und Mykorrhizierung der Feinwurzeln von Fichten. Forstwiss Centralbl 105: 324–329

Block J, Bartels U (1983) Hohe Konzentration von Luftverunreinigungen in den Waldniederschlägen. Mitt LÖLF 8(4): 19–34

Bloemen GW (1964) Hydraulic device for weighing large lysimeters. Trans Am Soc Agric Eng 7: 297–299

Blohm R, Schönwald B (1983) Bestimmung von Temperaturprofilen mit einem mobilen Mikrowellen-Radiometer. Meteorol Rundsch 36: 201–204

Blüthgen J (1966) Allgemeine Klimageographie Walter de Gruyter Berlin

Blum U, Tingey DT (1977) A study of the potential ways in which ozone could reduce root growth and nodulation of soybean. Atmos Environ 11: 737–739

Blume HP (1981) Alarmierende Versauerung Berliner Forsten. Berl Naturschutzbl 1981: 713–715

Blume HP (1988) Düngung schleswig-holsteinischer Böden in ihrer Bedeutung für Boden- und Wasserschutz. In: Landesnaturschutzverband Schleswig-Holstein: Grüne Mappe 1988: 18–23

Blume HP (ed) (1990) Handbuch des Bodenschutzes. ecomed, Landsberg Lech

Blume HP, Brümmer G (1987a) Prognose des Verhaltens von Schwermetallen in Böden. Mitt Dtsch Bodenkdl Ges 53: 111–117

Blume HP, Brümmer G (1987b) Prognose des Verhaltens von Pflanzenbehandlungsmitteln in Böden mittels einfacher Feldmethoden. Landwirtsch Forsch 40: 41–50

Blume HP, Drewes H, Horn R (1978) Bodenwasserbilanzen mittels Tracer. Mitt Dtsch Bodenkdl Ges 26: 103–114

Blume HP, Litz N, Dörung HW (1983) Adsorption, percolation, and decomposition methods for forecasting the behaviour of organic chemicals in soils. Ecotoxicol Environ Safety 7: 204–215

Blumer M (1976) Polycyclic aromatic compounds in nature. Sci Am 234: 35–45

Blumer M, Blumer W, Teich T (1977) Polycyclic hydrocarbons in soils of a mountain valley: correlation with highway traffic and cancer incidence. Environ Sci Technol 11: 1082–1084

Bobrowski U (1982) Pflanzengeographische Untersuchungen der Vegetation des

Bornhöveder Seengebietes auf quantitativ-soziologischer Basis. Kiel Geogr Schr 56: 1–168

Boehm PD, Fiest DL (1982) Subsurface distributions of petroleum from an offshore well blowout. The Ixtoc I blowout, Bay of Campeche. Environ Sci Technol 16: 67–74

Boening D (1989) Untersuchungen zur Modellierung des Stofftransportes in einem durch Deponiesickerwasser verunreinigten Grundwassergerinne. Berichte – Rep Geol Paläont Inst Univ Kiel 30

Boersma L (1965) Field measurement of hydraulic conductivity above a water table. Agronomy 9: 234–252

Bolin B (ed) (1981) Carbon cycle modelling. SCOPE 16. Wiley, Chichester

Bolin B, Cook RB (1983) The major biogeochemical cycles and their interactions. SCOPE 21. Wiley, Chichester

Bolin B, Degens ET, Kempe S, Ketner P (eds) (1979) The global carbon cycle. SCOPE 13. Wiley, Chichester

Bolt GH, Bruggenwert MGM (1976/1982) Soil chemistry. A. Basic elements. B. Physicochemical models. Elsevier, Amsterdam

Bonazountas M (1987) Chemical fate modelling in soil systems: a state-of-the art review. In: Barth H, L'Hermite P (eds) Scientific basis for soil protection in the European Community. Elsevier, London, pp 487–566

Bonazountas M, Wagner J (1984) SESOIL: A seasonal soil compartment model. UESPA/OTS Contract 68-01-6271. Report by Arthur D Little Inc, Cambridge MA

Bonell M, Gilmour DA, Cassels DS (1983) A preliminary survey of the hydraulic properties of rainforest soils in tropical north–east Queensland and their implications for the runoff process. CATENA Suppl 4: 57–78

Bonse G, Metzler M (1978) Biotransformationen organischer Fremdsubstanzen. Thieme, Stuttgart

Bork HR (1980) Oberflächenabfluß and Infiltration – Qualitative and quantitative Analysen von 50 Starkregensimulationen in der Südheide (Ostniedersachsen) Landschaftsgenese und Landschaftsökologie 6: 1–104

Bork HR (1983) Die holozäne Relief- und Bodenentwicklung in Lößgebieten - Beispiele and dem sudöstlichen Niedersachsen. CATENA Suppl 3: 1–93

Borka G (1980) The effect of cement dust pollution on growth and metabolism of *Helianthus annuus*. Environ Pollut A 22: 75–80

Bos R, Goudena EJG, Guicherit R, Hoogeven A, de Vreede JAF (1978) Atmospheric precursors and oxidants concentrations in The Netherlands. In: Guicherit R (ed) Photochemical smog formation in The Netherlands. s'Gravenhage, pp 20–59

Bossel H (1986) Dynamics of forest dieback: Systems analysis and simulation. Ecol Modelling 34: 259–288

Bossel H (1990) Umweltwissen. Bossel, Kassel

Bossel H, Metzler W, Schäfer H (eds) (1985) Dynamik des Waldsterbens. Springer, Berlin Heidelberg New York

Bouwer H, Jackson RD (1974) Determining soil properties. Agronomy 17: 611–666

Bowen IS (1926) The ratio of heat losses by conduction and by evaporation from any water surface. Physiol Rev 27: 779–787

Boxman AW, Roelofs JGM (1986) Some physiological effects of NH_4^+ and Al^{3+} on pine forest ecosystems. Schriftenr "Texte" Umweltbundesamtes 19: 407–414

Boyd SA (1982) Adsorption of substituted phenols by soil. Soil Sci 134: 337–343

Boyd SA, King R (1984) Adsorption of labile organic compounds by soil. Soil Sci 137: 115–119

Braden H (1985) Ein Energiehaushalts- und Verdunstungsmodell für Wasser- und Stoffhaushaltsuntersuchungen landwirtschaftlich genutzter Einzugsgebiete. Mitt Dtsch Bodenk Ges 42: 294–298

Brain CK (1981) The hunters or the hunted. Chicago Univ Press, Chicago

Brandt J, Agger P (eds) (1984) Methodology in landscape ecological research and planning. Proc 1st Int Sem Int Assoc of Landscape Ecol Roskilde, Roskilde Univ Centre

Branson DR (1978) Predicting the fate of chemicals in the aquatic environment from laboratory data. In: Cairns J, Dickson KL, Maki AW (eds) Estimating the hazard of chemical substances to aquatic life. Am Soc Testing Materials, Philadelphia Pa, pp 55–70

Brdicka R (1971) Grundlagen der physikalischen Chemie. Deutscher Verlag der Wissenschaften, Berlin

Bressan M, Brunetti R, Casellato S, Fava GC, Giro P et al. (1989) Effects of linear alkylbenzene sulfonate (LAS) on benthic organisms. Tenside Surfact Deterg 26: 148–158

Bressan RA, Wilson LG, LeCureux L, Filner P (1978) Use of ethylene and ethane emission to assay injury by SO_2. Plant Physiol 61: S–59

Briggs GG (1973) A simple relationship between soil sorption of organic chemicals and their octanol/water partition coefficients. Proc Br Insect Fung Conf 7: 83–86

Brimblecombe P, Stedman DH (1982) Historical evidence for a dramatic increase in the nitrate component of acid rain. Nature 298: 460–462

Brintzinger H, Beier HG (1933) Die Beeinflussung der Löslichkeit schwerlöslicher Stoffe durch hydrophile Kolloide bzw. adsorbierende Stoffe. I. Löslichkeitsbeeinflussung durch Gelatine. Kolloid-Z 64: 160–172

Brown CC (1978) The statistical analysis of dose-effect relationships. In: Butler GC (ed) Principles of ecotoxicology. Wiley, Chichester, pp 115–148

Brown DH (1976) Mineral uptake by lichens. In: Brown DH et al. (eds) Lichenology: progress and problems. pp 419–439

Bruck RI, Shafer SR, Heagle AS (1981) Effects of simulated acid rain on the development of fusiform rust on loblolly pine. Phytopathology 71: 864

Brümmer G, Tiller KG, Herms U, Clayton PM (1983) Adsorption–desorption and/or precipitation–dissolution processes of zinc in soils. Geoderma 31: 337–354

Brümmer G, Gerth J, Herms U (1986) Heavy metal species, mobility and availability in soils. Z Pflanzenernähr Bodenkd 149: 382–398

Brümmer G, Fränzle O, Kuhnt G, Kukowski H, Vetter L (1987) Fortschreibung der OECD-Prüfrichtlinie 'Adsorption/Desorption' im Hinblick auf die Übernahme in Anhang V der EG-Richtlinie 79/831: Auswahl repräsentativer Böden im EG-Bereich und Abstufung der Testkonzeption nach Aussagekraft und Kosten. Forschungsbericht 106 02 045 im Umweltforschungsplan des BMU, Kiel 1987

Bruland KW, Bertine K, Koide M, Goldberg ED (1974) History of metal pollution in Southern California coastal zone. Environ Sci Technol 8: 425–431

Bryan RB, Yair A (eds) (1982) Badland geomorphology and piping. GeoBooks, Norwich

Bucher JB (1978) SO_2-induziertes Stress-Äthylen in den Assimilationsorganen von Waldbaumarten. Ber Int Union For Res Organis (Ljubljana) X: 93–102

Buck M, Ihlfeld H, Ellermann K (1982) Die Entwicklung der Immissionsbelastung in den letzten 15 Jahren in der Rhein–Ruhr–Region. LIS-Ber 18

Budyko MI (1968) Solar radiation and the use of it by plants. Agroclimatological methods. Proc of the Reading Symp, Unesco, Paris, pp 39–53

Büchel KH (1987) Forschung für den Umweltschutz. Bayer-Umweltperspektive: 56–77

Bücking W, Krebs A (1986) Interzeption und Bestandesniederschläge von Buche und Fichte im Schönbuch. In: Einsele G (ed) Das landsohaftsökologische Forschungsprojekt Naturpark Schönbuch, VCH, Weinheim, pp 113–131

Buijsman E, Maas HFM, Asman WAH (1987) Anthropogenic NH_3 emissions in Europe. Atmos Environ 21: 1009–1022

Bundesgesundheitsamt (1973) Schadwirkungen von Tensiden unter wasserwirtschaftlichen Gesichtspunkten. Bundesgesundheitsblatt 18: 258–263

Bundesminister für Ernährung, Landwirtschaft und Forsten (1985) Waldschadenserhebung 1985. BMELF, Bonn

Bundesminister für Forschung und Technologie (1985) Umweltforschung zu Waldschäden. 2. Bericht. BMFT, Bonn

Bundesminister des Innern (BMI) (1981a) Internationales Vergleichsprogramm mit dem Gaußschen Ausbreitungsmodell. Umwelt 86: 35–36

Bundesminister des Innern (ed) (1981b) Was Sie schon immer über Umweltschutz wissen wollten. Kohlhammer, Stuttgart

Bunting BT (1965) The geography of soil. Hutchinson, London

Burhenne W (ed) (1962) Umweltrecht – Raum und Natur – Systematische Sammlung der Rechtsvorschriften des Bundes und der Länder. 5 Bde. Schmidt, Berlin

Burns R, Hardy R (1975) Nitrogen fixation in bacteria and higher plants. Springer, Berlin Heidelberg New York

Butler GC (ed) (1978) Principles of ecotoxicology. Wiley, Chichester

Butler PF, Prescott JA (1955) Evapotranspiration from wheat and pasture in relation to available moisture. Aust J Agric Res 6: 52–61

Butzer K (1977) Environment, culture and human evolution. Am Sci 65: 572–584

Butzke H (1981) Versauern unsere Wälder? Erste Ergebnisse der Überprüfung 20 Jahre alter pH-Wert-Messungen in Waldböden Nordrhein-Westfalens. Forst- Holzwirt 36: 542–548

Cabridenc R, Chouroulinkov I (1977) Tests biologiques d'evaluation d'effets toxiques de substances chimiques dans l'environnement. Ministère de la culture et de l'environnement, Paris

Cabridenc R, Yana A (1978) Biological laboratory tests permitting evaluation of the effects of a chemical substance on aquatic flora and fauna. Berichte d Umweltbundesamtes 10/78: 189–199

Cairns J (1979) Hazard evaluation with microcosms. Int J Environ Stud 13: 95–99

Cairns J, Dickson KL, Maki AW (1978) Estimating the hazard of chemical substances to aquatic life. ASTM (American Soc for Testing and Material) Spec Techn Publ 657, Philadelphia

Campbell R (1980) Mikrobielle Ökologie. Verlag Chemie, Physik, Weinheim

Carr RA, Wilkniss PE (1973) Mercury in Greenland ice sheet: Further data. Science 181: 843–844

Carter WPL, Darnall KR, Lloyd AC, Winer AM, Pitts JN (1976) Evidence for alkoxy radical isomerization in photooxidations of C_4–C_6 alkanes under simulated atmospheric conditions. Chem Phys Lett 42: 22–27

Carter WPL et al. (1981) Major atmospheric sink for phenols and cresols. Reaction with nitrate radical. Environ Sci Technol 15: 829–831

CEC Directive (1986) Council Directive of 12 June 1986 on the protection of the environment, and in particular of the soil, when sewage sludge is used in agriculture. Off J Eur Commun L181/6-12, 4 July 1986

Cederberg GA et al. (1985) A groundwater mass transport and equilibrium chemistry model for multicomponent systems. Water Resour Res 21: 1095–1104

Chamberlain AC (1953) Aspects of travel and deposition of aerosols and vapour clouds. AERE HP/R 1261 – AERE Harwell

Chamberlain AC (1975) The movement of particles in plant communities. In: Monteith JL (ed) Vegetation and the atmosphere. Academic Press, London

Chamberlain AC, Chadwick RC (1953) Deposition of airborne radioiodine vapour. Nucleonics 11: 22–25

Chandler TJ (1970) The climate of London. Urban climatology: summary and conclusions of the symposium in Urban Climates. WMO Tech Note 108: 375–379

Chandra M, O'Driscoll KF, Rempel GL (1984) Ligand sorption of aromatic amines on resin-bound ferrous ion. React Polymers 2: 279–290

Chandra P, Bollen WB (1961) Effects of nabam and mylone on nitrification, soil respiration and microbial numbers in four Oregon soils. Soil Sci 92: 387–393

Chang JS, Penner JE (1978) Analysis of global budgets of halocarbons. Atmos Environ 12: 1867–1873

Chapman J, Shorter J (1972) Advances in linear free-energy relationships. Plenum Press, New York

Chapman TG (1957) Two-dimensional ground-water flow through a bank with vertical faces. Géotechnique 7: 35–40

Chapman TG (1964) Effects of ground-water storage and flow on the water balance. In: Water resources, use and management. Melbourne University Press, pp 290–301

Charlson RJ, Rodhe H (1982) Factors controlling the acidity of natural rain water. Nature 295: 683–685

Chater KWA, Somerville HJ (eds) (1978) The oil industry and microbial ecosystems. Heyden & Son, New York

Chen W, Li Y, vd Ploeg RR (1990a) Modelling multi-ion transport in saturated soil and parameter estimation. I: Theory. Z Pflanzenernähr Bodenkd 153: 167–173

Chen W, Li Y, vd Ploeg RR (1990b) Modelling multi-ion transport in saturated soil and parameter estimation. II: Experimental. Z Pflanzenernähr Bodenkd 153: 175–179

Chiou CT, Shoup TD (1985) Soil sorption of organic vapours and effects of humidity on sorptive mechanism and capacity. Environ Sci Technol 19: 1196

Chow VT (1964) Runoff. In: Chow VT (ed) Handbook of applied hydrology. New York, McGraw-Hill, pp 14-1–14-54

Chu KC, Cueto C, Ward JM (1981) Factors in the evaluation of 200 National Cancer Institute carcinogen bioassays. J Toxicol Environ Health 8: 251–280

CITEPA (1982) Etude de l'évolution de la pollution de l'air dans Paris et sa proche banlieue

Clague JJ, Evans SG, Fulton RJ, Ryder JM, Stryd AH (1987) XII INQUA Congress field excursion A-18: Quaternary geology of the southern Canadian Cordillera. National Research Council of Canada, Ottawa Ont

Clements JB (1978) Philosophy and principles of reference methods. In: Schneider T, De Koning HW, Brasser LY (eds) Air pollution reference measurements, methods and systems. Elsevier Sci Publ, pp 21–30

Clymo RS (1963) Ion exchange in Sphagnum and its relations to bog ecology. Ann Bot 27: 309–324

Cogbill CV (1976) The history and character of acid precipitation in eastern North America. In: Dochinger LS, Seliga TA (eds) Proc First Internt Symposium on acid precipitation and forest ecosystem. USDA Forest Service, General Tech Report NE-23, Upper Darby, PA

Cole LK, Metcalf RL, Sandborn JR (1976) Environmental fate of insecticides in terrestrial model ecosystems. Int J Environ Stud 10: 7–14

Collett AR, Johnston J (1926) Solubility relations of isomeric organic compounds IV. Solubility of the nitroanilines in various liquids. J Phys Chem 30: 70–82

Cooley RD, Manning WF (1987) The impact of ozone on assimilate partitioning in plants: a reviev. Environmental Pollution 47: 95–113

Cooper DG, Zajic JE (1980) Surface–active compounds from microorganisms. Adv Appl Microbiol 26: 229–253

Cordon TC (1966) Synthetic detergents: some aspects of their relation to agriculture. Soil Sci 102: 1–17

Costa LG, Galli CL, Murphy SD (eds) (1987) Toxicology of pesticides: Experimental, clinical and regulatory aspects. NATO ASI, Ser. vol H 13. Plenum, New York London

Cowan IR (1965) Transport of water in the soil–plant–atmosphere system. J appl Ecol 2: 221–239

Cowling EB (1978) Effects of acid precipitation and atmospheric deposition on terrestrial vegetation. In: Galloway et al. (eds) A national program for assessing the problem of atmospheric deposition (acid rain). A report to the Council on Environmental Quality, pp 47–63

Cox RA, Penkett SA (1983) Formation of atmospheric acidity. In: Beilke S, Elshout AJ (eds) Acid deposition. Reidel, Dordrecht, pp 56–81

Cox RA, Derwent RG, Sandalls FJ (1976) Some air pollution measurements made at Harwell, Oxfordshire during 1973–1975. AERE-R 8324. HMSO, London

Crafts AS (1949) Toxicity of ammonium dinitro-o-sec-butyl phenolate in California soils. Hilgardia 19: 159–169

Crafts AS (1961) Translocation in plants. Holt, Rinehart, and Winston, New York

Crafts AS, Reiber HG (1945) Studies of the activation of herbicides. Hilgardia 16: 487–500

Cramér H (1957) Mathematical methods of statistics. Princeton Univ Press, Princeton

Cramer HH (1986) Klimaeinflüsse und Waldschäden. Schriftenr "Texte" Umweltbundesamtes 19: 308–312, 423–426

Cryer R (1980) The chemical quality of some pipeflow waters in upland mid-Wales and its implications. Cambria 6: 1–19

Cupitt LT (1980) Fate of toxic and hazardous materials in the air environment. Report to Environmental Sciences Research Laboratory, US Environmental Protection Agency, EPA-600/3-80-084

Custodio E et al. (1987) Groundwater flow and quality modelling. NATO ASI Ser 224. Reidel, Dordrecht

Czaja AT (1961) Zementstaubwirkungen auf Pflanzen: Die Entstehung der Zementkrusten. Qual Plant Water Veg 7: 201–238

Czaja AT (1966) Über die Einwirkung von Stäuben, speziell von Zementofenstaub auf Pflanzen. Angew Bot 40: 106–120

Czeratzki W (1968) Mehrjährige Vergleichsuntersuchungen zwischen gravimetrischer Methode und Neutronenmessungen zur Kontrolle der Bodenfeuchte bei einem Beregnungsversuch. Landwirtsch Forsch 21: 292–305

Dämmgen M, Grünhage L, Jäger J (1985) System zur flächendeckenden Erfassung von luftgetragenen Schadstoffen und ihren Wirkungen auf Pflanzen. Landschaftsökol Messen Auswerten 1: 95–106

Dahmke A (1988) Lösungskinetik von feldspat-reichen Gesteinen und deren Bezug zu Verwitterung und Porenwasser-Chemie natürlicher Sander-Sedimente. Berichte – Reports Geol Paläont Inst Univ Kiel 20

Dahmke A, Mattheß G, Petersen A, Schenk D (1988) Gleichgewichts- und Ungleichge-
wichtsbeziehungen zwischen Porenlösungen und festen Substanzen der Verwitte-
rungszone quartärer Sedimente. Berichte – Reports Geol Paläont Inst Univ Kiel 25
Dana MT, Hales JM (1976) Statistical aspects of the washout of polydisperse aerosols.
Atmos Environ 10: 45–50
Daniele PG, de Robertis A, de Stefano C, Sammarfano S (1985) On the possibility of
determining the thermodynamic parameters of the formation of weak complexes using
a simple model for the dependence of low molecular weight ligands in aqueous
solution. J Chem Soc Dalton Trans 1985: 2353–2361
Darcy HPG (1856) Les fontaines publiques de la ville de Dijon. Delmont Paris
Darimont T (1985) Naturwissenschaftlich-technische Anforderungen an die Sanierung
kontaminierter Standorte – Teil I: Erarbeitung eines Schemas zur Bewertung des
Migrationsverhaltens von Stoffen im Untergrund. Forschungsbericht 102 03 405 im
UFO-Plan des BMI. Berlin
Darnall KR, Carter WPL, Winer AM, Lloyd AC, Pitts JN (1976a) Importance of RO_2
+ NO in alkyl nitrate formation from C4–C6 alkane photooxidations under simu-
lated atmospheric conditions. J Phys Chem 80: 1948–1950
Darnall KR, Lloyd AC, Winer AM, Pitts JN (1976b) Reactivity scale for atmospheric
hydrocarbons based on reaction with hydroxyl radical. Environ Sci Technol 10:
692–696
Davids P, Lange M (1986) Die TA Luft '86. Technischer Kommentar. VDI, Düsseldorf
Davidson CI, Wu YL (1990) Dry deposition of particles and vapors. In: Lindberg SE,
Page AL, Norton SA (eds) Acidic precipitation, vol 3. Springer, New York, pp 103–216
Davis DD, Coppolino JB (1974) Relationship between age and ozone sensitivity of
current needles of ponderosa pine. Plant Dis Rep 58: 660–663
Dean C, Davis DR (1967) Ozone and soil moisture in relation to the occurrence of
weather fleck on Florida cigar-wrapper tobacco in 1966. Plant Dis Rep 51: 72–75
Delfiner P (1975) Geostatistical estimation of hydrocarbon reserves. Ecole des Mines,
Fontainebleau
Delfs J (1955) Die Niederschlagszurückhaltung im Walde (Interception). Mitt Arbeits-
kreises Wald und Wasser 2, Koblenz
Delmas RJ, Gravenhorst G (1983) Background precipitation acidity. In: Beilke S, Elshout
AJ (eds) Acid deposition. Reidel, Dordrecht, pp 82–108
Delschen T, Werner W (1989) Zur Aussagekraft der Schwermetallgrenzwerte in klär-
schlammgedüngten Böden. Landwirtsch Forsch 42: 29–49
Delvigne J (1965) Pédogenèse en zone tropicale – la formation des minéraux secondaires
en milieu ferralitique. Mém ORSTOM 13: 1–177
Denmead OT, Simpson JR, Freney JR (1974) Ammonia flux into the atmosphere from a
grazed pasture. Science 185: 609–610
Der Rat der Sachverständigen für Umweltfragen (1983) Waldschäden und Luftverun-
reinigungen. Kohlhammer, Stuttgart
Deutsche Forschungsgemeinschaft (DFG) (1988) Polychlorierte Biphenyle. Mitt XIII
Senatskomm zur Prüfung von Rückständen in Lebensmitteln. VCH, Weinheim
Deutscher Wetterdienst (1982) Untersuchungen zur großräumigen Ausbreitung von
Luftbeimengungen. Forschungsbericht 104 04 105 des Umweltforschungsplanes des
Bundesministers des Innern im Auftrag des Umweltbundesamtes, Berlin
de Ploey J (ed) (1983) Rainfall simulation, runoff and soil erosion CATENA Suppl 4.
CATENA, Cremlingen
de Wit CT (1978) Simulation of assimilation, respiration and transpiration of crops.
PUDOC, Wageningen

DGMK (1991) Mikrobieller Abbau von Kohlenwasserstoffen und Kohlenwasserstoffverbindungen. DGMK Forschungsbericht 461–01

Dickson KL, Maki AW, Cairns J (eds) (1979) Analysing the hazard evaluation process. American Fisheries Society, Washington DC

Dierßen K (1990) Einführung in die Pflanzensoziologie. Wiss Buchges, Damstadt

Dietze G (1985) Bindungsformen und Gleichgewichte von Aluminium im Sickerwasser saurer Böden. Dissertation, Univ Göttingen

Diez T (1965) Feuchte- und Dichtebestimmung in Böden mit Hilfe von radioaktiven Strahlen. Z Kulturtech Flurbereinig 4: 12–35

Dikau R (1983) Der Einfluß von Niederschlag, Vegetations-bedeckung und Hanglänge auf Oberflächenabfluß und Bodenabtrag von Meßparzellen. Geomethodica 8: 149–177

Dilger H, Nester K (1975) Aufstellung und Vergleich verschiedener Schemata zur Bestimmung von Ausbreitungskategorien. Meteorol Rundsch 28: 12–17

Dilling WL, Tefertiller NB, Kalles GJ (1975) Evaporation rates and reactivities of methylene chloride, chloroform, 1,1,1-trichloroethylene tetrachloroethylene, and other chlorinated compounds in dilute aqueous solutions. Environ Sci Technol 9: 833–837

di Toro DM (1985) A particle interaction model of reversible organic chemical sorption. Chemosphere 14: 1503–1538

Dobbins KA (1979) Atmospheric motion and air pollution. Pergamon, New York

Dobbs RA, Cohen JM (1980) Carbon adsorption isotherms for toxic organics. EPA 600/8-80-023

Dörrhofer G, Josopait V (1980) Eine Methode zur flächendifferenzierten Ermittlung der Grundwasserneubildungsrate. Geol Jahrb C 27: 45–65

Domsch KH (1985) Funktionen und Belastbarkeit des Bodens aus der Sicht der Bodenmikrobiologie. Materialien zur Umweltforschung 13. Kohlhammer, Stuttgart

Domsch KH, Jagnow G, Anderson TH (1983) An ecological concept for the assessment of side-effects of agrochemicals on soil-microorganisms. Residue Rev 86: 65–101

Doner HE, Mortland MM (1969) Intermolecular interactions in montmorillonites: NH–CO systems. Clays and Clay Minerals 17: 265–270

Dong PH, Kramer H (1987) Zuwachsverlust in erkrankten Fichtenbeständen. Allg Forst-Jagdztg 158: 122–125

Donnelly T (1984) Kinetics and mathematical modelling of an anaerobic contact digester. PhD Thesis, University of Newcastle upon Tyne

Draggan S, Giddings JM (1978) Testing toxic substances for protection of the environment. Sci Total Environ 9: 63–74

Draxler R, Taylor D (1982) Horizontal dispersion parameters for long-range transport modeling. J Appl Meteorol 21: 367–372

Drescher J, Horn R, de Boodt M (eds) (1988) Impact of water and external forces on soil structure. CATENA Suppl 11

Drewes H, Blume H-P (1977) Abbau, Bewegung und Sorption von Herbiziden in Böden. Landwirtsch Forsch Sonderh 33: 104–113

Driscoll CT, Likens GE (1982) Hydrogen ion budget of an aggrading forested ecosystem. Tellus 34: 283–292

Duce RA, Hoffmann GL, Zoller WH (1975) Atmospheric trace metals at remote northern and southern hemispheric sites: pollution or natural? Science 187: 59–61

Duchaufour P (1960) Précis de pédologie. Masson, Paris

Duinker JC (1981) Chemical pollution in the Wadden Sea. In: Tougaard S, Helweg Ovesen C (eds) Environmental problems of the Waddensea region. Fiskeri- og Søfartsmuseet, Esbjerg, pp 77–85

Duinker JC, Hillebrand MTJ (1979) Behaviour of PCB, pentachlorobenzene, hexa-chlorobenzene, α-HCH, γ-HCH, β-HCH, dieldrin, endrin and p,p'-DDD in the Rhine/ Meuse estuary and the adjacent coastal area. Neth J Sea Res 13: 256–281

Dunin FX (1976) Infiltration: its simulation for field conditions. In: Rodda JC (ed) Facets of Hydrology. Wiley, London, pp 199–227

Dunn B, Stich H (1976) Monitoring procedures for chemical carcinogens in coastal waters. J Fish Res Board Can 33: 2040–2046

Dunn B, Young D (1976) Baseline levels of benzo(a)pyrene in Southern California mussels. Mar Pollut Bull 7: 231–234

Dutil P, Durand R (1974) Acquisition des caractères chimiques de l'eau en milieu calcaire: Passage du sol à la nappe de la Craie en Champagne (France). Mém Congr Montpellier, Tome X: 61–67

Duvigneaud P, Kestemont P (ed) (1977) Productivité biologique en Belgique. Trav Sect Belge Progr Biol Int SCOPE, Paris-Gembloux

Duxbury JM, Bouldin DR, Terry RE, Tate III RL (1982) Emissions of nitrous oxide from soils. Nature 298: 462–464

Duynisveld WHM (1983) Entwicklung von Simulationsmodellen für den Transport von gelösten Stoffen in wasserungesättigten Böden und Lockersedimenten. Umweltbundesamt "Texte" 17/83, Berlin

Duysings JJHM, Verstraten JM, Bruynzeel L (1983) The identification of runoff sources of a forested lowland and catchment: a chemical and statistical approach. J Hydrol 64: 357–375

DVWK (Deutscher Verband für Wasserwirtschaft und Kulturbau e.V.) (1989) Abschätzen des Verhaltens organischer Chemikalien in Böden. DVWK, Bonn

Dyer AI (1961) Measurements of evaporation and heat transfer in the lower atmosphere by an automatic eddy-correlation technique. Q J R Meteorol Soc 87: 401–412

Dzerdzeevskii BL (1958) On some climatological problems and microclimatological studies of arid and semiarid regions in U.S.S.R. Climatology and microclimatology. Proc Canberra Symp/Climat et microclimatol Actes du colloque de Canberra. Unesco (Arid zone research/Recherches sur la zone aride XI), Paris, pp 315–325

Dzubay TG, Stevens RK (1975) Ambient air analysis with dichotomous sampler and X-ray fluorescence spectrometer. Environ Sci Technol 9: 663–668

Eatough DJ, Smith TJ, Mangelson NF, Hansen LD, Jensen TE et al. (1975) Metals, S (IV), and S (VI) in smelter particulates. Abstr Int Conf Heavy metals in the Environment, Toronto 1975, pp D89–91

Eatough DJ, Eatough NL, Hill MW, Mangelson NF, Ryder J et al. (1979) The chemical composition of smelter fine dusts. Atmos Environ 13: 489–506

ECETOC (European Chemical Industry Ecology and Toxicology Centre) (1983a) Experimental assessment of the phototransformation of chemicals in the atmosphere. Technical Report 7. ECETOC, Brussels

ECETOC (1983b) Biodegradation testing: an assessment of the present status. Technical Report No 8. ECETOC, Brussels

ECETOC (1985) Harmonisation of ready biodegradability tests. Technical Report No 18. ECETOC, Brussels

ECETOC (1986a) Structure–activity relationships in toxicology and ecotoxicology: an assessment. ECETOC monograph 8, Brussels

ECETOC (1986b) Evaluation of the toxicity of substances to be assessed for biodegradability. Technical Report 23. ECETOC, Brussels

ECETOC (1986c) Biodegradation tests for poorly-soluble compounds. Technical Report No 20. ECETOC, Brussels

ECETOC (1988a) Nitrate and drinking water. Technical Report 27. ECETOC, Brussels

ECETOC (1988b) Evaluation of anaerobic biodegradation. Technical Report 28. ECETOC, Brussels

ECETOC (1988c) Concentrations of industrial organic chemicals measured in the environment: The influence of physico-chemical properties, tonnage and use pattern. Technical Report 29. ECETOC, Brussels

ECETOC (1988d) Evaluation of anaerobic biodegradation. Technical Report No 28. ECETOC, Brussels

ECETOC (1991) Biodegradation kinetics. Technical Report No 44. ECETOC, Brussels

Edwards CA, Thomson AR (1973) Pesticides and the soil fauna. Res Rev 45: 2–79

EG-Working Party on Motor Vehicles (1983) Report of the air quality subgroup on item 1.3 of the ERGA-programme. Comm Europ Communities, Brussels

Ehlers W (1975) Observations on earthworm channels and infiltration on tilled and untilled loess soil. Soil Sci 119: 242–249

Eichinger L, Merkel B, Nemeth G, Salvamoser J, Stichler W (1984) Seepage velocity determinations in the unsaturated quaternary gravel. A comparison of various methods. Proc Int Symp RIZA 1: 303–315, Munich

Eidmann FE (1960) Die Interception in Buchen- and Fichtenbeständen-Ergebnis mehrjähriger Untersuchungen im Rothaargebirge (Sauerland). Publ Nr 48, Ass Int Hydr Sci, Gentbrugge, pp 5–25

Eilenberger G (1986) Die Erforschung komplexer Systeme. Allg Forstzg 22: 537–542

Einsele G, Agster G, Elgner M (1986) Niederschlag-Bodenwasser-Abflußbeziehungen bei Hochwasserereignissen im Keuper-Lias-Bergland des Schönbuchs. In: Einsele G (ed) Das landschaftsökologische Forschungsprojekt Naturpark Schönbuch VCH, Weinheim, pp 209–234

Eisenreich SJ, Looney BB, Thornton JD (1981) Airborne organic contaminants in the Great Lakes ecosystem. Environ Sci Technol 15: 30–38

Ekhalt DH (1974) The atmosphere cycle of methane. Tellus 26: 58–70

Eliassen A, Hov Ø, Isaksen ISA, Saltbones J (1982) A Lagrangian long-range transport model with atmospheric boundary layer chemistry. In: Schneider T, Grant L (eds) Air pollution by nitrogen oxides. Elsevier, Amsterdam, pp 347–356

Elkiey T, Omrod DP (1980) Response of turfgrass cultivars to ozone, sulfur dioxide, nitrogen dioxide, or their mixtures. J Am Soc Hortic Sci 105: 664–668

Ellenberg H (1979) Vegetation Mitteleuropas mit den Alpen. Ulmer, Stuttgart

Ellenberg H, Fränzle O, Müller P (1978) Ecosystem research with a view to environmental policy and development planning. Federal Ministry of the Interior of the Federal Republic of Germany, Bonn

Ellenberg H, Meyer R, Schauermann J (ed) (1986) Ökosystemforschung. Ergebnisse des Solling-Projekts. Enke, Stuttgart

Eller BM (1977a) Beeinflussung der Energiebilanz von Blättern durch Straßenstaub. Angew Bot 51: 9–15

Eller BM (1977b) Road dust-induced increase of leaf temperature. Environ Pollut 13: 99–107

Elstner EF, Osswald W (1984) Polymer-Beschichtung von Blättern und Nadeln – ein neuer Weg im Pflanzenschutz. Forst- and Holzwirt 39: 86–88

Elstner EF, Osswald W, Youngman RY (1985) Basic mechanisms of pigment bleaching and loss of structural resistance in spruce (Picea abies) needles: advances in phytomedical diagnostics. Experientia 41: 591–597

Emeis S (1985) Subsynoptic vertical heat fluxes in the atmosphere over Europe. Bonn Meteorol Abh 32

Enfield CG, Carsel RF, Cohen SZ, Phan T, Walters DM (1980) Approximating pollutant transport to groundwater. USEPA, RSKERL, Ada OK

Eno CF (1958) Insecticides and the soil. J Agric Food Chem 6: 348–351

Enslein K, Craig PN (1978) A toxicity estimation model. J Environ Pathol Toxicol 2: 115–132

Erikson K-EL, Blanchette RA, Ander P (1990) Microbial and enzymatic degradation of wood and wood components. Springer, Berlin Heidelberg New York

Erlenkeuser H, Suess E, Willkomm H (1974) Industrialization affects heavy metal and carbon isotope concentrations in recent Baltic Sea sediments. Geochim Cosmochim Acta 38: 823–842

Ernst W (1982) Tiere als Monitororganismen für organische Schadstoffe. Decheniana Beih 26: 55–66

Ernst WHO (1982) Monitoring of particulate pollutants. In: Steubing L, Jäger JH (eds) Monitoring of air pollutants by plants, Junk, The Hague, pp 121–128

Eurocop-Cost (1984) Extracts from an inventory of organic pollutants which have been identified in various surface waters, effluent discharges, aquatic animals and plants, and bottom sediments. Cost-Project 64B. Comm of the European Communities, Brussels

Evans LS, Ting IP (1974) Ozone sensitivity of leaves: relationship to leaf water content, gas transfer resistance and anatomical characteristics. Am J Bot 61: 592–597

Evans LS, Lewin KF, Conway CA, Patti MJ (1981) Seed yields (quantity and quality) of field-grown soybeans exposed to simulated acidic rain. New Phytol 89: 459–470

Evans LS (1982) Biological effects of acidity on vegetation: a review. Environ Expt Bot 22: 155–169

Evans LS, Curry TM (1979) Differential responses of plant foliage to simulated acid rain. Amer J Bot 66: 953–962

Evans LS, Gmur NF, Dacosta F (1977) Leaf surface and histological perturbations of leaves of Phaseolus vulgaris and Helianthus annuus after exposure to simulated acid rain. Am J Bot 64: 903–913

Evans LS, Gmur NF, Dacosta F (1978) Foliar response of six clones of hybrid poplar to simulated acid rain. Phytopathology 68: 847–856

Evans LS, Gmur NF, Kelsch JJ (1977) Perturbations of upper leaf surface structures by acid rain. Environ Exp Bot 17: 145–149

Evans LS, Hendrey GR, Stensland GJ, Johnson DW, Francis AJ (1981) acid precipitation: Considerations for an air quality standard. Water, Air, Soil Pollut 16: 469–509

Evans SG, Buchanan RG (1976) Some aspects of natural slope stability in silt deposits near Kamloops, British Columbia. 29th Can Geotech Conf, Vancouver, Proc IV-1-32. National Res Council of Canada, Ottawa, Ont

Ewert E (1978) Vegetationsschäden in der Umgebung landwirtschaftlicher Tierproduktionsanlagen. Luft- Kältetech 74: 218–220

Ewert E (1979) Zur Phytotoxität von Ammoniak. Hercynia NF 16: 75–80

Fabrewitz S (1986) Die Simulation des Stoffumsatzes in einem Agrarökosystem. Dr Dissert, Univ Osnabrück

Farmer VC, Mortland MM (1966) An i.r. study of the coordination of pyridine and water to exchangeable cations in montmorillonite and saponite. J Chem Soc Am 88: 344–351

Farrell EP, Nelsson J, Tamm CO, Wiklander G (1980) Effects of artificial acidification with sulphuric acid on soil chemistry in a Scots pine forest. In: Drabløs D, Tollan A (eds) Proc Int Conf Ecol Impact Acid Precipitation. SNSF-Project, Norway, pp 186–187

Faulstich H (1986) Verursacht Triethylblei Waldschäden? Schriftenr "Texte" Umweltbundesamtes 19: 22–27, 381–382

Faust CR (1984) Transport of immiscible fluids within and below the unsaturated zone – a numerical model. Geotrans Report 84-01. Geotrans, Herdon VA

Faust SL, Hunter JV (eds) (1978) Kinetics of biologically mediated oxidation of organic compounds in aquatic environments. Marcel Dekker, New York

Favre A, Kovasznay LS, Dumas R et al. (1976) La turbulence en mécanique des fluides. Gauthiers Villars, Paris

Feicht PG (1980) Exposure of soybeans to ozone or simulated acid rain and interactions with *Glomus macrocarpus*. MSc Thesis, North Carolina State University, Raleigh NC

Ferenbaugh RW (1976) Effects of simulated acid rain on *Phaseolus vulgaris* L. (Fabaceae). Am J Bot 63: 183–188

Ferguson P, Lee JA (1983) The growth of *Sphagnum* species in the southern Pennines. J Bryology 12: 579–586

Fernley HN, Evans WC (1958) Oxydative metabolism of polycyclic hydrocarbons by soil pseudomonads. Nature 182: 373–375

Field B (1976) Forbruk av fossilt breusel i Europa og utslipp av SO_2 i perioden 1900–1972. Norwegian Institute for Air Research, Teknik Notat 1/76

Figge K, Klahn J, Koch J (1985) Chemische Stoffe in Ökosystemen. Fischer, Stuttgart

Figge K, Schöberl P (1989) LAS and the application of sewage sludge in agriculture. Tenside Surfact Deterg 26: 122–128

Filip Z (1975) Wechselbeziehungen zwischen Mikroorganismen und Tonmineralen und ihre Auswirkung auf die Bodendynamik. Habil-Schrift, Univ Gießen

Fitter AH, Hay RK (1987) Environmental physiology of plants. Academic Press, London

Fittkau EJ (1983) Flow of nutrients in a large open system: the basis of life in Amazonia. Environmentalist 3, Suppl 5: 41–49

Fitzgerald GP (1962) The control of the growth of algae with CMU. WISC Acad Sci, Arts Lett 46: 281–294

Flaig-Baumann R, Herrmann M, Boehm HP (1970) Reaktionen der basischen Hydroxylgruppen auf der Oberfläche von Titandioxid. Z Anorg Allg Chem 372: 296–307

Flegel R (1970) Ein mathematisches Modell für die Optimierung von hydrotechnischen Schutzmaßnahmen gegen Wassererosion und seine Verwendung bei Planung und Projektierung. Z Landeskultur 11: 97–114

Fleischer S, Hamrin S, Kindt T, Rydberg L, Stibe L (1987) Coastal eutrophication in Sweden: reducing nitrogen in land run-off. Ambio 16: 246–251

Fleischhauer WJ, Schütze W, Zeise R (1983) Ausbreitung und Immissionshöhe von Schadstoffen aus Kohlekraftwerken in Abhängigkeit von der Schornsteinhöhe. Luftverunreinigung 1983: 1–6

Flohn H (1968) Contributions to a meteorology of the Tibetan highlands. Atmospheric Science Paper 130, Colorado State Univeristy, Fort Collins, Col

Flügel WA (1979) Untersuchungen zum Problem des Interflow. Heidelberger Geogr Arb 56, Heidelberg

Flügel WA (1981) Grundwassererneuerung durch Interflow in Abhängigkeit von Bodenfeuchte und Niederschlag. Beitr Hydrogeol Sonderh 2: 13–35

Förster R (1986) Ein Konvektions–Diffusions–Transportmodell mit Multispezies-Kationenaustausch, Ionenkomplexierung und Aluminiumhydroxosulfat. Diss, Univ Göttingen, 146 pp

Förstner U, Müller G (1974) Schwermetalle in Flüssen und Seen als Ausdruck der Umweltverschmutzung. Springer, Berlin Heidelberg New York

Folkeson L (1979) Interspecies calibration of heavy-metal concentrations in nine mosses and lichens: applicability to deposition measurements. Water Air Soil Pollut 11: 253–260

Foote HP (1982) (Information available from: Battelle Pacific Northwest Laboratories, PO Box 999, Richland, VA)

Forbes EA, Posner AM, Quirk JP (1976) The specific adsorption of divalent Cd, Co, Cu, Pb, and Zn on goethite. J Soil Sci 27: 154–166

Foster SSD, Bridge LR, Geak AK, Lawrence AR, Parker JH (1986) The groundwater nitrate problem: a summary of research on the impact of agricultural land-use practices on groundwater quality between 1976 and 1985. Hydrogeol Rep Brit Geol Surv 86/2. Geol Survey, London

Fournier JC (1980) Enumeration of the soil micro-organisms able to degrade 2,4-D by metabolism or co-metabolism. Chemosphere 9: 169–174

Fournier JC et al. (1981) Soil-adaptation at 2,4-D degradation in relation to the application rates and the metabloic behaviour of the degrading microflora. Chemosphere 10: 977–984

Fowler D, Cape JN (1984) The contamination of rain samples by dry deposition on rain collectors. Atmos Environ 18: 183–189

Fränzle O (1971a) Untersuchungen zur altquartären Tonilluviation der Sables de Fontainebleau. In: Ters M (ed) Etudes sur le Quaternaire dans le Monde. Vol I Suppl Bull AFEQ, 1971: 351–358

Fränzle O (1971b) Die Opferkessel im quarzitischen Sandstein von Fontainebleau. Z Geomorph NF 15: 212–235

Fränzle O (1976a) Die Schwankungen des pleistozänen Hygroklimas in Südost-Brasilien und Südost-Afrika. Biogeographica 7: 143–162

Fränzle O (1976b) Der Wasserhaushalt des amazonischen Regenwaldes und seine Beeinflussung durch den Menschen. Amazoniana 6: 21–46

Fränzle O (1976c) Ein morphodynamisches Grundmodell der Savannen- und Regenwaldgebiete. Z Geomorph NF, Suppl 24: 177–184

Fränzle O (1977a) Biophysical aspects of species diversity in tropical rain forest eco-systems. Biogeographica 8: 69–83

Fränzle O (1977b) Hang- und Flächenbildung in den Tropen unter dem Einfluß der Eisen- und Aluminiumdynamik. Z Geomorph NF, Suppl 28: 62–80

Fränzle O (1978a) Die Struktur und Belastbarkeit von Ökosystemen. Amazoniana 6: 279–297

Fränzle O (1978b) The structure of soil associations and cenozoic morphogeny in Southeast Africa. In: Nagl H (ed) Beiträge zur Quartär- und Landschaftsforschung – Festschrift. Fink, Wien, pp 159–176

Fränzle O (1979) Comparative studies on species diversity of plant associations in the U.S.A. and Northern Germany. Biogeographica 16: 113–126

Fränzle O (1981) Vergleichende Untersuchungen über Struktur, Entwicklung und Standortsbedingungen von Biozönosen in den immerfeuchten Tropen und der gemäßigten Zone. Aach Geogr Arb 14(1): 167–191

Fränzle O (1982) Erfassung von Ökosystemparametern zur Vorhersage der Verteilung von neuen Chemikalien in der Umwelt. "Texte" des Umweltbundesamtes Berlin, pp XV + 347

Fränzle O (1983) Ökosystemforschung: allgemeine Grundlagen und Definition, trophische Strukturen, biozönotische Gesetze und Thermodynamik. In: DFG (ed) Ökosystemforschung als Beitrag zur Beurteilung der Umweltwirksamkeit von Chemikalien. Verlag Chemie, Weinheim, pp 21–29

Fränzle O (1984a) Regionally representative sampling. In: Lewis RA, Stein N, Lewis CW (eds) Environmental specimen banking and monitoring as related to banking. Ni jhoff, Boston, pp 164–179

Fränzle O (1984b) Die Bestimmung von Bodenparametern zur Vorhersage der potentiellen Schadwirkung von Umweltchemikalien. Angew Bot 58: 207–216

Fränzle O (1986) Statistische Analyse der Datenstruktur der 512-Stoffliste. In: Beratergremium für Umweltrelevante Altstoffe (BUA) der Ges Dt Chemiker (ed) Umweltrelevante Alte Stoffe. VCH, Weinheim, pp 195–211

Fränzle O (1988) Environmental impact and environmental protection in the Federal Republic of Germany. Geogr Rundsch, Spec Ed, pp 4–11

Fränzle O (1990) Ökosystemforschung und Umweltbeobachtung als Grundlage der Raumplanung. MAB- Mitteilungen 33: 26–39

Fränzle O (1991) Mikrobielle Sanierung kontaminierter Böden und Lockergesteine. Geogr Rundschau 43: 84–89

Fränzle O, Bobrowski U (1983) Untersuchungen zur ökologischen Aussagefähigkeit floristisch definierter Vegetationseinheiten. Verh Ges Ökol Göttingen 11: 101–109

Fränzle O, Kuhnt G (1983) Regional repräsentative Auswahl der Böden für eine Umweltprobenbank. Exemplarische Untersuchung am Beispiel der Bundesrepublik Deutschland. Forschungsbericht 106 05 028 im Umweltforschungsplan des Bundesministers des Innern/Umweltbundesamtes. Kiel

Fränzle O, Zilling L (1984) Abschätzung der Exposition von Umweltchemikalien: Flächenhafte Verteilung in Geoökosystemen. Forschungsbericht 106 05 029/01 im Umweltforschungsplan des Bundesministeriums des Innern, Kiel

Fränzle O, Killisch WF, Ingenpaß A, Mich K (1980) Die Klassifizierung von Bodenprofilen als Grundlage agrarer Standortplanung in Entwicklungsländern. Ein Beispiel aus dem Savannengebiet Nordost-Ghanas. CATENA 7: 353–381

Fränzle O, Schröder W, Vetter L (1985) Synoptische Darstellung möglicher Ursachen des Waldsterbens. Forschungsbericht 106 07 046/13 im Umweltforschungsplan des Bundesministers des Innern, Berlin

Fränzle O, Killisch W, Mich N (1986) Die regionale Differenzierung und zeitliche Veränderung der Emissionssituation in der Bundesrepublik Deutschland. In: Fränzle O (ed) Geoökologische Umweltbewertung. Kiel Geogr Schr 64: 31–77

Fränzle O, Elhaus D, Fröhling J (1987) Naturwissenschaftliche Anforderungen an die Sanierung kontaminierter Standorte. Standortspezifische Klassifikation im Hinblick auf die Migration von Stoffen im Untergrund. Forschungsbericht 102 03 405/2 im Umweltforschungsplan des Bundesministers für Umwelt, Naturschutz und Reaktorsicherheit, Kiel

Fränzle O, Bruhm I, Grünberg K-U et al. (1989) Darstellung der Vorhersagemönglichkeiten der Bodenbelastung durch Umweltchemikalien. Texte des Umweltbundesamtes 34/89 Berlin

Fränzle O, Zölitz-Möller R, Außenthal R et al. (1992) Erarbeitung und Erprobung einer Konzeption für die ökologisch orientierte Planung auf der Grundlage der regionalisierenden Umweltbeobachtung am Beispiel Schleswig-Holsteins. Texte d Umweltbundesamtes 20/1992

Frankenberg P (1984) Ähnlichkeitsstrukturen von Ernteertrag und Witterung in der Bundesrepublik Deutschland. Erdwissenschaftliche Forschung 17. Steiner, Wiesbaden

Frankenberger P (1958) Ein Meßgerät für vertikale atmosphärische Wärmeströme. Tech Mitt Instr Anst Hamb, Dtsch Wetterdienst, NF 4: 21–28

Frankenberger E (1961) Beiträge zum Int Geophys Jahr 1957/58. 1. Meßergebnisse und

Beobachtungen zum Wärmehaushalt der Erdoberfläche. Ber Dtsch Wetterdienstes 10, 73

Franz F (1983) Auswirkungen der Walderkrankungen auf Struktur und Wuchsleistung von Fichtenbeständen. Forstwiss Centralbl 102: 186–200

Fredericksen CF, Lilly JH (1955) Measuring wireworm reactions to soil insecticides by tagging with radioactive cobalt. J Econ Entomol 48: 438–442

Free SM, Wilson JW (1964) A mathematical contribution to structure–activity studies. J Med Chem 7: 395–399

Freedman HI (1980) Deterministic mathematical models in population ecology. Plenum, New York

Freeman LG (1973) The significance of mammalian faunes from Paleolithic occupations in Cantabrian Spain. Am Anthropol 38: 3–44

Freeze RA, Cherry JA (1979) Groundwater. Prentice Hall, Englewood Cliffs

Frehe C (1984) Eine Versuchsanlage zur Prüfung der Wirkung von Umweltchemikalien auf Organismengemeinschaften. Verh Ges Ökol 12: 507–510

Frenzel B (1986) Viren als mögliche Ursache der Walderkrankungen. Schriftenr "Texte" Umweltbundesamtes 19: 205–213

Freundlich H (1930) Kapillarchemie. Akad Verlagsges, Leipzig

Frevert T, Klemm O (1984) Wie ändern sich pH-Werte im Regen- und Nebelwasser beim Abtrocknen auf Pflanzenoberflächen? Arch Met Geophys Biocl Ser B34: 75–81

Friedel B (1976) Formen der Salzakkumulation an der Bodenoberfläche und deren Einfluß auf die Evaporation. Mitt Leichtweiß-Inst Wasserbau Tech Univ Braunschweig 50: 402–412

Friesel P, Hansen PD, Kühn R, Trénel J (1984) Überprüfung der Durchführbarkeit von Prüfungsvorschriften und der Aussagekraft der Stufe 1 and 2 des Chemikaliengesetzes – Teil VI. Umweltforschungsplan des Bundesministers des Innern 106 04 011/08

Frische R, Klöpffer W, Schönborn W (1979) Bewertung von organisch-chemischen Stoffen and Produkten in bezug auf ihr Umweltverhalten – chemische, biologische und wirtschaftliche Aspekte. Forschungsbericht 101 04 009/03 des Battelle Instituts Frankfurt/M im Umweltforschungsplan des BMI. Battelle, Frankfurt am Main

Frissel MJ (1981) Twenty models for the behaviour of nitrogen in soil. Commun XIIth Annu Meet European Society of Nuclear Methods in Agriculture, Aberdeen, UK, 28 Sept to 2 Oct, 1981.

Fuchs M, Stanhill G (1963) The use of class A evaporation pan data to estimate the irrigation water requirements of the cotton crop. Isr J Agric Res 13: 63–78

Fuchs NA (1964) The mechanics of aerosols. Pergamon Press, Oxford

Führ F, Biel HM, Thielert M (1983) Methoden zur toxikologischen Bewertung von Chemikalien. Band 2: Böden und Modellsysteme. Spez Ber KFA Jülich 224

Fuhs GW (1961) Der mikrobielle Abbau von Kohlenwasserstoffen. Arch Mikrobiol 39: 374–422

Fujita T, Iwasa J, Hansch C (1964) A new substituent constant, pi, derived from partition coefficients. J Am Chem Soc 86: 5175–5180

Fuzzi S (1978) Study of iron (III) catalysed sulfur dioxide oxidation in aqueous solution over a wide range of pH. Atmos Environ 12: 1439–1442

Gabriel KR (1971) The biplot graphic display of matrices with application to principal component analysis. Biometrika 58: 453–467.

Gadde RR, Laitinen HA (1974) Studies of heavy metal adsorption by hydrous iron and manganese oxides. Anal Chem 46: 2022–2026

Gaddo PP, Weaving JH (1982) Urban scale mathematical model for pollutants in a

European city. In: Schneider T, Grant L (eds) Air pollution by nitrogen oxides. Elsevier, Amsterdam, pp 289–307

Gäb S, Parlar H, Korte F (1974) Ultraviolet-irradiation reactions of photodieldrin as a solid on glass and adsorbed to silica gel. Chemosphere 3: 187–192

Galbally IE, Garland JA, Wilson HJG (1979) Sulphur uptake from the atmosphere by forest and farmland. Nature 280: 49–50

Galloway JN, Dillon PJ (1983) Effects of acid deposition: the importance of nitrogen. In: National Swedish Environment Protection Board, Report PM 1636: Ecological effects of acid deposition. pp 145–160

Galloway JN, Thornton JD, Norton SA, Volchok HL, McLean RAN (1982) Trace metals in atmospheric deposition: a review and assessment. Atmos Environ 16: 1677–1700

Garber K (1935) Über die Physiologie der Einwirkung von Ammoniakgasen auf die Pflanze. Landwirtsch Vers Stat 123: 277–344

Garber K, Schürmann B (1971) Wirkung and Nachweis von Ammoniak. Immissionen in der Nähe von Großstallungen. Z Landw Forsch 26, Sonderheft I: 36–40

Gardner WR (1956) Calculation of capillary conductivity from pressure plate outflow data. Soil Sci Soc Am Proc 20: 317–320

Gardner WR (1958) Some steady-state solutions of the unsaturated moisture flow equation with application to evaporation from a water table. Soil Sci 85: 228–232

Gardner WR, Israelsen OW, Edlefsen NE, Clyde H (1922) The capillary potential function and its relation to irrigation practice. Phys Rev 1922, 2nd ser 20: 196–202

Garland JA (1977) The dry deposition of sulphur dioxide to land and water surfaces. Proc R Soc Lond A354: 245–268

Garland JA (1978) Dry and wet removal of sulphur from the atmosphere. Atmos Environ 12: 349–362

Garland JA (1983) Principles of dry deposition: application to acidic species and ozone. VDI-Berichte 500, Saure Niederschläge. VDI, Düsseldorf, pp 83–95

Garland JA, Branson JR (1976) The mixing height and mass balance of SO_2 in the atmosphere above Great Britain. Atmos Environ 10: 353–362

Garland JA, Branson JR (1977) The deposition of sulphur dioxide to a pine forest assessed by a radioactive tracer method. Tellus 29: 445–454

Garland JA, Cox LC (1980) The absorption of tritium gas by English soils, plants and the sea. Water Air Soil Pollut 14: 103–114

Garland JA, Derwent RG (1979) Destruction at the ground and the diurnal cycle of concentration of ozone and other gases. Q J R Meteorol Soc 105: 169–183

Garrels RM, Christ CL (1965) Solutions, minerals and equilibria. Freeman, Cooper & Co, San Francisco, Cal

Garrels RM, Mackenzie FT, Hunt C (1975) Chemical cycles and the global environment: assessing human influences. Kaufmann, Los Altos, Calif

Garrett SD (1981) Soil fungi and fertility. Pergamon Press, Oxford

GDCh-Advisory Committee on Existing Chemicals of Environmental Relevance (1989) Existing chemicals of environmental relevance. Criteria and list of chemicals. VCH, Weinheim

Gehrmann J (1983) Zur Entwicklung von Buchenjungpflanzen auf unterschiedlich immissionsbelasteten Standorten. Forst-Holzwirt 38: 150–154

Geiger R (1961) Das Klima der bodennahen Luftschicht, 4. Aufl. Vieweg, Braunschweig

Georgii HW (1982) Global distribution of the acidity in precipitation. In: Georgii HW, Pankrath J (eds) Deposition of atmospheric pollutants. Reidel, Dordrecht, pp 55–66

Georgii HW, Pankrath J (eds) (1982) Deposition of atmospheric pollutants. Reidel, Dordrecht

Georgii HW, Perseke C, Rohbock E (1983) Trockene und nasse Deposition säurebildender Verbindungen. VDI-Berichte 500, Saure Niederschläge. VDI, Düsseldorf, pp 127–134

Gerike P (1978) The biodegradability testing of water-soluble environmental chemicals: a concept and its justification on the ground of experimental results. UBA-Ber 10/78: 139–143

Gerike P, Fischer WK, Holtman W (1980) Biodegradability determination in trickling filter units compared with the OECD Confirmatory Test. Water Res 14: 753

Gerlach SA (1976) Meeresverschmutzung. Diagnose und Therapie. Springer, Berlin Heidelberg New York

Gerth J (1985) Untersuchungen zur Adsorption von Nickel, Zink und Cadmium durch Bodentonfraktionen unterschiedlichen Stoffbestandes und verschiedene Bodenkomponenten. Diss, Univ Kiel

Gerth J, Brümmer G (1979) Quantitäts–Intensitäts-Beziehungen zwischen Cadmium, Zink und Nickel in Böden unterschiedlichen Stoffbestandes. Mitt Dtsch Bodenkdl Ges 29: 555–566

Gerth J, Brümmer G (1983) Adsorption und Festlegung von Nickel, Zink und Cadmium durch Goethit (α-FeOOH). Fresenius Z Anal Chem 316: 616–620

Giam CS, Atlas E, Chan HS, Neff GS (1980) Phthalate esters, PCB and DDT residues in the Gulf of Mexico atmosphere. Atmos Environ 14: 65–69

Gibson DT (1976) Microbial degradation of carcinogenic hydrocarbons and related compounds. In: Sources, effects and sinks of hydrocarbons in the aquatic environment. Am Inst Biol Sci, Washington DC, pp 224–238

Gibson DT (1977) Biodegradation of aromatic petroleum hydrocarbons. In: Wolfe DA (ed) Fate and effects of petroleum hydrocarbons in marine ecosystems and organisms. Pergamon Press, New York, pp 36–46

Gibson DT, Mahadevan V, Jerina DM, Yagi H, Yeh HJC (1975) Oxidation of the carcinogens benzo(a)pyrene and benzo(a)-anthracene to dihydrodiols by a bacterium. Science 189: 295–297

Gigon A (1983) Über das biologische Gleichgewicht und seine Beziehungen zur ökologischen Stabilität. Ber Geobot Inst Eidg Tech Hochsch, Stift Rübel Zürich 50: 149–177

Gilman CS (1964) Rainfall. In: Chow VT (ed) Handbook of applied hydrology. New York, Mc Graw-Hill, pp 9-1–9-68

Glansdorff P, Prigogine I (1971) Thermodynamic theory of structure, stability and fluctuations. Wiley-Interscience, New York

Glas TK, Klute A, McWorther DB (1979) Dissolution and transport of gypsum in soils: I. Theory. Soil Sci Soc Am J 43: 265–268

Godbold DL, Hüttermann A (1986) The uptake and toxicity of mercury and lead to spruce (*Picea abies* Karst.) Seedlings. Water, Air and Soil Pollution 31: 509–515

Golberg L (ed) (1983) Structure–activity correlation as a predictive tool in toxicology. Hemisphere, Washington

Goldstein BD (1983) Toxic substances in the atmospheric environment. A critical review. JAPCA 33: 454–467

Golterman HL (1975) Physiological limnology. Elsevier, Amsterdam

Gordon R (1976) Distribution of airborne polycyclic aromatic hydrocarbons throughout Los Angeles. Environ Sci Technol 10: 370–373

Graedel T (1978) Chemical compounds in the atmosphere. New York

Graedel TE, Graney JP (1974) Atmospheric aerosol size spectra: rapid concentration fluctuations and bimodality. J Geophys Res 79: 5643–5645

Graeves MP, Malkomes HP (1980) Effects on soil microflora. In: Hance RJ (ed) Interactions between herbicides and the soil. Academic Press, London, pp 223–254

Graf G, Lagaly G (1980) Interaction of clay minerals with adenosine-5-phosphates. Clays Clay Minerals 28: 12–18

Gravenhorst G, Beilke S, Betz M, Georgii (1978) Sulfur dioxide absorbed in rain water. In: Hutchinson TC, Havas M (eds) Effects of acid precipitation on terrestrial ecosystems. Plenum Press, New York, pp 41–45

Gravenhorst G, Hofken KD, Georgii H-W (1982) Acidic input to a beech and spruce forest. In: Beilke S, Elshout E (eds) Acid deposition. Reidel, Dordrecht, pp 161–177

Greenfelt P, Bengtson C, Skarby L (1983) Dry deposition of nitrogen dioxide to scotts pine needles. In: Pruppacher HR, Semonin RG, Slinn WGN (eds) Precipitation scavenging, dry deposition and resuspension, Elsevier, New York

Greenland DJ, Hayes MBH (eds) (1981) The chemistry of soil processes. Wiley, Chichester

Gregorius HR, Hattemer HH, Bergmann F, Müller-Strack G (1985) Umweltbelastung und Anpassungsfähigkeit von Baumpopulationen. Silvae Genet 34: 230–241

Gregory PH (1945) The dispersal of airborne spores. Trans Br Mycol Soc 28: 26–72

Griffiths RP, McNamara TM, Caldwell BA, Morita RY (1981) A field study of the acute effects of the dispersant Corexit 9527 on glucose uptake by marine microorganisms. Mar Environ Res 5: 83–91

Grimmer G, Jacob J, Hildebrand A (1972) Kohlenwasserstoffe in der Umgebung des Menschen. 9. Mitteilung: Der Gehalt polycyclischer Kohlenwasserstoffe in isländischen Bodenproben. Z Krebsforsch 78: 65–72

Groot SR de, Mazur P (1962) Non-equilibrium thermodynamics. North-Holland, Amsterdam

Grosch S (1989) Der atmosphärische Gesamteintrag auf natürliche Oberflächen, unter besonderer Berüchsichtigung der trockenen Depositionen in Waldgebieten. Diss Univ Frankfurt

Gross PM, Saylor JH (1931) The solubilities of certain slightly soluble organic compounds in water. J Am Chem Soc 53: 1744–1751

Grossmann W-D, Schaller J, Sittard M (1984) "Zeitkarten": eine neue Methodik zum Testen von Hypothesen und Gegenmaßnahmen bei Waldschäden. Allg Forstz 38: 837–843

Grove JH, Ellis BG (1980) Extractable chromium as related to soil pH and applied chromium. Soil Sci Soc Am J 44: 238–242

Grover R (1975) A method for determining the volatility of herbicides. Weed Sci 23: 529–532

Grünberg KU (1987) Untersuchungen zur Differenzierung des Mikroklimas und des Energiehaushaltes im Meßgebiet Rastorf. Dipl-Arb Univ Kiel

Gruttke H, Kielhorn U, Kratz W, Weigmann G, Hague A (1987) Verteilung von ^{14}C-Natriumpentachlorphenol in einem geschlossenen Modellökosystem unter besonderer Berücksichtigung der Detritusnahrungskette. Verh Ges Ökol 15: 351–357

Gschwend PM, Wu S-C (1985) On the constancy of sediment–water partition coefficients of hydrophobic organic pollutants. Environ Sci Technol 19: 90–96

Guderian R (1978) Wirkungen sauerstoffhaltiger Schwefelverbindungen. VDI Ber 314: 207–217

Guderian R (ed) (1985) Air pollution by photochemical oxidants. Springer, Berlin Heidelberg New York

Guderian R (1988a) Critical levels for effects of NOx. ECE Critical Levels Worksh Bad Harzburg March 14–18, 1988. Final Draft Report Part II, pp 79–104. UN Econ Comm Europe, Geneva

Guderian R (1988b) Critical levels for effects of ozone (O_3). ECE Critical Levels Worksh Bad Harzburg March 14–18, 1988. Final Draft Report Part II, pp 51–78. UN Econ Comm Europe, Geneva

Guderian R, Küppers K, Six R (1985) Wirkungen von Ozon, Schwefeldioxid und Stickstoffdioxid auf Fichte und Pappel bei unterschiedlicher Versorgung mit Magnesium and Kalzium sowie auf die Blattflechte Hypogymnia physoides. VDI Ber 560: 657–701

Guisti DM, Conway RA, Lawson CT (1974) Activated carbon adsorption of petrochemicals. JWPCF 46: 947–965

Gurr CG, Marshall TJ, Hutton TJ (1952) Movement of water in soil due to a temperature gradient. Soil Sci 74: 335–345

Gutenmann WH, Loos MA, Alexander M, Lisk DJ (1964) Beta-oxidation of phenoxyalkanoic acids in soil. Proc Soil Sci Soc Am 28: 205–207

Gydesen H, Rasmussen L, Pilegaard K, Andersen A, Hovmand MF (1981) Differences in the regional deposition of cadmium, copper, lead and zinc in Denmark as reflected in bulk precipitation, epiphytic cryptogams and animal kidneys. Ambio 10: 229–230

Haagen-Smit AJ, Darley EF, Zaitlin M, Hull H, Noble W (1952) Investigation on injury to plants from air pollution in the Los Angeles area. Plant Physiol 27: 18–34

Haan FAM de (1987) Effects of agricultural practices on the physical, chemical and biological properties of soils: part III – chemical degradation of soil as the result of the use of mineral fertilizers and pesticides: aspects of soil quality evaluation. In: Barth H, L'Hermite P (eds) Scientific basis for soil protection in the European Community. Elsevier, London, pp 211–236

Haase B, Fränzle O (1984) Die regionale und gruppenspezifische Differenzierung der Benzolemissionen in der Bundesrepublik Deutschland. Schriffenreihe "Betrifft" des Umweltbandesames

Haberland W, Fränzle O (1975) Untersuchungen zur Bildung von Verwitterungskrusten auf Sandsteinoberflächen in der nördlichen und mittleren Sahara (Libyen und Tschad). Würzb Geogr Arb 43: 148–163

Haberland W, Sperlich M (1983) Untersuchungen zur Verwitterung des Jakobsbrunnens auf der Pfaueninsel zu Berlin-Wannsee. Berl Beitr Archäometrie 8: 299–324

Hacker-Thomae S (1985) Ein eindimensionales numerisches Modell zur Nebelvorhersage. Ber Dt Wetterdienst 168. DWD, Offenbach am Main

Hadassah K, Dinar N (eds) (1984) Boundary layer structure – modeling application to air pollution and wind energy. Reidel, Dordrecht

Häberle M (1982) Stoffkreisläufe der Natur und Einfluß des Menschen. Umwelt 1/82: 15–22, 2/82: 76–88

Haeupler H (1982) Evenness als Ausdruck der Vielfalt in der Vegetation. Cramer, Vaduz

Haider K (1985) Der Einfluß von Bodenparametern auf den Abbau von Pflanzenschutzmitteln. Ber Landwirtsch, Sonderh 198: 81–92

Halling CS (1966) The strategy of building models of complex ecological systems. In: Watt KEF (ed) Systems analysis in ecology. New York, pp 195–214

Hamaker JW (1972) Decomposition: quantitative aspects. In: Goring CAI, Hamaker JW (eds) Organic chemicals in the soil environment, vol I. Marcel Dekker, New York, pp 342–399

Hamburger B (1982) Aktuelle Probleme der Fischteste. Decheniana-Beih 26: 78–81

Hamburger B, Häberling H, Hitz HR (1977) Vergleichende Prüfungen der Fischtoxizitäten an Elritzen, Forellen und Goldorfen. Arch Fisch Wiss 28: 45–55

Hammett LP, Pfluger HL (1933) The rate of addition of methyl esters to trimethylamine. J Am Chem Soc 55: 4079–4080

Hance JR (ed) (1980) Interactions between herbicides and the soil. Academic Press, London

Hance RJ (1969) An empirical relationship between chemical structure and the sorption of some herbicides by soils. J Agric Food Chem 17: 667–668

Handley R, Overstreet R (1968) Uptake of carrier free ^{137}Cs by Ramalina reticulata. Plant Physiol 43: 1401–1405

Hanna SR, Briggs GA, Hosker RP (1982) Handbook on atmospheric diffusion. Springfield

Hansch C, Fujita T (1964) ρ, σ, π analysis. A method for the correlation of biological activity and chemical structure. J Am Chem Soc 86: 1616–1626

Hansch C, Leo A (1979) Substituent constants for correlation analysis in chemistry and biology. Wiley, New York

Hansen LD, Whiting L, Bartholomew CH, Cluff CL, Izatt RM, Christensen JJ (1974) Transition metal–SO_3^{2-}–complexes: a postulated mechanism for the synergistic effects of aerosols and SO_2 on the respiratory tract. In: Hemphill DD (ed) Trace substances in environmental health VIII. Univ Missouri Press, Columbia, pp 393–397

Hansen LD, Whiting L, Eatough DJ, Jensen TE, Izatt RM (1975) Determination of sulfur (IV) and sulfate in aerosols by thermometric methods. Anal Chem 48: 634–638

Hantel M, Reimer E, Speth P (1984) ALPEX-diagnostics: quantitative synoptics over Europe. Beitr Phys Atmos 57: 477–494

Haque R (1980) Dynamics, exposure and hazard assessment of toxic chemicals. Ann Arbor Science, Ann Arbor

Hardey JL, Harley EL (1987) A check-list of mycorrhiza in the British flora. New Phytol 105 1–102

Hardy RWF, Havelka UD (1975) Nitrogen-fixation research – the key to world food. Science 188: 633–643

Harkor R (1982) Toxic air pollutants – assessing their importance. Sci Total Environ 26: 67–85

Harley JL, Smith SE (1983) Mycorrhizal symbiosis. Academic Press, New York

Hartge K (1966) Problematik und Fehlerquellen bei der Bestimmung der Durchlässigkeit an Stechzylinderproben. Z Wasser Boden 18: 19–22

Hartge KH (1978) Einführung in die Bodenphysik. Enke, Stuttgart

Hartmann L (1966) Effect of surfactants on soil bacteria. Bull Environ Contam Toxicol 1: 219–224

Harvey GW, Legge AH (1979) The effect of sulfur dioxide upon the metabolic level of adenosine triphosphate. Can J Bot 57: 759–764

Hassel MP, May RM (1974) Aggregation of predators and insect parasites and its effect on stability. J Anim Ecol 43: 567–594

Hassett JJ, Means JC, Banwart WL, Wood SG (1980) Sorption properties of sediments and energy-related pollutants. EPA-600/3-80-041

Hauffe K, Morrison SR (1974) Adsorption. De Gruyter, Berlin

Hay A (1983) Defoliants in Vietnam: the long-term effects. Nature 302: 208–209

Heagle AS, Johnston JW (1979) Variable responses of soybeans to mixtures of ozone and sulfur dioxide. J Air Pollut Control Assoc 29: 729–732

Hech WW, Dunning JA (1967) The effects of ozone on tobacco and pinto bean as conditioned by several ecological factors. J Air Pollution Control Assoc 17: 112–114

Heck WW, Brandt CS (1977) Effects on vegetation: Native crops, forests. In: Air pollution, vol II, Academic Press, London New York, pp 157–229

Heckmann HJ, Schreiber KF, Thöle R (1985) Ein Vergleich unterschiedlicher Verfahren zur flächenhaften Ermittlung der Grundwasserneubildungsrate. Mitt Dtsch Bodenkdl Ges 41: 353–356

Heggestad HE, Middleton JT (1959) Ozone in high concentrations as a cause of tobacco leaf injury. Science 129: 208–210

Heicklen J (1976) Atmospheric chemistry. Academic Press, New York

Heicklen J (1981a) The correlation of rate coefficients for H-atom abstraction by OH radicals with C–H bond dissociation enthalpies. Int J Chem Kinet: 651

Heicklen J (1981b) The removal of atmospheric gases by particulate matter. Atmos Environ 15: 781–785

Heidelberger C (1976) Studies on the mechanisms of carcinogenesis by polycyclic aromatic hydrocarbons and their derivatives. In: Freudenthal R, Jones PW (eds) Carcinogenesis – a comprehensive survey. Volume I. Polynuclear aromatic hydrocarbons: chemistry, metabolism and carcinogenesis. Raven Press, New York, pp 1–8

Heidland S (1986) Darstellung der Komponenten des Wirkungskomplexes "Akkumulationspotential" als Standortkriterium. Dipl Arb Univ Kiel

Heinrich G, Güsten H (1978) Belastung der Atmosphäre durch polycyclische aromatische Kohlenwasserstoffe und Blei im Raum von Karlsruhe. Staub-Reinhalt Luft 38: 94–100

Heit M (1977) Review of current information on some ecological and health-related aspects of the release of trace metals into the environment associated with the combustion of coal. Rep HASL-320, p 53. NTIS, New York

Helfferich F (1959) Ionenaustauscher. Verl Chemie, Weinheim

Helling C et al. (1974) Behaviour of pesticides in soils. Adv Agron 23: 147–240

Hem JD (1970) Study and interpretation of the chemical characteristics of natural water. Geol Surv Water Suppl 1473, Washington DC

Henderson-Sellers B (1987) Modeling of plume rise and dispersion – The University of Salford model: U.S.P.R. Springer, Heidelberg

Hendry DG, Mill T (1980) Oxidation chemistry of pollutants in the environment. In: Funcke W, König J, Stöber W et al. (eds) Proc Int Worksh on test methods and assessment procedures for the determination of the photochemical degradation behaviour of chemical substances. Dec 2–4, 1980, Reichstag, Berlin. FhG, UBA, Münster

Herms U (1982) Untersuchungen zur Schwermetallöslichkeit in kontaminierten Böden und kompostierten Siedlungsabfällen im Abhängigkeit von Bodenreaktion, Redoxbedingungen und Stoffbestand. Diss, Univ Kiel

Herms U, Brümmer G (1980) Einfluß der Bodenreaktion auf Löslichkeit und tolerierbare Gesamtgehalte an Nickel, Kupfer, Zink, Cadmium und Blei in Böden und kompostierten Siedlungsabfällen. Landwirtsch Forsch 33: 408–423

Herms U, Brümmer G (1984) Einflußgrößen der Schwermetallöslichkeit und -bindung in Böden. Z Pflanzenernähr Bodenkde 147: 400–424

Herrera R, Jordan CF, Medina E, Klinge H (1981) How human activities disturb the nutrient cycles of a tropical rainforest in Amazonia. Ambio 10: 109–114

Herrmann H (1985) Untersuchungen zur Anionenadsorption an Tonmineralen. Diss, Univ Kiel

Herrmann H, Lagaly G (1985) ATP-clay interactions. 5th meeting of the European clay groups, Prague 1983. Charles University, Prague, pp 269–277

Herrmann R (1978) Regional patterns of polycyclic aromatic hydrocarbons in NE-Bavarian snow and their relationships to anthropogenic influence and air flow. CATENA 5: 165–175

Herrmann R (1984) Atmosphärische Transporte und raumzeitliche Verteilung von Mikroschadstoffen (Spurenmetalle, Organochlorpestizide, polyzyklische aromatische Kohlenwasserstoffe) in Nordostbayern. Erdkunde 38: 55–63

Herrmann R, Neuland H, Buss G (1978) Zur Geschichte der Spurenmetallverunreinigung der Luft: Eine Zeitreihenanalyse der Metallgehalte in Baumringen. Staub-Reinhalt Luft 38: 366–369

Herron JT, Huie RE, Hodgeson JA (eds) (1979) Chemical kinetic data needs for modeling the lower troposphere. NBS Spec Publ 557

Herzel F, Schmitt G (1979) Prüfung des Versickerungsverhaltens in Lysimetern und Kleinsäulen. Inst Wasser-, Boden- und Lufthygiene des Bundesgesundheitsamtes Berlin-Dahlem, Ber 3

Hicks BB, Matt DR, McMillen RT (1988) A micrometeorological investigation of surface exchange of SO_3, SO_2 and NO_2: a case study. NOAA/ATDD Contr No 88/36

Hidy GM, Burton CS (1980) Atmospheric aerosol formation by chemical reactions. In: Hidy GM, Mueller PK, Grosjean D et al. (eds) The character and origin of smog aerosols. Wiley, New York, pp 385–433

Hidy GM, Friedlander SK (1971) The nature of the Los Angeles aerosol. In: Englund H, Burg W (eds) Proc 2nd Int Clean Air Congress. Academic Press, New York, pp 390–404

Hidy GM, Mueller PK (1980) Prologue. In: Hidy GM, Appel BR, Wisolowski JJ (eds) The character and origin of smog aerosols. Wiley, New York, pp 17–52

Hidy GM, Mueller PK, Grosjean D, Appel BR, Wesolowski JJ (eds) (1980) The character and origins of smog aerosols. Wiley, New York

Hill AC (1971) Vegetation: a sink for atmospheric pollutants. J Air Pollut Control Assoc 21: 341–346

Hill IR, Wright SJL (eds) (1978) Pesticide microbiology. Academic Press, London

Hill MO (1973) Diversity and evenness: a unifying notation and its consequences. Ecology 54: 427–432

Hillel D (1980) Fundamentals of soil physics. Academic Press, New York

Hindawi IJ, Rae JA, Griffis WL (1980) Response of bush bean exposed to acid mist. Am J Bot 67: 168–172

Ho WW, Hidy GM, Goran RM (1980) Microwave measurements of the liquid water content of atmospheric aerosols. In: Hidy GM, Mueller PK, Grosjean D, Appel BR, Wesolowski JJ (eds) The character and origins of smog aerosols. Wiley, New York, pp 215–236

Hock B, Elstner E (ed) (1984) Pflanzentoxikologie: Der Einfluß von Schadstoffen und Schadwirkungen auf Pflanzen. Bibliographisches Institut – Wissenschaftsverlag, Mannheim

Hoechst (1987) Finding new ways – research and development at Hoechst. Hoechst, Frankfurt am Main

Höfken DK, Georgii HW, Gravenhorst G (1981) Untersuchungen über die Deposition atmosphärischer Spurenstoffe am Buchen- und Fichtenwald. Ber Inst Meteorol Geophys Frankfurt, Nr. 46

Höfken KD, Gravenhorst G (1982) Deposition of atmospheric aerosol particles to beech and spruce forest. In: Georgii H-W, Pankrath J (eds) Deposition of atmospheric pollutants. Reidel, Dordrecht, pp 191–194

Höhne H (1970) Streunutzungsschäden, ihre Entstehung und Verbreitung sowie Möglichkeiten ihrer Behebung. Wiss Tech Z TU Dresden 19: 1047–1054

Hölscher J, Walther W (1986) Belastung von Wasser und Boden in der Bundesrepublik Deutschland durch Luftverunreinigungen. Mitt Niedersächs Landesamt Wasserwirtschaft 2 Hildesheim

Hoffmann G (1983) Abschätzung einer möglichen Gefährdung der Ackerböden in Baden-Württemberg durch Säureeintrag aus der Atmosphäre. VDI-Berichte 500, Saure Niederschläge. VDI Düsseldorf, pp 301–306

Hoffmann J, Pahlich E, Steubing L (1976) Enzymatisch–analytische Untersuchungen zum Adenosinphosphatgehalt SO_2-begaster Erbsen. Int J Environ Anal Chem 4: 183–196

Hoffman WA, Lindberg SE, Turner RR (1980) Some observations of organic constituents in rain above and below the forest canopy. Environ Sci Technol 14: 95–100

Hofius K (1977) Wasserflüsse im ungesättigten Bodenbereich. Beitr Hydrol 4: 81–115

Hofmann G (1961) Meßtechnische Hinweise für mikroklimatologische und mikrometeorologische Untersuchungen. In: Geiger R (ed) Das Klima der bodennahen Luftschicht. Vieweg, Braunschweig, pp 544–567

Høiland K (1986) Is the mycorrhizal symbiosis being damaged by pollution? Acid Mag 4: 30–31

Holling CS (1966) The strategy of building models of complex ecological systems. In: Watt KEF (ed) Systems analysis in ecology. McGraw-Hill, New York, pp 195–214

Holling CS (1968) The tactics of a predator. In: Southwood TRE (ed) Insect abundance. Blackwell, Oxford, p 47

Holling CS (1973) Resilience and stability of ecological systems. Annu Rev Ecol Syst 4: 1–23

Holling CS (1976) Resilience and stability of ecosystems. In: Jantsch E, Waddington CH (eds) Evolution and consciousness. Addison–Wesley, Reading Ma, pp 73–92

Hollstein W (1963) Bodenkarte der Bundesrepublik Deutschland 1: 1 000 000. Bundesanstalt für Bodenforschung Hannover

Horner J, Que Hee SS, Sutherland RG (1974) Esterification of 2,4-dichlorophenoxyacetic acid – a quantitative comparison of esterification techniques. Anal Chem 46: 110–112

Horowitz M, Givelberg A (1979) Toxic effects of surfactants applied to plant roots. Pestic Sci 10: 547–557

Horsman DC, Wellburn AR (1977) Effect of SO_2 polluted air upon enzyme activity in plants originating from areas with different annual mean atmospheric SO_2 concentrations. Environ Pollut 13: 33–39

Horvath L (1982) On the vertical flux of gaseous ammonia above water and soil surfaces. In: Georgii HW, Pankrath J (eds) Deposition of atmospheric pollutants. Reidel, Dordrecht, pp 17–22

Houten IG ten (1966) Bezwaren van luchtverontreiniging voor de landbouw. Landbouwkd Tijdschr 78: 2–13

Hoyer H, Peperle W (1958) Dampfdruckmessungen an organischen Substanzen und ihre Sublimationwärmen. Z Elektrochem Ber Bunsenges Phys Chemie 62: 61–66

Hoyningen-Huene Jv (1983) Die Interzeption des Niederschlags in landwirtschaftlichen Pflanzenbeständen. Schriftenr Dtsch Verb Wasserwirtsch Kulturbau 57: 1–53

Hoyningen-Huene Jv, Nasdalack S (1985) Einfache Methoden zur Messung der Niederschlagsinterzeption in landwirtschaftlichen Pflanzenbeständen. Landschaftsökol Messen Auswerten 1: 107–116

Hsu TS, Bartha R (1974) Biodegradation of chloroaniline–humus complexes in soil and in culture solution. Soil Sci 118: 213–220

Huber SJ (1982) Modelluntersuchungen zur Wirkung aromatischer Herbizidmetaboliten auf Bodenmikroorganismen und auf die symbiotische Stickstoff-Fixierung. Diss, Univ München

Huebert BJ (1983) Measurements of the dry deposition flux of nitric acid vapor to grasslands and forests. In: Pruppacher HR, Semonin RG, Slinn WGN (eds) Precipitation scavenging, dry deposition and resuspension. Elsevier, New York

Hueck HJ, Adema DMM, de Kock WC, Kuiper J (1978) Experiences with the validation of ecotoxicological tests. Umweltbundesamt, Ber 10/78: 159–167

Hüttermann A (1978) Symptome des Waldsterbens. Allg Forst Jagdztg 4: 67–70

Hüttermann A (1986) Der Einsatz neuartiger Methoden zur Überprufung von Hypothesen beim Waldsterben: Elektronenmikroskopie/Röntgenanalyse und Computertomographie. In: Texte des Umweltbundesamtes, Berl 19/86: 247–258

Huff DD (1977) TEHM: a terrestrial ecosystem hydrology model. Oak Ridge National Laboratory, Oak Ridge TN

Hughes RD, Gilbert N (1968) A model of an aphid population – a general statement. J. Anim Ecol 40: 525–534

Hunding C (1978) Effects of toxic concentrations of chemicals on aquatic communities. Ber Umweltbundesamtes Berl 10/78: 45–55

Hunger W (1978) Über Absterbeerscheinungen an älteren Fichtenbeständen in der Nähe einer Schweinemastanlage. Beitr Forstwirtsch 4: 188–189

Husar RB, White WH, Blumenthal DL (1976) Direct evidence of heterogeneous aerosol formation in the Los Angeles smog. Environ Sci Technol 10: 490–491

Hutchinson GL et al. (1972) Atmospheric ammonia: absorption by plant leaves. Science 175: 771–772

Hutton M (1983) A prospective atmospheric emission inventory for cadmium – the European Community as a study area. Sci Total Environ 29: 29–47

Hutzinger O (ed) (1980) The handbook of environmental chemistry. Springer, Berlin Heidelberg New York

Huygen C, van Ijssel FW (1981) Lúchtverontreiniging bij kassen door bodemontsmetting met methylbromide. Rep G 1048. IMG-TND, Delft

Imeson AC (1983) Studies of erosion thresholds in semi-arid areas: field measurements of soil loss and infiltration in northern Morocco. CATENA Suppl. 4: 79–89

Ingle SE, Schuldt MD, Schults DW (1978) A user's guide for REDEQL. US Environ Prot Agency Rep EPA 600/3-78-024, Corvallis ORE

IPS (Industrieverband Pflanzenschutz) (1982) Wirkstoffe in Pflanzenschutz- und Schädlingsbekämpfungsmitteln. Bintz, Offenbach

Irving PM (1979) Response of field-grown soybeans to acid precipitation alone and in combination with sulfur dioxide. PhD Dissertation, University of Wisconsin, Milwaukee

Irving PM (1983) Acid deposition effects on vegetation: a review and analysis of methodology. VDI-Berichte 500, Saure Niederschläge VDI, Düsseldorf, pp 215–223

Isermann K (1983) Bewertung natürlicher und anthropogener Stoffeinträge über die Atmosphäre als Standortfaktoren im Hinblick auf die Versauerung land- und forstwirtschaftlich genutzter Böden. VDI-Berichte 500, Saure Niederschläge. VDI, Düsseldorf, pp 307-335

Israel GW (1977) Differences in the accumulation of gaseous and particulate fluorine compounds by foliage and limed filter paper samplers. Atmos Environ 11: 183–188

Israelachvili JN (1985) Intermolecular and surface forces. With applications to colloidal and biological systems. Academic Press, London

Ivanov MV, Freney JR (eds) (1983) The global biogeochemical sulphur cycle. SCOPE 19. Wiley, Chichester

Jacobson J (1980) The influence of rainfall composition on the yield and quality of agricultural crops. In: Drabløs D, Tollan A (eds) Ecological impact on acid precipitation. Intern Conf, Sandefjord, Norway, pp 41–46

Jacobson JS (1983) Experimental acidification of natural and agro–ecosystems. VDI-Berichte 500, Saure Niederschläge. VDI, Düsseldorf, pp 205–209

Jäger H-J (1982) Biochemical indication of an effect of air pollution on plants. In: Steubing L, Jäger H-J (eds) Monitoring of air pollutants by plants. Junk, The Hague, pp 99–107

Jäger HJ, Grill D (1975) Einfluß von SO_2 und HF auf freie Aminosäuren der Fichte (*Picea abies* L. Karsten). Eur J For Pathol 5: 279–282

Jäger HJ, Meyer HR (1977) Effect of water stress on growth and proline metabolism of *Phaseolus vulgaris* L. Oecologia 30: 83–96

Jäger HJ, Pahlich E, Steubing L (1972) Die Wirkung von Schwefeldioxid auf den Aminosäure- und Proteingehalt von Erbsenkeimlingen. Angew Bot 46: 199–221

Jäger H-J, Schulze E (1988) Critical Levels for effects of SO_2. ECE Critical Levels Worksh Bad Harzburg, 14–18 March 1988, pp 15–50. UN Econ Comm Europe, Geneva

Jaeschke W (ed) (1986) Chemistry of multiphase atmospheric systems. Springer, Berlin Heidelberg New York

Jäkel D, Dronia H (1976) Ergebnisse von Boden- und Gesteinstemperaturmessungen in der Sahara. Berliner Geogr Abh 24: 55–64

Jakhola S, Katainen H, Kellomaki S, Saukkola P (1980) The effects of artificial acid rain on the spectral reflectance and photosynthesis of Scots pine seedlings. In: Drabløs D, Tollan A (eds) Ecological impact of acid precipitation. Proc Intern Conf, Sandefjord, Norway pp 172–173

James RV, Rubin J (1979) Applicabilities of local equilibrium assumption to transport through soils affected by ion exchange. In: Jenne EA (ed) Chemical modeling in aqueous systems. ACS Symp Ser 93. American Chemical Society , Washington DC

Jansen AE, Dighton J, Bresser AHM (eds) (1988) Ectomycorrhiza and acid rain. Comm European Communities, Brussels

Jantsch E, Waddington CH (1976) Evolution and consciousness. Addison-Wesley Reading, Mass

Jauregui E (1987) Urban heat island development in medium and large urban areas in Mexico. Erdkunde 41: 48–51

Jellinek HHG (1977) Aspects of degradation and stabilization of polymers. Elsevier, Amsterdam

Jenkins DW (1976) Flow of toxic metals in the environment. Proc Int Conf Environ Sensing Assess 1975, 1, 1, 5 pp. IEEE, New York

Jenkinson DS, Smith KA (eds) (1988) Nitrogen efficiency in agricultural soils. Elsevier, London

Jenne EA (1979) Chemical modeling – goals, problems, approaches and priorities. In: Jenne EA (ed) Chemical modeling in aqueous systems. ACS Symp Ser 93. American Chemical Society, Washington DC

Jennings AA, Kirkner DJ, Theis TL (1982) Multicomponent equilibrium chemistry in groundwater quality models. Water Resour Res 18: 1089–1096

Jensen HL (1957) Decomposition of chloro-substituted aliphatic acids by soil bacteria. Can J Microbiol 3: 151–164

Jensen HL, Petersen HI (1952) Decomposition of hormone herbicides by bacteria. Acta Agric Scand 2: 215–231

Jensen K (1985) Untersuchungen zum Schwermetallhaushalt eines Waldökosystems unter besonderer Berücksichtigung des Stammabflusses. Dipl.-Arbeit Universität Kiel

Jensen-Huß K (1990) Raumzeitliche Analyse atmosphärischer Stoffeinträge in Schleswig-Holstein und deren ökologische Bewertung. Diss, Univ Kiel

Jerina DM, Daly JW, Jeffrey AM, Gibson DT (1971) Cis-1,2-dihydroxy-1,2-dihydro-naphthalene: a bacterial metabolite from naphthalene. Arch Biochem Biophys 142: 394–396

Jochheim H (1985) Der Einfluß des Stammablaufwassers auf den chemischen Boden-zustand und die Vegetationsdecke in Altbuchenbeständen verschiedener Waldbe-stände. Ber Forschungszentrum Waldökosysteme/Waldsterben 13: 1–226

Jörg F, Schmitt D, Ziegahn K-F (1985) Materialschäden durch Luftverunreinigungen. Forschungsbericht 85 106 08 010 im Umweltforschungsplan des Bundesministers des Innern. Fraunhofer Institut, Pfinztal-Berghausen

Jørgensen SE (1988) Fundamentals of ecological modelling. Developments in environ-mental modelling, 9, Elsevier, Amsterdam

Johann HP (1979) Beitrag zur Strukturierung des Gegenstandsbereichs Umweltschutz aus hochschuldidaktischer Sicht. VDI, Düsseldorf

Johnson AH, Siccama TG, Wang D, Turner RS, Barringer TH (1981) Recent changes in patterns of tree growth rate in the New Jersey pinelands: a possible effect of acid rain. J Environ Qual 10 (4): 427–430

Johnson AH, Siccama TG, Turner RS, Lord DG (1983) Assessing the possibility of a link between acid precipitation and decreased growth rates of trees in the northeastern U.S. In: The effects of acidic deposition on vegetation. Am Chem Soc Meeting, Ann Arbor Science. Las Vegas, NV

Johnson B (1976) Soil acidification by atmospheric pollution and forest growth. In: Dochinger LS, Seliga TA (eds) Proc First Internt Symposium on acid precipitation and forest ecosystems. USDA Forest Service, General Tech Report NE-23, Upper Darby, PA

Johnson DW, Cole DW (1980) Anion mobility in soils: relevance to nutrient transport from forest ecosystems. Environ Int 3: 79–90

Johnson S (1981) Survey of legal and administrative powers to control existing chemicals. Proc Worksh control of exist chemicals under the patronage of the organisation for economic co-operation and development. Umweltbundesamt, Berlin, pp 97–120

Johnston AE (1989) Benefits from long-term ecosystem research; some examples from Rothamsted. MAB-Mitt 31: 288–312

Johnstone HF, Conghanour DR (1958) Absorption of sulfur dioxide from air. Oxidation in drops containing dissolved catalysts. Ind Eng Chem 50: 1169–1172

Jones JAA (1981) The nature of soil piping – a review of research. British Geomorpho-logical Research Group Research Monogr 3. Geobooks, Norwich

Jones JAA (1987) The effects of soil piping on contributing areas and erosion patterns. Earth Surf Process Landforms 12: 229–248

Jones KC, Symon CJ, Johnston AE (1987) Retrospective analysis of an archived soil collection. II. Cadmium. Sci Total Environ 67: 75–89

Jost D, Beilke S (1983) Trend saurer Depositionen. VDI-Ber 500: 135–139

Journel AG, Huijbregts CJ (1978) Mining geostatistics. Academic Press, London

JRB Associates (1980) Methodology for rating the hazard potential of waste disposal sites. JRB Associates, McLean VA

Juhrén M, Noble W, Went FW (1957) The standardization of *Poa annua* as an indicator of smog concentrations. I. Effects of temperature, photoperiod and light intensity during growth of the test plants. Plant Physiol 32: 576–586

Jung L (1980) Messungen von Oberflächenabfluß und Bodenabtrag auf verschiedenen Böden der Bundesrepublik Deutschland. Dtsch Verb Wasserwirtsch Kulturbau, Schr 48

Junge CE (1960) Sulfur in the atmosphere. J Geophys Res 65: 227–237

Junge CE (1963) Air chemistry and radioactivity. Academic Press, New York

Jurry WA (1982) Simulation of solute transport using a transfer function model. Water Resour Res 18: 363–368

Kadowski S (1976) Size distribution of atmospheric total aerosols, sulfate, ammonium, and nitrate particulates in the Nagoya area. Atmos Environ 10: 39

Kändler U, Ullrich H (1964) Nachweis von NO_2-Schäden an Blättern. Naturwissenschaften 51: 518

Kahl G (1973) Das Thermoklima der bodennahen Luftschicht über dem Blunker See. Dipl-Arbeit Univ Kiel

Kallend AS (1983) Trends in the acidity of rain in Europe: a reexamination of European atmospheric chemistry network data. In: Beilke S, Elshout AJ (eds) Acid deposition. Reidel, Dordrecht, pp 108–113

Karickhoff SW (1984) Organic pollutant sorption in aquatic systems. J Hydraul Eng 110: 707–735

Karickhoff SW, Brown DS, Scott TA (1979) Sorption of hydrophobic pollutants on natural sediments. Water Res 13: 241–248

Katen PC, Hubbe JM (1983) Size-resolved measurements of the dry deposition velocity of atmospheric aerosol particles. In: Pruppacher HR, Semonin HR, Slinn WGN (eds) Precipitation scavenging, dry deposition and resuspension. Elsevier, New York

Katlein R (1986) Bestimmung von Komplexkapazitäten mit der Mangandioxid-Methode. Diss, Univ Marburg

Kato T, Tachibana S, Inden T (1974) Studies on the injuries of crops by harmfull gases under covering; Part 2: On the mechanism of crop injury due to gaseous nitrogen dioxide. Environmental Control Biology 12: 103–107

Kaupenjohann M, Schneider BU, Hantschel R, Zech W, Horn R (1988) Sulfuric acid rain treatment of *Picea abies* (Karst. L.): effects on nutrient solution, throughfall chemistry, and tree nutrition. Z Pflanzenernähr Bodenkd 151: 123–126

Kayser D, Boehringer U, Schmidt-Bleek F (1982) The environmental specimen banking project of the Federal Republic of Germany. Environ Monitor Assess 1: 241–255

Kayser H (1977) Effect of zinc on the growth of mono and multispecies cultures of some marine plancton algae. Helgol Wiss Meeresunters 30: 682–696

Kearney PC, Wodson EA, Ellington CP (1972) Persistence and metabolism of chlorodioxines in soils. Environm Sci and Techn 6: 1017–1019

Kellog WW, Cadle RR, Allen ER, Lazarus AZ, Martell EA (1972) The sulfur cycle. Science 175: 587–596

Kemp PH (1986) Calciumcarbonat- und Gipsneubildungen in kapillarporösen Medien unter simulierten Sebkha-ähnlichen Bedingungen in der Klimakammer. Berl Geowiss Abh A 80

Kenaga EE, Goring CAI (1980) Relationship between water solubility, soil sorption, octanol–water partitioning, and concentration of chemicals in biota. In: Eaton JG,

Parrish PR, Hendricks AC (eds) Aquatic toxicology. American Society for Testing and Materials, Philadelphia Pa, pp 78–115

Kenk GK (1983) Zuwachsuntersuchungen in geschädigten Tannenbeständen in Baden-Württemberg. Allg Forstztg 38: 650–652

Kenk GK (1984) Zum Problemkreis Walderkrankungen und Waldwachstumsforschung. Forst- Holzwirt 39: 435–438

Kenneweg H, Kramer H (1984) Großräumige Erfassung von Veränderungen des Holzzuwachses und der Bestandesstrukturen in immissionsexponierten Waldgebieten unter Einbeziehung von Methoden der Luftbildauswertung, der Dendrometrie und der Jahresringanalyse. Ber Forschungszentr Waldökosysteme/Waldsterben 2: 268–280

Kessler A (1968) Globalbilanzen von Klimaelementen. Ein Beitrag zur allgemeinen Klimatologie der Erde. Ber Inst Meteorol Klimatol Tech Hochsch Hannover 3

Kharaka YK, Barnes I (1973) SOLMNEQ: Solution–mineral equilibrium computations. NTIS Tech Rep PB 214–899, Springfield VA

Kimbrough RD, Jensen AA (1989) Halogenated biphenyls, terphenyls, naphthalenes, dibenzodioxins and related products. Elsevier, Amsterdam

Kimerle RA (1989) Aquatic and terrestrial ecotoxicology of linear alkylbenzene sulfonate. Tenside Surface Deterg 26: 169–176

Kimminich O, v Lersner H, Storm P-C (1986) Handwörterbuch des Umweltrechts, I. Band. Erich Schmidt, Berlin

Kincaid CT, Morrey JR, Rogers JE (1984) Geochemical models for solute migration, vol 1. Process description and computer code selection. Rep EA–3417. Electric Power Research Institute, Palo Alto CA

Kinniburgh DG, Jackson ML, Syers JK (1976) Adsorption of alkaline earth, transition, and heavy metal cations by hydrous oxide gels of iron and aluminium. Soil Sci Soc Am J 40: 796–799

Kins L (1982) Temporal variation of chemical composition of rainwater during individual precipitation events. In: Georgii HW, Pankrath J (eds) (1982) Deposition of atmospheric pollutants. Reidel, Dordrecht, pp 87–96

Kinzelbach W (1983) Modellierung des Transports von Schadstoffen im Grundwasser. Wasser Boden 9: 410–415

Kinzelbach W (1987) Numerische Methoden zur Modellierung des Transports von Schadstoffen im Grundwasser. Schriftenr Wasser–Abwasser 21. Oldenbourg, München

Kirkham D, Powers WL (1972) Advanced soil physics. Wiley-Interscience, London

Kirkner DJ, Theis TL, Jennings AA (1984) Multicomponent solute transport with sorption and soluble complexation. Adv Water Resour 7: 120

Kirschmer P, Reuter U, Ballschmiter K (1983) Die Belastung der unteren Toposphäre mit C_1–C_6 Organohalogenen. Chemosphere 12: 225–230

Kitano M (1983) Chemical structure and biodegradability. JETOC (Japan Chemical Industry Ecology, Toxicology & Information Center) Newsletter July 1983, 1

Kittredge J (1948) Forest influences. McGraw-Hill, New York

Klaer W, Krieter M (1982) Über die Bedeutung des Humus für Bodenerosion und Hangstabilität in den feuchten und wechselfeuchten Tropen von Papua Neuguinea. Erdkunde 36: 153–160

Klausing O (1970) Das Hessische Lysimeterprogramm. Verdunstungs- und Versickerungsmessungen in einem Netz von Lysimeterstationen. Dtsch Gewässerkd Mitt 14: 7–10

Klausing O, Salay G (1976a) Die Messung des Wasserumsatzes im Felde, Teil I. Dtsch Gewässerkd Mitt 20: 1–7

Klausing O, Salay G (1976b) Die Messung des Wasserumsatzes im Felde, Teil II. Dtsch Gewässerkd Mitt 20: 70–79

Klausing O, Salay G (1976c) Die Messung des Wasserumsatzes im Felde, Teil III. Reale Wasserbilanz. Dtsch Gewässerkd Mitt 20: 100–111

Klemme J-H (1988) Konzentrierung von Pestiziden im Nebel. Naturwiss Rundsch 41: 67

Klinge H (1983) Forest structures in Amazonia. Environmentalist 3, Suppl 5: 13–23

Klinge H, Furch K, Junk WJ, Irmler U (1981) Fundamental ecological parameters in Amazonia, in relation to the potential development of the region. In: Lal R, Russel EW (eds) Tropical agricultural hydrology. Wiley, Chichester, pp 19–36

Klingmüller W (1985) Genetik und Gentechnologie. Naturwiss Rundsch 3: 83–91

Klöpffer W, Rippen G (1983) Assessment of environmental hazard of substances and mixtures – integration of the final reports of the three working parties within the OECD hazard assessment project. Final Rep 106 04 014 Umweltbundesamt, Berlin

Kloke A (1972) Zur Anreicherung von Cadmium in Böden und Pflanzen. Landwirtsch Forsch 27: 200–206

Klug W (1969) Ein Verfahren zur Bestimmung der Ausbreitungs-bedingungen aus synoptischen Beobachtungen. Staub-Reinhalt Luft 29: 142–147

Klug W (1982) Physical transport or the problem how to model air pollution. In: Schneider T, Grant L (eds) Air pollution by nitrogen oxides. Elsevier, Amsterdam, pp 243–248

Klute A, Peters DB (1962) A recording tensiometer with a short response time. Soil Sci Soc Am Proc 26: 87–88

Knackmuss HJ (1979) Halogenierte und sulfonierte Aromaten – eine Herausforderung für Aromaten abbauende Bakterien. Forum Mikrobiol 6: 311–317

Knap AH, Iliffe TM, Butler JN (1980) Has the amount of tar on the open ocean changed in the past decade? Mar Pollut Bull 11: 161–164

Knapp BJ (1979) Soil processes. Allen & Unwin, London

Knie J (1978) Der Dynamische Daphnientest – ein automatischer Biomonitor zur Überwachung von Gewässern und Abwässern. Wasser Boden 12: 310–312

Knie J (1982) Der Daphnientest. Decheniana-Beih 26: 82–86

Koch R (1989) Umweltchemikalien–physikalisch–chemische Daten, Toxizitäten, Grenz- und Richtwerte, Umweltverhalten. VCH, Weinheim

Koeman IH (1982) Ecotoxicological evaluation: the eco-side of the problem. Ecotoxicol Environ Safety 6: 358–362

Koeman IH, van Genderen H (1972) Tissue levels in animals and effects caused by chlorinated hydrocarbon insecticides, chlorinated biphenyls and mercury in the marine environment along The Netherlands coast. In: Marine pollution and sea life, Fishing News. Ltd, Surrey, pp 428–435

Könemann H (1981a) QSAR in fish toxicity studies. Pt 1: Relationship for 50 industrial pollutants. Toxicology 19: 209–221

Könemann H (1981b) Fish toxicity tests with mixtures of more than two chemicals: a proposal for a quantitative approach and experimental results. Toxicology 19: 229–238

Könemann H, Musch A (1981) QSAR in fish toxicity studies. Pt 2: The influence of pH on the QSAR of chlorophenols. Toxicology 19: 223–228

König LA (1986) Gibt es einen Zusammenhang zwischen Umweltradioaktivität und Waldschäden? Schriftenr "Texte" Umweltbundesamtes 19: 163–165, 400

König N, Baccini P, Ulrich B (1986) Der Einfluß der natürlichen organischen Substanzen auf die Metallverteilung zwischen Böden und Bodenlösung in einem sauren Waldboden. Z Pflanzenernähr Bodenk 149: 68–82

Kohlmaier GH, Plöchl M (1985) Das Waldsterben in Mitteleuropa unter chronischer Luftschadstoffbelastung: Ein dynamisches Modell mit nicht-linearen Dosis–Wirkungsbeziehungen. Umweltbundesamt: IMA-Querschnittsseminar zur Waldschädenforschung "Belastung und Schäden auf Ökosystemebene und ihre Folgen"

Kokke R, Winteringham FPW (1980) Labelled substrate techniques as indicators of agrochemical residue-biota interactivity in soil and aquatic ecosystems. IAEA Panel Proc Ser, STI/PUB/548, pp 23–33, Vienna

Koltzenburg CH, Knigge W (1987) Holzeigenschaften von Buchen aus immissionsgeschädigten Beständen. Holz Roh- Werkstoff 45: 81–87

Kondratyev KY (1972) Radiation processes in the atmosphere. WMO, Genève

Kopczinski SL (1964) Photooxidation of alkylbenzene–nitrogen dioxide mixtures in air. Int J Air Water Pollut 8: 107–120

Korte F (1978) Abiotic processes. In: Butler GC (ed) Principles of Ecotoxicology. Wiley, Chichester, pp 11–35

Korte F (ed) (1987) Lehrbuch der ökologischen Chemie, 2. Aufl. Thieme, Stuttgart

Kramer H, Dong PH (1985) Kronenanalyse für Zuwachsuntersuchungen in immissionsgeschädigten Nadelholzbeständen. Forst- Holzwirt 40: 115–118

Kramer JR, Gleed J, Turner L, Pulfer K (1983) Aluminium speciation and analysis. Environmental Geochemical Report 1983/3. McMasters Univ, Ontario

Kramer W, Ulrich B (1985) Ergebnisse eines Kalksteigerungs-versuchs im Forstamt Syke. Forst- Holzwirt 40: 147–154

Krause GHM, Kaiser H (1977) Plant response to heavy metals and SO_2. Environ Pollut 12: 63–71

Kraybill HF (1980) Evaluation of public health aspects of carcinogenic/mutagenic biorefractories in drinking water. Prev Med 9: 212–218

Kraybill HF, Helmes CT, Sigman CC (1978) Biomedical aspects of biorefractories in water. In: Hutzinger O et al. (eds) Aquatic pollutants. Pergamon Press, New York, pp 419–459

Kremer BP (1989) Verölung und Ölabbau im Lebensraum Meer. Naturwiss Rundsch 42: 303–308

Kremling K (1983) The behavior of Zn, Cd, Cu, Ni, Co, Fe and Mn in anoxic Baltic waters. Mar Chem 13: 87–108

Kress LW, Skelly JM (1982) Response of several eastern forest trees to chronic doses of ozone and nitrogen dioxide. Plant Dis 66: 1149–1152

Kretschmar R (1979) Kulturtechnisch-bodenkundliches Praktikum. Inst Wasserwirtsch Meliorationswesen Christian-Albrechts-Univ Kiel, Kiel

Kreutzer K, Strebel O, Renger M (1978) Die Untersuchung des Wasserflusses im Boden mit Hilfe der Tritium-Markierung im Vergleich zur Tensiometer-Methode. Mitt Dtsch Bodenkd Ges 26: 93–102

Kriebitzsch WU (1978) Stickstoffnachlieferung in sauren Waldböden Nordwest-Deutschlands. Scripta Geobotanica. Golke, Göttingen

Krotzky A, Berggold R, Jaeger D, Dart PJ, Werner D (1983) Enhancement of aerobic nitrogenase activity (acetylene reduction assay) by phenol in soils and the rhizosphere of cereals. Z Pflanzenernähr Bodenkd 14b: 634–642

Kruyt HR, Robinson C (1927) Über Lyotropie. Chem Zentralbl I: 1117

Kucera V (1988) Studying effects on materials. Acid Mag 6: 17

Kübler W (1965) Methämoglobinämie im Säuglingsalter nach Spinatfütterung. D med Wschr 90: 1881–1982

Kühn H (1966) Absterbeerscheinungen an Koniferen in der Nähe von Hühnerställen mit

Entlüftung durch Ventilatoren. Nachrichtenbl Dtsch Pflanzenschutzdienstes (Braunschw) 18: 121–123

Külske S (1975) Der Stand der Anwendungstechnik von mathematisch–meteorologischen Ausbreitungsmodellen in der Praxis der Luftreinhaltung. Schriftenr Landesanst Immissions- und Bodennutzungsschutz Nordrhein- Westfalen 35: 69–112

Külske S (1986) Smogalarm. Geowiss Zeit 4(1): 1–9

Küppers K, Klumpp G (1987) Effects of ozone, sulfur dioxide and nitrogen dioxide on gas exchange and starch economy in Norway spruce (*Picea abies* L. Karst.). Vortrag, XIV Int Bot Cong, Berlin (West)

Kuhnt D (1987) Untersuchungen zum Sorptionsverhalten anionisch reagierender Umweltchemikalien in schlewsig-holsteinischen Böden. Diss, Univ Kiel

Kunte H, Pfeiffer EH (1985) Isolation of organic material from water. Bull Environ Contam Toxicol 34: 650–655

Kuokol J, Dugger WM (1967) Anthocyanin formation as a response to ozone and smog treatment in *Rumex crispus* L. Plant Physiol 42: 1023–1024

Kuttler W (1979) Einflußgrößen gesundheitsgefährdender Wetterlagen und deren bioklimatische Auswirkungen auf potentielle Erholungsgebiete – dargestellt am Beispiel des Ruhrgebietes und des Sauerlandes. Bochumer Geogr Arb 36. Schöningh, Paderborn

Kuttler W (1985) Stadtklima – Struktur und Möglichkeiten zu seiner Verbesserung. Geogr Rundsch 37: 226–233

Laaksovirta K, Olkkonen H, Alakuijala P (1976) Observations on the lead content of lichen and bark adjacent to a highway in southern Finland. Environ Pollut 11: 247–255

Labeyrie J (1978) Les aérosols. Recherche 9: 209–218

Lachenbruch PA (1975) Discriminant analysis. Hafner, New York

Lacy GH, Chevone BI, Cannon NP (1981) Effects of simulated precipitation on *Erwinia herbicola* and *Pseudomonas syringae* populations. Phytopathology 71: 888

Laflamme R, Hites R (1978) The global distribution of polycyclic aromatic hydrocarbons in recent sediments. Geochim Cosmochim Acta 42: 289–303

Lagaly G (1984) Clay–Organic interactions. Phil Trans R Soc London, A 311: 315–332

Lahmann E, Fett W (1980) 25 Jahre Staubniederschlagsmessung in Berlin. Gesundheits-Ing 101: 149–155

Lailach GE, Brindley GW (1969) Specific co-adsorption of purines and pyridines by montmorillonite. Clays and Clay Minerals 17: 95–100

Lailach GE, Thompson RD, Brindley GW (1968) Adsorption of pyrimidines, purines, and nucleosides by Li-, Na-, Mg-, and Ca-montmorillonite. Clays and Clay Minerals 16: 285–293, 295–301

Lalubie C (1991) Stoffeinträge durch Streufall in verschiedenen Waldökosystemen des Untersuchungsgebietes "Bornhöveder Seenkette". Dipl-Arb, Univ Kiel

Lambert SM (1967) Functional relationship between sorption in soil and chemical structure. J Agric Food Chem 15: 572–576

Langbein WB (1960) Water levels as indicators of long-term precipitation or runoff. Int Assoc Sci Hydrol Publ 51: 517–525

Langmuir D (1979) Techniques in estimation of thermodynamic properties for some aqueous complexes of geochemical interest. In: Jenne EA (ed) Chemical modeling in aqueous systems. ACS Symp Ser 93. American Chemical Society, Washington DC

Lariland NI, Lovelius NB, Yatsenko-Khmelevskii AA (1978) The effect of dust pollution from cement factories on growth of oak. Bot Zh (Leningr) 63: 721–729

Lauten H (1964) Untersuchungen über den Einfluß synthetischer grenzflächenaktiver Stoffe auf Pflanzen und ihre Abbaubarkeit im Boden. Diss, Univ Bonn

Lautensach H, Bögel R (1956) Der Jahresgang des mittleren geographischen Höhengradienten der Lufttemperatur in der verschiedenen Klimagebieten der Erok. Erdkunde 10: 270–282

Law RJ (1981) Hydrocarbon concentrations in water and sediments from UK marine waters, determined by fluorescence spectroscopy. Mar Pollut Bull 12: 153–157

Le Blanc F, Robitaille G, Rao DN (1974) Biological response of lichens and bryophytes to environmental pollution in the Murdochville Copper mine area, Quebec. J Hattori Bot Lab 38: 405–433

Lee DO (1984) Urban climates. Prog Phys Geogr 8: 1–31

Lee JA, Press MC, Woodin SJ (1985) Effects of NO_2 on aquatic ecosystems. In: CEC, Study on the need for a NO_2 long-term limit value for the protection of terrestrial and aquatic ecosystems. Final Report, EUR 10 546 EN

Lee JJ, Weber DE (1982) Effects of sulfuric acid rain on major cation and sulfate concentrations of water percolating through two model hardwood forests. J Environ Qual 11: 57–64

Lee MC et al. (1979) Solubility of polychlorinated biphenyls and capacitor fluid in water. Water Res 13: 1249–1258

Lee R (1978) Forest microclimatology. Columbia University Press, New York

Leffler HR, Cherry JH (1974) Destruction of enzymatic activities of corn and soybean leaves exposed to ozone. Can J Bot 52: 1233–1238

LeGrand HE (1980) A standard system for evaluating waste disposal sites. National Water Well Association, Washington DC

Lemerle P (1981) Opening address of the workshop on the control of existing chemicals under the patronage of the Organisation for Economic Co-operation and Development, Berlin (West), June 10–12, 1981. In: Umweltbundesamt (ed) Proc Worksh control of existing chemicals under the patronage of the Organisation for Economic Co-operation and Development, 1981. Berlin pp 9–11

Leo A, Hansch C, Elkins D (1971) Partition coefficients and their uses. Chem Rev 71: 525

Leo R (1983) Ölwehrhandbuch. Bekämpfung von Ölunfällen im Inland und auf See. Parey, Hamburg

Lerman SL, Darley EF (1975) Particulates. In: Mudd JB, Kozlowski TT (eds) Responses of plants to air pollution. Academic Press, New York, pp 141–158

Leroi-Gourhan A (1983) Le fil du temps. Fayard, Paris

Leser H (1983) Das achte "Basler Geomethodische Colloquium": Bodenerosion als methodisch–geoökologisches Problem. Geomethodica (Basel) 8: 7–22

Lesieur M (1982) La turbulence développée. Recherche 13: 1412–1425

Lettau H (1954) Improved models of thermal diffusion in the soil. Trans Am Geophys Union 35: 121–132

Lettau HH, Davidson B (eds) (1957) Exploring the atmosphere's first mile, vols 1 and 2. Academic Press, New York

Levitt J (1980) Responses of plants to environmental stresses, vol I. Chilling, freezing, and high temperature stresses. Academic Press, New York

Levy H (1971) Normal atmosphere: large radical and formaldehyde concentrations predicted. Science 173: 141–143

Lévy P (1954) Théorie de l'addition des variables aléatoires. Gauthier-Villars, Paris

Lewis JS, Prinn RG (1984) Planets and their atmospheres. Academic Press, San Francisco

Lewis RA (1987) Guidelines for environmental specimen banking with special reference to the Federal Republic of Germany. Ecological and Managerial Aspects. U.S. MAB Rep 12, Washington DC

Lewis RA, Stein N, Lewis CW (eds) (1984) Environmental specimen banking and monitoring as related to banking. Nijhoff, The Hague

Lewontin RC (1969) The meaning of stability. In: National Bureau of Standards (ed) Diversity and stability in ecological systems. Springfield Va, pp 13–24

Lichtfuß R (1977) Schwermetalle in den Sedimenten schleswig-holsteinischer Fließgewässer – Untersuchungen zu Gesamtgehalten und Bindungsformen. Diss, Univ Kiel

Liebscher HJ (1970) Eine neue wägbare und registrierende Lysimeteranlage in Niedermendig (Eifel). Dtsch Gewässerkd Mitt 14: 10–13

Lieth H (ed) (1987) Studien zum Osnabrücker Agrarökosystem-Modell OAM für das landwirtschaftliche Intensivgebiet Südoldenburg. MAB-Mitt 26

Likens GE (ed) (1981) Some perspectives of the major biogeochemical cycles. Scope 17. Wiley, Chichester

Likens GE, Johnson NM, Galloway JN, Bormann H (1976) Acid precipitation: strong and weak acids. Science 194: 643–645

Lindberg E, Turner R, Lovett M (1983) Mechanisms of the flux of acidic compounds and heavy metals onto receptors in the environment. VDI-Ber 500: 165–171

Lindberg RD, Runnels DD (1984) Groundwater redox reactions: an analysis of equilibrium state applied to E_h measurements and geochemical modeling. Science 425: 925

Lindberg SE (1982) Factors influencing trace metal, sulfate, and hydrogen ion concentrations in rain. Atoms Environ 16: 1701–1717

Lindberg SE, Johnson DW (eds) (1989) 1988 annual report of the integrated forest study. Environmental Science Division , Oak Ridge National Laboratory

Lindberg SE, Harriss RC (1981) The role of atmospheric deposition in an eastern U.S. deciduous forest. Water Air Soil Pollut 16: 13–31

Lindberg SE, Page AL, Norton SA (1990) Sources, deposition, and canopy interactions. In: Adriano DC, Salomons W (eds) Acidic precipitation, vol 3. Springer, Berlin Heidelberg New York London Paris Tokyo Hong Kong Barcelona

Lindsay WL (1979) Chemical equilibria in soils. Wiley, New York

Linthurst R (1983) Effects of acid deposition on vegetation: an overview. VDI-Ber 500: 175–185

Linzon SN, Heck WW, MacDowall FDH (1975) Effects of photochemical oxidants on vegetation. In: Photochemical air pollution: Formation, transport, effects. Can Nat Res Council, pp 89–142

Linzon SN, Temple PJ (1980) Soil resampling and pH measurements after an 18-year period in Ontario. In: Drabløs D, Tollan A (eds) Proc Int Conf Ecol Impact Acid Precipitation. SNSF-Project Norway, pp 176–177

Liss PG, Slater PG (1974) Flux of gases across the air–sea interface. Nature 247: 181–184

Litz N, Thiele M, Döring HW, Blume H-P (1987) The behavior of linear alkylbenzenesulfonate in different soils: a comparison between field and laboratory studies. Ecotoxicol Environ Safety 14: 103–116

Lloyd M, May RM (1974) On the population dynamics of periodical cicadas. In: May RM (ed) Stability and complexity in model ecosystems, 2nd edn. Princeton University Press, Princeton

Löpmeier F-J (1987) Verdunstung und Energiehaushalt. Verh 45. Dtsch Geographentages. Steiner, Stuttgart, pp 434–438

Lovett GM, Reiners WA, Olson RK (1982) Cloud droplet deposition in subalpine alsam fir forests. Science 218: 1303–1304

Luck WAP (1984) Structure of water and aqueous systems. In: Belfort G (ed) Synthetic membrane processes. Academic Press, New York, pp 21–72

Luckat S (1981) Quantitative Untersuchung des Einflusses von Luftverunreinigungen bei der Zerstörung von Naturstein. Staub-Reinhalt Luft 41: 440–442

Lull HW (1964) Ecological and silvicultural aspects. In: Chow VT (ed) Handbook of applied hydrology. McGraw-Hill, New York, pp 6.1–6.30

Lundgren DA (1973) Mass distribution of large atmospheric particles. PhD Thesis, University of Minnesota, Minneapolis

Lyman W, Reehl WF, Rosenblatt DH (1982) Handbook of chemical properties estimation methods. McGraw-Hill, New York

Mabey W, Mill T (1978) Critical review of hydrolysis of organic compounds in water under environmental conditions. J. Phys Chem Ref Data 7: 383–415

MacArthur RH (1955) Fluctuations of animal populations and a measure of community stability. Ecology 36: 533–536

MacArthur RH (1972) Geographical ecology. Harper and Row, New York

Mackay DM (1979) Finding fugacity feasible. Environ Sci Technol 13: 1218–1223

Mackay D, Leinonen PJ (1975) Rate of evaporation of low-solubility contaminants from water bodies to atmosphere. Environ Sci Technol 9: 1178–1180

Mackay D, Paterson S, Cheung B, Neely WB (1985) Evaluating the environmental behaviour of chemicals with a level III fugacity model. Chemosphere 14: 335–375

MacLean AJ (1976) Cadmium in different plant species and its availability in soils as influenced by organic matter and additions of lime, P, Cd and Zn. Can J Soil Sci 56: 129–138

MacRae IC, Alexander M (1963) Metabolism of phenoxyalkyl carboxylic acids by a *Flavobacterium* species. J Bacteriol 86: 1231–1235

Maier R, Altgayer M, Punz W, Rammer C, Schinniger R et al. (1979) Wasserhaushalt und Produktivitat staubbelasteter Pflanzen in der Umgebung einer Zementfabrik in Kärnten. Carinthia II 169: 167–193

Maignien R (1966) Compte rendu de recherches sur les latérites. Unesco, Paris

Mainwairing SJ, Harsha S (1975) Size distribution of aerosols in Melbourne city air. Atmos Environ 10: 57–60

Maki AW (1979) Correlations between *Daphnia magna* and fathead minnow chronic toxicity values for several classes of test substances. J Fish Res Board Can 36: 411–421

Malhotra SS, Khan AA (1984) Biochemical and physiological impact of major pollutants. In: Treshow M (ed) Air pollution and plant life. John Wiley & Sons, New York, pp 113–157

Malo BA, Purvis ER (1964) Soil absorption and atmospheric ammonia. Soil Sci 97: 242–247

Manier G (1975) Vergleich zwischen Ausbreitungsklassen und Temperaturgradienten. Meteorol Rundsch 28: 6–11

Manion PD (1981) Tree disease concepts. Prentice-Hall, Englewood Cliffs, NJ

Mann H (1976) Fischtest mit Goldorfen zur vergleichenden Prüfung der akuten Toxizität von Wasserinhaltsstoffen und Abwässern – praktische Erfahrungen aus drei Ringtesten. Z Wasser- Abwasser-Forsch 9: 103–110

Mansfield TA, Posthumus AC (1985) Effects of NO_2 in combination with other pollutants on vegetation. In: Commission of the European Communities (ed) Study on the need of a NO_2 long-term limit value for the protection of terrestric and aquatic ecosystems, G1-G83. Final Rep XI/70/85. Brussels

March J (1977) Advanced organic chemistry. McGraw-Hill, New York

Margalef R (1968) Perspectives in ecological theory. Univ of Chicago, Chicago

Marschner H (1986) Mineral nutrition of higher plants. Academic Press, London

Mart L, Rützel H, Klahe P, Sipos L, Platzek U, Valenta P, Nürnberg HW (1982) Comparative studies on the distribution of heavy metals in the oceans and coastal waters. Sci Total Environ 26: 1–17

Martin JP (1963) Influence of pesticide residues on soil microbiological and chemical properties. Residue Rev 4: 96–129

Martin JP, Ervin JO (1952) Effect of fumigation on soil organisms. Calif Citrogr 38: 6

Martin JP, Baines R, Ervin JO (1957) Influence of soil fumigation for citrus replants on the fungus population of the soil. Proc Soil Sci Soc Am 21: 163–166

Martin JT, Juniper BE (1970) The cuticles of plants. St. Martins Press, New York

Martin RJ, Al-Bahrani KS (1977) Adsorption studies using gas liquid chromatography – 2. Competitive adsorption. Water Res 11: 991–999

Marty D, Bianchi A, Gatellier C (1979) Effects of three oil spill dispersants on marine bacterial populations. I. Preliminary study. Quantitative evolution of aerobes. Mar Pollut Bull 10: 285–287

Marx DH (1969) The influence of ectotrophic mycorrhizal fungi on the resistance of pine roots to pathogenic infections. Phytopathology 59: 153–163

Marx DH (1972) Ectomycorrhizae as biological deterrents to pathogenic root infections. Annu Rev Phytopathol 10: 429–454

Mathé P (1985) Mitwirkung von Epibiosen in belasteten Waldökosystemen. In: Umweltbundesamt, IMA-Querschnittsseminar zur Waldschädenforschung Bioindikation, pp 177–178

Matheron G (1963) Principles of geostatistics. Econ Geol 58: 1246–1266

Matsumaru T, Yoneyama T, Totsuka T, Shiratori K (1979) Absorption of atmospheric NO_2 by plants and soils, I: Quantitative estimation of absorbed NO_2 in plants by ^{15}N method. Soil Sci Plant Nutr 25: 255–265

Mattheß G (1990) Die Beschaffenheit des Grundwassers, 2. Aufl. Borntraeger, Stuttgart

Mattheß G, Ubell K (1983) Allgemeine Hydrogeologie – Grundwasserhaushalt. Borntraeger, Berlin

Mattheß G, Isenbeck M, Pekdeger A, Schenk D, Schröter J (1985) Der Stofftransport im Grundwasser und die Wasserschutzgebietsrichtlinie W 101 – Statusbericht und Problemanalyse. UBA Berichte 7/85: 1–181, Berlin

Matveev LT (1984) Cloud dynamics. Atmospheric Sciences Library. Reidel, Dordrecht

Matziris DE, Nakos G (1977) Effect of simulated acid rain on juvenile characteristics of Aleppo pine (Pinus halepensis Mill.). Forest Ecology and Management 1: 267–272

Matzner E (1985) Auswirkungen von Düngung und Kalkung auf den Elementumsatz und die Elementverteilung in zwei Waldökosystemen im Solling. Allg Forstztg 40: 1143–1147

Matzner E, Thoma E (1983) Auswirkungen eines saisonalen Versauerungsschubs im Sommer/Herbst 1982 auf den chemischen Bodenzustand verschiedener Waldökosysteme. Allg Forstztg 26/27: 677–682

Matzner E, Ulrich B (1981) Effect of acid precipitation on soil. In: Fazzolare RA, Smith CB (eds) Beyond the energy crisis. Opportunity and challenge, vol II. Pergamon Press, Oxford, pp 555–564

Matzner E, Ulrich B (1983) Bilanzierung jährlicher Elementflüsse in Waldökosystemen im Solling. Z Pflanzenernähr Bodenkd 144: 660–681

May RM (1974) Stability and complexity in model ecosystems, 2nd edn. Princeton University Press, Princeton

Mayer R (1978) Adsorptionsisothermen als Regelgrößen beim Transport von Schwermetallen in Böden. Z Pflanzenern Bodenkde 141: 11–28

Mayer R (1981) Natürliche und authropogene Komponenten des Schwermetallhaushalts von Waldökosystemen. Göttinger Bodenk Ber 70

Mayer R (1983) Schwermetalle in Waldökosystemen der Lüneburger Heide. Mitt Dtsch Bodenkd Ges 38: 251–256

Mayer R, Heinrichs H (1981) Gehalte von Baumwurzeln an chemischen Elementen einschließlich Schwermetalle aus Luftverunreinigungen. Z Pflanzenernähr Bodenkd 144: 637–646

Mayer R, Ulrich B (1982) Calculation of deposition rates from the flux balance and ecological effects of atmospheric deposition upon forest ecosystems. In: Georgii HW, Pankrath J (eds) Deposition of atmospheric pollutants. Reidel, Dordrecht, pp 195–200

Mayer R, Heinrichs H, Seekamp G, Faßbender H (1980) Die Bestimmung repräsentativer Mittelwerte von Schwermetall-Konzentrationen in den Niederschlägen und im Sickerwasser von Wald–Standorten des Solling. Z Pflanzenernähr Bodenkd 143: 221–231

Maynard Smith J (1971) Models in ecology. Cambridge University Press, Cambridge

McCaig M (1979) The pipeflow streamhead – a type description. Univ of Leeds, School of Geography, Working Pap 242

McCall PJ, Laskowski DA, Swann RL, Dishburger HJ (1980) Measurement of sorption coefficients of organic chemicals and their use in environmental fate analysis in test protocols for environmental fate and movement of toxicants. Proc Symp. Assoc Off Anal Chem 94: 89–109

McClenahen JR (1978) Community changes in a deciduous forest exposed to air pollution. Can J For Res 8: 432–438

McClure VE (1976) Transport of heavy chlorinated hydrocarbons in the atmosphere. Environ Sci Technol 10: 1223–1229

McColl JG, Firestone MK (1986) Cumulative effects of simulated acid rain on soil chemical and microbial characteristics and conifer seedling growth. Soil Sci Soc Am J 51: 794–800

McConnell G, Ferguson DM, Pearson CR (1975) Chlorinated hydrocarbons and the environment. Endeavour 34: 13–18

McFarlane JC, Berry RL (1974) Cation penetration through isolated leaf cuticles (apricot). Plant Physiol 53: 723–727

McGrath SP (1984) Metal concentrations in sludges and soil from a long-term field trial. J Agric Sci 103: 25–35

McKenzie RM (1980) The adsorption of lead and other heavy metals on oxides of manganese and iron. Aust J Soil Res 18: 61–73

McMahon TA, Densison PJ (1979) Empirical atmospheric deposition parameters – a survey. Atmos Environ 13: 571–585

McNaughton SJ (1978) Stability and diversity of ecological communities. Nature 274: 251–253

Means JC, Wijayaratne R (1982) Role of natural colloids in the transport of hydrophobic pollutants. Science 215: 968–970

Meetham AR (1950) Natural removal of pollution from the atmosphere. Q J R Meteorol Soc 76: 359–371

Meigs P (1953) World distribution of arid and semi-arid homoclimates. Reviews of research on arid zone hydrology. P203–210. Unesco (Arid zone research I), Paris

Mejnartowicz L (1980) A genetic basis for the resistance of forest trees to anthropo-pressure, with special study of the effect of some toxic gases. First Annu Rep, Oct 1979–Sept 80. Polish Academy of Science, Institute of Dendrology, Kórnik-Poland, pp 4–18

Melamed D, Hanks RJ, Willardson LS (1977) Model of salt flow in soil with a source-sink term. Soil Sci Soc Am J 41: 29–33

Mengel K, Lutz HJ, Breininger MT (1987) Auswaschung von Nährstoffen durch sauren Nebel aus jungen, intakten Fichten (*Picea abies*). Z Pflanzenernähr Bodenkd 150: 61–68

Mercer JW (1984) Miscible and immiscible transport in groundwater. EOS, US Geol Survey, Reston VA

Mercier MJ (1977) 2,3,7,8-tetrachlordibenzo-*p*-dioxin – an overview. Proc of TCDD Pollution, CEC 1977

Merkel B, Grimmeisen W (1985) Bauanleitung für ein kostengünstiges Druckaufnehmer-tensiometer. Landschaftsökolog Messen Auswerten 1: 125–132

Metcalf RL (1976) Laboratory model ecosystem evaluation of the chemical and bio-logical behaviour of radiolabelled micropollutants. Environ Qual Saf 5: 141–151

Metcalf RL, Lu PY (1973) Environmental distribution and metabolic fate of key industrial pollutants and pesticides in a model ecosystem. Rep UILU-WRC-0069, University of Illinois at Urbana, Water Resources Center

Metcalf RL, Sangha GK, Kapoor JP (1971) Model ecosystem for the evaluation of pesticide biodegradability and ecological magnification. Environ Sci Technol 5: 709–712

Metzner H (1986) Vegetationsschäden durch kerntechnische Anlagen? Schriftenr "Texte" Umweltbundesamtes 19: 148–152, 396–397

Meyer RA, Hidy GM, Davis JH (1973) Determination of water and volatile organics in filter collected aerosols. Environ Lett 4: 9–20

Middleton JT, Kendrick JB, Darley EF (1955) Air-borne oxidants as plant-damaging agents. Proc Natl Air Pollut Symp, Washington, DC

Mill T (1980) Data needed to predict the environmental fate of organic chemicals. In: Haque R (ed) Dynamics, exposure and hazard assessment of toxic chemicals. Ann Arbor Science, Ann Arbor, pp 297–322

Mill T (1981) Minimum data needed to estimate environmental fate and effects for hazard classification of synthetic chemicals. In: Umweltbundesamt (ed) Proc worksh control of existing chemicals under the patronage of the Organisation for Economic Co-operation and Development, June 10–12, 1981, Umweltbundesamt, Berlin (West). pp 207–227

Mill T, Mabey WR, Bomberger DC, Chou TW, Hendry DG, Smith JH (1981) Labora-tory protocols for evaluating the fate of organic chemicals in air and water. EPA Contract 68-03-2227

Miller CW, Benson LV (1983) Simulation of solute transport in a chemically reactive heterogeneous system: model development and application. Water Resour Res 19: 381–391

Miller MS, Friedlander SK, Hidy GM (1972) A chemical element balance for the Pasadena aerosol. J Colloid Interface Sci 39: 165–176

Millot G (1964) Géologie des argiles. Masson, Paris

Milthorpe FL (1962) Plant factors involved in transpiration. In: UNESCO (ed) Plant–Water relationships in arid and semi-arid conditions. UNESCO, Paris. pp 107–115

Mingelgrin U, Gerstl Z (1983) Reevaluation of partitioning as a mechanism of nonionic chemicals adsorption in soils. J Environ Qual 12: 1–11

Mingelgrin U, Tsvetkov F (1985) Surface condensation of organophosphate esters on smectites. Clays and Clay Minerals 33: 62–70

Ministerie van Volkshuisvesting, Ruimtelijke Ordning en Milieubeheer (1983) Leidraad Bodemsanering. Amsterdam

Ministerie van Volkshuisvesting etc (ed) (1984) De aantasting van materialen door NO_x. Rep XI/659/84-EN

Mircetich SM, Zentmyer GA, Kendrick JB (1968) Physiology of germination of chlamydospores of *Phytophthora cinnamoni*. Phytopathology 58: 666–671

Mishustin EN, Shilnikova VK (1969) The biological fixation of atmospheric nitrogen by free-living bacteria. In: Unesco (ed) Soil biology. Unesco, Paris, pp 65–124

Mitchell A, Ormrod DP, Dietrich HF (1979) Ozone and nickel effects on pea leaf cell ultrastructure. Bull Environ Contam Toxicol 22: 379–385

Moll W (1978) Taschenbuch für Umweltschutz, Bd 1. Chemische und technologische Informationen. Steinkopff, Darmstadt

Monteith JL (1965) Evaporation and environment. Symp Soc Exp Biol 29: 205–234

Monteith JL, Szeicz G (1960) The carbon dioxide flux over a field of sugar beet. Q J R Meteorol Soc 86: 205–214

Mooi J (1984) Wirkungen von SO_2, NO_2, O_3 und ihrer Mischungen auf Pappeln und andere Pflanzenarten. Forst- und Holzwirt 39: 438–444

Moore JA (1984) Proposed design for a retrospective study of PMN hazard predictions. Memo to Science Advisory Board, US EPA

Moore JW, Ramamoorthy S (1984) Organic chemicals in natural waters. Springer, Berlin Heidelberg New York

Moore L, Barth EF (1976) Degradation of nitrilo-tri-acetic acid during anaerobic digestion. J Water Pollut Control Fed 18: 2406

Moreale A, van Bladel R (1980) Behaviour of 2,4 -D in Belgian soils. J Environ Qual 9: 627–633

Morrill LG, Mahilum BC, Mohiuddin SH (1982) Organic compounds in soil: Sorption, degradation and persistence. Ann Arbor, Michigan

Morrison RT, Boyd RN (1973) Organic Chemistry. Allyna Bacon, Boston

Mortland MM (1966) Urea complexes with montmorillonite: an infrared absorption study. Clays und Clay Minerals 6: 143–156

Mosebach D (1974) Stationärer Energie- und Wassertransport in den oberen Bodenschichten bei Verdunstung unter dem Einfluß eines Temperaturgradienten. Mitt Dtsch Bodenkd Ges 18: 75–83

Mosimann T (1980) Boden, Wasser und Mikroklima in den Geosystemen der Löss–Sand–Mergel–Hochfläche des Bruderholzgebietes (Raum Basel). Physiogeographica, Baseler Beiträge zur Physiogeographie 3

Mudd JB (1982) Effects of oxidants on metabolic function. In: Unsworth MH, Ormrod DP (eds) Effects of gaseous air pollution in agriculture and horticulture. Butterworth, London, pp 189–203

Mückenhausen E (1985) Bodenkunde, 2. Aufl. DLG, Frankfurt Main

Müller F (1987) Geoökologische Untersuchungen zum Verhalten ausgewählter Umweltchemikalien. Diss, Univ Kiel

Müller J (1982) Residence time and deposition of particle-bound atmospheric substances.

In: Georgii HW, Pankrath J (eds) Deposition of atmospheric pollutants. Reidel, Dordrecht, pp 43–52

Müller P (1977a) Die Belastbarkeit von Ökosystemen. Mitt 8, Schwerpunkt für Biogeographie Univ Saarlandes, Saarbrücken

Müller P (1977b) Biogeographie und Raumbewertung. Wissenschaftliche Buchgesellschaft, Darmstadt

Müller P (1979) Basic ecological concepts and urban ecological systems. In: Luepke NP (ed) Monitoring environmental material and specimen banking. Martinus Nijhoff, The Hague, pp 430–449

Müller PH (1975) Wahrscheinlichkeitsrechnung und mathematische Statistik. Lexikon der Statistik. Wissenschaftliche Buchgesellschaft, Darmstadt

Müller R, Lingens F (1988) Der mikrobielle Abbau von chlorierten Kohlenwasserstoffen. Wasser Abwasser 129: 55–60

Müller W (1975) Filtereigenschaften der Böden und deren kartiertechnische Erfaßbarkeit. Mitt Dtsch Bodenkd Ges 22: 323–330

Müller-Starck G (1985) Genetic differences between "tolerant" and "sensitive" beeches (*Fagus sylvatica* L.) in an environmentally stressed adult forest stand. Silvae Genet 34: 241–247

Münnich KO, Weiss W (1986) Konzentrationen von ^{14}C, Tritium und radioaktiven Edelgasen in der Atmosphäre. Schriftenr "Texte" Umweltbundesamtes 19: 153–162, 398–399

Muljadi D, Posner AM, Quirk IP (1966a) The mechanism of phosphate adsorption by kaolinite, gibbsite and pseudoboehmite. Part I. The isotherms and the effect of pH on adsorption. J Soil Sci 17: 238–247

Muljadi D, Posner AM, Quirk IP (1966b) The mechanism of phosphate adsorption by kaolinite, gibbsite and pseudoboehmite. Part II. The effect of temperature on adsorption. J Soil Sci 17: 248–257

Munger JW, Eisenreich SJ (1983) Continental-scale variations in precipitation chemistry. Environ Sci Technol 17: 32A–42A

Munn RE (1981) The design of air quality monitoring networks. Macmillan, London

Murach D (1984) Du Reaktion der Feinwurzeln von Fichte (*Picea abies* Karst.) auf zunehmende Bodenversauerung. Göttinger Bodenkdl Ber 77

Murdy WH (1979) Effect of SO_2 on sexual reproduction in *Lepidium virginicum* L. originating from regions with different SO_2 concentrations. Bot Gaz 140: 299–303

Murozumi M, Chow TJ, Patterson C (1969) Chemical concentrations of pollutant lead aerosols, terrestrial dust and sea salts in Greenland and Antarctic snow strata. Geochim Cosmochim Acta 33: 1247–1292

Musgrave G (1955) How much of the rain enters the soil? In: Stefferud A (ed) Water. US Dept Agriculture, Washington, pp 155–159

Myttenaere C, Daoust C, Roucoux P (1980) Leaching of technetium from foliage by simulated rain. Environ Exp Bot 20: 415–419

Nakaya U (1954) Snow crystals, natural and artificial. Harvard University Press, Cambridge, Mass

Nanson A (1962) Quelques éléments concernant le bilan d'assimilation photosynthétique en hêtraie ardennaise. Bull Inst Agron Stn Rech Gembloux 30: 320–331

NATO – Committee on the challenge of modern society (1982) Impact of air pollutants on materials. NATO CCMS pilot study on air pollution control strategies and impact modelling 139

Naturwissenschaftliche Rundschau (1990) Der bakterielle Abbau von Dioxinen. Nat Rd 43(5): 226

Naujokat D (1991) Modellierung von Oberflächenwiderständen der trockenen Deposition von SO_2 im Bereich der Bornhöveder Seenkette. Dipl-Arb Univ Kiel

Neff JM (1979) Polycyclic aromatic hydrocarbons in the aquatic environment. Sources, fates and biological effects. Applied Science Publishers, London

Nelson RW, Schur JA (1980) Assessment of effectiveness of geologic oscillation systems: PATHS groundwater hydrologic model. Battelle, Richland WA

Newman AS (1947) The effect of certain plant-growth-regulators on soil microorganisms and microbial processes. Proc Soil Sci Soc Am 12: 217–221

Newman AS, Thomas JR, Walker RL (1952) Disappearance of 2,4-dichlorophenoxy-acetic acid and 2,4,5-trichlorophenoxyacetic acid from soil. Proc Soil Sci Soc Am 16: 21–24

Nguyen VD, Valenta P, Nürnberg HW (1979) Voltammetry in the analysis of atmospheric pollutants. The determination of toxic metals in rainwater and snow by differential pulse stripping voltammetry. Sci Total Environ 12: 151–167

Nieboer E, Puckett KJ, Richardson DHS, Tomassini FD, Grace B (1975) Ecological and physicochemical aspects of the accumulation of heavy metals and sulphur in lichens. Int Conf Heavy Metals in the Environment, Toronto, 1975, II: 331–351

Nieboer E, Richardson DHS, Tomassini FD (1978) Mineral uptake and release by lichens: an overview. Bryologist 81: 226–246

Nieder R, Richter J (1986) C- und N-Festlegung in Böden Südost–Niedersachsens nach Krumenvertiefung. Z Pflanzenernähr Bodenkd 149: 189–201

Nielsen DR, Biggar JW, Erh KT (1973) Spatial variability of field-measured soil–water properties. Hilgardia 42: 215–260

Nienhaus F (1986) Ansichten eines Virologen zur Kontamination von Viren und Mikroorganismen in Waldökosystemen. Schriftenr "Texte" Umweltbundesamtes 19: 214–219

Niestlé A (1987) Die Änderungen des anorganischen Stickstoffgehaltes in einem kleinen Fließgewässer bei abnehmenden Abflüssen im Frühjahr. Dipl-Arb Univ Kiel

Nieuwstadt FTM, van Dop H (eds) (1984) Atmospheric turbulence and air pollution modeling. Atmospheric Sciences Library. Reidel, Dordrecht

Nihlgård B (1985) The ammonium hypothesis – an additional explanation for the forest dieback in Europe. Ambio 14: 2–8

Niki H, Maker PD, Savage CM, Breitenbach LP (1978) Mechanism for hydroxyl radical initiated oxidations of olefin–nitric oxide mixtures in parts per million concentrations. J Phys Chem 82: 135–137

Nikitin DJ (1973) The microstructure of the elementary microbial ecosystems. Bull Ecol Comun 17: 357–366

Nilsson SI, Miller HG, Miller JD (1982) Forest growth as a possible cause of soil and water acidification: An examination of the concepts. Oikos 39: 40–49

Noble WM (1956) Smog damage to plants. Lasca Leaves 15: 1–24

Noble WM, Wright LA (1958) Air pollution with relation to agronomic crops. II. A bio-assay approach to the study of air pollution. Agron 50: 551–553

Nogler P (1981) Auskeilende und fehlende Jahrringe in absterbenden Tannen (Abies alba Mill.). Allg Forstztg 36: 709–711

Nordstrom DK, Munoz JL (1986) Geochemical thermodynamics. Blackwell, Oxford

Nordstrom DK, Plummer LN, Wigley TML, Wolery TT, Ball JW et al. (1979) A comparison of computerized chemical models for equilibrium calculations in aqueous systems. In: Jenne EA (ed) Chemical modeling in aqueous systems. ACS Symp Ser 93. American Chemical Society, Washington DC

North Atlantic Treaty Organization (1981) Removal and transformation processes in the atmosphere with respect to SO_2 and NO_x. NATO/CCMS-Doc 127

Nounou P (1979) La pollution pétrolière des océans. Recherche 19: 147–156

Novakov T, Chang SG, Harker AB (1974) Sulfates as pollution particulates: catalytic formation on carbon (soot) particles. Science 186: 259–261

Nowak A (1984) Ein mathematisches Modell zum Einfluß von Monolinuron auf die mikrobielle Biomasse des Bodens. Z Pflanzenkr Pflanzenschutz, Sonderh X: 203–210

Nriagu JO (1978) Sulfur in the environment. Wiley, New York

Nürnberg HW, Valenta P, Nguyen VD (1982) Wet deposition of toxic metals from the atmosphere in the Federal Republic of Germany. In: Georgii H-W, Pankrath J (eds) Deposition of atmospheric pollutants. Reidel, Dordrecht. pp 143–157

Nukem GmbH (1976) Räumliche Erfassung der Emissionen ausgewählter luftverunreinigender Stoffe aus Industrie, Haushalt und Verkehr in der Bundesrepublik Deutschland. Materialienband Teil 1: Industrie 1960–1980. Nukem, Hanau

Nusch EA (1982) Prüfung der biologischen Schadwirkungen von Wasserinhaltsstoffen mit Hilfe von Protozoentests. Decheniana-Beih 26: 87–98

Nye PH, Greenland DJ (1960) The soil under shifting cultivation. Tech Commun 51. Commonwealth Bureau of Soils, Harpenden, UK

Oblisami G, Padmanabhan G, Padmanabhan C (1978) Effect of particulate pollutants from cement-kilns on cotton plants. Ind J Air Pollut Control 1: 91–94

O'Connor GA, Anderson JA (1974) Soil factors affecting the adsorption of 2,4,5-T. Soil Sci Soc Am Proc 38: 433–436

O'Connor GA, Wierenga PJ (1973) The persistence of 2,4,5-T in greenhouse lysimeter studies. Soil Sci Soc Am Proc 37: 398–400

OECD (Organisation for Economic Co-operation and Development) (1979) The OECD programme on long-range transport of air pollutants – measurements and findings. OECD, Paris

OECD (1981a) Guidelines for testing of chemicals. OECD, Paris

OECD (1981b) Guideline for testing of chemicals 301A – ready biodegradability: modified AFNOR Test. OECD, Paris

OECD (1984) Data interpretation guides for initial hazard assessment of chemicals. Provisional. OECD, Paris

Ogner G, Teigen O (1980) Effects of acid irrigation and liming on two clones of Norway spruce. Expanded version with basic data included. The Norwegian Forest Research Institute, 1432 Aas-NLH, Norway

Ohlendorf MM (1979) Archiving wildlife specimens for future analysis. In: Luepke NP (ed) Monitoring environmental material and specimen banking. Martinus Nijhoff, The Hague, pp 491–504

Oke TR (1978) Boundary layer climates. Methuen, London

Oke TR (1982) The energetic basis of the heat island. J R Meteorol Soc 1982: 1–24

Oke TR, East C (1971) The urban boundary layer in Montreal. Boundary Layer Meteorol 1: 411–437

Oke TR, Maxwell GB (1975) Urban heat island dynamics in Montreal and Vancouver. Atmos Environ 9: 191–200

Olsthoorn AA, Thomas R (1986) Milieuverontreiniging door Cadmium 1981–2000. Vrije Universiteit, Amsterdam

Opara–Nadi OA (1979) A comparison of some methods for determining the hydraulic conductivity of unsaturated soil in the low suction range. Gött Bodenkd Ber 57

Orgell WH, Weintrauf RL (1975) Influence of some ions on foliar absorption of 2,4-D. Bot Gaz 119: 88–93

Ottow JCG (1976) Mikroorganismen als Indikatoren unbelasteter, fäkalverschmutzter und biozidbelasteter Böden und Gewässer. Daten Dok Umweltschutz: 29–41

Ottow JCG (1982) Pestizide – Belastbarkeit, Selbstreinigungsvermögen und Fruchtbarkeit von Böden. Landwirtsch Forsch 35: 238–256

Ottow JCG (1983) Kann die Bodenfruchtbarkeit durch Pflanzenschutzmittel beeinträchtigt werden? Mitt Dtsch Landwirtsch Ges 98: 462–464

Ottow JCG (1985) Einfluß von Pflanzenschutzmitteln auf die Mikroflora von Böden. Naturwiss Rundsch 38: 181–189

Ouellet M, Jones HG (1983) Limnological evidence for the long-range atmospheric transport of acidic pollutants and heavy metals into the Province of Quebec, Eastern Canada. Can J Earth Sci 20: 23–36

Pankrath J (1987) Ausbreitungsrechnungen nach TA Luft 86 mit dem bundeseinheitlichen Programmsystem AUSTAL 86. Staub Reinhaltung der Luft 47: 239–244

Pankrath J (1983) Großräumiger Transport von Luftverunreinigungen in Europa: Anforderungen an Modelle zur Simulation der Ausbreitung und Deposition von säurebildenden Luftverunreinigungen. VDI-Ber 500: 43–50

Parker HM (1956) Radiation exposure from environmental hazards. Proc Int Conf Peaceful Uses Atomic Energy 13: 360–363

Parkhurst DL, Thorstenson DC, Plummer LN (1980) PHREEQE – a computer program for geochemical calculations. US Geol Surv Water Resour Inv 80–96. US Geol Survey, Washington DC

Pattenden NJ Branson JR, Fisher EMR (1982) Trace element measurements in wet and dry deposition and airborne particulate at an urban site. In: Georgii H-W, Pankrath J (eds) Deposition of atmospheric pollutants. Reidel, Dordrecht, pp 173–184

Paul JS, Bassham JA (1978) Effects of sulfite on metabolism in isolated mesophyll cells from *Papaver somniferum*. Plant Physiol 62: 210–214

Pearson CR, McConnell (1975) Chlorinated C_1 and C_2 hydrocarbons in the marine environment. Proc R Soc Lond Ser B 189: 305–332

Peichl L, Reiml D, Ritzl J, Schmidt-Bleek F (1987) Übersicht biologischer Wirkungs-Testsysteme zur Beobachtung unerwarteter Umweltveränderungen. Biosonden. Teil I. GSF-Bericht 28/87, München

Peiser GD, Yang SF (1979) Ethylene and ethane production from sulfur dioxide-injured plants. Plant Physiol 63: 142–145

Pekdeger A, Schulz HD (1975) Ein Methodenvergleich zur Laborbestimmung des k_f-Wertes von Sanden. Meyniana 27: 35–40

Penkett SA, Jones BMR, Eggleton AEI (1979) A study of SO_2 oxidation in stored rain water samples. Atmos Environ 13: 123–137

Penman HL (1963) Vegetation and hydrology. Commonwealth Agricultural Bureau, Farnham Royal, (England)

Persaud N, Wierenga PJ (1982) A differential model for one-dimensional cation transport in discrete homoionic ion-exchange media. Soil Sci Soc Am J 46: 482–490

Perseke C (1982) Composition of acid rain in the Federal Republic of Germany – spatial and temporal variations during the period 1979–1981. In: Georgii HW, Pankrath J (eds) Deposition of atmospheric pollutants. Reidel, Dordrecht, pp 77–86

Pestemer W (1985) Herbiziddynamik im Boden. Ber Landwirtsch Sonderh 198: 69–80

Petersen A (1988) Laboruntersuchungen zum Einfluß organischer Komplexbildner auf die Kinetik der Feldspatverwitterung. Berichte – Rep Geol Paläont Inst Univ Kiel 21

Petzold E (1981) Grundwasserneubildungsrate und Bodenfeuchtervorrat. Ein Vergleich unterschiedlicher Betrachtungsverfahren. Mitt Dtsch Bodenkd Ges 32: 211–218

Petzold E (1984) Erfassung und Bilanzierung des Wasserdargebotspotentials. Verh Ges Ökol 12: 487–498

Phadke SR, Gokhale SD, Phalnikar NL, Bhide BV (1945) Dipole moments from dielectric constants of liquids. Indian Chem Soc 22: 235–238

Philip JR (1957) Transient fluid motions in saturated porous media. Aust J Phys 10: 43–53

Philip JR (1964) The gain, transfer and loss of soil water. In: Hills ES (ed) Water resources, use and management. Melbourne University Press, Melbourne, pp 257–275

Philip JR, de Vries DA (1957) Moisture movement in porous materials under temperature gradients. Trans Am Geophys Union 38: 222–232

Pitts JN, Winer AM, Fitz DR, Knudsen AK, Atkinson R (1981) Experimental protocol for determining absorption cross-sections of organic compounds. EPA-600/3-81-05

Plummer LN, Parkhurst DL, Kosiur DR (1975) MIX2: a computer program for modeling chemical reactions in natural waters. Nat Tech Inform Serv Rep PB-251688, Springfield VA

Plummer LN, Jones BF, Truesdell AH (1976) WATEQF – a fortran IV version of WATEQ, a computer program for calculating chemical equilibrium of natural waters. US Geol Surv Water Res Invest 76/13, Washington DC

Pochon J, Tardieux P, d'Aguilar J (1969) Methodological problems in soil biology. Unesco Rev Res Paris IX: 13–63

Postgate JR (1982) The fundamentals of nitrogen fixation. Cambridge University Press, Cambridge

Posthumus AC (1977) Experimentelle Untersuchungen der Wirkung von Ozon und Peroxyacetylnitrat (PAN) auf Pflanzen. VDI-Ber 270: 153–161

Posthumus AC (1988) Critical levels for effects of NH_3 and NH_4^+. ECE Critical Levels Workshop Bad Harzburg, 14–18 March 1988. Final Draft Rep Part II, pp 117–127, UN Econ Comm Europe, Geneva

Posthumus AC, Tonneijck AEG (1982) Monitoring of effects of photo-oxidants on plants. In: Steubing L, Jäger HJ (eds) Monitoring of air pollutants by plants. Junk, The Hague, pp 115–119

Prance GT (1978) The origin and evolution of the Amazon flora. Interciencea 3: 207–222

Precht H (1967) A survey of experiments on resistance adaptation. In: Troshin AS (ed) The cell and environmental temperature. Pergamon Press, Oxford, pp 307–321

Prescott JA, Thomas JA (1949) The length of the growing season. Proc R Geogr Soc Aust 50: 42–46

Prickett TA, Naymik TG, Lonnquist CG (1981) A random walk solute transport model for selected groundwater quality evaluations. Ill State Water Surv Bull 65

Prigogine I (1967) Thermodynamics of irreversible processes, 3rd edn. Wiley-Interscience, New York

Prigogine I (1976) Order through fluctuation: Self-organization and social system. In: Jantsch E, Waddington CH (eds) Evolution and consciousness. Addison-Wesley, Reading Ma, pp 93–133

Prinz B (1984) Woran sterben unsere Wälder? Umschau 18: 544–549

Prinz B, Krause GHM, Stratmann H (1982) Waldschäden in der Bundesrepublik Deutschland. LIS Ber 28, Essen

Pruppacher HR, Semonin RG, Slinn WGN (eds) (1983) Precipitation scavenging, dry deposition and resuspension. Elsevier, Amsterdam

Puckett KJ, Niboer E, Gorzynski MJ, Richardson DHS (1973) The uptake of metal ions by lichens: a modified ion-exchange process. New Phytol 72: 329–342

Purcell WP, Bass GE, Clayton JM (1973) Strategy of drug design: a guide to biological activity. Wiley Interscience, New York

Ramade F (1979) Ecotoxicologie. Masson, Paris

Rambal S, Ibrahim M, Rapp M (1984) Variabilité spatiale des variations du stock d'eau du sol sous forêt. Application à l'optimisation d'un dispositif de mesure du bilan hydrique. CATENA 11: 177–186

Rappe C, Buser HR (1989) Chemical and physical properties, analytical methods, sources and environmental levels of halogenated dibenzodioxins and dibenzofurans. In: Kimbrough R, Jensen A A (eds) Halogenated biphenyls, terphenyls, naphtalenes, dibenzodioxins and related products. Elsevier, Amsterdam New York Oxford, pp 71–102

Rasmussen RA (1972) What do the hydrocarbons from trees contribute to air pollution? J Air Pollut Control Assoc 22: 537

Rasmussen RA, Went F (1965) Volatile organic material of plant origin in the atmosphere. Proc Natl Acad Sci 5: 215

Rasool SI (ed) (1973) Chemistry of the lower atmosphere. Wiley, New York

Raynal DJ, Leaf AL, Manion PD, Wang DJ (1980) Actual and potential effects of acid precipitation on a forest ecosystem in the Adirondack Mountains. New York State Energy Research and Development Authority, Research Report 80–28

Rehfuess KE, Bosch C, Pfankuch E (1982) Nutrient imbalances in coniferous stands in southern Germany. Int workshop on growth disturbances of forest trees. Oct 10–13, IUFRO/FFRJ-Jyvaskyla, Finland

Reiche EW (1990) Entwicklung, Validierung und Anwendung eines Modellsystems zur Beschreibung und flächenhaften Bilanzierung der Wasser- und Stickstoffdynamik in Böden. Diss, Univ Kiel

Reiter H et al. (1986) Einfluß von saurer Beregnung und Kalkung auf austauschbare und gelöste Ionen im Boden. Forstwiss Centralbl 105: 300–309

Rejment-Grochowska I (1976) Concentration of heavy metals, lead, iron, manganese, zinc and copper, in mosses. J Hattori Bot Lab 41: 225–230

Rekker RF (1977) The hydrophobic fragmental constant. Elsevier, Amsterdam

Renger M, Strebel O (1980) Jährliche Grundwasserneubildung in Abhängigkeit von Bodennutzung und Bodeneigenschaft. Wasser Boden 32: 362–366

Renk W (1977) Die räumliche Struktur und Genese der Bodendecke im Bereich der Großen Randstufe Transvaals und Swazilands. Diss, Univ Kiel

Rényi A (1961) On measures of entropy and information. In: Neyman J (ed) 4th Berkeley symp mathematical statistics and probability. Univ California Press, Berkeley, pp 547–561

Richards GA, Mulchi CL, Hall JR (1980) Influence of plant maturity on the sensitivity of turfgrass species to ozone. J Environ Qual 9: 49–53

Richards LA (1941) A pressure membrane extraction apparatus for soil solution. Soil Sci 51: 277–286

Richards LA (1942) Soil moisture tensiometer. Materials and construction. Soil Sci 53: 241–248

Richter DD, Johnson DW, Todd DE (1983) Atmospheric sulfur deposition, neutralization and ion leaching in two deciduous forest ecosystems. J Environ Qual 12: 263–268

Richter G (1978) Bodenerosion in den Reblagen an Mosel-Saar-Ruwer. Wirth E, Heinritz G (Hrg) Verh 41 Dtsch Geographentag Mainz 1977. Steiner, Wiesbaden, pp 371–389

Richter J (1986) Der Boden als Reaktor. Enke, Stuttgart

Riederer J (1973) Bibliographie der deutschsprachigen Literatur über die Verwitterung und Konservierung von Naturstein. Dtsch Kunst- Denkmalpflege 1973: 106–118

Riederer J (1977) Kunst und Chemie. Berlin

Riederer J (1985) Untersuchung der Gefährdung von kunst- und kulturgeschichtlichen Objekten in Museen und Archiven durch luftverunreinigende Stoffe. Umweltbundesamt, Berlin

Rijtema PE (1959) Calculation of capillary conductivity from pressure plate outflow data with non-negligible membrane impedance. Neth J Agric Sci 7: 209–215

Rijtema PE (1961) Evapotranspiration in relation to suction and capillary conductivity. In: Unesco (ed) Plant–water relationships in arid and semi-arid conditions. Unesco, Paris, pp 99–106

Rippen G Frank R, Zietz E, Hachmann R (1987a) Produktionsmengen und Verwendung chemischer Stoffe. Forschungsbericht 87–106 01 025 im Umweltforschungsplan des Bundesministers für Umwelt, Naturschutz und Reaktorsicherheit. Battelle, Frankfurt

Rippen G, Zietz E, Frank R, Knacker T, Klöpffer W (1987b) Do airborne nitrophenols contribute to forest decline? Environ Technol Lett 8: 475–482

Robbins CW, Jurinak JJ, Wagenet RJ (1980a) Calculating cation exchange in a salt transport model. Soil Sci Soc Am J 44: 1195–1199

Robbins CW, Wagenet RJ, Jurinak JJ (1980b) A combined salt transport–chemical equilibrium model for calcareous and gypsiferous soils. Soil Sci Soc Am J 44: 1191–1194

Robbins RC, Borg KM, Robinson E (1968) Carbon monoxide in the atmosphere. J Air Pollut Control Assoc 18: 106–110

Robin MJL, Elrick DE (1985) Effect of cation exchange on calculated hydrodynamic dispersion coefficients. Soil Sci Soc Am J 49: 39–45

Robinson DC, Wellburn AR (1983) Light-induced changes in the quenching of 9-amino-acridene fluorescence by photosynthetic membranes due to atmospheric pollutants and their products. Environmental Pollution (Series A) 32: 109–120

Robinson E, Rasmussen RA, Westberg HH, Holdren MW (1973) Nonurban nonmethane low molecular weight hydrocarbon concentration. J Geophys Res 78: 5345–5351

Rodhe H, Crutzen P, Vanderpol A (1981) Formation of sulfuric and nitric acid in the atmosphere during long-range transport. Tellus 33: 132–141

Roelofs JGM, Boxman AW (1986) The effect of air-borne ammoniumsulphate deposition on pine forests. Schriftenr "Texte" Umweltbundesamtes 19: 414–422

Roelofs JGM, Kempers AJ, Haudijk ALFM, Jansen J (1985) The effect of air-borne ammonium sulphate on Pinus nigra var. maritima in The Netherlands. Plant Soil 84: 45–56

Römpp H (1977) Römpps Chemie Lexikon. Franck'sche Verlagsbuchhandlung, Stuttgart

Roeschmann G (1986) Karte der Bundesrepublik Deutschland 1: 1 000 000. Bodenkarte. Bundesanstalt für Geowisseuscherffen und Rohstoffe. Hannover

Rogowski AS (1972) Watershed physics: soil variability criteria. Water Resour Res 8: 1015–1023

Rohbock E (1982) Atmospheric removal of airborne metals by wet and dry deposition. In: Georgii HW, Pankrath J (eds) Deposition of atmospheric pollutants. Reidel, Dordrecht, pp 159–171

Rohde G (1962) Flugasche und Pflanzenwachstum. Z Gesamte Hyg Grenzgeb 8: 333–339

Rohde K (1980) Warum sind ökologische Nischen begrenzt? Naturwiss Rundsch 33: 98–102

Roman JR, Raynal DJ (1980) Effect of acid precipitation on vegetation. In: Actual and potential effects of acid precipitation in the Adirondack Mountains. New York State Energy Research and Development Auth Rept ERDA 80–28, pp 1–63

Rose CW, Stern WR (1965) The drainage component of the water balance equation. Aust J Soil Res 3: 95–100

Roth R (1975) Der vertikale Transport von Luftbeimengungen in der Prandtl-Schicht und die Deposition-Velocity. Meteorol Rundsch 28: 65–71

Ruck A (1989) Vorliegende Konzepte für die Ableitung von Schwellenwerten zur Beurteilung von Schadstoffen im Boden – Bestandsaufnahme und Bewertung. Forschungsbericht 107 01 009 im Umweltforschungsplan des Bundesministers für Umwelt, Naturschutz und Reaktorsicherheit. Umweltbundesamt, Berlin

Rüstmann ML (1990) Risikoabschätzung des Pflanzenschutzmitteleinsatzes in Schleswig-Holstein mit Hilfe eines geographischen Informationssystems. Dipl-Arb Univ Kiel

Ruetz WF (1973) The seasonal pattern of CO_2 exchange of *Festuca rubra* L. in a montane meadow community in Northern Germany. Oecologia (Berl) 13: 247–269

Runge M (1973) Energieumsätze in den Biozönosen terrestrischer Ökosysteme. Scr Geobot 4

Sager H (1983) Mathematische Grundwassermodelle als Entscheidungshilfe bei Wassererschließungen. Informationsber Teil 1: 59–72. Bayer LA Wasserwirtschaft, München

Saltzman SB, Yaron B, Mingelgrin Y (1974) The surface-catalyzed hydrolysis of parathion on kaolinite. Soil Sci Soc Am Proc 38: 231–234

Samii AM, Lagaly G (1987) Adsorption of nuclein bases on smectites. In: Schultz LG, Mumpton FA (eds) Proc Int Clay Conf, Denver, 1985. The Clay Minerals Society, Bloomington Ind, pp 363–369

Sanders MO (1969) Toxicity to the crustacean *Gammarus lacustris*. Techn Pap US Bur Sport Fish Wildl 25: 1–18

Sandstede G (1961) Thermodynamische Betrachtungen zur BET-Theorie und zur Polanyi-Theorie der Gasadsorption. Z Phys Chem 29: 120–133

Sauerbeck D (1985) Funktionen, Güte und Belastbarkeit des Bodens aus agrikulturchemischer Sicht. Material zur Umweltforschung 10. Kohlhammer, Stuttgart

Sauerbeck D (1986) Vorkommen, Verhalten und Bedeutung von anorganischen Schadstoffen in Böden. In: Bodenschutz. Tagung über Umweltforschung an der Universität Hohenheim. Ulmer, Stuttgart, pp 77–96

Sauerbeck D (1987) Effects of agricultural practices on the physical, chemical and biological properties of soils: Part II. Use of sewage sludge and agricultural wastes. In: Barth H, L'Hermite PL (eds) Scientific basis for soil protection in the European Community. Elsevier, London, pp 181–210

Sax NI (1975) Dangerous properties of industrial materials. Van Nostrand Reinhold, New York

SCAS Test (1965) A procedure and standards for the determination of the biodegradability of alkyl benzene sulphonate and linear alkylate sulphonate. J Am Chem Soc 42: 986

Schachtschabel P, Blume H-P, Brümmer G, Hartge K-H, Schwertmann U (1989) Lehrbuch der Bodenkunde, 12. Aufl. Enke, Stuttgart

Schaefer R (1969) Amylose et amylopectine: recherches sur la dynamique de leur dégradation dans les sols d'une succession d'assèchement. Ann Inst Pasteur 116: 83–98

Schaller J (1988) Das geographische Informationssystem ARC/INFO. In: Faulbaum F, Uehlinger H-M (eds) Fortschritte der Statistik-Software 1. Fischer, Stuttgart, pp 503–514

Schams H (1967) Die Problematik der Wasserdurchlässigkeitsmessungen in Labor- und Felduntersuchungen auf Mineralböden. Diss, Tech Univ, Berlin

Schang J, Haussen JE, Nodop K, Dovland H (1984) Report to the steering body on the work at the chemical coordinating centre from 1 Oct 1983–30 Sept 1984. EMEP/CCC-Note 5/84. Norwegian Institute for Air Research, LillestrΦm

Scharpenseel HW (1971) Radiocarbon dating of soils – problems, troubles, hopes. In: Yaalon DH (ed) Paleopedology – origin, nature and dating of paleosols. Int Soc Soil Sci, Israel Univ Press, Jerusalem, pp 77–88

Scheerer E (1988) Toxizität von Pestizid-Stoffwechselprodukten. Naturwiss Rundsch 41: 370–371

Scheidegger AE (1957a) On the theory of flow of miscible phases in porous media. Proc IUGG, General Assembly Toronto 2: 236–242

Scheidegger AE (1957b) The physics of flow through porous media. Univ of Toronto Press, Toronto

Schernewski G, Matthies M, Litz N (1990) Untersuchung zur Anwendbarkeit von Sorptionskoeffizienten für die Simulation der Verlagerung von 2,4,5-T und LAS in Böden. Z Pflanzenernähr Bodenkd 153: 141–148

Schick AP, Hassan MA, Lekach J (1988) A vertical exchange model for coarse bedload movement: numerical considerations. CATENA Suppl 10: 73–83

Schidlowski M (1971) Probleme der atmosphärischen Evolution im Präkambrium. Geol Rundsch 60: 1351–1384

Schladot JD, Nürnberg HW (1982) Atmosphärische Belastung durch toxische Metalle in der Bundesrepublik Deutschland. Emission und Deposition. Ber Kernforschungsanlage Jülich 1776, Jülich

Schmidt K, Schöttler U, Zullei-Seibert N, Krutz H (1983) Die Ausbreitung von Stoffen im Grundwasser unter besonderer Berücksichtigung chemischer und biologischer Einflußgrößen. DVGW-Schriftenr Wasser 34: 197–224

Schmidt W (1925) Der Massenaustausch in freier Luft und verwandte Erscheinungen. Grand, Hamburg

Schmidt-Bleek F, Peichl L, Behling G, Müller KW, Reiml D (1987) A concept for early recognition and assessment for environmental changes. GSF-Bericht 21/87, München

Schmitt G (1982) Seasonal and regional distribution of polycyclic aromatic hydrocarbons in precipitation in the Rhein-Main-area. In: Georgii H-W, Pankrath J (eds) Deposition of atmospheric pollutants. Reidel, Dordrecht, pp 133–142

Schmölling J, Jörß KE (1983) Räumliche Verteilung und zeitliche Entwicklung von Emissionen der Vorläufer saurer Niederschläge und Oxidantien. VDI-Ber 500: 13–19

Schneider T, Grant L (eds) (1982) Air pollution by nitrogen oxides. Elsevier Sci, Amsterdam

Schoental R (1964) Carcinogenesis by polycyclic aromatic hydrocarbons. In: Clar E (ed) Polycyclic hydrocarbons, vol 1. Academic Press, London, pp 134–160

Schönwald B (1980) Ableitung des vertikalen Temperaturprofils der atmosphärischen Grenzschicht aus Strahldichtemessungen unter verschiedenen Zenitdistanzen auf den Frequenzen 54.5 GHz und 58 GHz. Hamb Geophys Einzelschr A 48

Schofield RL (1935) The pF of the water in soil. Trans III Int Congr Soil Sci, vol 2. Oxford, pp 37–48

Scholl G (1975) Positive und negative Wirkungen von Stickstoffverbindungen im Einwirkungsbereich einer Düngemittelfabrik. Staub – Reinhaltung der Luft 35: 201–205

Scholles U (1985) Die Kennzeichnung des Oberflächenabflusses an Hängen mit Hilfe von Tracerkurven. Landschaftsökol Messen Auswerten 1: 179–189

Scholles U, Rohdenburg H, Bork H-R (1985) Ein Versuchsgerinne zur Simulation von Oberflächenabfluß und Bodenerosion von Hangausschnitten. Landschaftsökol Messen Auswerten 1: 159–168

Scholz F (1981a) Genökologische Wirkungen von Luftverunreinigungen aufgrund von Expositionsunterschieden im Bestand. Forstarchiv 52: 58–61

Scholz F (1981b) Genecological aspects of air pollution effects on northern forests. Silva Fenn 15: 384–391

Scholz F, Geburek Th (1987) Untersuchungen über genetisch-ökologische Auswirkungen simultaner Immissionsbelastungen des Sproß- und Wurzelsystems der Baumart Fichte. 8. Seminarbericht Waldschäden/Luftverunreinigungen zum 9. Statusseminar am Fraunhofer-Institut für Umweltchemie und Ökotoxikologie in Schmallenberg-Grafschaft 1987. pp 301–316

Scholz F, Gregorius H-R, Rudin D (1989) Genetic effects of air pollutants in forest tree populations. Springer, Berlin Heidelberg

Schorling M (1990) Anwendung des Lagrange-Ausbreitungsmodells zur Berechnung des Mittelwertes und der Streuung der Konzentration. VDI Berichte 837: 151–164

Schrimpf E (1980) Zur zeitlichen und räumlichen Belastung des Fichtelgebirges mit Spurenmetallen: Analysen von Baumringen und Schnee. Nat Landschaft 55: 460–462

Schrimpf E, Herrmann R (1978) Spurenmentalle im Schnee Nordostbayerns. Gesundheits-Ing 99: 70–74

Schrimpf E, Thomas W, Herrmann R (1979) Regional patterns of contaminants (PAH, pesticides and trace metals) in snow of northeast Bavaria and their relationship to human influence and orographic effects. Water Air Soil Pollut 11: 481–497

Schrimpf R, Klemm O, Eiden R, Frevert T, Herrmann R (1984) Anwendung eines GRUNOW-Nebelfängers zur Bestimmung von Schadstoffen in Nebelniederschlägen. Staub-Reinhalt Luft 14: 72–75

Schroder H (1985) Nitrogen losses from Danish agriculture – trends and consequences. Agric Ecosyst Environ 14: 279–289

Schroeder M (1976) Grundsätzliches zum Einsatz von Lysimetern. – Erfahrungen aus Nordrhein-Westfalen. Dtsch Gewässerkd Mitt 20: 8–13

Schroeder M (1983) Neue Werte zur Grundwasserneubildung unter Wald für das Münsterland. Dtsch Gewässerkd Mitt 27: 121–124

Schröder W (1989) Ökosystemare und statistische Untersuchungen zu Waldschäden in Nordrhein-Westfalen: Methodenkritische Ansätze zur Operationalisierung einer wissenschaftstheoretisch begründeten Konzeption. Diss, Univ Kiel

Schröder W, Fränzle O, Vetter L (1986) 1st eine synoptische Darstellung von standörtlichen Randbedingungen der Waldschäden möglich? Allg Forstz 22: 543–544

Schrödter H (1985) Verdunstung: anwendungsorientierte Meßverfahren und Bestimmungsmethoden. Springer, Berlin

Schröter J (1983) Der Einfluß von Textur- und Struktureigenschaften poröser Medien auf die Dispersivität. Diss, Univ Kiel

Schubach K (1959) Wasserhaushaltsuntersuchungen nach Beobachtungen an wägbaren Lysimetern. Dtsch Gewässerkd Mitt 14: 1–7

Schulten HR, Rump HH, Simmleit N, Müller R (1986) Untersuchungen über immissionsbedingte Veränderungen ausgewählter Pflanzenstoffe in und an Koniferennadeln. Schriftenr "Texte" Umweltbundesamtes 18: 13–23

Schultz D (1982) Land disposal of hazardous waste. Proc 8th Annu Res Symp, March 8–10. US EPA, Cincinnati OH

Schultz R (1987) Vergleichende Betrachtung des Schwermetallhaushalts verschiedener Waldökosysteme Norddeutschlands. Berichte des Forschungszentrums Waldökosysteme/Waldsterben, Reihe A, Bd 32

Schwartz FW, Growe A (1980) A deterministic probabilistic model for contaminant transport. US NRC, NUREG/CR-1609, Washington, DC

Schwarz O (1986) Zum Abflußverhalten von Waldböden bei künstlicher Beregnung. In: Einsele G (ed) Das landschaftsökologische Forschungsprojekt Naturpark Schönbuch. VCH, Weinheim, pp 161–179

Schwarzenbach RP, Westall J (1981) Transport of nonpolar organic compounds from surface water to groundwater. Laboratory sorption studies. Environ Sci Technol 15: 1360–1367

Schwille F (1981) Groundwater pollution in porous media by fluids immiscible with water. Sci Total Environ 21: 173–185

Schwille F (1982) Die Ausbreitung von Chlorkohlenwasserstoffen im Untergrund, erläutert anhand von Modellversuchen. DVGW-Schriftenr Wasser 31: 203–234

Sehmel GA (1980) Particle and gas dry deposition. Atmos Environ 14: 983–1011

Seibt G, Wittich W, Reemtsma JB (1977) Ertragskundliche und bodenkundliche Ergebnisse langfristiger Kalkdüngungsversuche im nord- und westdeutschen Bergland. Schrift Forstl Fak Univ Göttingen Bd 50

Sevruk B (1981) Methodische Untersuchungen des systematischen Meßfehlers des Hellmann-Regenmessers im Sommerhalbjahr in der Schweiz. Diss, ETH Zürich 6798

Seymour J, Girardet H (1985) Fern vom Garten Eden. Fischer, Frankfurt

Shaffer MJ, Gupta SC (1981) Hydrosalinity models and field validation. In: Iskandar IK (ed) Modeling wastewater renovation. Wiley, New York, pp 136–181

Sheals JG (1955) The effects of DDT and BHC on soil Collembola and Acarina. Soil Zoology, Proc Nottingham School Agric Sci. Nottingham, pp 241–252

Sheehan PJ, Miller DR, Butler GC, Bourdeau P (eds) (1984) Effects of pollutants at the ecosystem level. SCOPE 22, Wiley, Chichester

Sheridan RP, Rosenstreter R (1973) The effect of hydrogen ion concentrations in simulated rain on the moss Tortula ruralis (Hedw.) Bryologist 76: 168–173

Shriner DS (1974) Effects of simulated rain acidified with sulfuric acid on host–parasite interactions. PhD Thesis, North Carolina State University, Raleigh, NC

Shriner DS (1977) Effects of simulated rain acidified with sulfuric acid on host-parasite interactions. Water, Air, and Soil Pollut 8: 9–14

Shriner DS (1981) Terrestrial vegetation air pollutant interactions: nongaseous pollutants, wet deposition. Paper presented at Inter Conf on air pollutants and their effects on terrestrial ecosystems. Banff, Alberta, Canada

Siccama TG, Bliss M, Vogelmann HW (1982) Decline of red spruce in the Green Mountains of Vermont. Bull Torrey Bot Club 109: 162–168

Silvius JE, Ingle M, Baer CH (1975) Sulfur dioxide inhibition of photosynthesis in isolated spinach chloroplasts. Plant Physiol 56: 434–437

Simon K (1983) Evolution der Evolution. Naturwiss Rundsch 36(4): 166–168

Simon-Sylvestre G, Beaumont A (1982) Effets de quelques pesticides sur les azotobacters. Agrochimica 26: 157–166

Singh SN, Rao DN (1981) Certain responses of wheat plants to cement dust pollution. Environ Pollut 24: 75–81

Sinn JP, Pell EJ (1984) Impact of repeated nitrogen dioxide exposures on composition and yield of potato foliage and tubes. J Am Soc Hortic Sci 109: 481–484

Sioli H (1980a) 40 Jahre Amazonasforschung. MPG Spiegel 1/80: 43–57

Sioli H (1980b) Foreseeable consequences of actual development schemes and alternative ideas. In: Barbira-Scazzocchio F (ed) Land, people and planning in contemporary Amazonia 3. Cambridge Univ, Centre of Latin America Studies, Cambridge pp 257–268

Sittig M (1980) Priority toxic pollutants. Health impacts and allowable limits. Noyes Data Corporation, New Jersey

Skiba U, Peirson-Smith TJ, Cresser MS (1986) Effects of simulated precipitation acidified with sulfuric and/or nitric acid on the throughfall chemistry of Sitka spruce (*Picea sitchensis*) and heather (*Calluna vulgaris*). Environ Pollut Ser B 11: 255–270

Skoulikidis TN (1983) Effects of primary and secondary air pollutants and acid depositions on (ancient and modern) buildings and monuments. In: Ott H, Stangl H (eds) Acid deposition, a challenge for Europe. Commission of the European Communities, Brussels, pp 193–226

Slanina J (1983) Collection and analysis of precipitation. Methods, data evaluation and interpretation. VDI-Ber 500: 117–124

Slanina J, Asman WAH (1981) Analytisch-chemisch onderzoek van neerslag in Nederland. Energiespectrum 1981: 292–299

Slatyer RO (1968) The use of soil water balance relationships in agroclimatology. Natural Resources Res VII: 73–87. UNESCO, Paris

Slinn WGN (1982) Predictions for particle deposition to vegetation. Atmos Environ 16: 1785–1794

Smith BJ (1977) Rock temperature measurements from the northwest Sahara and their implications for rock weathering. CATENA 4: 41–63

Smith FE (1970) Analysis of ecosystems. In: Ecological studies 1. Reichle ED (ed) Analysis of temperate forest ecosystems. Springer, Berlin Heidelberg New York, pp 7–18

Smith RE, Parlange JY (1978) A parameter efficient hydrologic infiltration model. Water Resour Res 14: 533–538

Sommer C (1980/81) A method for investigating the influence of soil water potential on water consumption, development and yield of plants. Soil Tillage Res 1: 163–172

Sommer C (1988) Soil compaction and water uptake of plants. CATENA Suppl 11: 107–111

Sontheimer R (1985) Nitrat im Grundwasser. Karlsruhe

Sørensen NA (1983) Ingen saerlig øking siden igoo. Miljøvernteknikk 3: 20–23, 38

Spearing AM (1972) Cation exchange capacity and galacturonic acid content on several species of *Sphagnum* in Sandy Ridge Bog, Central New York State. Bryologist 75: 154–158

Spencer WF, Farmer WJ (1980) Assessment of the vapor behavior of toxic organic chemicals. In: Haque R (ed) Dynamics, exposure and hazard assessment of toxic chemicals. Ann Arbor Science, Ann Arbor, pp 143–161

Spencer WF, Farmer W, Cliath MM (1973) Pesticide volatilization. Residue Rev 49: 1–47

Spierings FHFG (1971) Influence of fumigations with NO_2 on growth and yield of tomato plants. Neth J Plant Pathol 77: 194–200

Sposito G (1981) The thermodynamics of soil solutions. Clarendon Press, Oxford

Sposito G (1985) Chemical models of inorganic pollutants in soils. CRC Crit Rev Environ Control 15

Sposito G, Mattigod (1980) GEOCHEM: a computer program for the calculation of chemical equilibria in soil solutions and other natural water systems. Dep Soil Environ Sci, Univ Calif, Riverside

Spranger T (1992) Erfassung und ökosystemare Bewertung der atmosphärischen Deposition und weiterer oberirdischer Stofflüsse im Bereich der Bornhöveder, Seenkette. Diss, Univ Kiel

Srivastava HS, Ormrod DP (1984) Effects of nitrogen dioxide and nitrate nutrition on growth and nitrate assimilation in bean leaves. Plant Physiologist 76: 418–423

Stamm Sv (1992) Untersuchungen zur Primär produktion von *Corylus avellana* an einem Knickstandort in Schleswig-Holstein und Erstellung eines Produktionsmodells. Eco-Sys – Beiträge zur Ökosystemforschung, Suppl Bd 3. Kiel

Stand L (1980) The effect of acid precipitation on tree growth. In: Drabløs D, Tollan A (eds) Ecological impact of acid precipitation. Proc of an International Conf, Sandefjord, Norway

Stauffer TB, MacIntyre WG (1986) Sorption of low-polarity organic compounds on oxide minerals and aquifer material. Environ Toxicol Chem 5: 949–955

Stein N (1978) Coniferen im westlichen Malayischen Archipel. Biogeographica 11: 1–168

Step system group, OECD chemicals testing program (1979) Toxicology studies according to a hierarchical system, together with a presentation of the costs involved. Final report of the expert group physical chemistry, vol I, part 2: 2.5.4. OECD, Paris

Stephens ER, Darley EF, Taylor OC, Scott WE (1961) Photochemical reaction products in air pollution. Int J Air Water Pollut 4: 79–100

Stern AM (1980) Environmental testing of toxic substances. In: Haque R (ed) Dynamics, exposure and hazard assessment of toxic chemicals. Ann Arbor Science, Ann Arbor, pp 459–469

Steubing L (1987) Resistenz- und Konkurrenzverhalten der Waldvegetation unter dem Einfluß saurer Deposition. 8. Seminarbericht Waldschäden/Luftverunreinigungen zum 9. Statusseminar am Fraunhofer-Institut für Umweltchemie und Ökotoxikologie in Schmallenberg-Grafschaft 1987. pp 212–242

Steubing L, Fangmeier A (1987) SO_2-sensitivity of plant communities in a beech forest. Environm Pollution 44: 297–306

Steubing L, Jäger HJ (eds) (1982) Monitoring of air pollutants by plants. Methods and prblems. Junk, The Hague

Stigliani WM (1988) Changes in valued "capacities" of soils and sediments as indicators of nonlinear and time-delayed environmental effects. IIASA Ecoscript 35. Int Inst Applied Systems Analysis, Laxenburg

Stöcker G (1974) Zur Stabilität und Belastbarkeit von Ökosystemen. Arch Naturschutz Landschaftsforsch 14: 237–261

Strebel O, Giesel W, Renger M, Lorch S (1970) Automatische Registrierung der Bodenwasserspannung im Gelände mit dem Druckaufnehmertensiometer. Z Pflanzenernähr Bodenkd 126: 6–15

Stremme H (1981) Bodenkarte von Schleswig-Holstein. Geol Landesamt Schleswig-Holstein, Kiel

Strosher MT, Peake E (1979) Baseline states of organic constituents in the Athabasca River system upstream of Fort McMurray. The University of Calgary, AOSERP Rep 53

Stumm W, Morgan JJ (1981) Aquatic chemistry, 2nd edn. Wiley, New York

Stumm W, Kummert R, Sigg L (1980) A ligand exchange model for the adsorption of inorganic and organic ligands at hydrous oxide interfaces. Croat Chem Acta 48: 291–312

Stuper AJ, Jurs PC (1976) ADAPT: a computer system for automated data analysis using pattern recognition techniques. J Chem Inf Comput Sci 16

Sturm RN (1973) Biodegradability of nonionic surfactants: screening test for predicting rate and ultimate biodegradation. J Am Oil Chem Soc 50: 159–167

Suess E, Erlenkeuser H (1975) History of metal pollution and carbon input in Baltic Sea sediments. Meyniana 27: 63–75

Sundborg A (1951) Climatological studies in Uppsala with regard to temperature conditions in the urban area. Geographica 22. Geogr Institute Uppsala, Uppsala

Sverdrup GM (1977) Parametric measurement of submicron atmospheric aerosol size distributions. PhD Thesis, University of Minnesota, Minneapolis

Swartzendruber D (1962) Non-Darcy flow behaviour in liquid saturated porous mdeia. J Geophys Res 67: 5205–5213

Swinbank WC (1958) Turbulent transfer in the lower atmosphere. Climatology and microclimatology. Proc of the Canberra Symp. Unesco (Arid zone research XI), Paris, pp 35–37

Sykes P (1976) Reaktionsmechanismen der Organischen Chemie, 7. Aufl. VCH, Physik-Verlag, Weinheim

Taft RW (1956) The separation of polar, steric and resonance effects in rates of normal ester hydrolysis. In: Newman MS (ed) Steric effects in organic chemistry. Wiley, New York

Takken W, Balk F. Jansen RC, Koeman JH (1978) The experimental application of insecticides from a helicopter for the control of riverine populations of *Glossina tachinoides* in West Africa. VI Observations on side-effects. PANS 24: 455–466

Talsma T (1960) Comparison of field methods of measuring hydraulic conductivity. Int Comm on Irrigation and Drainage, 4th Congr, Madrid C11. pp 145–156

Talsma T (1969) In situ measurement of sorptivity. Aust J Soil Res 7: 269–276

Talsma T, Hallam PM (1980) Hydraulic conductivity measurement of forest catchments. Aust J Soil Res 18: 139–148

Tanji KK, Doneen LD, Ferry GV, Ayers RS (1972) Computer simulation analysis on reclamation of salt-affected soils in San Joaquin Valley, Calfornia. Soil Sci Soc Am Proc 36: 127–133

Tanner CB (1960) Energy balance approach to evapotranspiration from crops. Proc Soil Sci Soc Am 24: 1–9

Taylor GE, Murdy WH (1975) Population differentiation of an annual plant species *Geranium carolinianum* L. in response to sulfur dioxide. Bot Gaz 136: 212–215

Taylor GE, Pitelka LF, Clegg MT (1991) Ecological genetics and air pollution. Springer, Berlin Heidelberg New York

Taylor OC (1969) Importance of peroxyacetyl nitrate (PAN) as a phytotoxic air pollutant. J Air Pollut Control Assoc 19: 347–351

Taylor RI (1958) The automatic, direct measurement of natural evaporation. Climatology and microclimatology. Proc of the Canberra Symp. Unesco (Arid zone research XI), Paris, pp 42–44

Taylor SA (1962) The influence of temperature upon transfer of water in soil systems. Meded Landbouwhogesch Gent, Belgie 27, 2

Taylor SA, Cary JW (1960) Analysis of the simultaneous flow of water and heat or electricity with the thermodynamics of irreversible processes. Trans VIIIth Int Congr of Soil Science, Madison, Wis 1: 80–90

Taylor SA, Cary JW (1965) Soil–water movement in vapour and liquid phases. In: Eckardt FE (ed) Methodology of plant eco-physiology. Proc Montpellier Symp. Arid Zone Res XXV: 159–165

Taylor SA, Cavazza L (1954) The movement of soil moisture in response to temperature gradients. Proc Soil Sci Soc Am 18: 351–358

Taylor T (1989) Zum Transportverhalten der Schwermetalle Cadmium, Chrom, Kupfer und Zink in ausgewählten Böden und Sanden. Diss, Univ Kiel

Temple PJ (1982) Effects of peroxyacetyl nitrate (PAN) on growth of plants. Ph.D. Thesis, Univ California, Riverside

Temple PJ, Harper DS, Pearson RG, Linzon SN (1979) Toxic effects of ammonia on vegetation in Ontario. Environ Pollut 4: 297–301

Terhaar CJ, Ewell WS, Dziuba SP, White WW, Murphy PJ (1977) A laboratory model for evaluating the behaviour of heavy metals in an aquatic environment. Water Res 11: 101–110

Tesche M, Schmidtchen A (1978) Schädigungen an Koniferen in der Umgebung von Anlagen der industriemäßigen Hühnerhaltung. Arch Phytopathol Pflanzenschutz 14: 327–332

Theis TL, Kirkner DJ, Jenning AA (1982) Multi-solute subsurface transport modeling for energy solid wastes. Technical Progress Report. Dept Civil Engineering, University of Notre Dame, Notre Dame, Indiana

Thibodeaux LJ (1974) A test method for volatile component stripping of waste water. EPA-660/2-74-044. EPA, Washington DC

Thibodeaux LJ (1979) Chemodynamics. Environmental movement of chemicals in air, water, and soil. Wiley, New York

Thöle R, Schreiber KF (1985) Grundwasserneubildungsrate und Filterpotential. Landsch Stadt 17: 61–65

Thom NS, Agg AR (1975) The breakdown of synthetic organic compounds in biological processes. Proc R Soc Lond Ser B, 189: 347–357

Thom R (1969) Topological models in biology. Topology 8: 313–335

Thomas MD, Hendricks RH, Collier TR, Hill GR (1943) The utilisation of sulfate and sulfur dioxide for the sulfur nutrition of Alfalfa. Plant Physiol 18: 345–371

Thomas W, Herrmann R (1980) Nachweis von Chlorpestiziden, PCB, PCA und Schwermetallen mittels epiphytischer Moose als Biofilter entlang eines Profils durch Mitteleuropa. Staub-Reinhalt Luft 40: 440–444

Thomas W, Riess W, Herrmann R (1983) Processes and rates of deposition of air pollutants in different ecosystems. In: Ulrich B, Pankrath J (eds) Effects of accumulation of air pollutants in forest ecosystems. Reidel, Dordrecht

Thompson CR, Taylor OC (1969) Effects of air pollutants on growth, leaf drop, fruit drop and yield of citrus trees. Environ Sci Technol 3: 934–940

Thompson TD, Brindley GW (1969) Adsorption of pyrimidines, purines, nucleotides by Na-, Mg-, and Cu(III)-illite. Amer Mineral 54: 858–868

Thornthwaite CW (1948) An approach toward a rational classification of climate. Geogr Rev 38: 85–94

Thornthwaite CW, Holzman B (1942) Measurement of evaporation from land and water surfaces. USDA Techn Bull No 817, Washington

Tiller KG, Gerth J, Brümmer G (1984) The relative affinities of Cd, Ni and Zn for different soil clay fractions and goethite. Geoderma 34: 17–35

Tingey DT, Standley C, Field RW (1976) Stress ethylene evolution: A measure of ozone effects on plants. Atmos Environ 10: 969–974

Tingey DT, Taylor GE (1982) Variation in plant response to ozone: a conceptual model of physiological events. In: Unsworth MH, Ormrod DP (eds) Effects of gaseous air pollution in agriculture and horticulture. Butterworth, London, pp 111–138

Tjepkema JD, Cartica RJ (1981) Atmospheric concentration of ammonia in Massachusetts and deposition on vegetation. Nature 294: 445–446

Topp E (1986) Aufnahme von Umweltchemikalien durch die Pflanze in Abhängigkeit von physikalisch-chemischen Stoffeigenschaften. Diss, TU München

Tou JC, Westover LB, Sonnabend LF (1974) Kinetic studies of bis-(chloromethyl)ether hydrolysis by mass spectrometry. J Phys Chem 78: 1096–1098

Townsend AM, Dochinger LS (1974) Relationship of seed source and developmental stage to the ozone tolerance of *Acer rubrum* seedlings. Atmos Environ 8: 957–964

Trampisch HJ (1986) Zuordnungsprobleme in der Medizin: Anwendung des Lokalisationsmodells. Springer, Berlin Heidelberg New York

Traxler RW, Bhattacharya LS (1978) Effects of a chemical dispersant on microbial utilization of petroleum hydrocarbons. In: McCarthy LT, Lindblom GP, Walters HF (eds) Chemical dispersants for the control of oil spills. American Soc for Testing and Materials, ASTM STP 659: 181–187

Treshow M (ed) (1984) Air pollution and plant life. Wiley, Chichester

Troedsson T (1980) Long-term changes of forest soils. Ann Agric Fenn 19: 81–84

Trost A (1966) Feuchtemessung mit Neutronen. Informationsheft Büro Euvisotop, Ser: Monogr 3, Wildbad

Troyanowsky C (1985) Air pollution and plants. VCH, Weinheim

Tschapek MW (1959) El agua en el suelo. Instituto Nacional de Tecnología Agropecuaria, Buenos Aires

Tukey JW (1977) Exploratory data analysis. Addison–Wesley, Reading Mass

Tuominen Y (1967) Studies on the strontium uptake of the *Cladonia alpestris* thallus. Ann Bot Fenn 4: 1–28

Turc L (1954) Le bilan d'eau des sols. Relations entre les précipitations, l'évaporation et l'écoulement. Ann Agron 5: 491–596

Turner BD (1964) A diffusion model for an urban area. J Appl Meteorol 3: 83–91

Turner JC (1965) Some energy and microclimate measurements in a natural arid zone plant community. In: Eckardt FE (ed) Methodology of plant eco-physiology. UNESCO, Paris. pp 63–70

Tyler G (1972) Heavy metals pollute nature, may reduce productivity. Ambio 1: 52–59

UK-DOE (1986) Nitrate in water. Dep Environ Pollut Pap 26. HMSO, London

Ulrich B (1972) Chemische Wechselwirkungen zwischen Wald-Ökosystemen und ihrer Umwelt. Forstarchiv 43: 41–43

Ulrich B (1981a) Ökologische Gruppierung von Böden nach ihrem chemischen Bodenzustand. Z Pflanzenernähr Bodenkd 144: 289–305

Ulrich B (1981b) Eine ökosystemare Hypothese über die Ursachen des Tannensterbens (*Abies alba* Mill.). Forstwiss Centralbl 100: 228–236

Ulrich B (1982) Gefahren für das Waldökosystem durch saure Niederschläge. Mitt

Landesanst Ökologie, Landschaftsentwicklung und Forstplanung Nordrhein-Westfalen, 2. Aufl, Sonderh 1982: 9–25

Ulrich B (1983a) Interaction of forest canopies with atmospheric constituents: SO_2, alkali and earth alkali cations and chloride. In: Ulrich B, Pankrath J (eds) Effects of accumulation of air pollutants in forest ecosystems. Reidel, Dordrecht. pp 33–45

Ulrich B (1983b) Stabilität von Waldökosystemen unter dem Einfluß des "sauren Regens". Allg Forstztg 26/27: 670–677

Ulrich B (1984) Waldsterben durch saure Niederschläge. Umschau 11: 348–355

Ulrich B (1987a) Effects of air pollutants on the soil. In: Barth H, L'Hermite P (eds) Scientific basis for soil protection in the European Community. Elsevier, London, pp 299–311

Ulrich B (1987b) Stabilität, Elastizität und Resilienz von Waldökosystemen unter dem Einfluß saurer Deposition. Forstarchiv 58: 232–239

Ulrich B (ed) (1990) Internationalet Kongreß Waldschadensforschung: Wissensstand und Perspektiven. Kernforschungszentrum Karlsruhe, Karlsruhe

Ulrich B, Matzner E (1983) Raten der ökosysteminternen H^+-Produktion und der sauren Deposition und ihre Wirkungen auf Stabilität, Elastizität von Waldökosystemen. VDI-Berichte 500, Saure Niederschläge. VDI, Düsseldorf, pp 289–300

Ulrich B, Mayer R, Khanna PK (1979) Deposition von Luftverunreinigungen und ihre Auswirkungen in Waldökosystemen im Solling. Schr Forstl Fak Univ Göttingen 58: 1–291. Sauerländer, Frankfurt

Ulrich B, Sumner ME (1991) Soil acidity, Springer, Berlin Heidelberg

Umweltbundesamt (1977) Materialien zum Immissionsschutzbericht 1977 der Bundesregierung an den Deutschen Bundestag. Schmidt, Berlin

Umweltbundesamt (1982) Collection of minimum pre-marketing sets of data including environmental residue data on existing chemicals. Umweltbundesamt, Berlin

Umweltbundesamt (1983) Jahresbericht 1983. Umweltbundesamt, Berlin

Umweltbundesamt (1984) Daten zur Umwelt 1984. Schmidt, Berlin

Umweltbundesamt (1986a) Daten zur Umwelt 1986/87. Schmidt, Berlin

Umweltbundesamt (ed) (1986b) Wissenschaftliches Symposium zum Thema Waldschäden "Neue Ursachenhypothesen". Schriftenr "Texte" 19/86. Umweltbundesamtes, Berlin

Umweltbundesamt (1986c) Umweltforschungskatalog 1985/86 (UFOKAT '85/86). Schmidt, Berlin

Umweltbundesamt (ed) (1986d) IMA-Querschnittsseminar zur Waldschädenforschung "Belastung und Schäden auf Ökosystemebene und ihre Folgen". Schriftenr "Texte" Umweltbundesamtes, Berlin

Umweltbundesamt (1989) Daten zur Umwelt 1988/89. Schmidt, Berlin

UN Economic Commission for Europe (ed) (1984) Air-borne sulphur pollution – effects and control. United Nations, New York

Urland K, Bork H-R, Rohdenburg H (1985) Übertragung von Punktmessungen des Grundwasserstandes auf ein regelmäßiges Gebietsraster für die Kalibrierung eines deterministischen Gebietsmodells der Wasserflüsse. Landschaftsökol Messen Auswerten 1: 169–178

Utermann J, Richter J (1988) Die Verlagerung physikalisch wechselwirkender Ionen in Böden – Modellentwicklung und -kalibrierung. Z Pflanzenernähr Bodenkd 151: 165–170

van Aalst RM (1984) Depositie van verzurende stoffen in Nederland. In: Adema EH, van Ham J (eds) Zure regen, oorzaken, effecten en beleid. PUDOC, Wageningen, p 66

van Bavel CHM, Fritschen LJ (1965) Energy balance of bare surfaces in an arid climate.

In: Eckardt FE (ed) Methodology of plant eco-physiology. UNESCO, Paris. pp 99–107

van der Eerden LJ (1978) Invloed van stallucht afkomstig van intensieve veehouderijbedrijven en van ammoniak op planten. IPO-Rep R 203, 50 pp

van der Eerden LJ (1982) Toxicity of ammonia to plants. Agric Environm 7: 223–235

van der Eerden LJ, Wit AKH (1987) Effecten van NH_3 en NH_4^+ op planten en vegetaties; relevantie van effectgrenswaarden. In: Boxman AW, Geelen JFM (eds) Acute en chronische effecten van NH_3 (en NH_4^+) op levende organismen; Proceedings van de BEL-studiedag op 12 december 1986 te Nijmegen. Lab. voor Aquatische Oecologie, KU Nijmegen, pp 45–51

van der Maarel E (1976) On the establishment of plant community boundaries. Ber Dtsch Bot Ges 89: 415–443

van der Sloot HA, de Groot GH, Eggenkamp HGM et al. (1988) Versatile method for the measurement of (trace) element mobilities in waste material, soil and bottom sediments. Environ Sci Technol

van Egmond ND, Kesseboom H (1982) Modelling of mesoscale transport of NO_x and NO_2; concentration levels and source contributions. In: Schneider T, Grant L (eds) Air pollution by nitrogen oxides. Elsevier, Amsterdam, pp 327–346

van Eimern J, Ehrhardt O (1985) Zur räumlichen und zeitlichen Variabilität der Strahlung und Temperatur an der Boden- und Kronenoberfläche eines Buchenwaldes. Forstarchiv 56: 181–186

van Eimern J, Ehrhardt O (1986) Zur tages- und jahreszeitlichen Veränderung der Globalstrahlung und ihrer Komponenten über und in einem Buchenwald. Forstarchiv 57: 86–92

van Genuchten MT (1980) A closed-form equation for predicting the hydraulic conductivity of unsaturated soils. Soil Sci Soc Am J 44(5): 892–898

van Genuchten MT, Alves WJ (1982) Analytical solutions of the one-dimensional convective-dispersive solute transport equation. USDA-ARS Tech Bull 1661. US Government Printing Office, Washington DC

van Genuchten MT, Wierenga PJ (1977) Mass transfer studies in sorbing porous media. II. Experimental evaluation with tritium. Soil Sci Soc Am J 41: 272–278

van Genuchten MT, Wierenga PJ, O'Connor GA (1977) Mass transfer studies in sorbing porous media. III. Experimental evaluation with 2,4,5-T. Soil Sci Soc Am J 41: 278–280

van Haut H, Stratmann H (1967) Experimentelle Untersuchungen über die Wirkung von Stickstoffdioxid auf Pflanzen. Schriftenreihe der Landesanstalt für Immissions- und Bodennutzungsschutz des Landes NW 7: 50–70

van Hove LWA (1987) De opname van atmosferisch ammoniak door bladeren. In: Boxman AW, Geelen JFM (eds) Acute en chronische effecten van NH_3 (en NH_4^+) op levende organismen; Proceedings van de BEL-studiedag op 12 december 1986 te Nijmegen. Lab. voor Aquatische Oecologie, KU Nijmegen, pp 35–44

van Steenis CGGJ (1948/49) General considerations. Flora Malesiana Ser I, 4

van Wambeke A (1978) Properties and potentials of soils in the Amazon Basin. Interciencia 3: 233–242

van Wonderen JJ, Sage RC (1985) Implications of the use of regional groundwater models. Groundwater Water Resour Planning 2: 949–958

VCI (Verband der Chemischen Industrie) (1988) VCI-Altstoffliste. Chem Ind 4/1988: 102–109

VDI (Verein Deutscher Ingenieure) (1985) Materialkorrosion durch Luftverunreinigungen. VDI-Ber 530. VDI, Düsseldorf

VDI (1988) Stadtklima und Luftreinhaltung. Springer, Berlin-Heidelberg

Veith CD, De Foe DL, Bergstedt BV (1979) Measuring and estimating the bioconcentration factor for chemicals in fish. J Fish Res Board Can 36: 1040–1048

Verly G, David M, Journel AG, Maréchal A (eds) (1984) Geostatistics for natural resources characterization. NATO Advanced Science Institutes Series. Reidel, Dordrecht

Verschueren K (1983) Handbook of environmental data on organic chemicals. Van Nostrand Reinhold, New York

Versino B, Ott H (eds) (1982) Physico-chemical behaviour of atmospheric pollutants. Proc 1st Eur Symp, Ispra, Italy, 29 Sept. –1 Oct. 1981. Reidel, Dordrecht

Verwey EJW, Overbeek JThG (1948) Theory of the stability of lyophobic colloids. Elsevier, Amsterdam

Visser WC (1958) Soil science and sprinkler irrigation. Rep Supplemental Irrigation, Comm VI, ISSS; 51–63

Waldron JK (1978) Effects of soil drenches and simulated "rain" applications of sulfuric acid and sodium sulfate on the nodulation and growth of legumes. MSc Thesis, North Carolina State University, Raleigh NC

Walker A, Smith AE (1979) Persistence of 2,4,5-T in a heavy clay soil. Pestic Sci 10: 151–159

Walker JD, Calomiris JJ, Herbert TL, Colwell RR (1976a) Petroleum hydrocarbons: degradation and growth potential for Atlantic Ocean sediment bacteria. Mar Biol 34: 1–9

Walker JD, Seesman PA, Herbert TL, Colwell RR (1976b) Petroleum hydrocarbons: degradation and growth potential of deepsea sediment bacteria. Environ Pollut 10: 89–99

Wallace A, Romney EM, Alexander GV, Soufi SM, Patel PM (1977) Some interactions in plants among cadmium, other heavy metals, and chelating agents. Agron J 69: 18–20

Wallnöfer P, Koniger M, Engelhardt G (1975) The behaviour of xenobiotic chlorinated hydrocarbons (HCB and PCBs) in plant and soils. Z Pflanzenkr Pflanzenschutz 82: 91–100

Wang HF, Anderson MP (1982) Introduction to groundwater modeling. Finite difference and finite element methods. Freeman, San Francisco

Ward CT, Matsumura F (1978) Fate of 2,3,7,8-tetrachlorodibenzo-p-dioxin (TCDD) in a model aquatic environment. Arch Envrionm Contam Toxicol 7: 349–357

Wassink EC (1959) Efficiency of light energy conversion in plant growth. Plant Physiol 34: 356–361

Webber MD, Kloke A, Tjell JC (1984) A review of current sludge use guidelines for the control of heavy metal contamination in soils. In: L'Hermite P, Oh H (eds) Processing and use of sewage sludge. Reidel, Dordrecht, pp 371–386

Weber DE, Lee JJ (1979) The effects of simulated acid rain on seedling emergence and growth of eleven woody species. Forest Science 25: 393–398

Wedding JB, Carlson RW, Stukel JJ, Bazzaz FA (1975) Aerosol deposition on plant leaves. Environ Sci Technol 9: 151–153

Wehrmann J, Scharpf HC (1988) Nitrat in Grundwasser und Nahrungspflanzen. Auswertungs- und Informationsdienst für Ernährung Landwirtschaft und Forsten (AID), Bonn

Weischet W (1977) Die ökologische Benachteiligung der Tropen. Teubner, Stuttgart

Weischet W (1979) Klimatologische Interpretation von METEOSAT-Aufnahmen. Einführung und Interpretation der Bilder vom 21. August 1978. Geogr Rundsch 31: 337–339

Weischet W (1980) Klimatologische Interpretation von METEOSAT-Aufnahmen. Teil XII, Juli 1979. Geogr Rundsch 32: 378–380

Wellburn AR, Higginson C, Robinson D, Walmsley C (1981) Biochemical explanations of more than additive inhibitory effects of low atmospheric levels of sulphur dioxide plus nitrogen dioxide upon plants. New Phytol 88: 223–237

Wellburn AR, Wilson J, Aldrige PH (1980) Biochemical responses of plants to nitric oxide polluted atmospheres. Environmental Pollution (Series A) 22: 219–228

Welp G (1987) Einfluß des Stoffbestandes von Böden auf die mikrobielle Toxizität von Umweltchemikalien. Diss, Univ Kiel

Wendling U (1967) Bemerkungen zur Eichung von Neutronensonden. Alb-Thaer-Arch 11: 1105–1116

Wenzel B, Ulrich B (1988) Kompensationskalkung – Risiken und ihre Minimierung. Forst – Holzwirt 43: 12–16

Werner D (1987) Pflanzliche und mikrobielle Symbiosen. Thieme, Stuttgart

Werner D, Krotzky A, Berggold R, Thierfelder H, Preiß M (1982) Enhancement of specific nitrogenase activity in *Azospirillum brasilense* and *Klebsiella pneumoniae*, inhibition in *Rhizobium japonicum* under air by phenol. Arch Microbiol 132: 51–56

Wesely ML (1983) Turbulent transport of ozone to surfaces common in the eastern half of the United States. In: Schwartz SE (ed) Trace atmospheric constituents. Wiley, New York, pp 345–370

Wesely ML (1989) Parametrization of surface resistances to gaseous dry deposition in regional-scale numerical models. Atmosph Environ 23: 1293–1304

Wesely ML, Hicks BB, Dannevik WP, Frisella S, Husar B (1977) An eddy-correlation measurement of particulate deposition from the atmosphere. Atmos Environ 11: 561–563

Wesely ML, Eastman JA, Cook DR, Hicks BB (1978) Daytime variations of ozone eddy fluxes to maize. Boundary Layer Meteorol 15: 361–373

Wesely ML, Eastman JA, Stedman DH, Yalvac ED (1982) An eddy-correlation measurement of NO_2 flux to vegetation and comparison to O_3 flux. Atmos Environ 16: 815–820

Wessolek G, Renger M, Facklam M, Strebel O (1985) Einfluß von Standortnutzungsänderungen auf die Grundwasserneubildung(srate). Mitt Dtsch Bodenkd Ges 41: 357–364

Westall JC, Zachary JL, Morel FMM (1976) MINEQL – a computer program for the calculation of chemical equilibrium composition of aqueous systems. Dep Civil Eng Mass Inst Technol, Tech Note 18, Cambridge MASS

Weyman DR (1975) Run-off processes and streamflow modelling. Oxford University Press, Oxford

Whelpdale DM, Bottenheim JW (1982) Recent Canada–USA transboundary air pollution studies. In: Schneider T, Grant L (eds) Air pollution by nitrogen oxides. Elsevier, Amsterdam, pp 357–363

Whitby KT (1977) The physical characteristics of sulfate aerosols. Atmos Environ 12: 135–159

Whitby KT, Sverdrup GM (1980) California aerosols: Their physical and chemical characteristics. In: Hidy GM, Mueller PK, Grosjean D, Appel BR, Wesolowski JJ (eds) The character and origins of smog aerosols. Wiley, New York, pp 477–517

White ID, Mottershead ND, Harrison SJ (1984) Environmental systems. Allen and Unwin, London

Whiteside JS, Alexander M (1960) Measurement of microbiological effects of herbicides. Weeds 8: 204–213

Whitmore ME, Mansfield TA (1983) Effects of long-term exposures to SO_2 and NO_2 on *Poa pratensis* and other grasses. Environmental Pollution (Series A) 31: 217–235

Wigley TML (1977) WATSPE – a computer program for determining the equilibrium speciation of aqueous solutions. Br Geomorph Res Group, Tech Bull 20, Norwich

Wiklander L (1964) Cation and anion exchange phenomena. In: Bear FE (ed) Chemistry of the soil. Reinhold, New York, pp 163–205

Wiklander L (1973/74) The acidification of soil by acid precipitation. Grundförbättring 26: 155–164

Wiklander L (1980) The sensitivity of soils to acid precipitation. In: Hutchinson TC, Havas M (eds) Effects of acid precipitation on terrestrial ecosystems. NATO Conference Series, Ser I, Ecology, vol 4, Plenum, New York. pp 553–567

Wiklander L, Andersson A (1972) The replacing efficiency of hydrogen ion in relation to base saturation and pH. Geoderma 7: 159–165

Wildförster E (1985) Untersuchungen zur standörtlichen Variabilität des mikrobiellen Abbaus ausgewählter Herbizide. Dipl Arb, Univ Kiel

Wilke B-M (1988) Langzeitwirkungen potentieller anorganischer Schadstoffe auf die mikrobielle Aktivität einer sandigen Braunerde. Z Pflanzenernähr Bodenkd 151: 131–136

Williams CH, David DJ (1976) The accumulation in soils of cadmium residues from phosphate fertilizers and their effect on the cadmium content of plants. Soil Sci 121: 86–93

Wind GP (1955) A field experiment concerning capillary rise of moisture in a heavy clay soil. Neth J Agric Sci 3: 60–69

Windhorst HW (1984) Das agrarische Intensivgebiet Südoldenburg. Geowiss Zeit 2: 181–193

Winger RJ (1960) In-place permeability tests and their use in subsurface drainage. Int Comm on Irrigation and Drainage, 4th Congr, Madrid, pp 48

Winkler P (1982) Zur Trendentwicklung des pH-Wertes des Niederschlages in Mitteleuropa. Z Pflanzenernähr Düng Bodenkd 145: 576–583

Winkler P (1983a) Saurer Niederschlag – eine Trendanalyse. Ann Meteorol NF 20: 117–118

Winkler P (1983b) Der Säuregehalt von Aerosol, Nebel und Niederschlägen. VDI-Be 500: 141–147

Winteringham FPW (1977) Comparative ecotoxocology of halogenated hydrocarbon residues. Ecotoxicol Environ Safety 1: 407–425

Winteringham FPW (1981) Comparative aspects of classical toxicology and ecotoxicology. VIIth Int Symp Chemical and Toxicological Aspects of Environmental Quality, London, 7–10 Sept. 1981. Abstract No. 4

Wishart D (1975) CLUSTAN 1C user manual. Academic Press, London

Wishart D (1984) CLUSTAN 1C user manual. Academic Press, London

Wittig R, Ballach HJ, Brandt CJ (1985) Increase of number of acid indicators in the herb layer of the millet grass-beech forest of the Westphalian Bight. Angew Bot 59: 219–232

Woese CR, Magrum LJ, Fox GE (1978) Archaebacteria. J Mol Evol 11: 245–252

Wohlrab B (1983) Zur Grundwasserneubildung – Menge und Qualität. Arb Dtsch Landwirtschaftsges 177: 54–68

Wolery TJ (1979) Calculations of chemical equilibrium between aqueous solution and minerals: The EQ 3/6 software package. NTIS UCRL 52658, Washington, DC

Wolfe NL (1980) Determining the role of hydrolysis in the fate of organics in natural waters. In: Haque R (ed) Dynamics, exposure and hazard assessment of toxic chemicals. Ann Arbor Science, Ann Arbor, pp 163–178

Wolkewitz H (1959/60) Die Weiterentwicklung des Verfahrens der pF-Untersuchung zur Feststellung der Bindungsintensität des Wassers im Boden. Z Kulturtech 1959/60: 37–50

Wood JM (1976) The biochemistry of toxic elements in aqueous systems. Biochem Biophys Perspect Mar Biol 1976: 407–431

Wood T, Bormann FH (1974) The effects of an artificial acid mist upon the growth of *Betula alleghaniensis* Britt. Environ Pollut 7: 259–268

Wood T, Bormann FH (1977) Short-term effects of a simulated acid rain upon the growth and nutrient relations of *Pinus strobus* L. Water Air Soil Pollut 7: 479–488

Woodwell GM (ed) (1984) The role of terrestial vegetation in the global carbon cycle. SCOPE 23, Wiley, Chichester

Woodwell GM, Whittaker RH, Reiners WA et al. (1978) The biota and the world carbon budget. Science 199: 141–146

World Health Organization (1971) International standards for drinking-water. WHO, Geneva

World Meteorological Organization (WMO) (1956) International cloud atlas. WMO, Geneva

World Meteorological Organization (WMO) (1965) Guide to hydrometeorological practices. WMO-No 168. TP 82, Geneva

Wright EA, Lucas PW, Cottam DA, Mansfield TA (1986) Physiological responses of plants to SO_2, NO_x and O_3: implications for drought resistance. In: CEC Air Poll Res Rep 4: 187–200

Wuebbles DJ, Cornell PS (1981) A screening methodology for assessing the potential impact of surface releases of chlorinated halocarbons on stratospheric ozone. Rep US Dep of Energy by the Lawrence Livermore Lab under Contract W-7405-Eng-48

Wuhrmann K, Eichenberger E (1980) Künstliche Bäche als Hilfsmittel der experimentellen Fließwasserökologie. Vom Wasser 54: 1–8

Wyckhoff RD, Botset HG, Muskat M (1932) Flow of liquids through porous material under the action of gravity. Phys Rev 40: 1027

Wyndham RC, Costerton JW (1981a) In vitro microbial degradation of bituminous hydrocarbons and in situ colonization of bitumen surfaces within the Athabasca oil sands deposit. Appl Environ Microbiol 41: 791–800

Wyndham RC, Costerton JW (1981b) Heterotrophic potentials and hydrocarbon degradation potentials of sediment microorganisms within the Athabasca oil sands deposit. Appl Environ Microbiol 41: 783–790

Yanagita T (1990) Natural microbial communities. Springer, Berlin Heidelberg New York

Yariv S, Heller-Kallai L (1975) Comments on the paper: the adsorption of aromatic, heterocyclic and cyclic ammonium cations by montmorillonite. Clay Miner 10: 479–481

Yeh GT, Ward DS (1981) FEMWASTE: A finite-element model of waste transport through saturated-unsaturated porous media. Oak Ridge National Laboratory, Publ 1462

Yoneyama T, Totsuka T, Hayakawa N, Yasaki J (1980) Absorption of atmospheric NO_2 by plants and soils, II: Nitrite accumulation, nitrite reductase activity and diurnal change of NO_2 absorption in leaves. Soil Sci Plant Nutrit 25: 267–275

Yoshino MM (1975) Climate in a small area. An introduction to local meteorology. Univ of Tokyo Press, Tokyo

Youngblood W, Blumer M (1975) Polycyclic aromatic hydrocarbons in the environment: homologous series in soils and recent marine sediments. Geochim Cosmochim Acta 39: 1303–1314

Zahn R, Wellens M (1974) Ein einfaches Verfahren zur Prüfung der biologischen Abbaubarkeit von Produkten und Abwasserinhaltsstoffen. Chemiker-Z 98: 228–232

Zanger CN (1953) Theory and problems of water percolation. Engineering Monographs 8. Bureau of Reclamation, Denver, pp 48–71

Zech W, Popp E (1983) Magnesiummangel, einer der Gründe für das Fichten- und Tannensterben in NO-Bayern. Forstwiss Centralbl 102: 50–55

Zeevaart AJ (1976) Some effects of fumigating plants for short periods with NO_2. Environmental Pollution 11: 97–108

Zellner R (1980) The reactive removal of anthropogenic emissions from the atmosphere. In: Funcke W, König J, Stöber W, Klein AW, Schmidt-Bleek F (eds) Proc Int worksh test methods and assessment procedures for the determination of the photochemical degradation behaviour of chemical substances, December 2–4 1980, Reichstag, Berlin. Fraunhofer-Gesellschaft, Umweltbundesamt, Berlin, pp 30–37

Ziechmann W (1980) Huminstoffe. Verlag Chemie, Weinheim

Zilling L (1987) Ölunfallbekämpfung auf Sylt–Ansätze zur Entwicklung eines lokalen Vorsorgekonzeptes. Schr Naturwiss Ver Schleswig-Holstein 57: 1–18

Zölitz R (1985) Spatial validity of the EMEP monitoring net: geostatistical investigations of the 1981 sulphate deposition in Europe. In: Grosch W (ed) Advancements in air pollution monitoring equipment and procedures. Federal Ministry of the Interior, Bonn, pp 200–210

Zoetman BCJ, Harmsen K, Linders JBHJ, Morra CFH, Sloff W (1980) Persistent organic pollutants in river water and groundwater of The Netherlands. Chemosphere 9: 231–249

Zsoldes F, Hannold E (1979) Effects of pH changes in ion and 2,4-D uptake of wheat roots. Physiol Plant 47: 77–80

Zwerver S (1982) An air quality management system as a tool for establishing a NO_x policy. In: Schneider T, Grant L (eds) Air pollution by nitrogen oxides. Elsevier, Amsterdam, pp 929–950

Subject Index

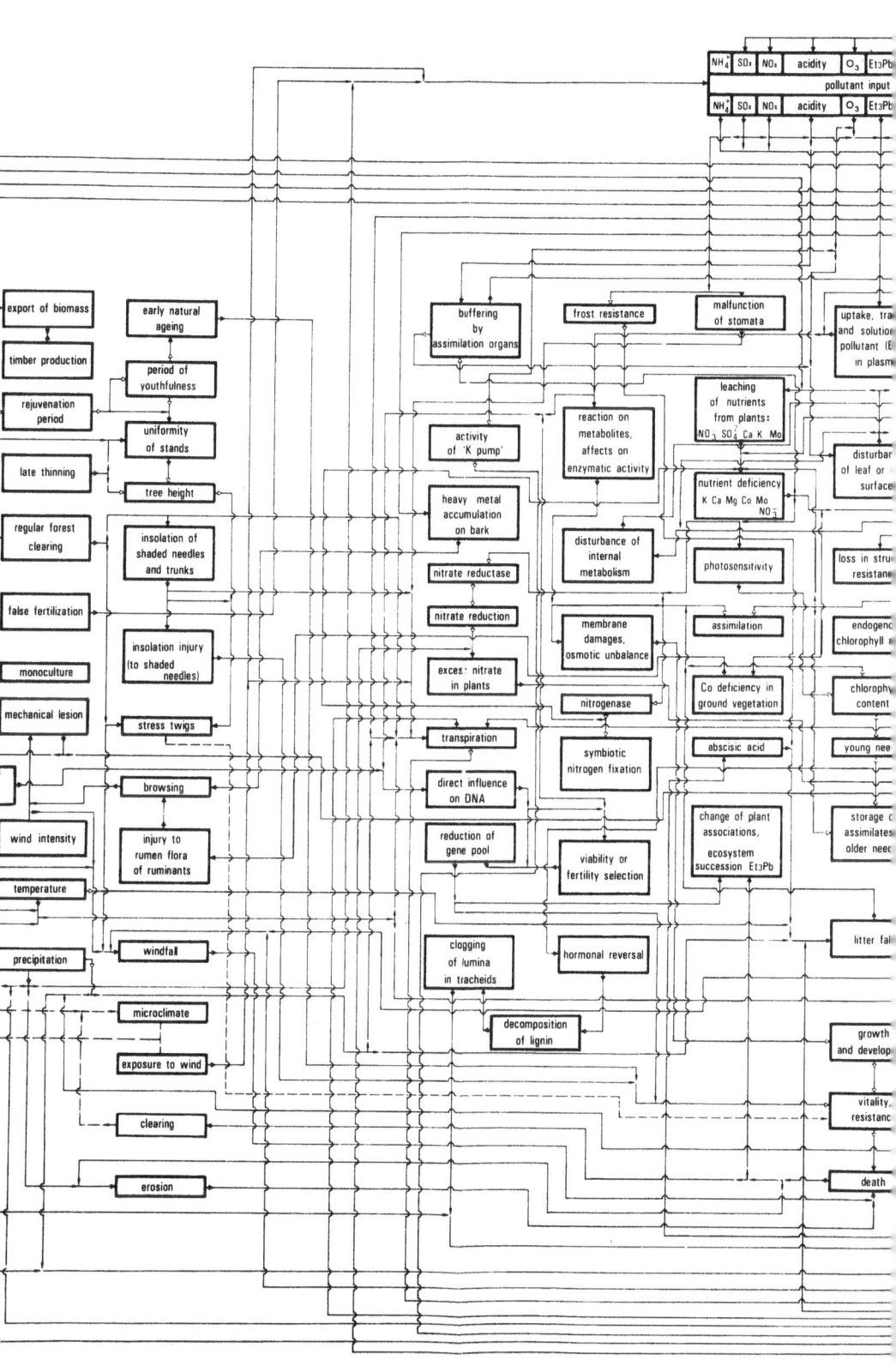

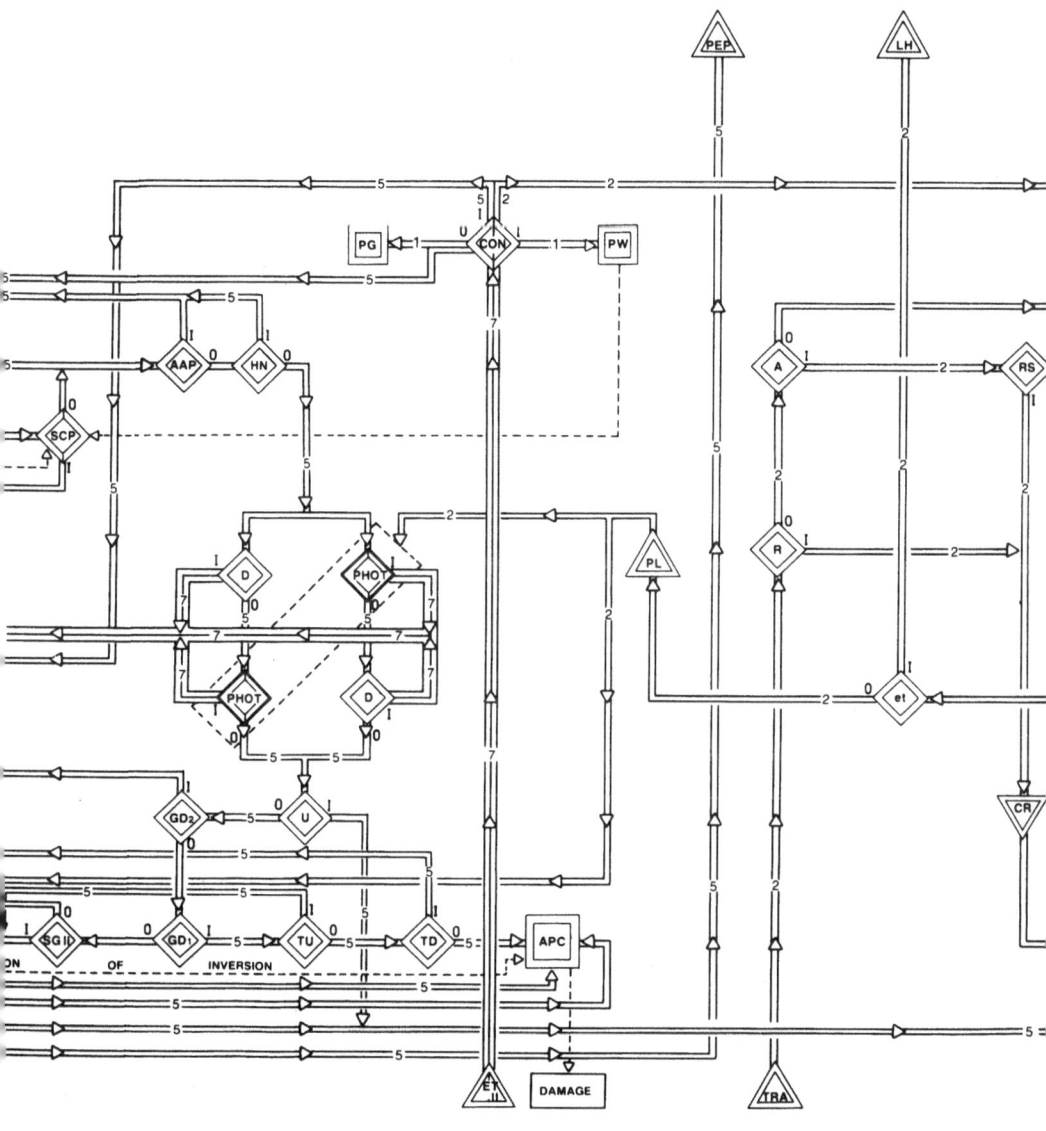

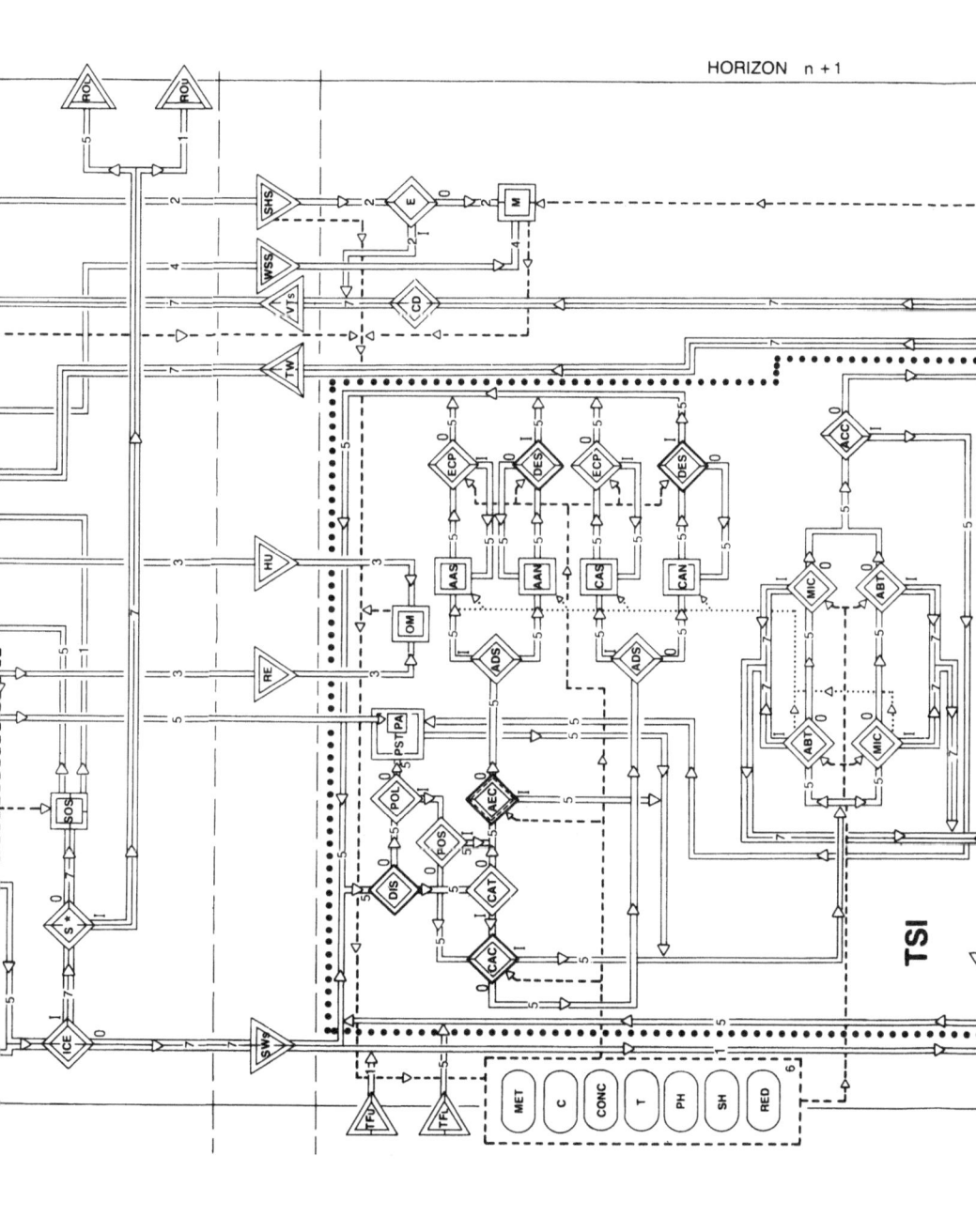

TSI